ANIMAL STRUCTURE AND FUNCTION

CECIE STARR / RALPH TAGGART

BIOLOGY

THE UNITY AND DIVERSITY OF LIFE

EIGHTH EDITION

WADSWORTH PUBLISHING COMPANY

I(T)P® AN INTERNATIONAL THOMSON PUBLISHING COMPANY

Belmont, CA • Albany, NY • Bonn • Boston
Cincinnati • Detroit • Johannesburg • London • Madrid
Melbourne • Mexico City • New York
Paris • San Francisco • Singapore
Tokyo • Toronto • Washington

BIOLOGY PUBLISHER: Jack C. Carey

ASSISTANT EDITOR: Kristin Milotich

MEDIA PROJECT MANAGER: Pat Waldo

MARKETING MANAGER: Halee Dinsey

DEVELOPMENTAL EDITOR: Mary Arbogast

PROJECT EDITOR: Sandra Craig

EDITORIAL ASSISTANT: Michael Burgreen

PRINT BUYER: Karen Hunt

PRODUCTION: Mary Douglas, Rogue Valley Publications

TEXT AND COVER DESIGN, ART DIRECTION: Gary Head, Gary Head Design

ART COORDINATOR: Myrna Engler-Forkner

EDITORIAL PRODUCTION: Mary Roybal, Karen Stough, Susan Gall

PRIMARY ARTISTS: Raychel Ciemma; Precision Graphics (Jan Troutt, J.C. Morgan)

ADDITIONAL ARTISTS: Robert Demarest, Darwen Hennings, Vally Hennings, Betsy Palay, Nadine Sokol, Kevin Somerville, Lloyd Townsend

PHOTO RESEARCH, PERMISSIONS: Stephen Forsling, Roberta Broyer

COVER PHOTOGRAPH: *Minnehaha Falls*, © Richard Hamilton Smith

COMPOSITION: Precision Graphics (Jim Gallagher, Kirsten Dennison)

COLOR PROCESSING: H&S Graphics (Tom Anderson, Nancy Dean, John Deady, Rich Stanislawski)

PRINTING AND BINDING: World Color, Versailles

For more information, contact Wadsworth Publishing Company, 10 Davis Drive, Belmont, California 94002, or electronically at http://www.thomson.com/wadsworth.html

International Thomson Publishing Europe
Berkshire House 168-173, High Holborn
London, WC1V7AA, England

Thomas Nelson Australia
102 Dodds Street
South Melbourne 3205, Victoria, Australia

Nelson Canada
1120 Birchmount Road
Scarborough, Ontario, Canada M1K 5G4

International Thomson Editores
Campos Eliseos 385, Piso 7
Col. Polanco, 11560 México D.F. México

International Thomson Publishing GmbH
Königswinterer Strasse 418
53227 Bonn, Germany

International Thomson Publishing Asia
221 Henderson Road, #05-10 Henderson Building
Singapore 0315

International Thomson Publishing Japan
Hirakawacho Kyowa Building, 3F
2-2-1 Hirakawacho, Chiyoda-ku, Tokyo 102, Japan

International Thomson Publishing Southern Africa
Building 18, Constantia Park
240 Old Pretoria Road
Halfway House, 1685 South Africa

CONTENTS IN BRIEF

Highlighted chapters are included in ANIMAL STRUCTURE AND FUNCTION.

Highlighted chapters are included in ANIMAL STRUCTURE AND FUNCTION.

DETAILED CONTENTS

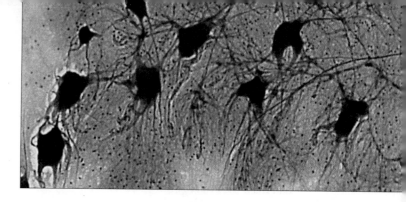

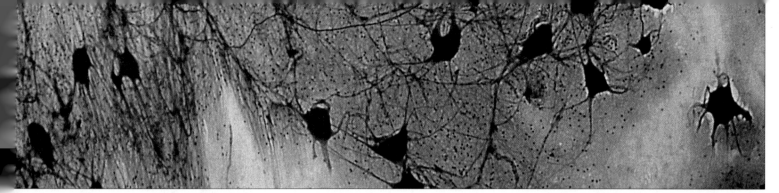

SOLDIER TERMITES DEFENDING THEIR COLONY

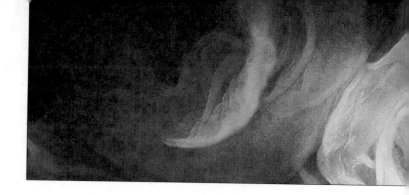

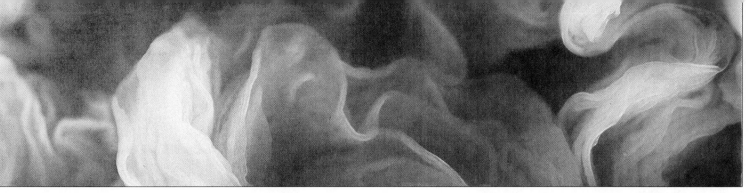

BILLOWING ENTRANCE TO AN OVIDUCT

PREFACE

Not too long from now we will cross the threshold of a new millennium, a rite of passage that invites reflection on where biology has been and where it might be heading. About 500 years ago, during an age of global exploration, naturalists first started to systematically catalog and think about the staggering diversity of organisms all around the world. Less than 150 years ago, just before the start of a civil war that would shred the fabric of a new nation, the naturalist Charles Darwin shredded preconceived notions about life's diversity. It was only about 50 years ago that biologists caught their first glimpse of life's unity at the molecular level. Until that happened a biologist could still hope to be a generalist—someone who viewed life as Darwin did, without detailed knowledge of mechanisms that created it, that perpetuate it, that change it.

No more. Biology grew to encompass hundreds of specialized fields, each focused on one narrow aspect of life and yielding volumes of information about it. Twenty years ago I wondered whether introductory textbooks could possibly keep up with the rapid and divergent splintering of biological inquiry. James Bonner, a teacher and researcher at the California Institute of Technology, turned my thinking around on this. He foresaw that authors and instructors for introductory courses must become the new generalists, the ones who give each generation of students broad perspective on what we know about life and what we have yet to learn.

And we must do this, for the biological perspective remains one of the most powerful of education's gifts. With it, students who travel down specialized roads can sense intuitively that their research and its applications may have repercussions in unexpected places in the world of life. With that perspective, students in general might cut their own intellectual paths through social, medical, and environmental thickets. And they might come to understand the past and to predict possible futures for ourselves and all other organisms.

CONCERNING THE EIGHTH EDITION

Like earlier editions, this book starts with an overview of the basic concepts and scientific methods. Three units on the principles of biochemistry, inheritance, and evolution follow. The principles provide the conceptual background necessary for deeper probes into life's unity and diversity, starting with a richly illustrated evolutionary survey of each kingdom. Units on the comparative anatomy and physiology of plants, then animals, follow. The last unit focuses on the patterns and consequences of organisms interacting with one another and with their environment. Thus the organization parallels the levels of biological organization, from cells through the biosphere. We adhere to this traditional approach for good reason: it works.

As before, we identify and highlight the key concepts, current understandings, and research trends for the major fields of inquiry. Through examples of problem solving and experiments, we give ample evidence of "how we know what we know" and thus demonstrate the power of critical thinking. We explain the structure and functioning of a broad sampling of organisms in enough detail so that students can develop a working vocabulary about life's parts and processes. We also updated the glossary.

CONCEPT SPREADS

In the first chapter, an overview of the levels of biological organization kicks off a story that continues through the rest of the book. Telling such a big, complex story might be daunting unless you remind yourself of the question *"How do you eat an elephant?"* and its answer, *"One bite at a time."* We who have told the story again and again know how the parts fit together, but many students need help to keep the story line in focus within and between chapters. And they need to chew on concepts one at a time.

In every chapter we present each concept on its own table, so to speak. That is, we organize the descriptions, art, and supporting evidence for it on two facing pages, at most. Think of this as a concept spread, as in Figure *A*. Each starts with a numbered tab and ends with boldface statements to summarize the key points. Students can use these cues as reminders to digest one topic before starting on another. Well-crafted transitions between spreads help students focus on where topics fit in the larger story and gently discourage memorization for its own sake. The clear demarcation also gives instructors greater flexibility in assigning or skipping topics within a chapter.

By restricting the space available for each concept, we force ourselves to clear away the clutter of superfluous detail. Within each concept spread, we block out headings and subheadings to rank the importance of its various parts. Any good story has such a hierarchy of information, with background settings, major and minor characters, and high points and an ending where everything comes together. Without a hierarchy, a story has all the excitement, flow, and drama of an encyclopedia. Where details are useful as expansions of concepts, we integrate them into suitable illustrations to keep them from disrupting the text flow.

Not all students are biology majors, and many of them approach biology textbooks with apprehension. If the words don't engage them, they sometimes end up hating the book, and the subject. It comes down to line-by-line judgment calls. During twenty-two years of authorship, we developed a sense of when to leave core material alone and when to loosen it up to give students breathing room. Interrupting, say, an account of mitotic cell division with a distracting anecdote does no good. Plunking a humorous aside into a chapter that ties together the evolution of the Earth and life trivializes a magnificent story. Including an entertaining story is fine, provided that doing so reinforces a key concept. Thus, for example, we include the story of a misguided species introduction that resulted in wild European rabbits running amok through Australia.

BALANCING CONCEPTS WITH APPLICATIONS

Each chapter starts with a lively or sobering application that leads into an adjoining list of key concepts. The list is an advance organizer for the chapter as a whole. At strategic points, examples of applications parallel the core material—not so many as to be distracting, but enough to keep minds perking along with the conceptual development. Many brief applications are integrated in the text. Others are in *Focus* essays, which give more depth on medical, environmental, and social issues but do not interrupt the text flow.

FOUNDATIONS FOR CRITICAL THINKING

To help students develop a capacity for critical thinking, we walk them through experiments that yielded evidence in favor of or against hypotheses being discussed. The main index for the book will give you a sense of the number and types of experiments used (see the entry *Experiments*).

We use certain chapter introductions as well as entire chapters to show students some of the productive results of critical thinking. Among these are the introductions to the chapters on Mendelian genetics (11), DNA structure and function (14), speciation (19), immunology (40), and animal behavior (51).

Many *Focus on Science* essays provide more detailed, optional examples of how biologists apply critical thinking to problem solving. For example, one of these describes RFLP analysis (Section 16.3) and a few of its more jarring applications. Another essay (Section 21.4) helps convey to students that biology is not a closed book. Even when new research brings a sweeping story into sharp focus—in this case, the origin of the great prokaryotic and eukaryotic lineages—it also opens up new roads of inquiry.

This edition has *Critical Thinking* questions at the end of chapters. Katherine Denniston of Towson State University developed these thought-provoking questions. Chapters 11 and 12 also include a large selection of *Genetics Problems* that help students grasp the principles of inheritance.

To keep readers focused, we cover each concept on one or two facing pages, starting with a numbered tab . . .

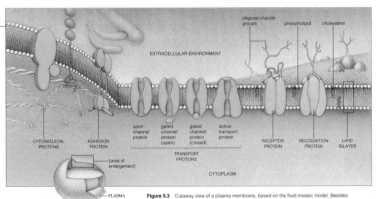

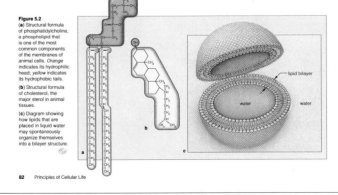

FIGURE A   *A concept spread from this edition.*

. . . and ending with one or more summary statements.

VISUAL OVERVIEWS OF MAJOR CONCEPTS

While writing the text, we simultaneously develop the illustrations as inseparable parts of the same story. This integrative approach appeals to students who are visual learners. When they can first work their way through a visual overview of some process, then reading through the corresponding text becomes less intimidating. Over the years, students have repeatedly thanked us for our hundreds of overview illustrations, which contain step-by-step, written descriptions of biological parts and processes. We break down the information into a series of illustrated steps that are more inviting than a complex, "wordless" diagram. Figure *B* is a sample. Notice how simple descriptions, integrated with the art, take students through the stages by which mRNA transcripts become translated into polypeptide chains, one step at a time.

Similarly, we continue to create visual overviews for anatomical drawings. The illustrations integrate structure and function. Students need not jump back and forth from the text, to tables, to illustrations, and back again in order to comprehend how an organ system is put together and what its parts do. Even individual descriptions of parts are hierarchically arranged to reflect the structural and functional organization of that system.

COLOR CODING

In line illustrations, we consistently use the same colors for the same types of molecules and cell structures. Visual consistency makes it easier for students to track complex parts and processes. Figure C is the color coding chart.

ZOOM SEQUENCES

Many illustrations in the book progress from macroscopic to microscopic views of the same subject. Figure 7.2 is an example; this zoom sequence shows where the reactions of photosynthesis proceed, starting with a plant growing by a roadside. As another example, Figures 38.19 and 38.20 move down through levels of skeletal muscle contraction, starting with a muscle in the human arm.

ICONS

Within the text, small diagrams next to an illustration help relate the topic to the big picture. For instance, in Figure *A*, a simple representation of a cell subtly reminds students of the location of the plasma membrane relative to the cytoplasm. Other icons serve as reminders of the location of reactions and processes in cells and how they

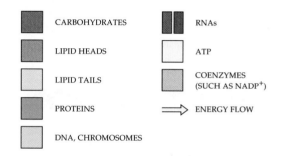

FIGURE C   *Color coding chart for the diagrams of biological molecules and cell structures.*

Step-by-step art with simple descriptions helps students visualize a process before reading text about it.

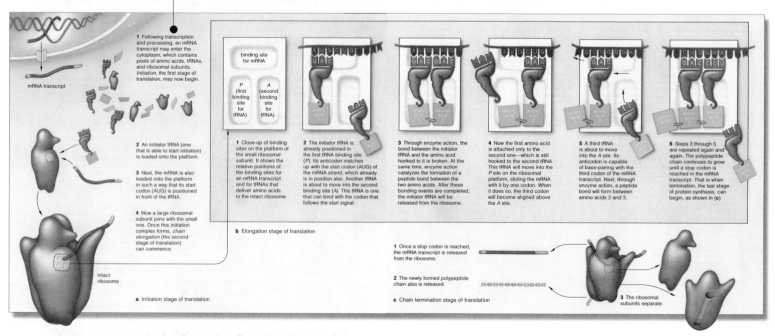

FIGURE B   *A visual overview from this edition.*

interrelate to one another. Still others remind students of the evolutionary relationships among groups of organisms, as in Chapter 26.

New to this edition are icons that invite students to use multimedia. One icon directs them to art in the CD-ROM enclosed with each student copy, another to supplemental material on the Web, and a third to InfoTrac:

CD-ROM ICON: WEB ICON: INFOTRAC ICON:

END-OF-CHAPTER STUDY AIDS

Figure *D* shows a sampling of our end-of-chapter study aids, which reinforce the key concepts. Each chapter ends with a summary in list form, review questions, a self-quiz, critical thinking questions, selected key terms, and a list of

readings. Italicized page numbers tie the review questions and key terms to relevant text pages.

END-OF-BOOK STUDY AIDS

At the book's end, the detailed classification scheme in Appendix I is helpful for reference purposes. Appendix II includes metric-English conversion charts. Appendix III has answers to the self-quizzes at the end of each chapter.

A Glossary includes boldfaced terms from the text, with pronunciation guides and word origins to make the formidable words less so. The Appendixes as well as the Glossary are printed on paper that is tinted different colors to preclude frustrating searches for where one ends and the next begins. The Index is detailed enough to help readers find doors to the text more quickly.

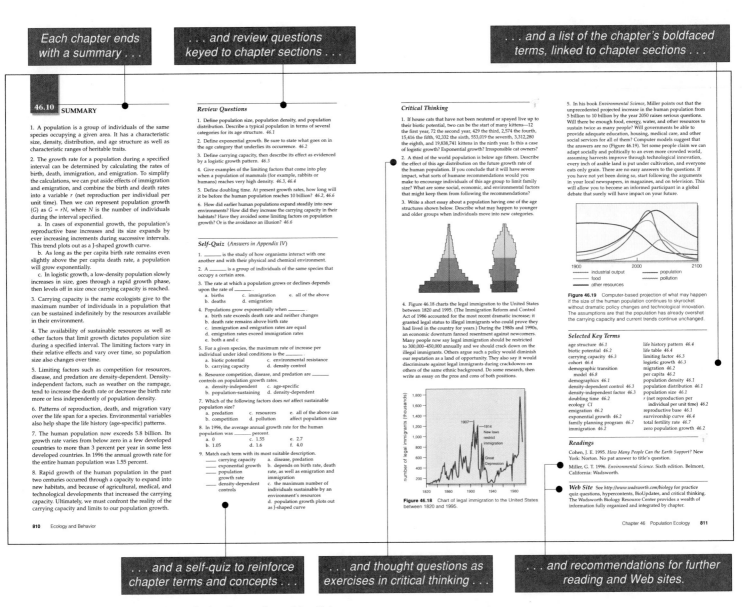

FIGURE D   *Study aids at the end of a chapter from this edition.*

CONTENT REVISIONS

UNIT VI. ANIMAL STRUCTURE AND FUNCTION   The unit takes a more comparative approach, with selections from numerous invertebrate and vertebrate groups. Chapter 34 has a less abstract model of the sodium-potassium pump as well as reorganized treatment of chemical synapses and paths of information flow. In Chapter 35, sections on the vertebrate brain now adjoin. They describe the reticular formation, blood-brain barrier, and limbic system early on. The section on memory is updated. Chapter 36 has a better classification of sensory receptors, and new, updated sections on somatic sensations and special senses. Chapter 37 now organizes the description of endocrine functions in sections on feedback controls, responses to local chemical changes, and responses to environmental cues. Chapter 38 includes a new introduction and two revised essays. Chapter 39 has refined art, a new section on blood cell disorders, and new descriptions of arterial blood pressure as well as capillary function. Chapter 40 has more on inflammation, new art on the generation of antibody diversity, and updates on HIV replication. Chapter 41 has separated sections on the invertebrates and vertebrates. Its sections on breathing, gas exchange and transport, acclimatization, hypoxia, and the physiology of diving are rewritten and updated.

Chapter 42 opens with a new story on obsessions with dieting. A new essay traces how molecular detectives used obese mice to isolate the gene for leptin, a hormone involved in appetite suppression and higher rates of metabolism. Chapter 43 has better delineation between sections on kidney structure and function, and tightened the sections on temperature regulation. Chapter 44 is overhauled to reflect extensive new research and to present a coherent picture of the vital topic of development. It includes a new essay on *Drosophila* development, which is yielding astounding insights into development in general. We rewrote much of Chapter 45 and tightened definitions to improve clarity.

ANIMAL STRUCTURE AND FUNCTION VI

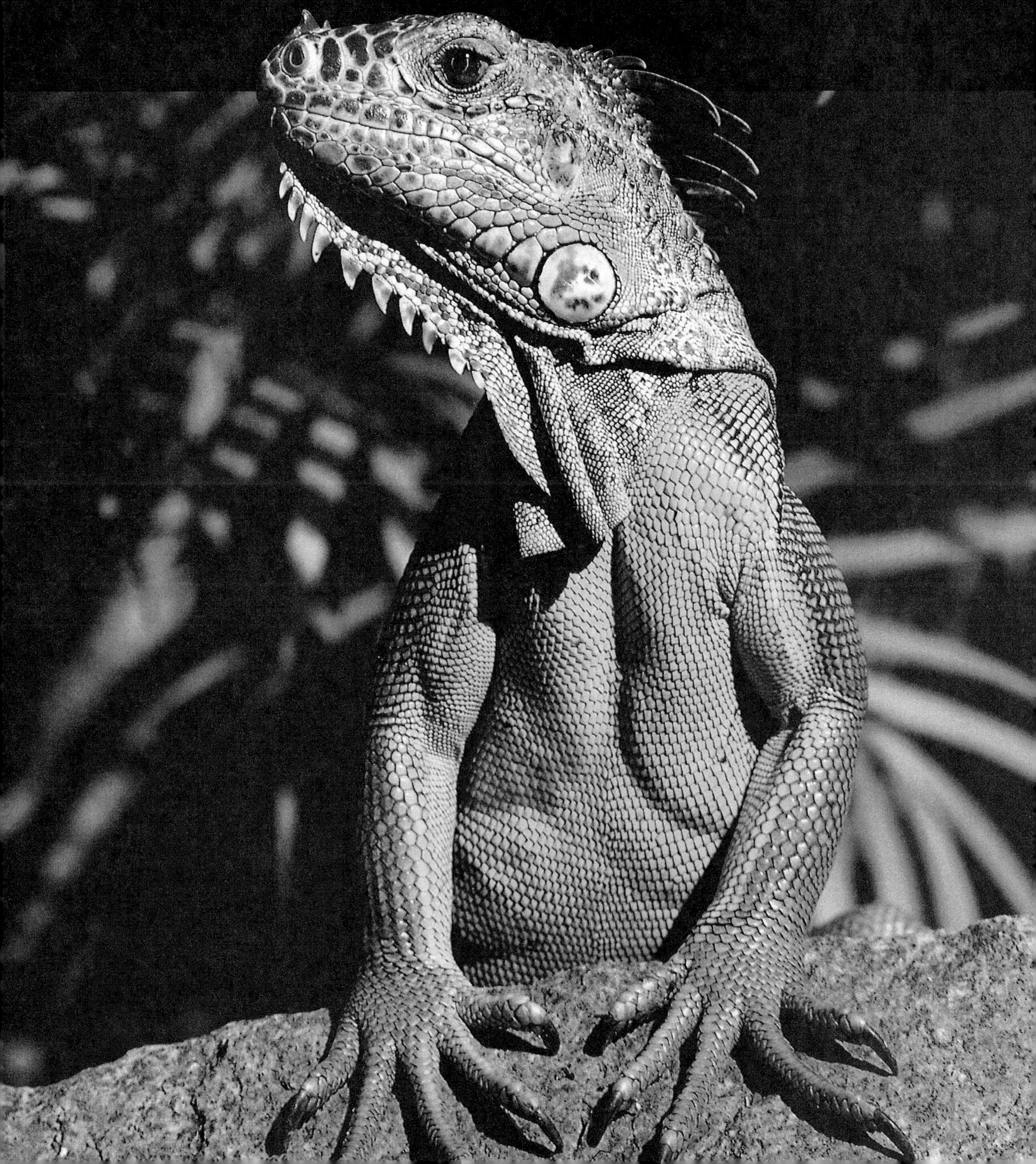

33

TISSUES, ORGAN SYSTEMS, AND HOMEOSTASIS

Meerkats, Humans, It's All the Same

After a cold night in Africa's Kalahari Desert, animals small enough to fit inside a coat pocket emerge stiffly from their burrows. These "meerkats" are a type of mongoose. They stand on their hind legs and face east, exposing their chilled bodies to the warm rays of the morning sun (Figure 33.1). Meerkats don't know it, but sunning behavior helps their enzymes. If the internal temperature of their body were to fall below a tolerable range, the activity of countless enzyme molecules in their cells would falter. With such a change in enzyme activity, metabolism would suffer.

Once meerkats warm up, they fan out from their burrows and look for food. Into the meerkat gut go insects and an occasional lizard. These are pummeled, dissolved, and then digested into glucose and other nutritious tidbits small enough to move across the gut wall, into the bloodstream, and on to cells throughout the body. In cells, aerobic machinery cracks molecules of glucose and other organic compounds apart and so releases vital energy. A respiratory system works with the bloodstream to supply the machinery with oxygen and take away its carbon dioxide leftovers.

All of this activity changes the composition and volume of the **internal environment**, which consists of interstitial fluid (tissue fluids) and blood that bathes the living cells of any complex animal. Drastic changes in those fluids would kill the cells, but a urinary system works to keep this from happening. Governing this system and all others are the body's central command posts—a nervous system and an endocrine system. The two systems work together and mobilize the body as a whole for everything from simple housekeeping tasks to heart-thumping flights from predators.

And so meerkats start us thinking about this unit's central topics: how the animal body is structurally put together (its *anatomy*) and how the body functions (its *physiology*). This chapter is an overview of the animal tissues and organ systems that we will be considering. It also introduces the central concept of **homeostasis**. With respect to the animal body, the word refers to stable operating conditions in the internal environment, as brought about by the coordinated activities of cells, tissues, organs, and organ systems.

Amazingly, the body of all complex animals consists of only four basic types of tissues. These are epithelial, connective, muscle, and nervous tissues. A **tissue** is an interacting group of cells and intercellular substances that take part in one or more particular tasks. As one example, muscle tissue takes part in contraction. An **organ** consists of different tissues that are organized in

specific proportions and patterns. Thus every vertebrate heart has predictable proportions and arrangements of epithelial, connective, muscle, and nervous tissues. An **organ system** consists of two or more organs that are interacting physically, chemically, or both in a common task, as when interconnected arteries and other vessels transport blood through the body under the driving force of the beating heart.

Cells, tissues, organs, and organ systems split up the work, so to speak, in ways that contribute to the survival of the body as a whole. This is sometimes known as a **division of labor**. By the end of this unit, you may have an abiding appreciation of the sheer magnitude of the division of labor among the separate parts and of the extent to which their activities are integrated. As you will see, regardless of whether you look at a flatworm or salmon, a meerkat or human, the body of every complex animal is structurally and physiologically adapted to perform four overriding tasks:

1. *Maintain conditions in the internal environment within ranges that are most favorable for cell activities.*

2. *Acquire nutrients and other raw materials, distribute them through the body, and dispose of wastes.*

3. *Afford protection against injury or attack from viruses, bacteria, and other agents of disease.*

4. *Reproduce, then often help nourish and protect the new individuals during their early growth and development.*

Figure 33.1 In the Kalahari Desert, gray meerkats (*Suricata suricatta*) face the warming rays of the morning sun, just as they do every morning. This simple behavior helps maintain internal body temperature. How animals function in their environment is the subject of this unit.

KEY CONCEPTS

1. The cells of most animals interact at three levels of organization—in tissues, many of which are combined in organs, which are components of organ systems.

2. Most animals are constructed of only four types of tissues, which are called epithelial, connective, muscle, and nervous tissues.

3. Each animal cell engages in basic metabolic activities that assure its own survival. At the same time, animal cells of a given tissue perform one or more activities that contribute to the survival of the animal as a whole.

4. The body's internal environment consists of all fluids that are not inside cells—that is, blood and interstitial fluid.

5. The combined contributions of cells, tissues, organs, and organ systems help maintain stability in the internal environment, which is required for the survival of each individual cell. This concept helps us understand the functions of any organ or organ system.

6. Homeostasis is the formal name for stable operating conditions in the internal environment.

General Characteristics

We commonly refer to an epithelial tissue as **epithelium** (plural, epithelia). This tissue has a free surface, which faces either a body fluid or the outside environment. *Simple* epithelium, with a single layer of cells, functions as a lining for body cavities, ducts, and tubes. *Stratified* epithelium, which has two or more cell layers, typically functions in protection, as it does in skin. Figure 33.2 shows a few examples of this type of animal tissue.

All cells in epithelium are close together, with little intervening material. As is true of nearly every animal tissue, specialized junctions provide both structural and functional links between its individual cells, which absorb, synthesize, and secrete substances.

Cell-to-Cell Contacts

Figure 33.3 shows three kinds of cell junctions that occur in epithelium and other tissues. **Tight junctions** are strands of proteins that help stop substances from leaking across a tissue. **Adhering junctions** cement cells together. **Gap junctions** help cells communicate by promoting the rapid transfer of ions and small molecules among them.

Consider the lining of your stomach. If highly acidic, gastric fluid in the stomach were to leak across this epithelium, it would digest proteins of your own body instead of those brought in with meals. (As described in Section 42.4, this actually is an outcome of a peptic ulcer.) Tight junctions in this epithelial lining and others form a leakproof barrier between the cells near their free surface. Other junctions serve as spot welds and as open channels between cells.

Figure 33.2 (a) Some basic characteristics of epithelium. Epithelia have a free surface exposed to a body fluid or to the environment. Between the epithelium's other surface and the connective tissue on which it rests is a basement membrane. The diagram below this athlete shows this arrangement for a simple epithelium, which has a single layer of cells. The light micrograph shows the upper portion of stratified epithelium. This type of epithelium has more than one layer of cells, which are flattened out near the surface.

(b) Light micrographs and sketches of three simple epithelia, which will give you an idea of the three basic shapes of cells in this type of tissue.

a Some characteristics of epithelium

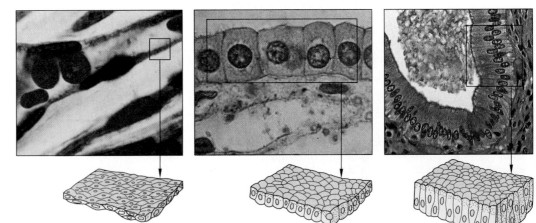

TYPE: Simple squamous

DESCRIPTION: Single layer of flattened cells

COMMON LOCATIONS: Blood vessel walls; air sacs of lungs

FUNCTION: Diffusion

TYPE: Simple cuboidal

DESCRIPTION: Single layer of cubelike cells; free surface may have microvilli (absorptive structures)

COMMON LOCATIONS: Glands and nephrons (slender tubes) in kidneys

FUNCTION: Secretion, absorption

TYPE: Simple columnar

DESCRIPTION: Single layer of tall, slender cells; free surface may have microvilli

COMMON LOCATIONS: Part of lining of gut and respiratory tract

FUNCTION: Secretion, absorption

b Examples of cell shape in simple epithelium

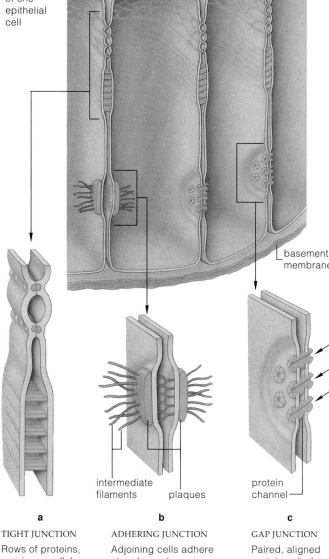

free surface of one epithelial cell

basement membrane

intermediate filaments   plaques

protein channel

a

TIGHT JUNCTION

Rows of proteins, running parallel to the tissue's free surface, form strands that prevent leaks between adjacent cells.

b

ADHERING JUNCTION

Adjoining cells adhere at a plaque (a mass of proteins) that is firmly anchored just inside the plasma membrane of each by intermediate filaments of the cytoskeleton.

c

GAP JUNCTION

Paired, aligned protein cylinders form channels that span the membranes of adjoining cells and interconnect their cytoplasm.

Figure 33.3 *Left:* Examples of cell junctions.

(**a**) In some epithelia, protein strands that ring each cell form tight seals and fuse it with its neighbors. The tight junctions prevent substances from leaking across the epithelium's free surface. Substances reach the tissues below only by passing *through* the epithelial cells. Built-in controls make the plasma membrane of these cells selectively permeable. At a given time, they are allowing some substances but not others to move across, through the interior of transport proteins (Section 5.3).

(**b**) Adhesion junctions are spot welds that hold the cells of epithelium (and all other tissues) together, so that they function as a unit. They are profuse in the skin's surface layer and in other tissues subjected to abrasion.

(**c**) Gap junctions are channels across the plasma membrane that allow the cytoplasm of adjoining cells to interconnect. They promote the diffusion of ions and small molecules from cell to cell. They are abundant in the heart, liver, and other organs in which cell activities must be rapidly coordinated.

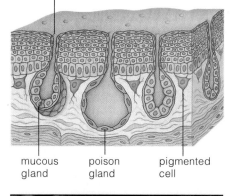

pore opening at surface of skin

mucous gland   poison gland   pigmented cell

Figure 33.4
Section through a frog's glandular epithelium. The photograph shows a frog of the genus *Dendrobates*, which produces one of the most lethal glandular secretions known. Some natives of a tribe in Colombia use this exocrine gland secretion to poison dart tips, which they shoot through a blowgun. Pigment-rich cells that produce the surface coloration branch into lower epithelial layers of skin. The showy coloration of all poisonous frogs serves as a strong warning signal to potential predators.

Glandular Epithelium

Glands are secretory cells or multicelled structures that are derived from epithelium and often remain linked to it. **Exocrine glands** secrete mucus, saliva, earwax, milk, oil, digestive enzymes, and other cell products. Usually the products are released onto a free epithelial surface through ducts or tubes. Figure 33.4 shows an example.

By contrast, **endocrine glands** have no ducts. Their products are hormones, which are secreted directly into the fluid bathing the gland. Typically, the bloodstream picks up the hormone molecules and distributes them to target cells elsewhere in the body.

Epithelia are sheetlike tissues that have one free surface. Different types of epithelia line the body's surface and its cavities, ducts, and tubes.

Profuse cell-to-cell contacts bind the cells closely together, with little intercellular material between them.

Glands are secretory cells or multicelled structures derived from epithelium and often connected to it.

CONNECTIVE TISSUE

Of all tissues in complex animals, connective tissues are the most abundant and widely distributed. They range from connective tissue proper to specialized types, which include cartilage, bone, adipose tissue, and blood (Table 33.1). In all types except blood, fibroblasts and other kinds of cells secrete fibers of collagen or elastin, which are structural proteins. (This is the collagen that plastic surgeons use to plump wrinkled skin, sunken acne scars, and lips.) Fibroblasts also secrete modified polysaccharides. Secreted material accumulates between cells and fibers, as the tissue's "ground substance."

Connective Tissue Proper

All of these tissues have mostly the same components but in different proportions. **Loose connective tissue** has its fibers and cells loosely arranged in a semifluid ground substance (Figure 33.5a). Often it serves as a support framework for epithelium. Besides fibroblasts, it contains infection-fighting white blood cells. When small cuts or other wounds allow bacteria to breach the skin or lining of the digestive, respiratory, or urinary tracts, these cells mount an early counterattack.

Dense, irregular connective tissue contains fibers, mostly collagen-containing ones, and a few fibroblasts. This tissue forms protective capsules around organs that do not stretch much. It also is present in the deeper

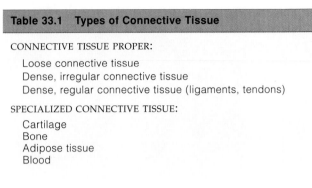

Table 33.1 Types of Connective Tissue

CONNECTIVE TISSUE PROPER:

    Loose connective tissue
    Dense, irregular connective tissue
    Dense, regular connective tissue (ligaments, tendons)

SPECIALIZED CONNECTIVE TISSUE:

    Cartilage
    Bone
    Adipose tissue
    Blood

part of skin (Figure 33.5b). **Dense, regular connective tissue**, which has parallel bundles of many collagen fibers, resists being torn apart. Rows of fibroblasts often intervene between the bundles. This is true of tendons, which attach skeletal muscle to bones (Figure 33.5c), and of elastic ligaments, which attach bones to each other.

Specialized Connective Tissue

Intercellular material of **cartilage** is solid yet pliable, like solid rubber, and resists compression. The material is produced by cells that in time become imprisoned in small cavities in their own secretions (Figure 33.5d). Cartilage elements in vertebrate embryos are structural models for bones that replace most of them. Cartilage

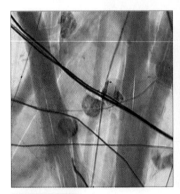

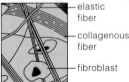

a

TYPE: Loose connective tissue
DESCRIPTION: Fibroblasts, other cells, plus fibers loosely arranged in semifluid ground substance
COMMON LOCATIONS: Under the skin and most epithelia
FUNCTION: Elasticity, diffusion

— elastic fiber
— collagenous fiber
— fibroblast

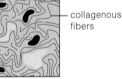

b

TYPE: Dense, irregular connective tissue
DESCRIPTION: Collagenous fibers, fibroblasts, less ground substance
COMMON LOCATIONS: In skin and capsules around some organs
FUNCTION: Support

— collagenous fibers

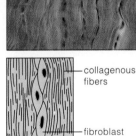

c

TYPE: Dense, regular connective tissue
DESCRIPTION: Collagen fibers in parallel bundles, long rows of fibroblasts, little ground substance
COMMON LOCATIONS: Tendons, ligaments
FUNCTION: Strength, elasticity

— collagenous fibers
— fibroblast

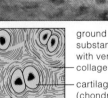

d

TYPE: Cartilage
DESCRIPTION: Cells embedded in pliable, solid ground substance
COMMON LOCATIONS: Ends of long bones, nose, parts of airways, skeleton of vertebrate embryos
FUNCTION: Support, flexibility, low-friction surface for joint movement

ground substance with very fine collagen fibers
— cartilage cell (chondrocyte)

Figure 33.5 Examples of connective tissue proper and of specialized connective tissue.

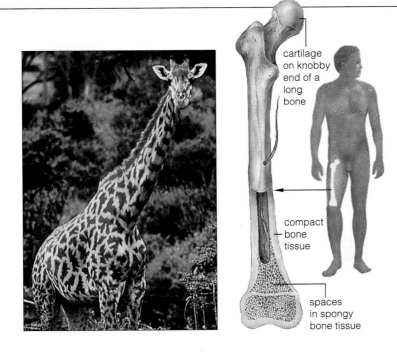

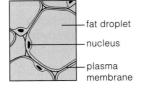

cartilage on knobby end of a long bone

compact bone tissue

spaces in spongy bone tissue

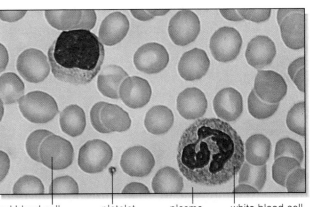

red blood cell   platelet   plasma   white blood cell

Figure 33.7 Some components of human blood, a vascular tissue. Its straw-colored, liquid matrix (plasma) is mostly water in which nutrients, diverse proteins, oxygen, carbon dioxide, ions, and other substances are dissolved.

Figure 33.6 Cartilage and bone tissue. Spongy bone tissue has tiny, needlelike hard parts with spaces in between. Compact bone tissue is more dense. Bone is a load-bearing tissue that resists compression. Over time, it was the basis of increases in the body size of many land vertebrates, including giraffes. It gives large animals selective advantages. Among other things, they can ignore most predators with impunity, roam farther for food and water, and heat up and cool off more slowly than small animals (because of a lower surface-to-volume ratio and greater heat production).

compact bone tissue

space that contained living bone cell (osteocyte)

fat droplet

nucleus

plasma membrane

TYPE: Bone tissue

DESCRIPTION: Collagen fibers, ground substance hardened with calcium

COMMON LOCATIONS: Bones of vertebrate skeleton

FUNCTION: Movement, support, protection

e

TYPE: Adipose tissue

DESCRIPTION: Large, tightly packed fat cells occupying most of ground tissue

COMMON LOCATIONS: Under skin, around heart, kidneys

FUNCTION: Energy reserves, insulation, padding

f

maintains the shape of the nose, outer ear, and other body parts. It cushions joints between adjacent bones of the vertebral column, limbs, hands, and elsewhere.

Bone (Figures 33.5*e* and 33.6) is the weight-bearing tissue of vertebrate skeletons, which support or protect softer tissues and organs. Minerals harden this tissue; its collagen fibers and ground substance are loaded with calcium salts. Limb bones, such as the long bones of your legs, interact with the skeletal muscles that are attached to them to bring about movements. Tissues in some bones also are production sites for blood cells.

Adipose tissue is so chockful of large fat cells, it no longer looks like a connective tissue (Figure 33.5*f*). The body's excess carbohydrates and proteins are converted to storage fats and tucked away in this tissue. Adipose tissue is richly supplied with blood, which serves as an immediately accessible "highway" along which fats can move to and from the tissue's individual cells.

Blood, derived mainly from connective tissue, has transport functions. Circulating within *plasma*, its fluid medium, are a great many red blood cells, white blood cells, and platelets (Figure 33.7). Red blood cells deliver oxygen to metabolically active tissues, and carry carbon dioxide and other wastes away from them. Plasma is largely water, but it has a great number of different kinds of proteins, ions, and other substances dissolved in it. Section 39.2 describes this complex tissue.

Diverse types of connective tissues bind together, support, strengthen, protect, and insulate other tissues in the body.

Most connective tissues consist of protein fibers and a variety of cells in a ground substance. One type, blood, is a fluid tissue. Another type, adipose tissue, serves as a reservoir of stored energy.

In muscle tissue alone, cells *contract* (that is, shorten) in response to stimulation, then lengthen and so return to their uncontracted state. Many long, cylindrical cells are arranged in parallel in these tissues. Their coordinated contraction and relaxation help move the body through the environment and maintain or change the position of

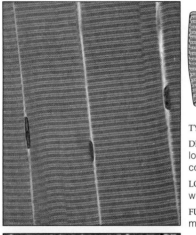

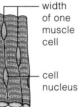

TYPE: Skeletal muscle

DESCRIPTION: Bundles of long, cylindrical, striated, contractile cells

LOCATION: Associated with skeleton

FUNCTION: Locomotion, movement of body parts

a

width of one muscle cell

cell nucleus

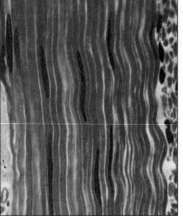

cells, teased apart for clarity

TYPE: Smooth muscle

DESCRIPTION: Contractile cells with tapered ends

LOCATION: Wall of internal organs, such as stomach

FUNCTION: Movement of internal organs

b

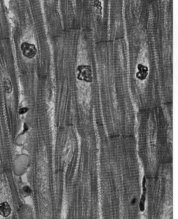

junction between adjacent cells

TYPE: Cardiac muscle

DESCRIPTION: Cylindrical, striated cells that have specialized end junctions

LOCATION: Wall of heart

FUNCTION: Pump blood within circulatory system

c

Figure 33.8   Characteristics and examples of skeletal muscle, smooth muscle, and cardiac muscle tissues.

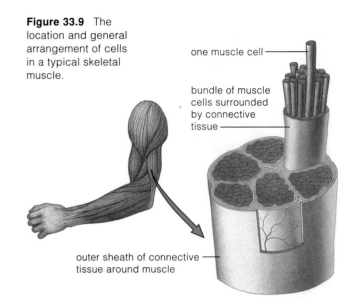

Figure 33.9   The location and general arrangement of cells in a typical skeletal muscle.

one muscle cell

bundle of muscle cells surrounded by connective tissue

outer sheath of connective tissue around muscle

its individual parts. The three types of muscle tissue are called skeletal, smooth, and cardiac muscle tissues.

The muscles that are connected to the bones of your skeleton consist of **skeletal muscle tissue** (Figure 33.8*a*). In a typical muscle, such as the biceps, striated skeletal muscle cells are bundled together in parallel. (*Striated* means striped.) A sheath of tough connective tissue encloses several bundles of the muscle cells, as Figure 33.9 indicates. The structure and function of skeletal muscle tissue are topics of Chapter 38.

The contractile cells of **smooth muscle tissue** taper at both ends (Figure 33.8*b*). Cell junctions hold them together, and a connective tissue sheath encloses them. The wall of internal organs, such as blood vessels, the stomach, and the intestines, contains this type of muscle tissue. Smooth muscle action is sometimes said to be "involuntary," because we usually are not able to make it contract merely by thinking about it (as we can do with skeletal muscle).

Cardiac muscle tissue is a contractile tissue that is present only in the heart (Figure 33.8*c*). Cell junctions fuse together the plasma membranes of cardiac muscle cells. Junctions at some fusion points allow the cells to contract as a unit. When one cell receives a signal to contract, its neighbors are stimulated to contract, also.

Muscle tissue, which can contract (shorten) in response to stimulation, helps move the body and specific body parts.

Skeletal muscle is the only muscle tissue attached to bones. Smooth muscle is a component of internal organs. Cardiac muscle alone makes up the contractile walls of the heart. Connective tissue sheathes all three types of tissues.

NERVOUS TISSUE

Of all tissues, **nervous tissue** exerts the greatest control over the body's responsiveness to changing conditions. In a human nervous system, much of the tissue consists of **neurons**, which are a type of excitable cell. More than half of it consists of **neuroglia**, diverse cells that protect and structurally and metabolically support the neurons.

When a neuron is suitably stimulated, an electrical disturbance swiftly travels along its plasma membrane. Arrival of the disturbance at the neuron's endings, or output zone, triggers events that may cause stimulation or inhibition of adjacent neurons and other cells.

cell body of one motor neuron

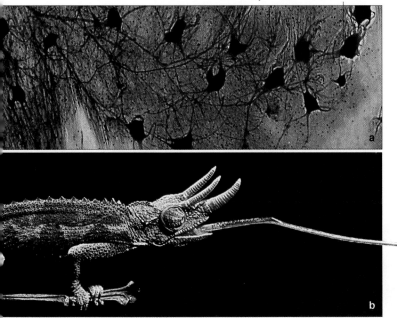

Figure 33.10 (**a**) A few motor neurons that form part of the human nervous system. This type relays signals from the brain or spinal cord to muscles and glands. Collectively, different types of neurons detect and process a great number and variety of signals about environmental changes, and initiate suitable responses. (**b**) Without neurons, for instance, this chameleon could not detect an edible insect, calculate its distance, and command a long, sticky, prey-capturing tongue to uncoil with stunning speed.

For example, more than a hundred billion neurons are organized as communication lines throughout your body. Some detect specific changes in environmental conditions. Others coordinate immediate and long-term responses to change. The type shown in Figure 33.10*a* relays signals from the brain to muscles and glands. How these cells function is a topic of later chapters.

Neurons are the basic units of communication in nervous tissue. Different kinds detect specific stimuli, integrate information, and issue or relay commands for response.

FRONTIERS IN TISSUE RESEARCH

As you probably will gather after thinking about the many tissues that are used as examples throughout this unit, a tissue is more than a sum of its cells. As each new animal grows and develops, its cells interact and become organized in particular ways to give rise to the body's diverse tissues. And as each new tissue develops, its cells synthesize specific gene products that are vital for normal body functioning.

For many decades, medical researchers have attempted to find a way to construct artificial tissues in quantity in the laboratory. Currently, they can grow extensive sheets of epidermis from a patient and use it to regenerate skin that was destroyed through third-degree burns and other types of severe damage. In some laboratories, researchers are taking small sections of epidermis from the skin of patients and exposing them to a culture medium that contains nutrients and growth factors. The cells proliferate and form *laboratory-grown epidermis*. When surgeons place an epidermal sheet over a wound, the cells in the sheet interact biochemically and structurally with underlying cells. As an outcome of the interactions, the damaged or missing tissue is regenerated.

On the horizon are *designer organs*. These encapsulated, selected groups of living cells might synthesize specific hormones, enzymes, and other substances. The idea is to surgically snip a bit of epithelium from a patient and then use it to   enclose the cells. Because the capsule is derived   from epithelial cells of the patient's own body, it will not be chemically recognized as "foreign" and attacked by the patient's immune system. As you will see in Chapter 40, rejection of tissue and organ implants can have serious medical consequences.

Currently, biotechnologists are close to understanding how to synthesize molecular cues that will allow designer organs to stick to appropriate sites in the body. Once the synthetic organs stick, they may become integral parts of normal body functioning.

Ultimately, the goal of this research is to put together packages of cells capable of producing specific life-saving substances that are absent in patients who suffer from severe genetic disorders or chronic diseases. For example, imagine the potential for people who have type I *diabetes mellitus*. This metabolic disorder results in an elevated concentration of glucose in the blood. Affected people produce little if any insulin, which is the hormone that signals cells to take up glucose from the blood. Unless they receive insulin injections on a regular basis, they will die. However, if a customized, insulin-secreting organ can be successfully installed inside the body of such individuals, their daily injections of insulin might be a thing of the past.

Overview of the Major Organ Systems

Figure 33.11 gives an overview of the organ systems of a typical vertebrate, the adult human. Figure 33.12 lists some terms that are used when describing the positions of the various organs. It also shows major body cavities in which they are located.

Each organ system contributes to the survival of all living cells in the animal body. You may think this is stretching things a bit. For example, how could muscles and bones be helping each microscopically small cell stay alive? And yet interactions between the skeletal and muscular systems allow us to move about—toward sources of nutrients and water, for example. Some parts of the two organ systems help keep blood circulating to cells, as when leg muscle contractions help move blood in veins back to the heart. Blood inside the circulatory system rapidly transports oxygen, nutrients, and other substances to cells, and transports products and wastes away from them. The respiratory system imports and exports the gases, skeletal muscles assist the respiratory system—and so it goes, throughout the entire body.

Figure 33.12 (**a**) The major cavities in the human body. (**b,c**) Directional terms and planes of symmetry for the vertebrate body. Notice how the *midsagittal* plane divides the body into right and left halves. Most vertebrates, such as fishes and rabbits, move with the main body axis parallel with Earth's surface. For them, *dorsal* pertains to their back or upper surface, and *ventral* pertains to the opposite, lower surface.

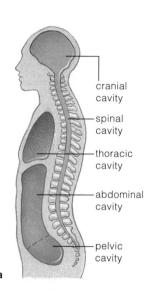

cranial cavity
spinal cavity
thoracic cavity
abdominal cavity
pelvic cavity

a

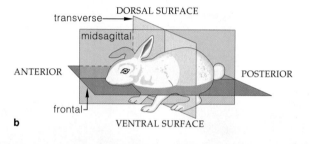

DORSAL SURFACE
transverse
midsagittal
ANTERIOR
POSTERIOR
frontal
VENTRAL SURFACE

b

Figure 33.11 *Below*: Human organ systems and their functions.

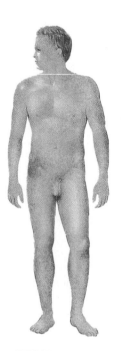
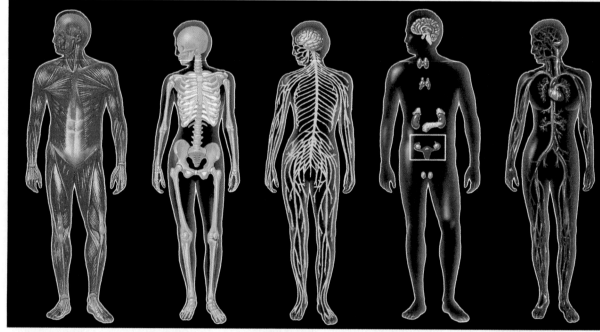

INTEGUMENTARY SYSTEM	MUSCULAR SYSTEM	SKELETAL SYSTEM	NERVOUS SYSTEM	ENDOCRINE SYSTEM	CIRCULATORY SYSTEM
Protect body from injury, dehydration, and some pathogens; control its temperature; excrete some wastes; receive some external stimuli	Move body and its internal parts; maintain posture; produce heat (by high metabolic activity)	Support and protect body parts; provide muscle attachment sites; produce red blood cells; store calcium, phosphorus	Detect both external and internal stimuli; control and coordinate responses to stimuli; integrate all organ system activities	Hormonally control body function; work with nervous system to integrate short-term and long-term activities	Rapidly transport many materials to and from cells; help stabilize internal pH and temperature

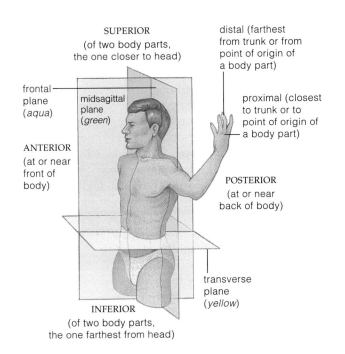

SUPERIOR
(of two body parts,
the one closer to head)

distal (farthest
from trunk or from
point of origin of
a body part)

frontal plane
(*aqua*)

midsagittal plane
(*green*)

proximal (closest
to trunk or to
point of origin of
a body part)

ANTERIOR
(at or near
front of
body)

POSTERIOR
(at or near
back of body)

transverse plane
(*yellow*)

INFERIOR
(of two body parts,
the one farthest from head)

c Unlike quadrupedal animals, humans walk upright, with their main body axis perpendicular to the ground. *Anterior* refers to the front of the body; it corresponds to ventral, as shown in (**b**). *Posterior* refers to the back; it corresponds to dorsal.

Tissue and Organ Formation

Where do the tissues of organ systems come from? To get a sense of how they originate, start with a sperm and egg. Recall that sperm and eggs develop from germ cells, which are immature reproductive cells. (All other cells in the body are "somatic," after the Greek word for body.) After a zygote forms at fertilization, mitotic cell divisions form an early embryo. In vertebrates, the cells become arranged as three primary tissues—ectoderm, mesoderm, and endoderm. The three are the embryonic forerunners of all tissues in the adult. **Ectoderm** gives rise to the skin's outer layer and to tissues of a nervous system. **Mesoderm** gives rise to tissues of the muscles, bones, and most of the circulatory, reproductive, and urinary systems. **Endoderm** gives rise to the lining of the digestive tract and to organs derived from it.

In general, vertebrates have the same kinds of organ systems. Each organ system serves specialized functions, such as gas exchange, blood circulation, and locomotion.

Vertebrate tissues, organs, and organ systems arise from three primary tissues in the developing embryo. These primary tissues are ectoderm, mesoderm, and endoderm.

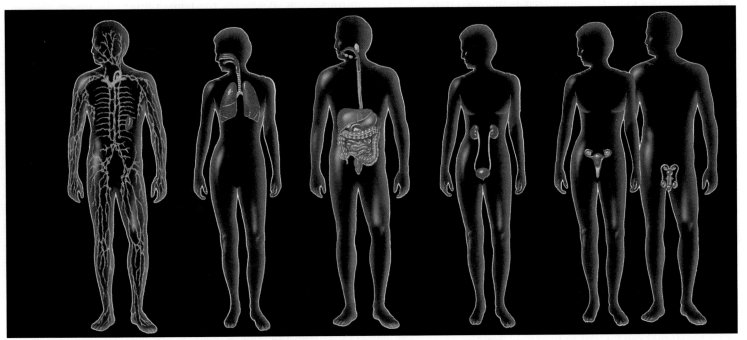

LYMPHATIC SYSTEM	RESPIRATORY SYSTEM	DIGESTIVE SYSTEM	URINARY SYSTEM	REPRODUCTIVE SYSTEM
Collect and return some tissue fluid to the bloodstream; defend the body against infection and tissue damage	Rapidly deliver oxygen to the tissue fluid that bathes all living cells; remove carbon dioxide wastes of cells; help regulate pH	Ingest food and water; mechanically, chemically break down food, and absorb small molecules into internal environment; eliminate food residues	Maintain the volume and composition of internal environment; excrete excess fluid and blood-borne wastes	*Female:* Produce eggs; after fertilization, afford a protected, nutritive environment for the development of new individual. *Male:* Produce and transfer sperm to the female. Hormones of both systems also influence other organ systems.

Concerning the Internal Environment

To stay alive, your cells must remain bathed in a fluid that offers nutrients and carries away metabolic wastes. In this they are no different from an amoeba or some other free-living, single-celled organism. The difference is, trillions of cells coexist in your body. They must draw nutrients from and dump wastes into the same fifteen liters of fluid, which is less than sixteen quarts.

The fluid *not* inside cells is **extracellular fluid**. Much of it is *interstitial*, meaning it occupies spaces between cells and tissues. The remainder is *plasma*, which is the fluid portion of blood. The interstitial fluid exchanges substances with the cells it bathes and with blood.

In functional terms, extracellular fluid is continuous with the fluid in cells. That is why drastic changes in the composition and volume of extracellular fluid have drastic effects on cell activities. The type and number of ions are especially crucial, for they must be maintained at concentrations that are compatible with metabolism. Otherwise, the animal itself cannot survive.

It makes no difference whether an animal is simple or complex. *The component parts of any animal work together to maintain the stable fluid environment required by all of its living cells.* This concept is absolutely central to understanding the structure and function of animals, and its key points may be summarized as follows. First, each cell of the animal body engages in basic metabolic activities that ensure its own survival. Second, the cells of a given tissue also perform one or more activities that contribute to the survival of the whole organism. Third, the combined contributions of individual cells, organs, and organ systems help maintain the stable internal environment—that is, the extracellular fluid—required for individual cell survival.

Mechanisms of Homeostasis

Homeostasis, recall, refers to stable operating conditions in the internal environment. Three components interact to maintain this state. They are called sensory receptors, integrators, and effectors. **Sensory receptors** are cells or cell parts that can detect a **stimulus**, which is a specific change in the environment. When someone kisses you, for example, there is a change in pressure on your lips. Receptors in the skin of your lips translate the stimulus into a signal that can be sent to the brain. Your brain is an **integrator**, a central command post where different bits of information are pulled together in the selection of a response. The brain sends signals to your muscles or glands (or both). Muscles and glands are **effectors**— they carry out the response. In this particular

case, the response might include flushing with pleasure and kissing the person back. Of course, you cannot engage in a kiss indefinitely, for this would prevent you from eating and carrying out other activities necessary to maintain operating conditions inside your body.

So how does your brain reverse the physiological changes induced by the kiss? Receptors only provide it with information about how things *are* operating. The brain also receives information about how things *should be* operating—that is, information from "set points." When physical or chemical conditions deviate sharply from a set point, the brain functions to bring them back to an effective operating range. It does this by way of signals that cause specific muscles and specific glands to increase or decrease their activity.

NEGATIVE FEEDBACK Feedback mechanisms are among the controls that operate to keep physical and chemical aspects of the body within tolerable ranges. As one example, with a **negative feedback mechanism**, some activity alters a condition in the internal environment, and this triggers a response that reverses the altered condition (Figure 33.13).

Think of a furnace with a thermostat. A thermostat senses the air temperature and "compares" it against a preset point on a thermometer built into the furnace's control system. Whenever the temperature falls below the preset point, the thermostat signals a switching mechanism that can turn on the furnace. When the air becomes heated enough to match the prescribed level, the thermostat signals the switching mechanism, which shuts off the furnace.

Similarly, feedback mechanisms help keep the body temperature of meerkats, humans, huskies, and many other animals near 37°C (98.6°F), even during hot or cold weather. Visualize a young husky running around on a hot summer day. Soon its body gets hot, and receptors trigger events that slow down the whole dog *and* its cells. The husky searches for shade and rests under a tree. Moisture from its respiratory system evaporates from the tongue and carries away some body heat with

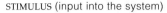

STIMULUS (input into the system)

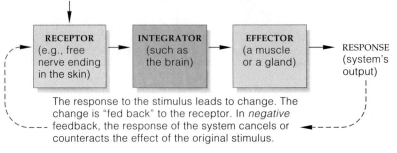

The response to the stimulus leads to change. The change is "fed back" to the receptor. In *negative* feedback, the response of the system cancels or counteracts the effect of the original stimulus.

Figure 33.13 Components necessary for negative feedback at the organ level.

STIMULUS:

The husky is overactive on a hot, dry day and its body surface temperature rises.

RECEPTORS
in skin and elsewhere detect the temperature change.

An INTEGRATOR
(the hypothalamus, a brain region) compares input from the receptors against a set point.

Some EFFECTORS
(pituitary gland and thyroid gland) trigger widespread adjustments.

RESPONSE:

Temperature of circulating blood starts decreasing.

Many **EFFECTORS** *carry out specific responses:*

SKELETAL MUSCLES	SMOOTH MUSCLE IN BLOOD VESSELS	SALIVARY GLANDS	ADRENAL GLANDS
Husky rests, starts to pant (behavioral changes).	*Blood carrying metabolically generated heat shunted to skin, some heat lost to surroundings.*	*Salivary secretions increase, evaporation from tongue has cooling effect.*	*Output drops, husky is less stimulated.*

Activity of the body in general slows down (behavioral change).

The overall slowdown in activities results in less metabolically generated heat.

Figure 33.14 Homeostatic controls over the internal temperature of a husky's body. *Blue* arrows indicate the main control pathways. The *dashed line* shows how a feedback loop is completed.

it, as shown in Figure 33.14. These control mechanisms and others counter overheating by curbing activities that naturally generate metabolic heat and by giving up the body's excess heat to the surrounding air.

POSITIVE FEEDBACK In some cases, **positive feedback mechanisms** operate. These controls set in motion a chain of events that *intensify* a change from an original condition—and after a limited time, the intensification reverses the change. Positive feedback is associated with instability in a system. For example, during sexual intercourse, chemical signals from a female's nervous system can induce her to make intense physiological responses to her sexual partner. Her responses stimulate changes in her partner that stimulate the female even more—and so on until an explosive, climax level of excitation is reached. Normal conditions now return, and homeostasis prevails.

As another example, at childbirth, a fetus exerts pressure on the wall of its mother's uterus. Pressure stimulates the production and secretion of oxytocin, a hormone. Oxytocin causes wall muscles to contract and exert pressure on the fetus, which exerts more pressure on the wall, and so on until the fetus is expelled.

What we have been describing is a general pattern of monitoring and responding to a constant flow of information about an animal's internal and external environments. During all of this activity, organ systems operate together in astoundingly coordinated fashion. Throughout this unit, we will be asking the following questions about their operation:

1. *What physical or chemical aspects of the internal environment are organ systems working to maintain as conditions change?*

2. *By what means are organ systems kept informed of the various changes?*

3. *By what means do they process incoming information?*

4. *What mechanisms are set in motion in response?*

As you will see in later chapters, operation of all organ systems is under neural and endocrine control.

Each living cell of an animal body engages in basic metabolic activities that ensure its own survival. Concurrently, the cells of any given tissue are performing one or more activities that contribute to the survival of the whole animal.

The combined contributions of cells, organs, and organ systems help maintain the stable internal environment (the extracellular fluid) required for individual cell survival.

Homeostatic control mechanisms help maintain physical and chemical aspects of the body's internal environment within ranges that are most favorable for cell activities.

SUMMARY

1. A tissue is an aggregation of cells and intercellular substances that perform a common task. An organ is a structural unit of different tissues combined in definite proportions and patterns that allow them to perform a common task. An organ system has two or more organs interacting chemically, physically, or both in ways that contribute to the survival of the body as a whole.

2. Epithelial tissues cover external body surfaces and line internal cavities and tubes. These tissues have one free surface exposed to body fluids or the environment.

3. A great variety of connective tissues bind together, support, strengthen, protect, and insulate other tissues. Most types are composed of fibers of structural proteins (especially collagen), fibroblasts, and other cells within a ground substance.

   a. Loose connective tissue, with a semifluid ground substance, is present under skin and most epithelia.

   b. Dense, irregular connective tissue contains mostly collagen fibers and fibroblasts. It is present in skin, and it forms protective capsules around a number of organs.

   c. Dense, regular connective tissue, with its parallel bundles of collagen fibers, provides structural support for tendons, skin, and other organs.

   d. Cartilage, with its solid yet pliable intercellular material, has structural and cushioning roles. Bone, the weight-bearing tissue of vertebrate skeletons, interacts with skeletal muscle to bring about movement.

   e. Blood, a specialized connective tissue, consists of plasma, cellular components, and dissolved substances. Adipose tissue, another specialized connective tissue, is a reservoir of energy; it consists mainly of fat cells.

4. Muscle tissues contract (shorten), then return to the resting position. They help move the body or parts of it. The three types of muscle tissue are skeletal muscle, smooth muscle, and cardiac muscle tissue.

5. Nervous tissue intercepts and integrates information about internal and external conditions, and governs the body's responses to change. Neurons of this tissue are the basic units of communication in nervous systems.

6. Tissues, organs, and organ systems work together to maintain the stable internal environment (that is, the extracellular fluid) required for individual cell survival. At homeostasis, conditions in the internal environment are balanced at levels most favorable for cell activities.

7. Feedback controls help maintain internal operating conditions for the body's cells. With negative feedback, for example, a change in a particular condition triggers a response that results in a reversal of the change.

8. Homeostasis depends on receptors, integrators, and effectors. Receptors detect stimuli, or specific changes in the environment. Integrating centers, such as a brain, process the information and direct muscles and glands (the body's effectors) to carry out responses.

Review Questions

1. Describe the characteristics of epithelial tissue in general. Then describe the various types of epithelial tissues in terms of specific characteristics and functions. *33.1*

2. List the major types of connective tissues; add the names and characteristics of their specific types. *33.2*

3. Identify and describe the following tissues: *33.1–33.3*

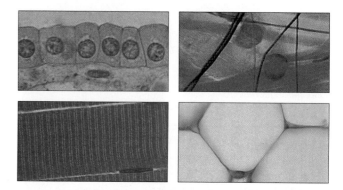

4. Identify this category of tissue and its characteristics. *33.4*

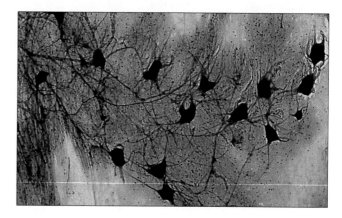

5. What type of cell serves as the basic unit of communication in nervous systems? *33.4*

6. Define animal tissue, organ, and organ system. List and define the functions of the eleven major organ systems of the human body. *CI, 33.6*

7. Define extracellular fluid, interstitial fluid, and plasma. *33.7*

8. Define homeostasis. *CI, 33.7*

9. Briefly describe two major categories of the homeostatic mechanisms operating in the human body. *33.7*

Self-Quiz (*Answers in Appendix IV*)

1. _____ tissues have closely linked cells and one free surface.
   a. Epithelial   b. Connective   c. Nervous   d. Muscle

2. In most _____ , cells secrete fibers of collagen and elastin.
   a. epithelial tissue   c. muscle tissue
   b. connective tissue   d. nervous tissue

3. _____ has a semifluid ground substance and occurs under skin and most epithelia.
   a. Dense, irregular connective tissue
   b. Loose connective tissue
   c. Dense, regular connective tissue
   d. Cartilage

4. _____, a specialized connective tissue, is mostly plasma with cellular components and various dissolved substances.
   a. Irregular connective tissue   c. Cartilage
   b. Blood   d. Bone

5. After you eat too many carbohydrates and proteins, your body converts the excess to storage fats, which accumulate in _____ .
   a. connective tissue proper   c. adipose tissue
   b. dense connective tissue   d. both b and c

6. Components of _____ detect and coordinate information about changes and control responses to those changes.
   a. epithelial tissue   c. muscle tissue
   b. connective tissue   d. nervous tissue

7. In your own body, _____ can shorten (contract).
   a. epithelial tissue   c. muscle tissue
   b. connective tissue   d. nervous tissue

8. Cells of complex animals _____ .
   a. survive by their own metabolic activities
   b. contribute to the survival of the whole animal
   c. help maintain extracellular fluid
   d. all of the above

9. At _____, physical and chemical conditions in the internal environment are being kept within tolerable ranges.
   a. positive feedback   c. homeostasis
   b. negative feedback   d. metastasis

10. With negative feedback mechanisms, _____ .
    a. a stimulus brings about a response that tends to return internal operating conditions to the original state
    b. a stimulus suppresses internal operating conditions to levels below a set point for the body
    c. a stimulus raises internal operating conditions to levels above a set point for the body
    d. fewer solutes are fed back to the affected cells

11. Of the components that exert feedback control of organ activity, _____ detect specific changes in the environment, an _____ pulls together different bits of information and selects a suitable response, and _____ carry out the response.

12. Match the terms with the suitable description.
    ____ epithelium   a. rather pliable, like rubber
    ____ cartilage   b. covers or lines body surfaces
    ____ homeostasis   c. stable internal environment
    ____ muscles and glands d. integrating center
    ____ positive feedback   e. the most common type of
    ____ negative feedback   homeostatic control mechanism
    ____ brain   f. effectors
                             g. chain of events intensifies original condition in body

Critical Thinking

1. *Anhidrotic ectodermal dysplasia*, a genetic disorder described in Section 15.4, has been associated with a recessive allele on the mammalian X chromosome. Among other symptoms of this disorder, affected males and females have no sweat glands in the tissues where the recessive allele is expressed. What type of tissue are we talking about?

2. Adipose tissue and blood are often said to be "atypical" connective tissues. Compared with other connective tissues, which of their features are *not* typical?

3. After graduating from high school, Jeff and Ryan set out on a trip through the desert roads of California and Arizona. One hot, dry morning in Joshua Tree National Monument, they saw an unusual rock formation that didn't appear to be too far from the road. They left the car and started hiking toward it. Their destination turned out to be farther away than they thought and the sun's rays were more relentless than they had anticipated. They reached the shade of the rocks in the early afternoon. Their canteen was nearly empty, and the physiological meaning of "thirst" made itself known to them in a scary way. They knew they had to locate and drink water (or some other fluid), which is what the brain usually tells us to do when the body starts to get dehydrated.

From what you read in this chapter, would you suspect that Ryan and Jeff's *thirst behavior* is part of a positive or negative feedback control mechanism?

4. *Porphyria* is a genetic disorder that shows up in about 1 in every 25,000 individuals. Affected individuals lack certain enzymes that are part of a metabolic pathway leading to formation of heme, which is the iron-containing group of hemoglobin. An accumulation of porphyrins, which are intermediates of the pathway, causes awful symptoms, especially after exposure to sunlight. Lesions and scars form on the skin. Hair grows thickly on the face and hands. As gums retreat from teeth, the canines take on a fanglike appearance. Symptoms worsen upon exposure to a variety of substances, including garlic and alcohol. Affected individuals avoid sunlight and aggravating substances, and get injections of heme from normal red blood cells.

If you are familiar with vampire stories, which date from the Middle Ages or earlier, speculate on how they may have evolved among superstitious folk who did not have medical knowledge of porphyria.

Selected Key Terms

adhering junction *33.1*	homeostasis *CI*
adipose tissue *33.2*	integrator *33.7*
blood *33.2*	internal environment *CI*
bone *33.2*	loose connective tissue *33.2*
cardiac muscle tissue *33.3*	mesoderm *33.6*
cartilage *33.2*	negative feedback mechanism *33.7*
dense, irregular connective tissue *33.2*	nervous tissue *33.4*
	neuroglia *33.4*
dense, regular connective tissue *33.2*	neuron *33.4*
	organ *CI*
division of labor *CI*	organ system *CI*
ectoderm *33.6*	positive feedback mechanism *33.7*
effector *33.7*	sensory receptor *33.7*
endocrine gland *33.1*	skeletal muscle tissue *33.3*
endoderm *33.6*	smooth muscle tissue *33.3*
epithelium *33.1*	stimulus *33.7*
exocrine gland *33.1*	tight junction *33.1*
extracellular fluid *33.7*	tissue *CI*
gap junction *33.1*	

Readings

Bloom, W., and D. W. Fawcett. 1995. *A Textbook of Histology.* Twelfth edition. Philadelphia: Saunders.

Leeson, C. R., T. Leeson, and A. Paparo. 1988. *Textbook of Histology.* Philadelphia: Saunders.

Ross, M., L. Romrell, and G. Kaye. 1988. *Histology: A Text and Atlas.* Baltimore: Williams & Wilkins.

Web Site See *http://www.wadsworth.com/biology* for practice quiz questions, hypercontents, BioUpdates, and critical thinking. The Wadsworth Biology Resource Center provides a wealth of information fully organized and integrated by chapter.

INFORMATION FLOW AND THE NEURON

TORNADO!

Figure 34.1 A terrifying view across the Kansas prairie—a tornado about to touch down. Imagine yourself alone in the prairie when you first see a tornado, knowing you have only minutes to remove yourself from harm's way. By what means do you conceive of plans of action? Can the plans be put together and evaluated quickly enough? The answers begin with the functioning of neurons in your nervous system.

It is spring in the American Midwest, and you are fully engrossed in photographing wildflowers in an expanse of shortgrass prairie. So intent are you on capturing all the species on film that you fail to notice the rapidly darkening sky. What's that rumbling you hear in the distance? It sounds something like a freight train. You wonder how can that be, when there is no train track, anywhere, in this part of the prairie. You turn to identify the source of the sound. Then you see it, but you don't want to believe your eyes. A dark funnel cloud is advancing across the prairie and heading right for you! *TORNADO!* The ominous image rivets your attention as nothing else has ever done (Figure 34.1). You know that you cannot remain where you are and survive. Suddenly you remember that hiding in a low area is better than standing out in the open. Commands rush from your brain to your four limbs: *GET MOVING OUT OF HERE!* With heart thumping, you run along a path, looking frantically for safety. You're in luck! Just ahead is a steep-banked creek. With a stunning burst of speed you reach the creek in less than a minute and scramble down the muddy bank. There you find a small ledge over the water. You quickly wedge yourself under it, wishing wildly to be inconspicuous, to be overlooked by a force of nature on the rampage.

It takes a few minutes before you realize the tornado has roared past the creek. You

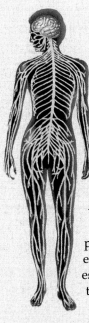

remain motionless, fingers clutching mud. Finally you do sit up and look around. Some distance away, you see a swath of twisted grass that marks the tornado's path. Your heart no longer feels as if it is slamming against your chest wall, but your legs feel like rubber when you stand up.

Thank your nervous system for every perception, every memory, and nearly every action that collectively helped you escape the tornado. Thank the information that traveled swiftly, in suitable directions, among interacting cells called **neurons**.

These are the cells that work together to monitor conditions in and around the body and to issue commands for responsive actions that benefit the body as a whole. They represent the communication lines of your brain, spinal cord, and nerves.

This chapter begins with the structure and function of neurons. In chapters to follow, you will consider how these cells interact with one another in nervous systems and sensory systems. Through such interactions, you detect tornadoes and other events that have bearing on whether you survive from one day to the next.

There are three classes of neurons. Different types of **sensory neurons** are adapted to respond to specific stimuli and relay information about them to the spinal cord and brain. A **stimulus** is a specific form of energy, such as light and pressure. In the spinal cord and brain you find **interneurons**. These receive sensory input, integrate it with other incoming information and with stored information, then influence the activity of other neurons. **Motor neurons** relay information from the brain and spinal cord to effectors—muscles or glands— that carry out the specified responses:

INPUT *(stimulus)*   OUTPUT *(response)*

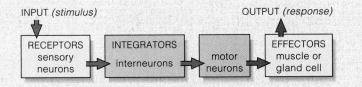

Neurons are not the only cells in vertebrate nervous systems. As you will see in this chapter and the next, a variety of cells, collectively called neuroglia, make up more than half the volume of the vertebrate systems. Different kinds of neuroglial cells metabolically assist, protect, and structurally support the neurons.

KEY CONCEPTS

1. Neurons are basic units of communication in nervous systems. Collectively, many neurons interact to detect and integrate information about external and internal conditions, then select or control muscles and glands in ways that produce suitable responses.

2. Neurons are a type of excitable cell, which means a suitable stimulus can disturb the distribution of electric charge across their plasma membrane. Such disturbances are the basis of messages that travel from the input zone of a neuron to its output zone near a neighboring cell.

3. In an undisturbed neuron, the cytoplasm is negatively charged with respect to the fluid just outside the plasma membrane. With suitable stimulation, the polarity of charge across the membrane may undergo an abrupt, brief reversal. Such reversals are action potentials.

4. Information flow through the nervous system starts with action potentials, which are self-propagating along the plasma membrane of individual neurons.

5. Action potentials start and end at small gaps between neurons. Chemical signals released from one neuron bridge the gap and stimulate or inhibit the adjoining neuron, muscle cell, or gland cell.

6. Information flow through nervous systems depends on moment-by-moment integration of excitatory and inhibitory signals that act upon the neurons of given pathways.

Functional Zones of a Neuron

To understand how your own nervous system works, start with how its neurons function. Neurons consist of a nucleated cell body and cytoplasmic extensions that differ in number and length (Figure 34.2). Typically, the cell body and the slender extensions called **dendrites** are *input* zones, where the neuron receives information. Another slender but often longer extension, called an **axon**, is a *conducting* zone. Axons rapidly propagate signals that arise at a neuron's trigger zone. Except in sensory neurons, the *trigger* zone is a patch of plasma membrane at the junction between the cell body and an axon. The branched endings of an axon are *output* zones, where messages are sent to other cells.

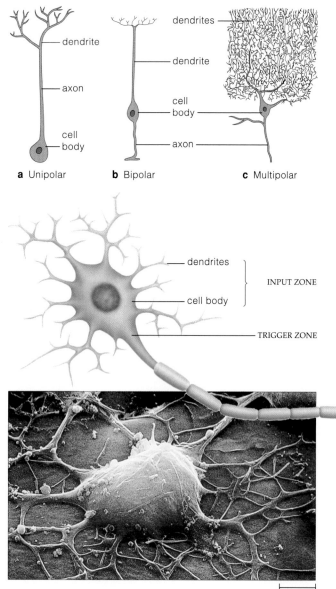

a Unipolar **b** Bipolar **c** Multipolar

dendrites

dendrite

cell body

axon

dendrite

axon

cell body

dendrites

cell body

INPUT ZONE

TRIGGER ZONE

A Neuron at Rest, Then Moved to Action

Different forms of signals arise in the nervous system. Let's start with one that begins and ends on the same neuron and is not transferred to another cell.

When a neuron is not being bothered, a difference in electric charge is being maintained across its plasma membrane. The cytoplasmic fluid next to the membrane remains negatively charged, compared to the interstitial fluid right outside. We measure these charges in units called millivolts. The amount of energy inherent in the steady voltage difference across the plasma membrane is the **resting membrane potential**. For many neurons, that amount is about −70 millivolts.

Suppose a weak signal reaches a patch of membrane in a neuron's input zone. The voltage difference across the patch changes only slightly, if at all. By contrast, a strong signal might trigger an **action potential**, which is an abrupt, short-lived reversal in the voltage difference across a plasma membrane. For a fraction of a second, the inside becomes positive with respect to the outside. The localized reversal invites an action potential at the adjoining membrane patch, which invites another at the next patch, and so on away from the initiation point. In short, *stimulation of a neuron disturbs the distribution of electric charge across its plasma membrane.*

Restoring and Maintaining Readiness

How does the neuron restore the voltage difference across each patch of plasma membrane and maintain it between action potentials? It relies on two membrane properties. First, a membrane has a lipid bilayer, which bars the passage of potassium ions (K^+), sodium ions (Na^+), and other charged substances. Thus, the neuron can build up differences in ion concentrations across the membrane. Second, ions can flow from one side to the other through the interior of transport proteins that span the bilayer, and that flow is controlled (Figure 34.3).

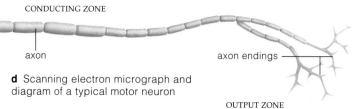

CONDUCTING ZONE

axon

axon endings

d Scanning electron micrograph and diagram of a typical motor neuron

OUTPUT ZONE

Figure 34.2 (**a–c**) Classification of neurons based on the number of cytoplasmic extensions of the cell body. *Unipolar* cells have a single axon with dendritic branchings. *Bipolar* cells have one dendrite and one axon. Many sensory neurons are like this. *Multipolar* cells, with one axon and many dendrites, predominate in vertebrate nervous systems. (**d**) Functional zones of a motor neuron, a multipolar cell.

10 μm

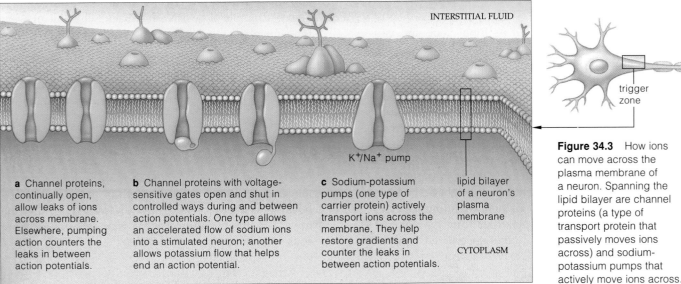

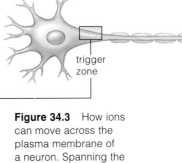

Figure 34.3 How ions can move across the plasma membrane of a neuron. Spanning the lipid bilayer are channel proteins (a type of transport protein that passively moves ions across) and sodium-potassium pumps that actively move ions across.

INTERSTITIAL FLUID

K⁺/Na⁺ pump

a Channel proteins, continually open, allow leaks of ions across membrane. Elsewhere, pumping action counters the leaks in between action potentials.

b Channel proteins with voltage-sensitive gates open and shut in controlled ways during and between action potentials. One type allows an accelerated flow of sodium ions into a stimulated neuron; another allows potassium flow that helps end an action potential.

c Sodium-potassium pumps (one type of carrier protein) actively transport ions across the membrane. They help restore gradients and counter the leaks in between action potentials.

lipid bilayer of a neuron's plasma membrane

CYTOPLASM

Suppose a motor neuron has 15 sodium ions inside the membrane for every 150 outside. Suppose it has 150 potassium ions inside for every 5 on the outside. You can depict each ion's concentration gradient in this way (from the large to the small letters):

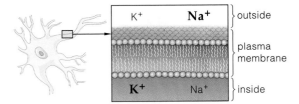

Gradients such as these determine the net direction in which sodium and potassium ions diffuse. The ions can diffuse through the interior of channel proteins, a type of transport protein in the membrane. Some of the channels never shut, so ions leak (diffuse) through them all of the time. Others have molecular gates, which can open only after the neuron is adequately stimulated.

Suppose that motor neuron is not being stimulated. Its sodium channels are shut, so sodium ions can't rush inside. Some potassium is leaking out through a few open channels and making the cytoplasm a bit more negative, so some potassium is attracted back in. When the inward pull of electric charge balances the outward force of diffusion, there is no more net movement of potassium. The concentration and electric gradients now existing across the plasma membrane will permit the neuron to respond to stimulation.

After the gradients have reversed during an action potential, **sodium-potassium pumps** restore them. These are transport proteins that span the plasma membrane (Section 5.5). When they get an energy boost from ATP,

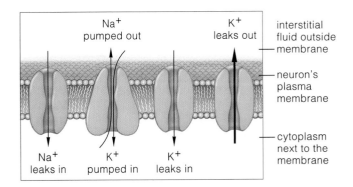

Figure 34.4 Pumping and leaking processes that influence the distribution of sodium and potassium ions across the plasma membrane of a neuron at rest. Arrow widths indicate the magnitude of the movements. Notice how the total inward and outward movements for each kind of ion are balanced.

the pumps actively transport potassium into the neuron and sodium out at the same time. Sodium-potassium pumps must maintain as well as restore the gradients. Why? Even in a resting neuron, a tiny fraction of the outward-leaking potassium isn't attracted back in. Also, a tiny fraction of sodium leaks in, through a few open channels (Figure 34.4). If the leaks went unattended, the crucial gradients would gradually disappear.

An undisturbed neuron is maintaining a resting membrane potential—a voltage difference across its plasma membrane. An action potential is an abrupt, short-lived reversal in that voltage difference in response to adequate stimulation.

After an action potential, sodium-potassium pumps restore and maintain the resting membrane potential.

A CLOSER LOOK AT ACTION POTENTIALS

Approaching Threshold

Weakly stimulate a neuron at its input zone and you disturb the ion balance across the membrane, but not much. Suppose your toes gently tap a cat and put a bit of pressure on its skin. Tissues beneath the skin surface have receptor endings—input zones of sensory neurons. Patches of plasma membrane at the receptor endings deform under the pressure. Some ions now flow across, so the voltage difference across the membrane changes slightly. The pressure produced a graded, local signal.

Graded means that signals arising at an input zone can vary in magnitude. Such signals might be large or small, depending on the intensity of the stimulus. *Local* means the signals do not spread far from the point of stimulation. It takes specialized types of

and so on. The ever increasing inward flow of sodium is a good example of **positive feedback**, whereby an event intensifies as a result of its own occurrence:

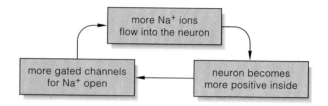

At threshold, the opening of more sodium gates no longer depends upon the strength of the stimulus. The positive-feedback cycle is now under way, so that the inward-rushing sodium itself is enough to cause more sodium gates to open.

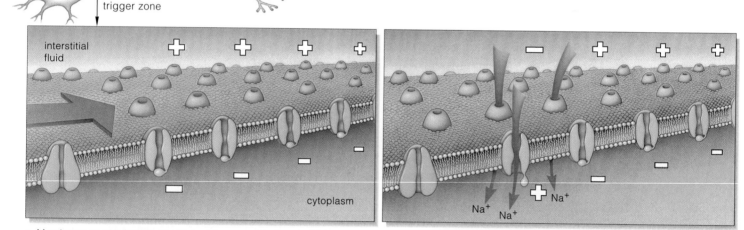

a Membrane at rest (inside negative with respect to the outside). An electrical disturbance (*red arrow*) spreads from an input zone to an adjacent trigger region of the membrane, which has a great number of gated sodium channels.

b A strong disturbance initiates an action potential. Sodium gates open. The sodium inflow decreases the negativity inside the neuron. The change causes more gates to open, and so on, until threshold is reached and the voltage difference across the membrane reverses.

Figure 34.5 Propagation of an action potential along the axon of a motor neuron.

ion channels to propagate a signal along the membrane, and input zones simply don't have them.

However, when a stimulus is intense or long lasting, graded signals can spread out from the input zone and into an adjacent trigger zone. At this zone, a certain minimum amount of change in the voltage difference across the plasma membrane can trigger an action potential. That amount is known as the **threshold level**. *Threshold can be reached at any membrane patch that has voltage-sensitive, gated channels for sodium ions.*

As Figure 34.5 shows, the stimulus causes sodium ions to flow across the membrane, into the neuron. With the influx of positively charged ions, the cytoplasmic side of the plasma membrane becomes less negative. This causes more gates to open, more sodium to enter,

An All-or-Nothing Spike

Figure 34.6 shows a recording of the voltage difference across the plasma membrane before, during, and after an action potential. Notice how the membrane potential spikes once threshold is reached. Every single action potential in the neuron spikes to the same level above threshold as an *all-or-nothing* event. That is, once the positive-feedback cycle starts, nothing will stop the full spiking. If the threshold is not reached, however, the disturbance to the plasma membrane will subside when the stimulus is removed.

Each spike lasts only for a millisecond or so. Why? At the membrane site of the charge reversal, the gated sodium channels closed and shut off the sodium inflow.

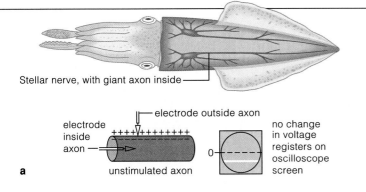

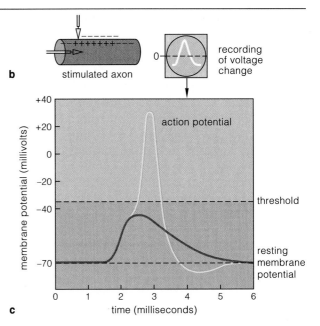

Figure 34.6 Action potentials. (**a**) When early researchers studied neural function, the squid *Loligo* provided them with evidence of action potential spiking. The squid's "giant" axons were large enough to slip electrodes inside. (**b**) When such an axon was stimulated, electrodes positioned inside and outside detected voltage changes, which showed up as deflections in a beam of light across the screen of an oscilloscope. (**c**) This is a typical waveform (*yellow* line) for an action potential on an oscilloscope screen. The *red* line represents a recording of a local signal that did not reach the threshold of an action potential; spiking did not occur.

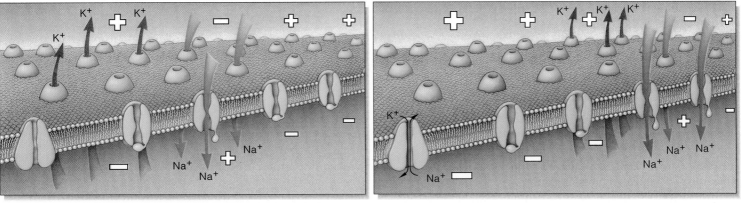

c With the reversal, sodium gates shut and potassium gates open (*pink arrows*). Potassium follows its gradient out of the neuron. Voltage is restored. The disturbance triggers an action potential at the adjacent site, and so on, away from the point of stimulation.

d Following each action potential, the inside of the plasma membrane becomes negative once again. However, the sodium and potassium concentration gradients are not yet fully restored. Active transport at sodium-potassium pumps restores them.

Also, about halfway through the reversal, potassium channels opened, so many more potassium ions flowed out and restored the original voltage difference across the membrane. And sodium-potassium pumps restored the ion gradients. Later on, after the resting membrane potential has been restored, most potassium gates close and sodium gates are in their initial state, ready to be opened with the arrival of a suitable disturbance.

Propagation of Action Potentials

The membrane disturbances leading up to an action potential are self-propagating, and they do not diminish in magnitude. As they spread to an adjacent membrane patch, an equivalent number of gated channels open. With this new disturbance, gated channels open in the *next* adjacent patch, then the next, and so on. For a brief period after each membrane patch has been excited, it is insensitive to stimulation. Sodium gates in the patch are inactivated and ions cannot move through them. This is the reason why action potentials do not spread back to the trigger zone (the site where they were initiated) but rather are self-propagating away from it.

The cytoplasm next to the plasma membrane of a neuron at rest is more negative than the interstitial fluid just outside the membrane.

During an action potential, the inside of a disturbed patch of membrane becomes more positive than the outside.

After an action potential, resting conditions are restored at the membrane patch.

When action potentials reach a neuron's output zone, they usually do not proceed farther than this. But their arrival may induce the neuron to release one or more **neurotransmitters**, which are signaling molecules that diffuse across chemical synapses. A **chemical synapse** is a narrow junction, or cleft, between the output zone of a neuron and an input zone of an adjacent cell (Figure 34.7). Some clefts intervene between two neurons, and others between a neuron and a muscle cell or gland cell.

At each chemical synapse, one of the two cells stores neurotransmitter molecules inside synaptic vesicles in its cytoplasm. Think of it as the *pre*synaptic cell. Here, gated channels for calcium ions span the membrane, and they open when an action potential arrives. There are more calcium ions outside the cell. When they flow into the cell (down their gradient), the synaptic vesicles are induced to fuse with the plasma membrane, so that neurotransmitter is released into the synaptic cleft.

The neurotransmitter molecules diffuse through the cleft. On the *post*synaptic cell membrane are protein receptors that bind specific neurotransmitters. Binding changes the receptor shape and creates a passageway through its interior. Ions cross the plasma membrane by diffusing through the passageway (Figure 34.8).

A postsynaptic cell's response depends on the type and concentration of neurotransmitter in the cleft, what kinds of receptors the cell bears, and the number and responsiveness of gated channels in its membrane. Such factors influence whether a neurotransmitter will have an *excitatory* effect and help drive the postsynaptic cell's membrane toward the threshold of an action potential. They also influence whether it will have an *inhibitory* effect and drive the membrane away from threshold.

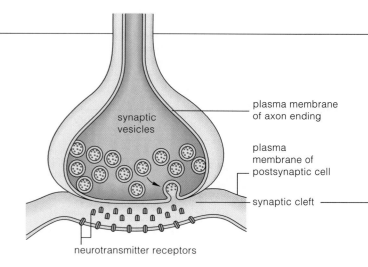

a

Figure 34.8   Closer look at a chemical synapse. (**a**,**b**) The plasma membranes of a motor neuron and a muscle cell face each other across a cleft between them. The arrival of an action potential at an axon ending triggers the release of molecules of neurotransmitter from the neuron. (**c**) These molecules diffuse through the cleft and then bind to receptors on gated channel proteins of the muscle cell membrane. The gates open; ions flow in and trigger a graded potential at the membrane site.

Consider **acetylcholine** (ACh), a neurotransmitter. It has excitatory and inhibitory effects on the brain, spinal cord, glands, and muscles. For example, it acts at the chemical synapse between a motor neuron and muscle cell, as in Figure 34.7. ACh released from the motor neuron diffuses across the cleft and binds to receptors on the muscle cell membrane. In this kind of cell it has excitatory effects; it can trigger action potentials, which in turn initiate muscle contraction (Section 38.9).

A Smorgasbord of Signals

Acetylcholine is only one of a veritable smorgasbord of signals that neurons deliver to target cells. For example, serotonin acts on neurons in brain regions that govern sleeping, sensory perception, temperature control, and emotions. Norepinephrine works in brain regions that control emotions, dreaming, and waking up. Dopamine also works in brain regions that deal with emotions. GABA (gamma aminobutyric acid) is the most common inhibitory signal in the brain. Antianxiety drugs, such as Valium, may exert their effects by enhancing GABA's effects. Except for ACh, these neurotransmitters and the other types are amino acids or are derived from them.

Signaling molecules known as **neuromodulators** can magnify or reduce the effects of a neurotransmitter on neighboring or distant neurons. They include substance P, which induces pain perception, and endorphins—the natural pain killers that inhibit the release of substance P from sensory nerves. Neuromodulators might also influence memory and learning, sexual activity, control of body temperature, and emotional states.

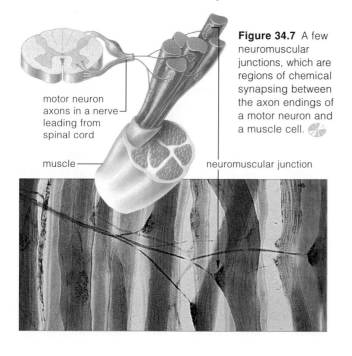

Figure 34.7 A few neuromuscular junctions, which are regions of chemical synapsing between the axon endings of a motor neuron and a muscle cell.

motor neuron axons in a nerve leading from spinal cord

muscle

neuromuscular junction

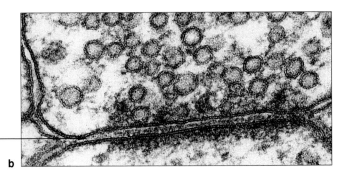

b

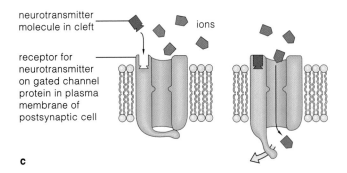

neurotransmitter molecule in cleft — ions

receptor for neurotransmitter on gated channel protein in plasma membrane of postsynaptic cell

c

Synaptic Integration

Anywhere from 1,000 to 10,000 communication lines form synapses with a typical neuron in your brain. And your brain contains at least 100 billion neurons. As long as you are alive, those neurons continually hum with messages about doing what it takes to be a human.

At any moment, a great number of excitatory and inhibitory signals may be washing over the input zones of a postsynaptic cell. Some signals drive its membrane closer to threshold; others maintain the resting level or drive it away from threshold. Said another way, signals compete for control of the neuron's membrane.

All synaptic signals are graded potentials. The ones we call **EPSPs** (for excitatory postsynaptic potentials) have a *depolarizing* effect. This simply means they bring the membrane closer to threshold. The **IPSPs** (inhibitory postsynaptic potentials) might have a *hyperpolarizing* effect (drive the membrane away from threshold) or might help maintain the membrane at its resting level.

With **synaptic integration**, competing signals that reach an input zone of a neuron at the same time are summed. By this process, two or more signals arriving

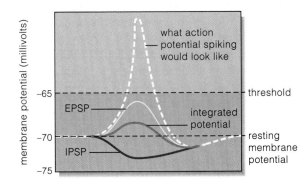

Figure 34.9 Example of synaptic integration. The *yellow* line shows how an EPSP of a certain magnitude would register on an oscilloscope screen if it were acting alone. The *purple* line shows the effect of an IPSP if *it* were acting alone. Suppose both signals arrive at a postsynaptic cell membrane at the same time. The *red* line shows their effect. In this case, when the two signals are integrated, threshold is not reached. Thus an action potential cannot be initiated in the target cell.

at a neuron may be dampened, suppressed, reinforced, or sent onward to other cells in the body.

Figure 34.9 shows what two separate recordings of an EPSP and an IPSP might look like, and it includes a recording of their summation. Such integration occurs when neurotransmitter from several presynaptic cells reaches the input zone of a neuron at the same time. It also occurs after neurotransmitter is released swiftly and repeatedly from a single presynaptic cell that has been whipped into a frenzy of excitability by a rapid series of action potentials.

Removing Neurotransmitter From the Synaptic Cleft

The flow of information through the nervous system depends on the prompt, precisely controlled removal of neurotransmitter molecules from synaptic clefts. Some amount of these molecules simply diffuses out of the cleft. Enzymes cleave others right in the cleft, as when acetylcholinesterase breaks apart ACh. Also, transport proteins actively pump the molecules back into the presynaptic cells or into neighboring neuroglial cells.

What happens if neurotransmitter accumulates in the cleft? As one example, cocaine blocks the uptake of dopamine. Molecules of this neurotransmitter linger in synaptic clefts and just keep on stimulating target cells. At first the stimulation produces euphoria (intense pleasure). Later on, it has disastrous effects, as you will read in the chapter to follow.

Neurotransmitters are signaling molecules that bridge a synaptic cleft, which is a tiny gap between two neurons or between a neuron and a muscle cell or gland cell.

Neurotransmitters have excitatory or inhibitory effects on different kinds of receiving cells. Synaptic integration is the moment-by-moment combining of excitatory and inhibitory signals acting on a postsynaptic cell.

With the summation process, messages traveling through the nervous system can be reinforced or downplayed, sent onward or suppressed. The process is essential for normal body functioning.

Blocks and Cables of Neurons

Through synaptic integration, messages arriving at any neuron in the body might be reinforced and sent on to neighboring neurons. What determines the direction in which a given message will travel? That depends on the organization of neurons in different body regions.

For example, your brain deals with its staggering numbers of neurons in a manner analogous to block parties. Regional blocks of hundreds or thousands of neurons receive excitatory and inhibitory signals. They integrate signals entering the block, then send out new ones in response. Some regions have neurons organized as *divergent* circuits, as when their processes fan out from one block and form connections with many others. Different regions have neurons arranged in *convergent* circuits, with signals from many sent on to just a few. In still other regions, neurons synapse back on themselves and repeat signals like gossip that just won't go away. Such neurons form *reverberating* circuits. They include the circuits that make your eye muscles rhythmically twitch while you sleep.

In the cablelike **nerves**, long axons of many sensory neurons, motor neurons, or both permit long-distance communication between the brain or spinal cord and the rest of the body. Connective tissue bundles most of the axons in parallel array (Figure 34.10). Each axon has a **myelin sheath** that enhances the rate of action potential propagation. The sheath is a series of **Schwann cells**, a type of neuroglial cell, wrapped like jellyrolls around the long axon. An exposed node, or gap, separates each cell from adjacent ones. There, voltage-sensitive, gated sodium channels pepper the plasma membrane (Figure 34.11). The sheathed regions in between nodes hamper ion movements across the membrane. Ion disturbances tend to flow along the membrane until the next node in line. At each node, ion flow can produce a new action potential. In large sheathed axons, action potentials are propagated at a remarkable 120 meters per second.

With *multiple sclerosis*, myelin sheaths around axons in the spinal cord degenerate slowly but inevitably. Gene mutation may predispose a person to the disease, but viral infection may be the actual trigger. Symptoms include weakening of muscles, fatigue, and numbness.

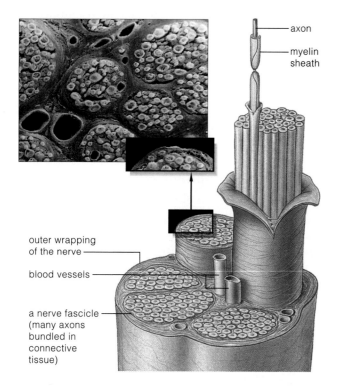

Figure 34.10 Structure of a nerve. Axons in the nerve are bundled together inside wrappings of connective tissue.

axon
myelin sheath
outer wrapping of the nerve
blood vessels
a nerve fascicle (many axons bundled in connective tissue)

Reflex Arcs

Figure 34.12 provides a specific example of the direction of information flow through nervous systems. It shows how sensory and motor neurons of certain nerves take part in a path known as the stretch reflex. **Reflexes** are simple movements made in response to specific sensory stimuli. In the simplest reflex arcs, sensory neurons synapse directly on motor neurons.

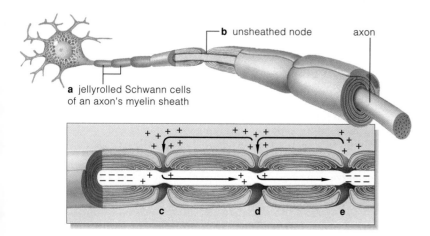

a jellyrolled Schwann cells of an axon's myelin sheath
b unsheathed node
axon
c d e

Figure 34.11 High-speed travel in the nervous system—how an action potential is propagated along sheathed neurons. (**a**) A myelin sheath is a series of Schwann cells, each wrapped like a jellyroll around an axon. (**b**) Each jellyroll blocks ion movements across the membrane, but ions can cross at nodes in between. The unsheathed nodes have very dense arrays of gated sodium channels. (**c,d**) A disturbance caused by an action potential spreads down the axon. When it reaches a node, sodium gates open, sodium ions rush inward, and another action potential results. (**e**) The new disturbance spreads rapidly to the next node and triggers another action potential, and so on down the line.

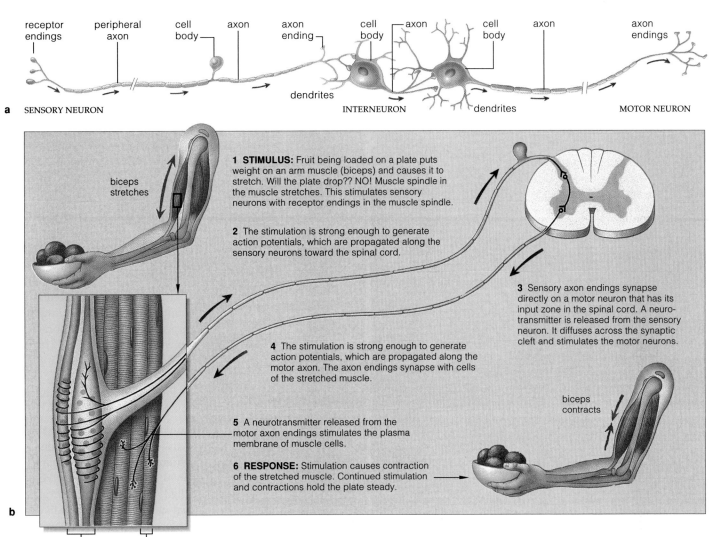

a SENSORY NEURON — receptor endings — peripheral axon — cell body — axon — axon ending — dendrites — INTERNEURON — cell body — axon — cell body — dendrites — axon — MOTOR NEURON — axon endings

b

biceps stretches

1 STIMULUS: Fruit being loaded on a plate puts weight on an arm muscle (biceps) and causes it to stretch. Will the plate drop?? NO! Muscle spindle in the muscle stretches. This stimulates sensory neurons with receptor endings in the muscle spindle.

2 The stimulation is strong enough to generate action potentials, which are propagated along the sensory neurons toward the spinal cord.

3 Sensory axon endings synapse directly on a motor neuron that has its input zone in the spinal cord. A neurotransmitter is released from the sensory neuron. It diffuses across the synaptic cleft and stimulates the motor neurons.

4 The stimulation is strong enough to generate action potentials, which are propagated along the motor axon. The axon endings synapse with cells of the stretched muscle.

5 A neurotransmitter released from the motor axon endings stimulates the plasma membrane of muscle cells.

6 RESPONSE: Stimulation causes contraction of the stretched muscle. Continued stimulation and contractions hold the plate steady.

biceps contracts

muscle spindle   muscle cell

Figure 34.12 (**a**) General direction of information flow in nervous systems. Sensory neurons relay information *into* the spinal cord and brain, where they synapse with interneurons. Interneurons *within* the spinal cord and brain integrate signals. Many synapse with motor neurons, which carry signals *away* from the spinal cord and brain. (**b**) Organization of nerves in a reflex arc that deals with muscle stretching. In a skeletal muscle, stretch-sensitive receptors of a sensory neuron are located in muscle spindles. The stretching generates action potentials, which reach axon endings in the spinal cord. These synapse with a motor neuron that carries signals to contract, from the spinal cord back to the stretched muscle.

The stretch reflex works to contract a muscle after gravity or some other load has caused the muscle to stretch. Suppose you hold out a large bowl and keep it stationary as someone puts several peaches into it. The peaches add weight to the bowl, and when your hand starts to drop, a muscle in your arm (the biceps) is stretched.

In the muscle, stretching activates receptor endings that are a part of muscle spindles—sensory organs in which specialized cells are enclosed in a sheath that runs parallel with the muscle. The receptor endings are the input zones of sensory neurons, the axons of which synapse with motor neurons within the spinal cord (Figure 34.12). Axons of the motor neurons lead back to the stretched muscle. Action potentials that reach the axon endings trigger the release of ACh, which initiates contraction. As long as receptor activity continues, the motor neurons are excited further, and this allows them to maintain your hand's position.

In the vast majority of reflexes, sensory neurons also make connections with a number of interneurons, which then activate or suppress all the motor neurons necessary for a coordinated response.

Vertebrates have interneurons organized in information-processing blocks. Cablelike nerves that have long axons of sensory neurons, motor neurons, or both connect the brain and spinal cord with the rest of the body.

Reflex arcs, in which sensory neurons synapse directly on motor neurons, are the simplest paths of information flow.

What happens when something disrupts information flow in the nervous system? To give one example, think about *Clostridium botulinum*. This anaerobic, endospore-forming bacterium normally lives in soil, but it can cause the disease *botulism*. When its endospores contaminate improperly stored, preserved, or canned food, they can germinate and produce a dangerous toxin, which people might ingest and absorb. The toxin can bind to neurons that synapse with muscle cells and block their release of acetylcholine (ACh). Once that happens, muscles cannot contract. The muscles become progressively flaccid and paralyzed. Death can follow within ten days, usually from respiratory and cardiac failure. Recovery is possible if antitoxins are quickly administered.

A related bacterium, *C. tetani*, lives in the gut of horses, cattle, and other grazing animals, even many people. Its endospores survive in soil, especially manure-rich ones. They persist for years if sunlight and oxygen do not reach them. They resist disinfectants, heat, and boiling water. When they enter the body through a deep puncture or a deep cut, they germinate in anaerobic, dead tissues. These bacteria do not spread from the dead tissue. They produce a toxin that the blood or nerves deliver to the spinal cord and brain. In the spinal cord, the toxin affects interneurons that help control motor neurons. It blocks the release of inhibitory neurotransmitters (GABA and glycine) and frees the motor neurons from normal inhibitory control. Symptoms of the disease *tetanus* are about to begin.

Muscles overstimulated for four to ten days stiffen and go into spasm. They cannot be released from contraction, and prolonged, spastic paralysis follows. Fists and jaws may remain clenched (the disease is also called lockjaw). The back may arch permanently. Spasms may be strong enough to break bones. When respiratory and cardiac muscles become paralyzed, death nearly always follows.

Vaccines were not available for soldiers of early wars, when dead cavalry horses and manure littered battlefields (Figure 34.13). Today, vaccines have all but eradicated tetanus in the United States.

Figure 34.13 Painting of a young victim of a contaminated battle wound, as he lay dying in a military hospital.

1. The nervous system senses, interprets, and issues commands for responses to specific aspects of the environment. In nearly all animals, its communication lines consist of sensory neurons, interneurons, and motor neurons, which activate muscle and gland cells.

a. Sensory neurons are receptors that detect specific stimuli (specific forms of energy, such as light).

b. Interneurons are integrators in the brain and spinal cord. They receive signals and process signals from sensory receptors, and then issue commands for suitable responses.

c. Motor neurons carry commands away from the brain and spinal cord to the body's effectors, which are muscle cells and gland cells that carry out responses.

2. A neuron's dendrites and cell body are input zones. If arriving signals spread to a trigger zone (such as the start of an axon), they may trigger an action potential that can be propagated to an output zone (axon endings).

a. Transport proteins pepper the plasma membrane from the trigger zone to axon endings of a neuron. They serve as gated or open channels for the passage of ions, Na^+ and K^+ especially, across the membrane.

b. The controlled flow of ions across the membrane is the basis of message propagation along a neuron.

3. For a neuron at rest, the ion distribution between the cytoplasm and extracellular fluid results in a difference in electric charge (voltage difference) across the plasma membrane. The cytoplasmic fluid is more negatively charged, by an amount called the resting membrane potential. An action potential is an abrupt, short-lived reversal in the voltage difference across the membrane in response to adequate stimulation.

a. Stimuli give rise to local, graded potentials, which vary in their magnitude and do not spread very far. A disturbance caused by a number of graded potentials may spread to a trigger zone and drive the membrane of a neuron to threshold of an action potential.

b. A strong disturbance makes many gated channels for sodium ions open in an ever accelerating, self-propagating way. As a result, the voltage difference across the membrane reverses abruptly. Accelerated openings proceed in patch after patch of membrane, to the neuron's output zone.

c. In between action potentials, sodium-potassium pumps counter small ion leaks across the membrane and so maintain the gradients. After an action potential, they restore the gradients.

4. Action potentials arriving at an output zone trigger the release of neurotransmitters. These diffuse across chemical synapses, where the neuron forms a junction with another neuron, a muscle cell, or a gland cell. They may excite the cell's plasma membrane (drive it closer to threshold) or inhibit it (drive it away from threshold).

5. Integration is the moment-by-moment summation of excitatory and inhibitory signals reaching all of the synapses on a neuron. It is a means of playing down, suppressing, reinforcing, or sending on information to other neurons of the nervous system.

Review Questions

1. Describe sensory neuron, interneuron, and motor neuron in terms of their structure and functions. *CI, 34.1, 34.4*

2. Label the functional zones of this motor neuron. *34.1*

3. Define resting membrane potential, graded potential, and action potential. *34.1, 34.2*

4. A neuron at rest is controlling the ion distribution across its plasma membrane. Identify the two major kinds of ions. Do they leak across the membrane, are they pumped across, or both? As part of your answer, describe gated and ungated transport proteins embedded in the neural membrane. *34.1*

5. With respect to action potentials, explain what is meant by threshold level, by all-or-nothing spikes, and by self-propagation of an action potential. *34.2*

6. Define chemical synapse and neurotransmitter. Choose an example of a neurotransmitter and state where it acts. *34.3*

7. Define synaptic integration. Include definitions of EPSPs and IPSPs as part of your answer. *34.3*

8. What is a myelin sheath? Do all neurons have one? *34.4*

9. How do divergent, convergent, and reverberating circuits among blocks of neurons differ from one another? *34.4*

10. Define reflex, then give an example of a reflex arc. *34.4*

Self-Quiz *(Answers in Appendix IV)*

1. Action potentials occur when _____ .
   a. a neuron receives adequate stimulation
   b. sodium gates open in an ever accelerating way
   c. sodium-potassium pumps kick into action
   d. both a and b

2. Compared with interstitial fluid near the plasma membrane of a neuron at rest, the cytoplasm near the membrane's inner surface carries a slight _____ charge.
   a. positive   c. graded and local
   b. negative   d. both b and c

3. The resting membrane potential is maintained by _____ .
   a. ion leaks   c. neurotransmitters
   b. ion pumps   d. both a and b

4. Neurotransmitters diffuse across a _____ .
   a. chemical synapse   c. myelin sheath
   b. channel protein   d. both a and b

5. An action potential lasts only briefly because _____ at the membrane region where it occurred.
   a. gates for sodium open, gates for potassium close
   b. gates for sodium close, gates for potassium open
   c. sodium-potassium pumps restore gradients
   d. both b and c

6. A nerve may consist of bundled-together axons of _____ .
   a. sensory neurons   c. sensory and motor neurons
   b. motor neurons   d. all are correct

7. Integration is the _____ .
   a. self-propagation of ion flow along a neuron
   b. trigger that releases neurotransmitters
   c. summation of signals acting at all synapses on a neuron
   d. collection of all information about several stimuli

8. Match the terms with their most suitable description.
   ____ synaptic integration   a. occurs at neuron's input zone
   ____ muscle spindle   b. spatial, temporal summation
   ____ graded, local   of all signals arriving at neuron
      potential   c. arises at trigger zone
   ____ action potential   d. stretch-sensitive receptor

Critical Thinking

1. With *multiple sclerosis*, recall, myelin sheaths around axons in the spinal cord slowly degenerate. Affected individuals progressively lose the ability to control movements of their skeletal muscles. Reflect on the role of myelin sheaths in the nervous system, then formulate a hypothesis of how their degeneration results in a loss of control over skeletal muscles.

2. *Epilepsy* is a neurological disorder characterized by short but recurring episodes of sensory and motor malfunctioning. The episodes, or epileptic seizures, might arise when reverberating circuits involving millions of interneurons in the brain become abnormally activated. Affected individuals typically experience involuntary muscle contractions and often sense lights, sounds, and odors even if receptors in the eyes, ears, and nose have not been stimulated. The drug valproic acid can eliminate or lessen the severity of the seizures by stimulating the body's synthesis of GABA. Why would stepping up GABA production help?

3. A mud-covered nail punctured Evita's foot. To protect Evita from tetanus, her doctor administered an antitoxin. This one contained enough protein molecules that specifically bind with and neutralize the *C. tetanus* toxin molecules. Will Evita now have long-lasting protection against future attacks by this bacterium?

Selected Key Terms

acetylcholine (ACh) *34.3*   neuron *CI*
action potential *34.1*   neurotransmitter *34.3*
axon *34.1*   positive feedback *34.2*
chemical synapse *34.3*   reflex *34.4*
dendrite *34.1*   resting membrane potential *34.1*
EPSP *34.3*   Schwann cell *34.4*
interneuron *CI*   sensory neuron *CI*
IPSP *34.3*   sodium-potassium pump *34.1*
motor neuron *CI*   stimulus *CI*
myelin sheath *34.4*   synaptic integration *34.3*
nerve *34.4*   threshold level *34.2*
neuromodulator *34.3*

Readings

Dunant, Y., and M. Israel. April 1985. "The Release of Acetylcholine." *Scientific American* 252(4): 58–83.

Nicholls, J., A. Martin, and B. Wallace. 1992. *From Neuron to Brain*. Third edition. Sunderland, Massachusetts: Sinauer.

Web Site See *http://www.wadsworth.com/biology* for practice quiz questions, hypercontents, BioUpdates, and critical thinking. The Wadsworth Biology Resource Center provides a wealth of information fully organized and integrated by chapter.

INTEGRATION AND CONTROL: NERVOUS SYSTEMS

Why Crack the System?

Suppose your biology instructor asks you to volunteer for an experiment. You will get a microchip implanted in your brain. It will make you feel really good. But it may mess up your health, lop ten years off your life, and destroy a good part of your brain. Your behavior will change for the worse, so you might have trouble completing school, getting or keeping a job, or even having a normal family life.

The longer the chip is implanted, the less you will want to give it up. You won't get paid. You will pay the experimenter—first at bargain rates, then a little more each week. The chip is illegal. If you get caught using it, you and the experimenter will go to jail.

Sometimes Jim Kalat, a professor at North Carolina State University, proposes this experiment, which of course is hypothetical. Hardly any students volunteer.

Figure 35.1 Owners of an evolutionary treasure—a complex brain that is the foundation for our memory and reasoning, and our future.

Then he substitutes *drug* for microchip and *dealer* for experimenter, and an amazing number of students come forward! Like 30 million other Americans, the "volunteers" seem ready to engage in self-destructive uses of drugs that alter emotional and behavioral states.

The destruction shows up in unexpected places. Each year, for instance, about 300,000 newborns are already addicted to crack, thanks to their addicted mothers. *Crack* is a cheap form of cocaine. It causes relentless stimulation of brain regions that govern the sense of pleasure. It dampens normal urges to eat and to sleep, and blood pressure rises. Elation and sexual desire intensify. In time, however, the brain cells that produce the stimulatory chemicals cannot keep up with the abnormal demand. The chemical vacuum makes crack users frantic and then profoundly depressed. Only crack makes them feel good again.

Addicted babies cannot know all of this. They can only quiver with "the shakes" and respond to the world

with chronic irritation. And they are abnormally small. While they were developing inside their mother, their body tissues simply were not provided with enough oxygen and nutrients. As one of its side effects, crack causes blood vessels to constrict, and maternal blood vessels are the only supply lines that the developing individual has.

Paradoxically, crack babies are exceptionally fussy, yet they cannot respond to rocking and other forms of stimulation that normally have soothing effects. It may be a year or more before they recognize even their own mother. Without treatment, they are likely to grow up as emotionally unstable children, prone to aggressive outbursts and stony silences. Why? Their mother's drug habit crippled their nervous system.

Think about it. The nervous system evolved as a way to sense and respond, with exquisite precision, to changing conditions inside and outside the body. Awareness of sounds and sights, of odors, of hunger and passion, fear and rage—all such things begin with the flow of information along the communication lines of the nervous system. Do those lines remain silent until they receive outside signals, much as telephone lines wait to carry calls from all over the country? Absolutely not. Even before you were born, excitable cells called neurons became organized into extensive gridworks in your newly forming tissues and started chattering among themselves. All through your life, in moments of danger or reflection, supreme excitement or sleep, their chattering has never ceased and will not cease until the time you die.

Each of us possesses a body of great complexity. Its architecture, its functioning are legacies of millions of years of evolution. Its nervous system is unparalleled in the living world. One of its most astonishing products is language, the encoding of shared experiences of groups of individuals in time and space. Through the evolution of our nervous system, the sense of history was born, and the sense of destiny. Through this system we can ask how we have come to be what we are, and where we are headed from here. Perhaps the sorriest consequence of drug abuse is its implicit denial of this legacy—the denial of self when we cease to ask, and cease to care.

This chapter can give you insight into the structure and function of the nervous system. Use these insights to think about what can happen when its operation falters, whether by disease or mutation, or by deliberate and abusive use of drugs.

KEY CONCEPTS

1. Nervous systems are composed of neurons and a variety of cells called neuroglia, which structurally and functionally support the neurons. The neurons interact in the detection and integration of information about external and internal conditions, and in commanding muscles and glands to carry out suitable responses.

2. The simplest nervous systems of all are the nerve nets of radial animals, such as sea anemones. The nervous systems of most animals show pronounced cephalization and bilateral symmetry.

3. Vertebrate nervous systems are functionally divided into central and peripheral regions. The brain and spinal cord make up the central nervous system. Paired nerves that thread through the rest of the body are the key components of the peripheral nervous system.

4. The somatic nerves of the peripheral nervous system deal with skeletal muscles. The autonomic nerves deal with the heart, lungs, and other internal organs.

5. The vertebrate brain is functionally divided into three regions, called the hindbrain, midbrain, and forebrain. Its most ancient parts deal with reflex control over breathing, blood circulation, and other basic functions that are essential for staying alive.

6. Long ago, in certain vertebrate lineages, the brain expanded in volume and complexity. Its newer regions appropriated more and more control over the ancient reflex functions.

7. In birds and mammals especially, the portion of the forebrain called the cerebrum contains the most complex centers for receiving, integrating, storing, and responding to sensory information.

8. The hypothalamus, another part of the forebrain, is the main homeostatic control center over the internal environment and the functioning of internal organs. Being part of the limbic system, the hypothalamus also influences emotional states and memory.

Life-styles and Nervous Systems

Even if you stare at it for hours on end, a sea urchin will never dazzle you with spectacular bursts of speed or precision acrobatics. Because it seems only a bit livelier than the pincushion it resembles, you might suspect this animal is brainless, and in fact it is. However, the sea urchin does have a nervous system, even if it is a decentralized one.

At this point in the book, you know that all animals except sponges have some type of a **nervous system** in which nerve cells, such as neurons, are oriented relative to one another in signal-conducting and information-processing pathways. At the minimum, communication lines of its component cells receive information about changing conditions outside and inside the body, then elicit suitable responses from muscle and gland cells.

To appreciate any animal's nervous system, you have to ask, *What does the animal do?* Many sea urchins spend most of their lives moving about slowly on little tube feet, scraping bits of algae off rocks with a toothy feeding apparatus (Section 26.20 and Figure 35.2). Their nervous system includes three sets of nerves. A **nerve**, recall, is like a cable in which sensory axons, motor

a

b

Figure 35.2 A spiny sea urchin, (**a**) right-side up and (**b**) flipped over to show its toothy, all-important mouth.

axons, or both are bundled together inside a sheath of connective tissue (Section 34.4). The ones in sea urchins deal mainly with sampling the environment, moving through it, and (you guessed it) getting food into the gut. They interconnect with one another by a **nerve net**, a loose mesh of nerve cells intimately associated with epithelial tissue. Information flow through the nerve net is not highly directional. Signals travel diffusely, in all directions, from a point of stimulation. You might not find this type of nervous system impressive, but if you did little more than slowly inch about eating algae, it would be quite sufficient.

Regarding the Nerve Net

Animals first evolved in the seas, and it is in the seas that we still find the animals with the simplest nervous systems. They are sea anemones, jellyfishes, and other cnidarians (Sections 26.4 and 26.5). These invertebrates show *radial* symmetry. Their body parts are arranged about a central axis, much like spokes of a bike wheel.

Like sea urchins, cnidarians have a nerve net (but no nerves). It extends throughout the body and is equally responsive to food or potential danger coming from any direction. Its nerve cells interact with sensory cells and contractile cells of the epithelium in **reflex pathways**. In such pathways, remember, sensory stimulation directly triggers simple, stereotyped movements (Section 34.4).

For example, in jellyfishes, reflex pathways permit slow swimming movements and keep the body right-side up. In all cnidarians, a reflex pathway concerned with feeding behavior extends from sensory receptors present in the tentacles, along nerve cells, to contractile cells around the mouth. Figure 35.3 illustrates the cells involved in such a pathway.

On the Importance of Having a Head

Flatworms are the simplest animals having a bilateral nervous system (Figure 35.4). *Bilateral* symmetry means having equivalent body parts on the left and right sides of the body's midsagittal plane. (Imagine sliding down a staircase banister that turns into a razor and you may never forget the location of the midsagittal plane.) Both sides have the same array of muscles that function in moving the body forward. Both have the same array of nerves to control the muscles, and so on.

The ladderlike nervous system of the flatworm has two cordlike nerves running longitudinally through the body. Besides the two nerve cords, some systems have **ganglia** (singular, ganglion), which are clusters of nerve cell bodies that serve as integrating centers. At the head end of this animal, ganglia form a two-part brainlike structure that coordinates signals from paired sensory

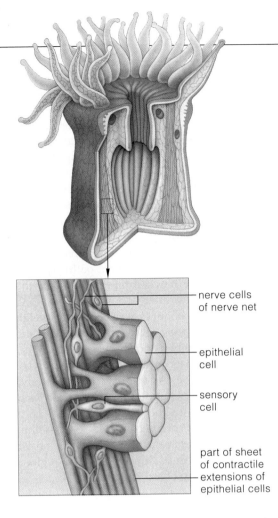

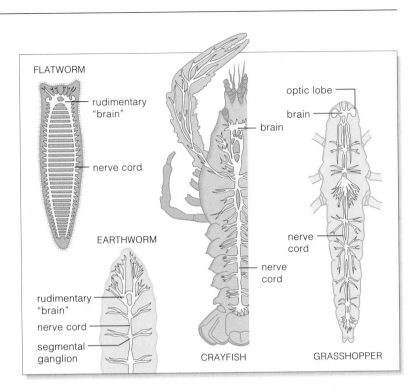

Figure 35.4 Bilateral nervous systems of a few invertebrates. The sketches are not to scale relative to one another.

Figure 35.3 Nerve net of a sea anemone, one of the cnidarians. The nerve cells interact with sensory cells and with contractile cells of a sheetlike epithelium between the outer epidermis and a jellylike middle layer (mesoglea) of the body wall.

organs, including two eyespots, and provides a degree of control over the nerve cords.

Did bilateral nervous systems evolve from nerve nets? Maybe. Consider that some cnidarian life cycles include a self-feeding larval stage, called a **planula**. Like flatworms, a planula has a flattened body and uses cilia to swim or crawl about:

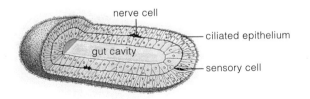

Imagine a few ancient planulas crawling about on the seafloor in Cambrian times. By chance, mutations of regulatory genes blocked their transformation into adults but had no effect on maturation of reproductive organs. (As you read in Section 27.6, this does happen in the larvae of certain salamanders and other animals.)

Such planulas kept on crawling, they reproduced, and so they passed on the mutated genes responsible for the capacity for forward mobility. The planula-like animals must have crawled into different, potentially dangerous or splendid situations. There, they would not have been helped much by sensory cells located at the trailing end of their body. But if such cells became concentrated at the leading end, rapid, effective responses to stimuli would have been possible. Probably natural selection did favor animals with concentrations of sensory cells at the leading end. Cephalization (formation of a head) and bilateral symmetry may have started this way. Both features do occur in most invertebrate lineages (Figure 35.4). And as you will see in the next section, *patterns of bilateral symmetry and cephalization also are evident in paired nerves and muscles, paired sensory structures, and paired brain centers of yourself and all other vertebrates.*

In nervous systems, nerve cells receive and process sensory input, then elicit responses from muscle and gland cells.

Radial animals have a nerve net, a diffuse mesh of nerve cells that take part in simple reflex pathways involving sensory cells and contractile cells of an epithelial tissue.

Nervous systems of cephalized, bilateral animals include a brain (or ganglia at the head end) as well as paired nerves and paired sensory structures.

Nerves are cablelike communication lines of sensory axons, motor axons, or both bundled in a connective tissue sheath.

Evolutionary High Points

Hundreds of millions of years ago, the earliest fishlike vertebrates were evolving (Section 27.3). A column of bony segments was taking over the function of the notochord, a long rod of stiffened tissue that worked with muscles to bring about movements. Above the notochord, a hollow, tubular nerve cord was evolving also. It was the forerunner of the spinal cord and brain. Remember, these developments were a foundation for new life-styles. Not long afterward, genetic divergences from lineages of filter-feeding, scavenging fishes gave rise to swift, jawed predators of the seas.

At first, simple reflex pathways predominated. The sensory neurons that were tripped into action synapsed on motor neurons, which directly stimulated muscles to contract. No other neurons altered the flow of signals from reception of a stimulus to the response. However, in the changing world of the fast-moving vertebrates,

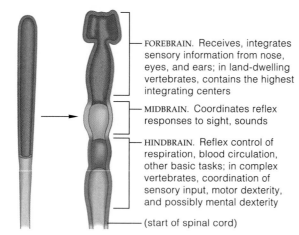

FOREBRAIN. Receives, integrates sensory information from nose, eyes, and ears; in land-dwelling vertebrates, contains the highest integrating centers

MIDBRAIN. Coordinates reflex responses to sight, sounds

HINDBRAIN. Reflex control of respiration, blood circulation, other basic tasks; in complex vertebrates, coordination of sensory input, motor dexterity, and possibly mental dexterity

(start of spinal cord)

a Diagram of how the anterior end of the dorsal, hollow nerve cord of vertebrates expanded into functionally distinct regions, which also increased in complexity in certain lineages

predators and prey that were better equipped to sense one another's presence had the competitive edge.

The senses of smell, hearing, and balance became even keener among vertebrates that invaded land. Bones and muscles evolved in ways that enhanced specialized motor activities. The brain became variably thickened with nervous tissue that could integrate rich sensory information and issue orders for complex responses.

The oldest parts of the vertebrate brain still deal with reflex coordination of breathing and other vital functions. Now, however, many interneurons synapse on sensory and motor neurons of ancient pathways and with one another in the newer brain regions. In the most complex vertebrates, interneurons receive, store, and compare information about experiences. They weigh possible responses. And they give our own species the capacity to reason, remember, and learn.

The nerve cord persists in all vertebrate embryos. We call it the **neural tube**. As the embryo grows and develops, the neural tube undergoes expansion and regional modifications into the brain and spinal cord (Figure 35.5a). The spinal cord becomes enclosed within the vertebral column. Adjacent tissues in the embryo give rise to nerves that thread through all body regions and connect with the spinal cord and brain.

Functional Divisions of the Vertebrate Nervous System

Figure 35.6 gives you a general sense of the expansions of nervous tissue in the human nervous system. This diagram shows the major paired nerves of this bilateral system. What it cannot possibly show are the 100 billion interneurons in the brain alone. To be sure, humans do have the most intricately wired nervous system in the animal world. Even so, you find similar patterns among other vertebrates.

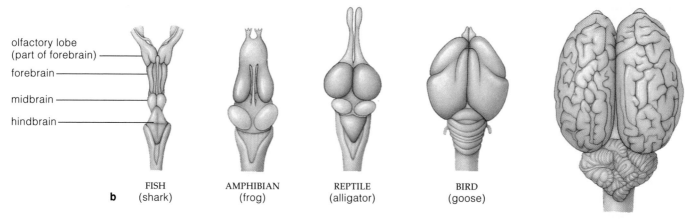

olfactory lobe (part of forebrain)

forebrain

midbrain

hindbrain

b

| FISH (shark) | AMPHIBIAN (frog) | REPTILE (alligator) | BIRD (goose) | MAMMAL (horse) |

Figure 35.5 Evolutionary trend toward an expanded, more complex brain, as suggested by comparing the brains of some existing vertebrates. These dorsal views are not to the same scale.

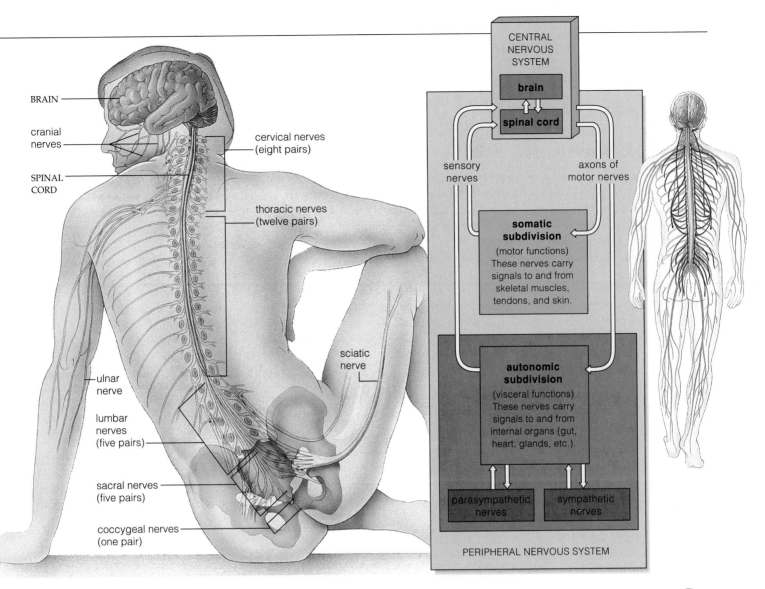

BRAIN

cranial nerves

SPINAL CORD

cervical nerves (eight pairs)

thoracic nerves (twelve pairs)

sciatic nerve

ulnar nerve

lumbar nerves (five pairs)

sacral nerves (five pairs)

coccygeal nerves (one pair)

CENTRAL NERVOUS SYSTEM

brain

spinal cord

sensory nerves

axons of motor nerves

somatic subdivision

(motor functions) These nerves carry signals to and from skeletal muscles, tendons, and skin.

autonomic subdivision

(visceral functions) These nerves carry signals to and from internal organs (gut, heart, glands, etc.).

parasympathetic nerves

sympathetic nerves

PERIPHERAL NERVOUS SYSTEM

Figure 35.6 Sketch of the human nervous system, showing the brain, spinal cord, and some of the major peripheral nerves. This nervous system also includes twelve pairs of cranial nerves that originate from the brain. Other vertebrates have a similar system.

Figure 35.7 Functional divisions of the nervous system. The central nervous system is color-coded *blue*, the somatic nerves *green*, and the autonomic nerves *red*. Sometimes the nerves carrying sensory input to the central nervous system are said to be *afferent* (a word meaning "to bring to"). The ones carrying motor output away from the central nervous system to muscles and glands are *efferent* ("to carry outward").

Investigators typically approach the complexity of the vertebrate nervous system by functionally dividing it into central and peripheral regions (Figure 35.7). All of the interneurons are confined to the **central nervous system**, or the spinal cord and brain. The **peripheral nervous system** consists mainly of nerves that thread through the rest of the body and carry signals into and out of the central nervous system.

Inside the brain and spinal cord, the communication lines are called **tracts**, not nerves. The tracts making up **white matter** contain axons with glistening white myelin sheaths and specialize in rapid signal transmission. By contrast, **gray matter** consists of unmyelinated axons, dendrites, and nerve cell bodies and neuroglial cells. **Neuroglial cells**, remember, protect or structurally and functionally support neurons. They make up more than half the volume of vertebrate nervous systems.

The coevolution of nervous, sensory, and motor systems made possible more complex life-styles among vertebrates.

The vertebrate nervous system has become so intricately wired that the structure and functions of its central and peripheral regions are described separately.

The central nervous system consists of the brain and spinal cord. The peripheral nervous system consists of nerves that thread through the rest of the body and carry signals into and out of the central region.

Myelinated axons of tracts make up the white matter of the brain and spinal cord. Neuroglia as well as unmyelinated axons, dendrites, and cell bodies of neurons make up their gray matter.

THE MAJOR EXPRESSWAYS

Let's now take a look at the peripheral nervous system and the spinal cord, which interconnect as the major expressways for information flow through the body.

Peripheral Nervous System

SOMATIC AND AUTONOMIC SUBDIVISIONS   In humans, the peripheral nervous system includes thirty-one pairs of *spinal* nerves, which connect with the spinal cord. The system also includes twelve pairs of *cranial* nerves, which connect directly with the brain.

Cranial and spinal nerves can be further classified according to function. The ones that carry signals about moving your head, trunk, and limbs are the **somatic nerves**. The sensory axons inside these nerves deliver information from receptors in the skin, skeletal muscles, and tendons to the central nervous system. Their motor axons deliver commands from the brain and spinal cord to the body's skeletal muscles.

By contrast, spinal and cranial nerves dealing with smooth muscle, cardiac (heart) muscle, and glands are the **autonomic nerves**. They deal with the visceral parts of the body—in other words, with its internal organs and structures. Autonomic nerves carry signals to and from these organs and structures.

SYMPATHETIC AND PARASYMPATHETIC NERVES   Figure 35.8 shows the two categories of autonomic nerves. We call them parasympathetic and sympathetic. Normally they work antagonistically, with the signals from one opposing signals from the other. Both carry excitatory and inhibitory signals to internal organs. Often their signals arrive at the same time at muscle or gland cells and compete for control over them. In such instances, synaptic integration at the cellular level leads to minor adjustments in the organ's level of activity.

Parasympathetic nerves dominate when the body is not receiving much outside stimulation. They tend to

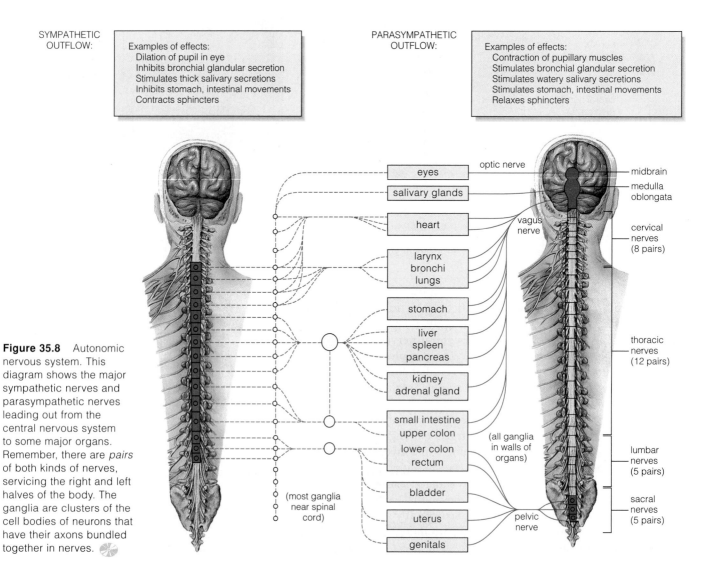

SYMPATHETIC OUTFLOW:

Examples of effects:
  Dilation of pupil in eye
  Inhibits bronchial glandular secretion
  Stimulates thick salivary secretions
  Inhibits stomach, intestinal movements
  Contracts sphincters

PARASYMPATHETIC OUTFLOW:

Examples of effects:
  Contraction of pupillary muscles
  Stimulates bronchial glandular secretion
  Stimulates watery salivary secretions
  Stimulates stomach, intestinal movements
  Relaxes sphincters

Figure 35.8 Autonomic nervous system. This diagram shows the major sympathetic nerves and parasympathetic nerves leading out from the central nervous system to some major organs. Remember, there are *pairs* of both kinds of nerves, servicing the right and left halves of the body. The ganglia are clusters of the cell bodies of neurons that have their axons bundled together in nerves.

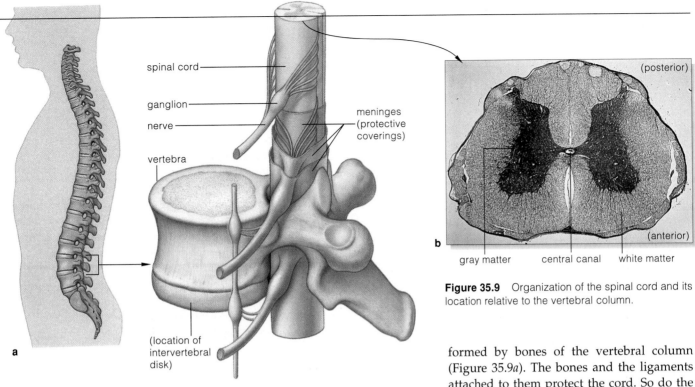

spinal cord

ganglion

nerve

vertebra

meninges (protective coverings)

(location of intervertebral disk)

a

(posterior)

b

(anterior)

gray matter   central canal   white matter

Figure 35.9 Organization of the spinal cord and its location relative to the vertebral column.

slow down the body overall and divert energy to basic "housekeeping" tasks, such as digestion.

Sympathetic nerves dominate in times of sharpened awareness, excitement, or danger. They tend to shelve housekeeping tasks. At the same time, they prepare the animal to fight or escape (when threatened) or to frolic (as in play or sexual behavior). Right now, sympathetic nerves are commanding your heart to beat a bit faster, and parasympathetic nerves are commanding it to beat a bit slower. Integration of the opposing signals adjusts the heart rate. If something scares or excites you, the parasympathetic input to the heart drops. Sympathetic signals cause the release of epinephrine, which makes your heart beat faster. It also makes you breathe faster and start to sweat. In this state of intense arousal, you are primed to fight (or play) hard or to get away fast. Hence the name **fight-flight response**.

Suppose the stimulus for the fight-flight response ends. Sympathetic activity may decrease abruptly, and parasympathetic activity may rise suddenly. You might observe this "rebound effect" after someone has been instantly mobilized to rush onto a highway to save a child from an oncoming car. The person may well faint as soon as the child has been swept out of danger.

The Spinal Cord

By definition, the **spinal cord** is a vital expressway for signals between the peripheral nervous system and the brain. Also, sensory and motor neurons make direct reflex connections in the spinal cord. (The stretch reflex, described in Section 34.4, arcs through the spinal cord this way.) The spinal cord threads through a canal

formed by bones of the vertebral column (Figure 35.9a). The bones and the ligaments attached to them protect the cord. So do the **meninges**, three tough, tubelike coverings around the spinal cord and brain. The coverings are not impervious to all attacks. *Meningitis*, an often-fatal disease, results from viral or bacterial infection. Disease symptoms include severe headaches, fever, a stiff neck, and nausea.

Signals travel rapidly up and down the spinal cord in bundles of axons with glistening myelin sheaths. The sheaths distinguish the cord's white matter from its gray matter (Figure 35.9b). Gray matter, recall, consists of dendrites and cell bodies of neurons, and neuroglial cells. It plays an important role in controlling reflexes for limb movements (as when you walk or wave your arms) and organ activity (such as bladder emptying).

Experiments with frogs provide evidence for these pathways. Between the frog's spinal cord and brain are neural circuits that deal with straightening legs after they have been bent. If these circuits are severed at the base of the brain, the legs become paralyzed—but only for about a minute. So-called extensor reflex pathways in the spinal cord recover quickly and have the frog hopping about in no time. Recovery is minimal after similar damage in humans and other primates—the vertebrates with the greatest cephalization.

The nerves of the peripheral nervous system connect the brain and spinal cord with the rest of the body.

Its somatic nerves deal with skeletal muscle movements. Its autonomic nerves deal with the functions of internal organs, such as the heart and glands.

The spinal cord is a vital expressway for signals between the brain and peripheral nerves. Some of its interneurons also exert direct control over certain reflex pathways.

The anterior end of the spinal cord merges with the **brain**, the body's master control center. The brain receives, integrates, stores, and retrieves information. It also coordinates responses to information by adjusting activities throughout the body. Like the spinal cord, the brain is protected by bones and membranes.

Reflect for a moment on Figures 35.5 and 35.10. The forebrain, midbrain, and hindbrain of vertebrates form from three successive portions of the neural tube. The nervous tissue that evolved first in all three regions is called the **brain stem**. The brain stem is still identifiable in an adult brain, and it still contains many simple, basic reflex centers. Over evolutionary time, expanded layers of gray matter developed from the brain stem. The more recent additions have been correlated with an increasing reliance on three major sensory organs: the nose, ears, and eyes. The forebrain and possibly parts of the cerebellum, which is part of the hindbrain region, contain the newest additions of gray matter. Figure 35.10 provides an overview of their main functions.

Hindbrain

The medulla oblongata, cerebellum, and pons are all components of the hindbrain. The **medulla oblongata** has reflex centers for vital tasks, such as respiration and blood circulation. It coordinates motor responses with complex reflexes, such as coughing. It influences other brain regions that help you sleep or wake up.

The **cerebellum** integrates sensory input from the eyes, ears, and muscle spindles with motor commands from the forebrain. It helps control motor dexterity. More recent expansions of the human cerebellum may be crucial in language and some other forms of mental dexterity.

Bands of many axons extend from both sides of the cerebellum to the **pons** (which means bridge). Although lodged in the brain stem, the pons is a major traffic center for information passing between the cerebellum and the higher integrating centers of the forebrain.

Midbrain

The midbrain coordinates reflex responses to sights and sounds. The **tectum** is a more recent layering of gray matter; it forms the roof of the midbrain. The fish and amphibian tectum is a center for coordinating nearly all sensory input and initiating motor responses. A frog surgically deprived of its highest brain center but with its tectum intact still does almost everything frogs do.

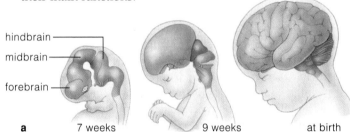

hindbrain
midbrain
forebrain

a   7 weeks   9 weeks   at birth

Division	Main Parts	Typical Functions
FOREBRAIN	Cerebrum	Localizes and processes sensory inputs; initiates and controls skeletal muscle activity; in the most complex vertebrates, roles in memory, mediating emotions, and abstract thought
	Olfactory lobes	Relaying of sensory input from the nose to olfactory centers of the cerebrum
	Thalamus	Relay stations for conducting sensory signals to and from cerebral cortex; role in memory
	Hypothalamus	Together with pituitary gland, homeostatic control center over the volume, composition, and temperature of the internal environment, influences behavior relating to organ functions (such as hunger and thirst) and physical expressions of emotions (such as sweating)
	Limbic system	A complex of structures that govern emotions and that have roles in memory
	Pituitary gland	With hypothalmus, endocrine control of metabolism, growth, and development
	Pineal gland	Control of some circadian rhythms; role in mammalian reproductive physiology
MIDBRAIN	Tectum	In fishes and amphibians, key coordinating center for sensory input and motor responses; in mammals, mainly reflex centers that rapidly relay sensory input to forebrain
HINDBRAIN	Pons	"Bridge" of tracts between cerebrum and cerebellum; its other tracts connect forebrain and spinal cord; works with medulla oblongata to control rate and depth of respiration
	Cerebellum	Coordinates motor activity for limb movements, maintaining posture, and spatial orientation
	Medulla oblongata	Contains tracts between pons and spinal cord; reflex centers involved in control of heart rate, blood vessel diameter, respiratory rate, vomiting, coughing, other vital functions

b

anterior end of spinal cord

Figure 35.10 (**a**) How the anterior end of a human embryo's neural tube develops into a brain. (**b**) Overview of the brain's subdivisions, correlated with the neural tube at an early developmental stage.

In most vertebrates (but not mammals), the midbrain includes a pair of brain centers called optic lobes, which deal with sensory input from eyes. Mammalian eyes deserted the tectum, so to speak; they formed important functional connections with integrating centers in the forebrain. The mammalian tectum ended up as a reflex center that swiftly relays sensory signals to those higher integrating centers.

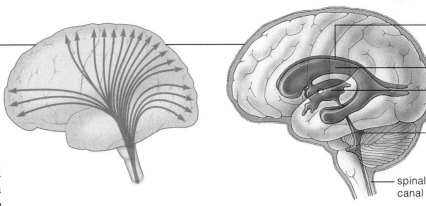

Figure 35.11 Communication pathways of the reticular formation, an evolutionarily ancient, diffuse network of neurons that extends from the anterior end of the spinal cord on up to the highest integrative centers of the cerebral cortex.

Figure 35.12 Cerebrospinal fluid (*blue*) in the human brain. This extracellular fluid surrounds and cushions the spinal cord and the brain. It also fills four interconnected cavities (cerebral ventricles) within the brain and the spinal cord's central canal.

right ventricle

left ventricle

third ventricle

fourth ventricle

spinal canal

Forebrain

For much of vertebrate history, chemical odors diffusing in aquatic environments were the major clues to survival. An olfactory lobe dealing mainly with odors from predators, prey, and potential mates was the main forebrain structure (Figure 35.5), along with paired outgrowths of the brain stem where olfactory input and responses to it were integrated. In time the outgrowths expanded greatly, especially during and after certain vertebrates invaded the land. We now call them the two hemispheres of the **cerebrum**. Another forebrain region, called the **thalamus**, evolved as a coordinating center for sensory input and as a relay station for signals to the cerebrum. Below this, the **hypothalamus** evolved as the main center for homeostatic control over the internal environment. It became central to behaviors related to internal organ activities, such as thirst, hunger, and sex, and to emotional expression, such as sweating with fear.

The Reticular Formation

By now, you may be thinking that the brain is tidily subdivided into three main regions. This is not the case. An evolutionarily ancient mesh of interneurons still extends from the uppermost part of the spinal cord, on through the brain stem, and on into higher integrative centers of the cerebral cortex (Figure 35.11). This major network of interneurons is the **reticular formation**. It persists as a low-level pathway to motor centers of the medulla oblongata and spinal cord. It also can activate centers in the cerebral cortex and thereby help govern the activities of the nervous system as a whole.

Brain Cavities and Canals

The hollow neural tube that first develops in vertebrate embryos persists in the adults, as a continuous system of fluid-filled cavities and canals. Within this system is the **cerebrospinal fluid**, a clear extracellular fluid that cushions the tissues of the brain and spinal cord from sudden, jarring movements (Figure 35.12).

A mechanism called the **blood-brain barrier** exerts some control over which solutes enter the cerebrospinal fluid and thereby helps to protect the brain and spinal cord. In no other portion of the extracellular fluid are solute concentrations maintained within such narrow limits. If the brain were exposed to the compositional shifts that typically accompany, say, digesting a meal or exercising, it would suffer. Why? Some of the hormones and amino acids in blood are signaling molecules that affect neurons. Also, shifting levels of certain ions (such as K^+) can skew the threshold for action potentials.

The barrier operates at the plasma membrane of cells making up the walls of capillaries that service the brain. In most brain regions, continuous tight junctions fuse together the abutting walls of these cells, so all water-soluble substances must move *through* the cells to reach the brain. Membrane transport proteins allow essential nutrients (such as glucose) and some ions to move into and out of the cells but bar urea and other metabolic wastes, certain toxins, and most drugs. The barrier does not bar fat-soluble substances, including oxygen, carbon dioxide, anesthetics, alcohol, caffeine, and nicotine. The barrier is nonexistent around the hypothalamus and the brain stem's vomiting center. (Can you guess why?)

The vertebrate brain develops from a hollow neural tube, which persists in adults as a system of cavities and canals filled with cerebrospinal fluid. The fluid cushions nervous tissue from sudden, jarring movements.

The vertebrate brain is functionally subdivided into three regions called the hindbrain, forebrain, and midbrain.

In all three regions, the nervous tissue that was the first to evolve is still recognizable and is called the brain stem. The brain stem affords reflex control over the most basic functions necessary for survival.

In the most complex vertebrates, the highest integrative centers reside in the forebrain, especially in its cerebral hemispheres.

Organization of the Cerebral Hemispheres

Study any nervous system or even the best computers and you realize the human brain is the most complex integration center of all. It only weighs 1,300 grams (3 pounds) in a person of average size, and more than half of that mass consists of neuroglial cells. But remember, the human brain also has at least *100 billion* interneurons!

The human cerebrum, housed in a chamber of hard skull bones, resembles the much-folded nut in a walnut shell (Figure 35.13*a*). Interneurons here contribute most to your humanness. A longitudinal fissure divides the cerebrum into left and right **cerebral hemispheres**. Most of the brain centers dealing with analytical skills,

temporal, parietal, and frontal lobes. Figure 35.14 gives a few examples. (*EEG*, short for electroencephalogram, is simply a recording of summed electrical activity in a given brain region. Section 2.2 describes PET scans.)

The *occipital* lobe at the back of each hemisphere has centers for vision. Severe blows to it may destroy part of the visual field, the outside world that the individual sees. The *temporal* lobe near each temple has centers that deal with hearing and visual associations. These centers have become highly developed among humans and other primates. A blow here won't leave you blind, but it might impair your ability to recognize complex visual patterns, such as faces. Centers in the temporal lobe also influence emotional behavior.

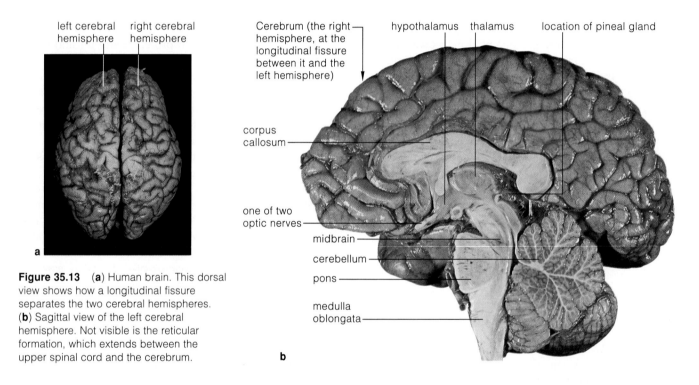

left cerebral hemisphere | right cerebral hemisphere

Cerebrum (the right hemisphere, at the longitudinal fissure between it and the left hemisphere)

hypothalamus   thalamus   location of pineal gland

corpus callosum

one of two optic nerves

midbrain

cerebellum

pons

medulla oblongata

Figure 35.13 (**a**) Human brain. This dorsal view shows how a longitudinal fissure separates the two cerebral hemispheres. (**b**) Sagittal view of the left cerebral hemisphere. Not visible is the reticular formation, which extends between the upper spinal cord and the cerebrum.

speech, and mathematics reside in the left hemisphere. The centers dealing with spatial relationships, music, and other nonverbal (abstract) skills reside mainly in the right hemisphere. Each half receives, processes, and coordinates responses to sensory input mostly from the opposite side of the body. For example, signals about pressure on the right arm reach the left hemisphere. A transverse band of nerve tracts, the corpus callosum, carries signals in both directions and coordinates the functioning of both hemispheres.

Above the brain's axons is the **cerebral cortex**, a thin layer of gray matter (the cell bodies of interneurons). Different parts receive and process different signals. EEGs as well as PET scans have localized the activities in each hemisphere to four tissue lobes: the occipital,

In the *parietal* lobe is the somatosensory cortex, the main receiving area for sensory input from the skin and joints. In the *frontal* lobe, the motor cortex deals with signals for motor responses. Thumb, finger, and tongue muscles get much of the brain's attention. This gives you an idea of how much control is required for hand movements and verbal expression (Figure 35.15).

The prefrontal cortex, a region in front of the motor cortex, is crucial in planning movements and inhibiting unsuitable behavior. As you will read later, it also has roles in some aspects of memory. PET scans suggest that this cortical region and part of the cerebellum (especially a structure called the dentate nucleus) probably interact to govern the motor abilities underlying language and some thought processes.

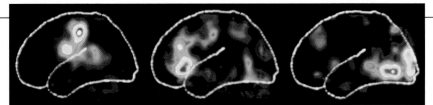

Motor cortex activity when speaking   Prefrontal cortex activity when generating words   Visual cortex activity when observing words

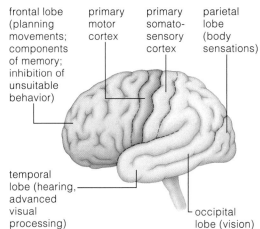

frontal lobe (planning movements; components of memory; inhibition of unsuitable behavior)   primary motor cortex   primary somato-sensory cortex   parietal lobe (body sensations)

temporal lobe (hearing, advanced visual processing)   occipital lobe (vision)

Figure 35.14 *Right:* Primary receiving and integrating centers for the human cerebral cortex. Primary cortical areas receive signals from receptors on the body's periphery. Association areas coordinate and process sensory input from different receptors. The three PET scans above identified which brain regions were active when a person performed three specific tasks: speaking, generating words, and observing words.

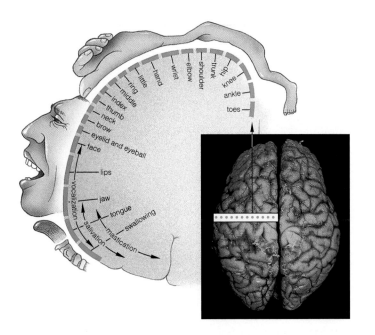

Figure 35.15 Body parts under control of the primary motor cortex. This diagram is a slice through the motor cortex of the left cerebral hemisphere. It controls muscles on the body's left side. The distortions to the human body that is draped over the diagram of the primary motor cortex indicate which body parts receive the most precise control.

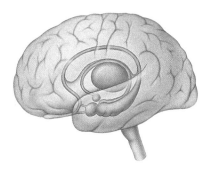

Figure 35.16 The location of the limbic system, which deals with memories as well as emotional reactions to stimuli. It consists of a complex group of nerve tracts and gray matter at the middle of the cerebral hemispheres. See also Section 35.7.

Many ingenious experiments have provided us with information on how the cerebral hemispheres function. Section 35.6 describes a classic example. For a lyrical example (no pun intended), consider reading "Music of the Hemispheres," an article in the October 1996 issue of *Discover* magazine that explores the cerebral circuits dealing with melodies and language.

Connections With the Limbic System

The **limbic system**, which is located at the middle of the cerebral hemispheres, governs our emotions and has roles in memory (Figure 35.16). It is distantly related to olfactory lobes and still deals with the sense of smell. That's one reason why you may feel warm and fuzzy

when your brain recalls the cologne of a special person who wore it. Connections from the cerebral cortex and other brain centers pass through this system, which includes the amygdala, hippocampus, hypothalamus, and parts of the thalamus. The connections enable us to correlate organ activities with self-gratifying behavior, such as eating and sex. They can make you feel that (say) your heart and stomach are on fire—with passion or with indigestion. Neural signals based on reasoning in the cerebral cortex often can override or dampen rage, hatred, and other "gut reactions."

The cerebrum of humans and other complex vertebrates is divided into hemispheres. Two-way signals along a nerve tract, the corpus callosum, afford communication between both hemispheres.

The cerebral cortex, the thin, outermost layer of gray matter of the hemispheres, contains the brain's primary receiving and integrating centers for sensory input.

The left hemisphere affords most of the control over speech, mathematics, and analytical skills. The right hemisphere affords most control over spatial abilities, music, and other abstract, nonverbal skills.

The cerebral cortex makes functional connections with the limbic system, which deals with emotions and memory.

SPERRY'S SPLIT-BRAIN EXPERIMENTS

Some time ago a neural surgeon, Roger Sperry, and his colleagues demonstrated some intriguing differences in perception between the two halves of the cerebrum of epileptics. Severe *epilepsy* is characterized by seizures, sometimes as often as every half hour. The seizures are analogous to an electrical storm in the brain. Sperry asked: Would *cutting* the corpus callosum of epileptics confine the electrical storm to one hemisphere, leaving at least the other hemisphere to function normally? Earlier studies of laboratory animals and humans whose corpus callosum had been damaged suggested this might be so.

Sperry performed the surgery on some patients. The electrical storms did subside in frequency and intensity. Cutting the neural bridge ended what must have been a positive feedback loop of ever intensifying electrical disturbances between the two hemispheres. The "split-brain" patients were able to lead what seemed, on the surface, entirely normal lives. But then Sperry devised some elegant experiments to test whether their conscious experience was indeed "normal." Given that the corpus callosum contains 200 million axons, surely *something* was different. Something was. "The surgery," Sperry later reported, "left these people with two separate minds, that is, two spheres of consciousness. What is experienced in the right hemisphere seems to be entirely outside the realm of awareness of the left."

Sperry presented the two hemispheres of split-brain patients with two different portions of the same visual stimulus. Researchers knew at the time that the visual connections to and from one hemisphere are mainly concerned with the opposite half of the visual field, as described in Figure 35.17. Sperry projected words—say, COWBOY—onto a screen so that COW fell in the left half of the visual field, and BOY fell in the right (Figure 35.18).

The subjects of this experiment reported *seeing* the word BOY. The left hemisphere, which controls language, perceived only the letters BOY. However, when asked to write the perceived word with the left hand—a hand that was deliberately blocked from a subject's view—the subject wrote COW. The right hemisphere "knew" the other half of the word (COW) and had directed the left hand's motor response. But it could not tell the left hemisphere what was going on because of the severed corpus callosum. The subject knew that a word was being written, but could not say what it was!

Thus Sperry showed that signals across the corpus callosum coordinate the functioning of the two cerebral hemispheres, each of which had responded to visual signals from the opposite side of the body.

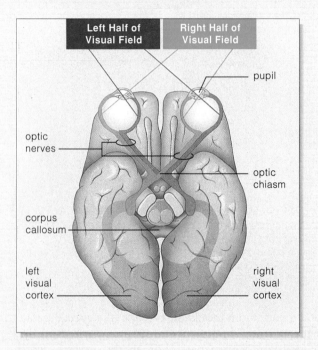

Figure 35.17 Pathway by which sensory input about visual stimuli reaches the visual cortex of the human brain.

Each eye gathers visual information at the retina, a layer of densely packed photoreceptors at the back of the eyeball (Section 36.8). Light from the *left* half of the visual field strikes receptors on the right side of both retinas. Parts of two optic nerves carry signals from the receptors to the right cerebral hemisphere. Light from the *right* half of the visual field strikes receptors on the left side of both retinas. Parts of the optic nerves carry signals from them to the left hemisphere.

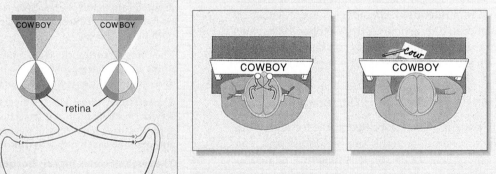

Figure 35.18 Response of a split-brain patient to different portions of a visual stimulus.

35.7 MEMORY

Memory is the capacity of an individual's brain to store and retrieve information about past sensory experience. Learning and adaptive modifications of our behavior would be impossible without it. Information is stored in stages. *Short-term* storage is a stage of neural excitation that lasts a few seconds to a few hours. It is limited to a few bits of sensory information—numbers, words of a sentence, and so on. In *long-term* storage, seemingly unlimited amounts of information get tucked away more or less permanently, as shown in Figure 35.19.

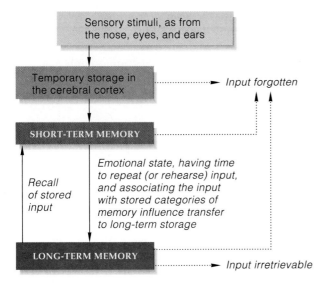

Figure 35.19 Stages of memory processing, starting with the temporary storage of sensory inputs in the cerebral cortex.

Not all of the sensory input bombarding the cerebral cortex ends up in memory storage. Only some is selected for transfer to brain structures involved in short-term memory. Information in these temporary holding bins is processed for relevance, so to speak. If irrelevant, it is forgotten; otherwise it is consolidated with the banks of information in long-term storage structures.

The human brain processes facts separately from skills. Dates, names, faces, words, odors, and other bits of explicit information are *facts*, soon forgotten or filed away in long-term storage, along with the circumstance in which they were learned. That is why you might associate, say, the smell of warm watermelon with a special picnic held long ago at the beach. By contrast, *skills* are gained by practicing specific motor activities. A skill such as slam-dunking a basketball or playing a piano concerto is best recalled by actually performing it, rather than by recalling the circumstances in which the skill was initially learned.

Separate memory circuits handle different kinds of input. A circuit leading to fact memory (Figure 35.20*a*) starts with inputs at the sensory cortex that flow to the

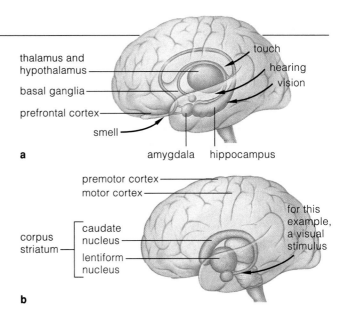

Figure 35.20 Possible circuits involved in (**a**) fact memory and (**b**) skill memory.

amygdala and hippocampus, which are structures in the limbic system. The amygdala is the gatekeeper; it connects the sensory cortex with parts of the thalamus and parts of the hypothalamus that govern emotional states. The hippocampus mediates learning and spatial relations. Information flows on to the prefrontal cortex, where multiple banks of fact memories are retrieved and used to stimulate or inhibit other parts of the brain. The new input also flows to basal ganglia, which send it back to the cortex in a feedback loop that reinforces the input until it can be consolidated in long-term storage.

Skill memory also starts at the sensory cortex, but this circuit routes sensory input to the corpus striatum, which promotes motor responses (Figure 35.20*b*). Motor skills involve muscle conditioning, and as you might suspect, the circuit extends to the cerebellum, the brain region that coordinates motor activity.

Amnesia is a loss of memory, the severity of which depends on whether the hippocampus, amygdala, or both are damaged, as by a severe head blow. Amnesia does not affect capacity to learn new skills. By contrast, basal ganglia are destroyed and learning ability is lost during *Parkinson's disease*, yet skill memory is retained. *Alzheimer's disease*, which usually has its onset later in life, is linked to structural changes in the cerebral cortex and hippocampus. Affected people often can remember long-standing information, such as their Social Security number, but they have difficulty remembering what has just happened to them. In time they become confused, depressed, and unable to complete a train of thought.

Memory, the storage and retrieval of sensory information, results from circuits between the cerebral cortex and parts of the limbic system, thalamus, and hypothalamus. Sensory input is processed through short-term and long-term storage.

35.8 STATES OF CONSCIOUSNESS

The spectrum of **consciousness** includes sleeping and aroused states, during which neural chattering shows up as wavelike patterns in EEGs. Specifically, EEGs are electrical recordings of the frequency and strength of membrane potentials at the surface of the brain (Figure 35.21). PET scans also can show the precise location of brain activity as it is proceeding.

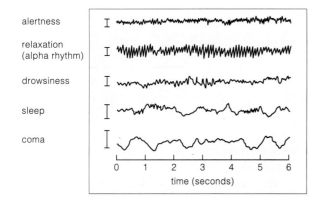

Figure 35.21 EEG patterns. Vertical bars indicate a fifty-range microvolt of electrical activity. The irregular horizontal graph lines indicate the electrical response over time.

The prominent wave pattern for someone who is relaxed, with eyes closed, is an *alpha rhythm*. The EEG waves are recorded in "trains" of one after the other, about ten per second. Alpha waves predominate during the state of meditation. During the transition to sleep, wave trains become larger, slower, and more erratic. This *slow-wave sleep* pattern dominates when sensory input is low and the mind is more or less idling. When subjects are awakened from slow-wave sleep, they usually report that they were not dreaming; rather, they often seemed to be mulling over recent, ordinary events.

Slow-wave sleep is punctuated by brief spells of *REM sleep*. *Rapid Eye Movements* accompany this pattern (the eyes jerk under closed lids), as do irregular breathing, faster heartbeat, and twitching fingers. Most people awakened from REM sleep report they were experiencing vivid dreams. The transition from sleep or deep relaxation into alert wakefulness is marked by a shift to low-amplitude, higher frequency wave trains. The transition, *EEG arousal*, occurs when conscious effort is made to focus on external stimuli or even on one's own thoughts.

Part of the reticular formation promotes chemical changes that influence whether you stay awake or fall asleep. Serotonin, a neurotransmitter released from one of its sleep centers, inhibits other neurons that arouse the brain and maintain wakefulness. At high levels, serotonin triggers drowsiness and sleep. Substances released from another brain center inhibit serotonin's effects and bring about wakefulness.

The spectrum of consciousness, which includes sleeping and states of arousal, is influenced by the reticular formation.

35.9 DRUGGING THE BRAIN

Broadly speaking, a **drug** is a substance introduced into the body to provoke a specific physiological response. Some drugs help a person cope with illness or emotional stress. Others act on the brain, artificially fanning pleasure associated with sex and other self-gratifying behaviors.

Many drugs are habit-forming. That is, even if the body is able to function well without them, a person continues to use drugs for the real or imagined relief they afford. Often the body develops tolerance of such drugs; it takes larger or more frequent doses to produce the same effect. Habituation and tolerance are signs of **drug addiction**, a chemical dependence on a drug (Table 35.1). *With an addiction, the drug has assumed an "essential" biochemical role in the body.* Abruptly deprive addicts of the drug, and they go through physical pain and mental anguish. The entire body goes through a period of biochemical upheavals. Stimulants, depressants, hypnotics, narcotic analgesics, hallucinogens, and psychedelics have such effects.

STIMULANTS Caffeine, nicotine, amphetamines, cocaine, and other stimulants increase alertness and body activity, then cause depression. Caffeine, a truly common stimulant, is present in coffee, tea, chocolate, and many soft drinks. Low doses arouse the cerebral cortex first, leading to increased alertness and restlessness. Higher doses acting at the medulla oblongata disrupt motor coordination and mental coherence. Nicotine, found in tobacco, has powerful effects on the nervous system. It mimics acetylcholine and directly stimulates a variety of sensory receptors. Its short-term effects include water retention, irritability, increased heart rate and high blood pressure, and gastric upsets.

Amphetamines, including "speed," mimic two of the neurotransmitters: dopamine and norepinephrine. In time, the brain produces less and less of these signaling molecules and depends more on artificial stimulation.

Table 35.1 Warning Signs of Drug Addiction*
1. Tolerance—it takes increasing amounts of the drug to produce the same effect.
2. Habituation—it takes continued drug use over time to maintain self-perception of functioning normally.
3. Inability to stop or curtail use of the drug, even if there is persistent desire to do so.
4. Concealment—not wanting others to know of the drug use.
5. Extreme or dangerous behavior to get and use the drug—as by stealing, asking more than one doctor to write drug prescriptions, or jeopardizing employment by drug use at work.
6. Deterioration of professional and personal relationships.
7. Anger and defensive behavior when someone suggests there may be a problem.
8. Preference of drug use over previously customary activities.

* Three or more of these signs may be cause for concern.

Possibly 2 million Americans are cocaine abusers. This drug gives a rush of pleasure by *blocking* the reabsorption of norepinephrine, dopamine, and some other neurotransmitters. Because these molecules accumulate in synaptic clefts, they incessantly stimulate postsynaptic cells for an extended period. Heart rate, blood pressure, and sexual appetite increase. In time the molecules diffuse away. However, neurons cannot synthesize replacements fast enough to counter the loss, and the sense of pleasure evaporates as the hypersensitized postsynaptic cells demand stimulation. After prolonged, heavy use of cocaine, "pleasure" can no longer be experienced. Addicts become anxious and depressed. They lose weight and cannot sleep properly. Their immune system weakens, and heart abnormalities set in.

Granular cocaine, which is inhaled (snorted), has been around for some time. Abusers burn crack cocaine and inhale the smoke, as in Figure 35.22. As suggested at the start of this chapter, crack is extremely addictive. Its highs are higher, the crashes are more devastating, and the social and economic tolls are extreme.

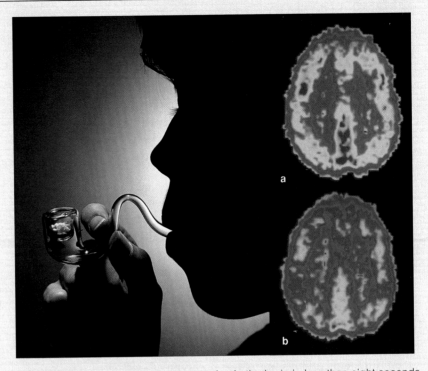

Figure 35.22 Smoking crack puts cocaine in the brain in less than eight seconds. (**a**) PET scan of a horizontal section of the brain, showing normal activity. (**b**) PET scan of a comparable section showing cocaine's effect. *Red* indicates greatest activity; *yellow, green,* and *blue* indicate successively inhibited activity.

DEPRESSANTS, HYPNOTICS   These drugs lower activity in nerves and parts of the brain. Some act at synapses in the reticular formation and thalamus. Responses depend on dosage and emotional and physiological states. They range from emotional relief through drowsiness, sleep, anesthesia, and coma to death. Low doses have the most effect on inhibitory synapses, so a person feels excited or euphoric at first. Increased doses suppress excitatory synapses as well and therefore lead to depression. Both the depressants and the hypnotics have additive effects. One amplifies another, as when alcohol plus barbiturates increases the depression.

Alcohol, or ethyl alcohol, acts directly on the plasma membrane to alter cell function. Like nicotine and cocaine, it is lipid soluble and easily crosses the blood-brain barrier to exert rapid effects. Some people mistakenly think of it as a harmless stimulant because of the initial "high" that it produces. However, alcohol is one of the most powerful psychoactive drugs and a major factor in many deaths. In the short term, small amounts cause disorientation and uncoordinated motor functions, and diminish judgment. Long-term addiction can lead to *cirrhosis.* Then, connective tissue permanently replaces damaged liver cells, so the self-regenerating capacity of liver tissue slowly diminishes. In time, the outcome is chronic liver failure and death.

ANALGESICS   When severe stress leads to physical or emotional pain, the brain produces natural analgesics, or pain relievers. Endorphins and enkephalins are two examples. They act on many parts of the nervous system, including brain centers that deal with emotions and pain. Narcotic analgesics, including codeine and heroin, sedate the body and relieve pain. They are among the most addictive substances known. Deprivation after massive doses of heroin results in hyperactivity and anxiety, fever, chills, violent vomiting, cramping, and diarrhea.

PSYCHEDELICS, HALLUCINOGENS   These drugs skew sensory perception by interfering with the activity of acetylcholine, norepinephrine, or serotonin. LSD (or lysergic acid diethylamide) alters serotonin's influence on inducing sleep, on temperature regulation, and on sensory perception. Even in small doses, LSD warps perceptions. For example, some users "perceived" they could fly and "flew" off buildings.

Marijuana, another hallucinogen, is made from crushed leaves, flowers, and stems of the plant *Cannabis.* In low doses it is like a depressant. It slows down but does not impair motor activity; it relaxes the body and elicits mild euphoria. It also causes disorientation, increased anxiety bordering on panic, delusions (including paranoia), and hallucinations.

Like alcohol, marijuana affects the performance of complex tasks, such as driving a car. In one study, pilots showed a marked deterioration in instrument-flying ability for more than two hours after smoking marijuana. Over time, marijuana smoking can suppress the immune system and impair mental functions.

SUMMARY

1. Nervous systems detect, interpret, and issue calls for response to stimuli in the animal's external and internal environments. At their most basic operating level are reflexes: simple, stereotyped movements made directly in response to stimuli.

   a. The simplest nervous systems are nerve nets, such as those of cnidarians and other radial animals. These meshworks of nerve cells form reflex connections with contractile and sensory cells of the epithelium.

   b. Most animals have a bilateral, cephalized nervous system, with a brain or ganglia at the anterior end. Such animals also have cordlike nerves and tracts (bundled axons of sensory neurons, motor neurons, or both).

2. The vertebrate central nervous system consists of the brain and spinal cord. The peripheral nervous system consists mainly of nerves that carry signals between all body regions and the spinal cord and brain.

   a. Somatic nerves of the peripheral nervous system deal with skeletal muscles. Autonomic nerves deal with the heart, lungs, glands, and other internal organs.

   b. Autonomic nerves are either parasympathetic or sympathetic. Parasympathetic nerves dominate when outside stimulation is low and tend to divert energy to basic housekeeping tasks. Overall, sympathetic nerves step up body activities during heightened awareness or danger. They govern the fight-flight response.

3. Reflex connections for limb movements and internal organ activity are made in the spinal cord. Many tracts in the spinal cord also carry signals between the brain and the peripheral nervous system.

4. During development, the brain develops by regional expansions of a hollow neural tube in the embryo. The tube's anterior end becomes the brain stem, which is still discernible in an adult brain. The brain's most recent expansions of gray matter deal with storing, comparing, and using experiences to initiate novel action.

   a. A neural tube's interior becomes a system of canals and cavities filled with cerebrospinal fluid. A protective mechanism called the blood-brain barrier exerts some control over which water-soluble substances enter the cerebrospinal fluid, which must be maintained within very narrow limits for neurons to function properly.

   b. The reticular formation, a mesh of interneurons extending from the top of the spinal cord to the cerebral cortex, is a low-level route to centers of motor activity.

5. The vertebrate brain has three main divisions:

   a. The hindbrain includes the medulla oblongata, pons, and cerebellum. It contains reflex centers for vital functions and muscle coordination.

   b. The midbrain coordinates and relays visual and auditory information to other brain regions.

   c. The forebrain includes the cerebrum, thalamus, and hypothalamus. The cerebral hemispheres contain the highest centers for receiving, processing, storing, and responding to sensory information. The thalamus relays signals and helps coordinate motor responses. The hypothalamus, the main homeostatic control center over the internal environment and internal organs, also governs thirst, sexual activity, and other behaviors related to organ functions. It is part of the limbic system, which governs emotions and memory.

Review Questions

1. Generally describe the type of nervous systems found in radially symmetrical animals and in bilaterally symmetrical animals. *35.1*

2. What constitutes the central nervous system? The peripheral nervous system? *35.2*

3. Distinguish between the following:
   a. central and peripheral nervous system *35.2*
   b. cranial and spinal nerves *35.3*
   c. somatic and autonomic nerves *35.3*
   d. parasympathetic and sympathetic nerves *35.3*

4. Distinguish between:
   a. white matter and gray matter *35.2*
   b. nerve and tract *35.2*

5. Describe the structure and function of the spinal cord. *35.3*

6. Label the major parts of the human brain. *35.4, 35.5*

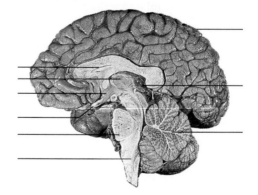

7. Explain the difference between the reticular formation and the limbic system in terms of their component parts and main functions. *35.4, 35.5*

8. Define cerebrospinal fluid and the blood-brain barrier. *35.4*

9. Briefly describe one of the brain centers of the cerebral cortex; mention the tissue lobe in which it is located. *35.5*

10. Distinguish between: *35.7*
    a. short-term memory and long-term memory
    b. fact memory and skill memory

11. Which part of the brain has major influence over states of consciousness, which include sleeping and arousal? *35.8*

12. What is a habit-forming drug? Briefly describe the effects of one such drug on the central nervous system. *35.9*

Self-Quiz *(Answers in Appendix IV)*

1. In sea anemones and jellyfishes, the _____ is a loose mesh of nerve cells that interact with sensory and contractile cells of an epithelium to bring about reflex responses to stimuli.

2. The oldest parts of the vertebrate brain provide _____ .
   a. reflex control of breathing, circulation, and other basic activities
   b. coordinating and relaying visual and auditory input
   c. storing, comparing, and retrieving sensory information
   d. both a and c

3. Is this statement true or false: White matter and gray matter are components of the spinal cord alone.

4. Is this statement true or false: The peripheral nervous system has nerves; the brain and spinal cord have tracts.

5. Overall, _____ nerves slow down the body and divert energy to digestion and other basic housekeeping tasks; and _____ nerves slow down housekeeping tasks and increase overall activity in times of heightened awareness or excitement.
   a. autonomic; somatic
   b. sympathetic; parasympathetic
   c. parasympathetic; sympathetic
   d. peripheral; central

6. Parasympathetic and sympathetic nerves _____ , to bring about minor adjustments in internal organ activity.
   a. service entirely different internal organs
   b. work antagonistically, most often
   c. come into play only during a fight-flight response
   d. none of the above

7. In tracts of the brain and spinal cord, _____ is associated with myelinated axons; _____ consists of neuroglia as well as unmyelinated axons, dendrites, and cell bodies of neurons.
   a. gray matter; white matter
   b. white matter; gray matter

8. In an adult brain, the brain stem is present in _____ .
   a. the hindbrain only
   b. the midbrain only
   c. the forebrain only
   d. all three divisions of the brain

9. The hindbrain, which includes the medulla oblongata, pons, and cerebellum, contains _____ .
   a. part of the reticular formation   c. the tectum
   b. the limbic system   d. all of the above

10. The _____ are located in the cerebrum.
    a. medulla oblongata, pons, and cerebellum
    b. thalamus, hypothalamus, and limbic system
    c. medulla oblongata, pons, and cerebral cortex
    d. cerebellum, medulla oblongata, pons, and limbic system
    e. hypothalamus, limbic system, pons, and cerebral cortex

11. The _____ is central to planning movements and blocking unsuitable behavior, and also has roles in memory.
    a. visual cortex
    b. motor cortex
    c. somatosensory cortex
    d. prefrontal cortex

12. Match the component with its main functions.
    ____ spinal cord   a. receives, integrates, issues
    ____ medulla   commands for response to sensory
         oblongata   information
    ____ hypothalamus   b. homeostatic control of internal
    ____ limbic   environment, internal organs
         system   c. deals with emotions, memory
    ____ cerebral   d. reflex control of respiration, blood
         cortex   circulation, other basic activities
                             e. expressway for signals between
                                brain and peripheral nervous system;
                                also, direct reflex connections

Critical Thinking

1. In human newborns and premature babies, the blood-brain barrier is not fully developed. Explain why this might be reason enough to pay careful attention to their diet.

2. When Jennifer was only six years old, a man lost control of his car and hit a tree in front of her house. She ran over to the car and screamed when she saw blood from a deep head wound dripping onto a huge bouquet of red roses. Thirty-five years later, someone gave Jennifer a bottle of *Tea Rose* perfume for a birthday present. When she sniffed the perfume, she became frightened and extremely anxious. A few minutes later she also had a vivid recollection of the accident. Explain this incident in terms of what you have learned about memory.

3. Eric typically drinks a cup of coffee nearly every hour, all day long. Today, by midafternoon, he is having trouble concentrating on his studies, and he feels tired and more than a little clumsy. Drinking still another cup of coffee didn't make Eric more alert. Explain how the caffeine in coffee might produce such symptoms.

Selected Key Terms

autonomic nerve 35.3
blood-brain barrier 35.4
brain 35.4
brain stem 35.4
central nervous system 35.2
cerebellum 35.4
cerebral cortex 35.5
cerebral hemisphere 35.5
cerebrospinal fluid 35.4
cerebrum 35.4
consciousness 35.8
drug 35.9
drug addiction 35.9
fight-flight response 35.3
ganglion (ganglia) 35.1
gray matter 35.2
hypothalamus 35.4
limbic system 35.5
medulla oblongata 35.4
memory 35.7

meninges (singular, meninx) 35.3
nerve 35.1
nerve net 35.1
nervous system 35.1
neural tube 35.2
neuroglia 35.2
parasympathetic nerve 35.3
peripheral nervous system 35.2
planula 35.1
pons 35.4
reflex pathway 35.1
reticular formation 35.4
somatic nerve 35.3
spinal cord 35.3
sympathetic nerve 35.3
tectum 35.4
thalamus 35.4
tract 35.2
white matter 35.2

Readings

Bloom, F., and A. Lazerson. 1988. *Brain, Mind, and Behavior.* Second edition. New York: Freeman.

Churchland, P., and P. Churchland. January 1990. "Could a Machine Think?" *Scientific American* 262(1): 32–37.

Fischbach, G. D. September 1992. "Mind and Brain." *Scientific American* 267(3): 48–57. This entire issue is devoted to the development, function, and certain disorders of the brain.

Romer, A., and T. Parsons. 1986. *The Vertebrate Body.* Sixth edition. Philadelphia: Saunders. Fine insights into the evolution of vertebrate nervous systems.

Shepherd, G. 1994. *Neurobiology.* Third edition. New York: Oxford. Paperback.

Web Site See *http://www.wadsworth.com/biology* for practice quiz questions, hypercontents, BioUpdates, and critical thinking. The Wadsworth Biology Resource Center provides a wealth of information fully organized and integrated by chapter.

36 SENSORY RECEPTION

Different Stokes for Different Folks

Even though you might be reluctant to pet a snake or scratch a bat behind the ears, you have to give them credit for being two of your vertebrate relatives with some special sensory traits.

Consider the python. Figure 36.1*a* shows one of these reptiles, which have thermoreceptors in rows of pits above and below the mouth. Thermoreceptors detect infrared energy—in this case, body heat of small, night-foraging mammals that are the snake's prey of choice. When stimulated, the receptors send messages to the brain, where several centers process the signals and issue commands to muscle cells. In no time at all, a strike is aimed and executed with stunning accuracy.

A motionless, edible frog would not have the same stimulatory effect on the receptors, so the snake would slither past it. The skin of a frog is not warm, and it blends with the colors of the frog's habitat. A python does not have the types of receptors that can detect a frog or a neural program for responding to it.

Or consider the mammals called bats. Nearly all species of bats sleep during the day and spread their webbed wings at dusk. Different kinds take to the air in search of nectar, fruit, frogs, or insects. Many sensory receptors that are located inside their eyes, nose, ears, mouth, and skin are not all that different from yours. However, other receptors provide bats with a sense of hearing that you cannot even begin to match. Even tiny-eyed, nearly blind species navigate and capture flying insects with precision in the dark!

Bats are masters of **echolocation**. They emit calls, as the bat in Figure 36.1*b* is doing. When sound waves of the calls bounce off insects, trees, and other objects, acoustical receptors inside the bat's ears detect the echoes and send signals about them to the bat brain.

As an echolocating bat flies, it emits a steady stream of about ten clicking sounds per second—sounds you cannot hear. The clicks are intense "ultrasounds," above the range of sound waves that receptors in human ears

a

Figure 36.1 Examples of sensory receptors. (**a**) Thermoreceptors in pits above and below this python's mouth detect body heat, or infrared energy, of nearby prey. (**b**) Some bats listen to echoes of their own high-frequency sounds. Their brain constructs a sound map, based on echoes bouncing back from objects in the surroundings. Such maps help this night-flying bat capture insects in midair, without the help of eyes.

are able to detect. When a bat hears a pattern of distant echoes from, say, an airborne mosquito or moth, it increases the rate of ultrasonic clicks to as many as 200 per second, which is faster than a machine gun fires bullets. In the few milliseconds of silence between the clicks, sensory receptors detect the echoes and deliver messages about them to the bat brain. The brain rapidly constructs a "map" of the sounds that the bat uses in its maneuvers through the night world.

b

With this chapter, we turn to the means by which animals receive signals from the external and internal environments, then decode signals in ways that give rise to awareness of sounds, sights, odors, pain, and other sensations. Sensory neurons, nerve pathways, and specific brain regions are required for these tasks. Together, they represent the portions of the nervous system called sensory systems.

KEY CONCEPTS

1. Sensory systems are portions of the nervous system. Each consists of specific types of sensory receptors, nerve pathways from receptors to the brain, and brain regions that receive and process the sensory information.

2. A stimulus is a form of energy that activates a specific type of sensory receptor, which is either a sensory neuron or a specialized cell adjacent to it. Photoreceptors detect light energy, thermoreceptors detect infrared energy, and so on.

3. A sensation is conscious awareness of change in some aspect of the external or internal environment. It begins when sensory receptors detect a specific stimulus. The stimulus energy becomes converted to a graded, local signal that may help initiate an action potential.

4. Information concerning the stimulus becomes encoded in the number and frequency of action potentials sent to the brain along particular nerve pathways. Then specific brain regions translate the information into a sensation.

5. The somatic sensations include touch, pressure, pain, temperature, and muscle sense.

6. The special senses include taste, smell, hearing, and vision.

SENSORY RECEPTORS AND PATHWAYS—AN OVERVIEW

Sensory systems, the front doors of the nervous system, receive information about specific changes inside and outside the body and notify the spinal cord and brain of what is going on. They consist of sensory receptors, nerve pathways leading from receptors to the brain, and brain regions where sensory information is translated into sensations. A **sensation** is conscious awareness of a stimulus. It is not the same as **perception** (understanding what a sensation means). *Compound* sensations arise when information about different stimuli is integrated at the same time. For example, "wetness" is not a single stimulus; our perception of it arises from simultaneous inputs concerning pressure, touch, and temperature.

Types of Sensory Receptors

Sensory receptors are the receptionists at the front door of the nervous system. As listed in Table 36.1, there are six major categories of sensory receptors, based on the type of stimulus energy that each type detects:

Mechanoreceptors detect forms of mechanical energy (changes in pressure, position, or acceleration).

Thermoreceptors detect infrared energy (heat).

Pain receptors (nociceptors) detect tissue damage.

Chemoreceptors detect chemical energy of specific substances dissolved in the fluid surrounding them.

Osmoreceptors detect changes in water volume (solute concentration) in the surrounding fluid.

Photoreceptors detect visible and ultraviolet light.

Depending upon the kinds and numbers of their sensory receptors, animals sample the environment in different ways and differ in their awareness of it. Unlike bees, for example, you have no receptors for ultraviolet light and do not see many flowers the way they do. The introduction to Chapter 31 provides an example of this. Unlike many bats, you and bees have no receptors for ultrasound. Unlike pythons, you, bees, and bats have no receptors for detecting warm-blooded prey in the dark.

Sensory Pathways

Recall, from Chapter 34, that sensory axons carry signals from receptors to the brain. Before they can do so, the stimulus energy must be converted to action potentials, the basis of neural messages. Briefly, when a stimulus disturbs the plasma membrane of a receptor's ending, certain ions flow across a local patch of the membrane. In other words, the disturbance triggers a local, graded potential. This type of signal does not spread far from the point of stimulation, and it can vary in magnitude.

When a stimulus is intense or when it is repeated fast enough for a summation of local signals, action potentials may be the result. Action potentials propagate themselves from the receptor to the axon endings of sensory neurons (Figure 36.2). There, neurotransmitter is released from the presynaptic cell, and it influences the electrical activity of the cell adjacent to it (another interneuron or a motor neuron). The disturbance may trigger action potentials in the postsynaptic cell, which is part of an information pathway leading to the brain.

Table 36.1 Major Categories of Sensory Receptors		
Category	Examples	Stimulus
MECHANORECEPTORS		
Touch, pressure	Certain free nerve endings and Pacinian corpuscles in skin	Mechanical pressure against body surface
Baroreceptor	Carotid sinus (Section 39.7)	Pressure changes in fluid that bathes them
Stretch	Muscle spindle in skeletal muscle	Stretching
Auditory	Hair cells in organ inside ear	Vibrations (sound or ultrasound waves)
Balance	Hair cells in organ inside ear	Fluid movement
THERMORECEPTORS	Certain free nerve endings	Change in temperature (heating, cooling)
PAIN RECEPTORS (NOCICEPTORS)*	Certain free nerve endings	Tissue damage (e.g., distortions, burns)
CHEMORECEPTORS		
Internal chemical sense	Carotid bodies in blood vessel wall	Substances (O_2, CO_2, etc.) dissolved in extracellular fluid
Taste	Taste receptors of tongue	Substances dissolved in saliva, etc.
Smell	Olfactory receptors of nose	Odors in air, water
OSMORECEPTORS	Hypothalamic osmoreceptors (Section 43.2)	Change in water volume (solute concentration) of fluid that bathes them
PHOTORECEPTORS		
Visual	Rods, cones of eye	Wavelengths of light

* Extremely intense stimulation of any sensory receptor also may be perceived as pain.

Action potentials that are traveling along a sensory neuron are not like a wailing ambulance siren. That is, *they do not vary in amplitude.* How, then, does the brain assess the nature of a given stimulus? The answer lies with (1) *which* nerve pathways happen to be carrying action potentials, (2) the *frequency* of action potentials traveling along each axon of the pathway, and (3) the *number* of axons that the stimulus has recruited.

First, the network of neurons in each animal's brain is genetically prescribed, and it can interpret action potentials only in certain ways. That is why you "see stars" when one of your eyes has been poked, even when it gets poked in a dark room. Photoreceptors inside the eye were mechanically disturbed enough to trigger messages that traveled along an optic nerve to your brain. And your brain always interprets any signals that are arriving from an optic nerve as "light."

Second, when a stimulus is strong, receptors fire action potentials more frequently than they do with a weak stimulus. The same receptor detects the sounds of a throaty whisper and a wild screech, and the brain senses the difference through frequency variations in signals that the receptor sends to it.

Third, a strong stimulus recruits more sensory receptors than a weaker stimulus does. Gently tap a spot of skin on one of your arms and you activate just a few receptors. Press hard on the same spot and you activate more receptors in a larger area. The increased disturbance translates into action potentials in many sensory axons at the same time. The brain interprets signals about the combined activity as

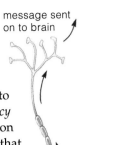

message sent on to brain

an interneuron inside the spinal cord

Figure 36.2 Example of a sensory nerve pathway that leads away from a sensory receptor to the brain. The sensory neuron is coded *red* and the interneurons are coded *yellow*.

Receptor endings of a sensory neuron are stimulated when a bare foot lands on a tack.

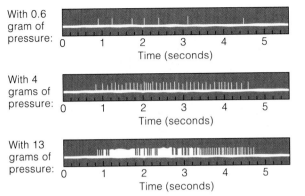

With 0.6 gram of pressure:

Time (seconds)

With 4 grams of pressure:

Time (seconds)

With 13 grams of pressure:

Time (seconds)

Figure 36.3 Recordings of action potentials from a single pressure receptor with endings in the skin of a human hand. The recordings correspond to variations in stimulus strength. The investigator pressed a thin rod against the skin with the amount of pressure indicated on the left side of each diagram. The vertical bars above each thick horizontal line represent individual action potentials. In each case, the increases in frequency correspond to increases in the strength of the applied stimulus.

an increase in the stimulus intensity. Figure 36.3 shows the effect of increases in stimulus strength.

In some cases, the frequency of action potentials decreases or stops even when a stimulus is being maintained at constant strength. A decrease in the response to a stimulus is known as **sensory adaptation**. After you put on clothing, for example, awareness of its pressure against your skin ceases. Some of the mechanoreceptors inside your skin adapt rapidly to the sustained stimulation; they are of a type that only signals a change in a stimulus (its onset and its removal). By contrast, other receptors adapt slowly or not at all; they help the brain monitor certain stimuli all the time. The stretch receptors you read about in Section 34.4 are like this. Many continually inform the brain about the length of particular muscles that help maintain balance and posture.

We turn next to some specific examples of sensory receptors. The types that are present at more than one body location contribute to **somatic sensations**. Other types are restricted to particular locations, such as inside the eyes or ears. They contribute to the **special senses**.

A sensory system has sensory receptors for specific stimuli, nerve pathways that conduct information from receptors to the brain, and brain regions that receive the information.

The brain assesses a given stimulus based on which nerve pathways are carrying action potentials, the frequency of action potentials traveling along each axon of that pathway, and the number of axons that have been recruited to action.

Somatic sensations begin with receptors in the body's surface tissues, skeletal muscles, and walls of internal organs. The receptors are most highly developed in birds and mammals; amphibians have few, and apparently fishes have none. Inputs from these receptors travel into the spinal cord and on to the **somatosensory cortex**, part of the surface layer of gray matter of the cerebral hemispheres (Section 35.5). Interneurons of this part of the brain are organized like maps, which correspond to particular parts of the body's surface. Map regions with the largest areas correspond to body parts that have the most sensory acuity and that require the most intricate control. Such body parts include the fingers, thumbs, and lips (Figure 36.4).

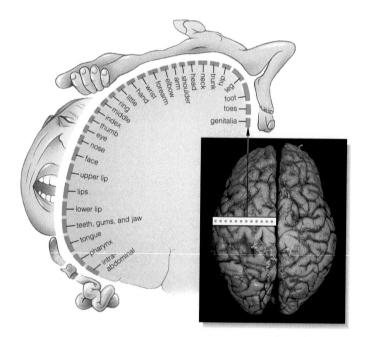

Figure 36.4 Differences in the representation of different body parts in the primary somatosensory cortex. This region is a strip of cerebral cortex, a little wider than 2.5 centimeters (about an inch), from the top of the head to above the ear.

Receptors Near the Body Surface

You and other mammals discern sensations of touch, pressure, cold, warmth, and pain near the body surface. Regions with the greatest number of sensory receptors, such as the fingertips and the tip of the tongue, are the most sensitive to stimulation. Other regions, such as the back of the hand and neck, do not have nearly as many receptors and are far less sensitive.

Free nerve endings are the simplest receptors. They are thinly myelinated or unmyelinated (naked) branched endings of sensory neurons in the epidermis or in the underlying dermis of skin. Different types function as mechanoreceptors, thermoreceptors, and pain receptors,

and all of them adapt very slowly to stimulation. One subpopulation gives rise to a sensation of prickling pain, as when you jab a finger with a pin. Another contributes to sensations of itching or warming that are elicited by chemicals, including histamine. Two thermoreceptive types have peak sensitivities that are higher and lower than the normal body temperature, respectively. One mechanoreceptive type coils around the hair follicle and detects movement of the hair inside (Figure 36.5).

Encapsulated receptors also are common near the body surface. A Meissner's corpuscle adapts slowly to vibrations of low frequencies. It is notably abundant in the lips, fingertips, eyelids, nipples, and genitals. The bulb of Krause is an encapsulated thermoreceptor that is activated by temperatures below 20°C (68°F). Below 10°C, it gives rise to painful freezing sensations. Ruffini endings are other slowly adapting encapsulated types. They are sensitive to steady touching and pressure, and to temperatures above 45°C (113°F).

The Pacinian corpuscle is an encapsulated receptor that is widely distributed in dermis, where it may have a role in perceptions of fine textures. It also is present deeper in the body, as in the membrane near freely movable joints and in some internal organs. Concentric, onionlike layers of membrane alternating with fluid-filled spaces surround this sensory ending (Figure 36.5). The construction enhances the detection of rapid pressure changes associated with touch and vibrations.

Muscle Sense

Sensing limb motions and the body's position in space requires mechanoreceptors in skeletal muscle, joints, tendons, ligaments, and skin. Examples include stretch receptors of muscle spindles. As described in Section 34.4, these sensory organs run parallel with skeletal muscle cells. Their responses to stimulation depend on how much and how fast the muscle stretches.

Regarding the Sensation of Pain

Pain is the perception of injury to some body region. The most important pain receptors are subpopulations of free nerve endings, several million of which are distributed throughout the skin and internal tissues. They trigger awareness of pain in the thalamus, but the type and intensity of pain are interpreted in the cerebral cortex.

Sensations of *somatic* pain start with pain receptors in skin, skeletal muscles, joints, and tendons. Sensations of *visceral* pain, which is associated with the internal organs, are related to excessive chemical stimulation, muscle spasms, muscle fatigue, inadequate blood flow to organs, and other abnormal conditions.

Figure 36.5 *Right:* A sampling of receptors in human skin. Subpopulations of free nerve endings function as mechanoreceptors (including those around hair follicles, which detect hair movements), thermoreceptors, and pain receptors. Pacinian corpuscles detect rapid pressure changes associated with touch and vibrations. Ruffini endings detect steady touching and pressure. The bulb of Krause is most sensitive to cold. Meissner's corpuscle detects vibrations of lower frequency than those detected by Pacinian corpuscles.

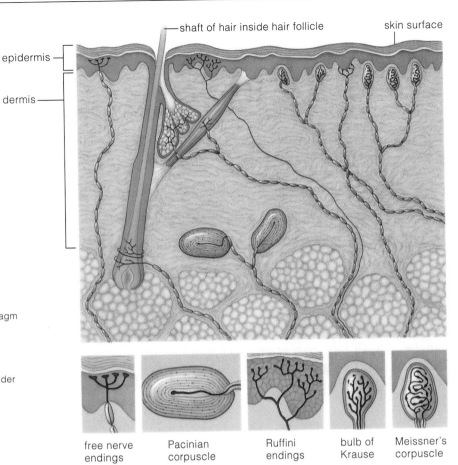

- shaft of hair inside hair follicle
- skin surface
- epidermis
- dermis

free nerve endings | Pacinian corpuscle | Ruffini endings | bulb of Krause | Meissner's corpuscle

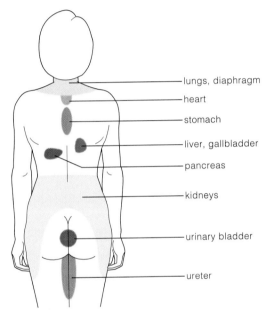

- lungs, diaphragm
- heart
- stomach
- liver, gallbladder
- pancreas
- kidneys
- urinary bladder
- ureter

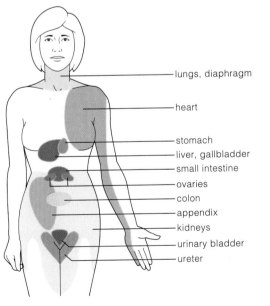

- lungs, diaphragm
- heart
- stomach
- liver, gallbladder
- small intestine
- ovaries
- colon
- appendix
- kidneys
- urinary bladder
- ureter

Figure 36.6 Referred pain. Receptors in some internal organs detect painful stimuli. Instead of localizing the pain at the organs, the brain projects the sensation to the skin areas indicated.

Responses to pain depend on the ability of the brain to identify the affected tissue and project the sensation back to it. For example, get smacked in the face with a snowball and you "feel" the contact on facial skin. However, sensations of pain from certain internal organs may be wrongly projected to part of the skin surface. All people have a capacity to experience this response, which is called *referred pain*, so it is related to the way the nervous system is constructed. Possibly, sensory inputs from the skin and from certain internal organs enter the spinal cord along common nerve pathways, so that the brain cannot accurately identify their source. For example, the brain commonly projects the pain of a heart attack to the skin above the heart and along the left shoulder and arm (Figure 36.6).

As a final example, amputees often sense *phantom pain* from a missing body part, as if it were still there. In some undetermined way, the brain projects pain past the healed regions where sensory nerves were severed.

Diverse sensory receptors near the body surface and in internal organs detect touch, pressure, temperatures, pain, motion, and positional changes of body parts. Their input travels through the spinal cord to the somatosensory cortex and other brain regions where somatic sensations arise.

SENSES OF TASTE AND SMELL

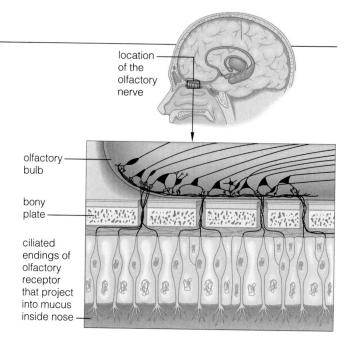

location of the olfactory nerve

We turn now to examples of the special senses, starting with those of taste and smell. Both are *chemical* senses; their sensory pathways start at chemoreceptors, which are activated when they bind a chemical substance that is dissolved in the fluid bathing them. The receptors themselves wear out and new ones replace them on an ongoing basis. In both pathways, sensory input travels from the receptors through the thalamus and on to the cerebral cortex, where perceptions of the stimulus take shape and undergo fine-tuning. The input also travels to the limbic system, which can integrate it with emotional states and stored memories (Sections 35.5 and 35.7).

Different animals taste substances with their mouth, antennae, legs, tentacles, or fins. It all depends on where the chemoreceptors called **taste receptors** are located. In your throat and mouth, especially the tongue's upper surface, they are located in about 10,000 sensory organs called taste buds (Figure 36.7). Each taste bud has a pore through which fluids in the mouth contact the surface of receptor cells. Of thousands of perceived tastes, all are combinations of four primary sensations: sweet (elicited by sucrose, glucose, and other simple sugars), sour (acids), salty (NaCl and other salts), and bitter (alkaloids and other potentially toxic plant substances).

olfactory bulb

bony plate

ciliated endings of olfactory receptor that project into mucus inside nose

Figure 36.8 Sensory pathway leading from the sensory endings of olfactory receptors in the nose to the cerebral cortex and limbic system. Axons from these receptors pass through holes in a bony plate that separates the nasal lining from the brain. In the earliest vertebrates, an olfactory bulb and an olfactory lobe dominated the forebrain. In certain lineages, the olfactory bulb became reduced in importance, and the cerebrum expanded greatly in size and functions, as described in Sections 35.2 and 35.4.

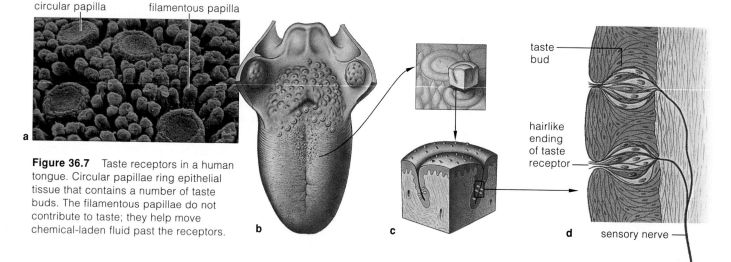

circular papilla   filamentous papilla

a

Figure 36.7 Taste receptors in a human tongue. Circular papillae ring epithelial tissue that contains a number of taste buds. The filamentous papillae do not contribute to taste; they help move chemical-laden fluid past the receptors.

b

c

taste bud

hairlike ending of taste receptor

d   sensory nerve

Olfactory receptors detect water-soluble or volatile (easily vaporized) substances. A bloodhound nose has more than 200 million; a human nose has about 5 million. Axons of such receptors lead into one of two small brain structures called olfactory bulbs (Figure 36.8). There they synapse with groups of cells that sort out the components of a given scent and relay the information onward, by way of an olfactory tract, for further processing.

A vomeronasal organ, or "sexual nose," is common among animals, including humans. Its receptors detect **pheromones**, a type of signaling molecule with roles in social aspects of reproduction. These exocrine gland secretions affect the behavior of other individuals of the same species, especially potential mates. Consider the effect of bombykol on the olfactory receptors of a male silk moth. Contact with merely one bombykol molecule per second sends action potentials to the moth brain. They help a male locate a female in the dark, even if she is more than a kilometer upwind from him.

The senses of taste and smell both start at chemoreceptors. Both involve sensory pathways that lead to processing regions in the cerebral cortex and in the limbic system.

36.4 SENSE OF BALANCE

Most animals assess and respond to displacements from their natural, *equilibrium* position, in which the body is balanced in relation to gravity, velocity, acceleration, and other forces that influence its positions and movement. For example, animals right themselves after tilting or turning upside-down. A variety of fishes, amphibians, and reptiles have paired **inner ears**, which evolved first as organs of equilibrium and had little, if anything, to do with hearing. These are systems of fluid-filled sacs and canals on both sides of the brain. Some parts detect rotational, accelerated motions of the head, as when you ride a looping roller coaster (Figure 36.9). Other parts detect linear motion of the head. In amphibians, birds, and mammals, the brain integrates sensory input from internal ears with input from receptors in the eyes, skin, and joints to arrive at sensations of balance.

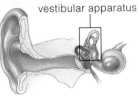

vestibular apparatus

Figure 36.9 Location of the internal ear in humans. Inside the vestibular apparatus (a system of fluid-filled sacs and canals), organs detect the head's linear, rotational, and accelerated motions.

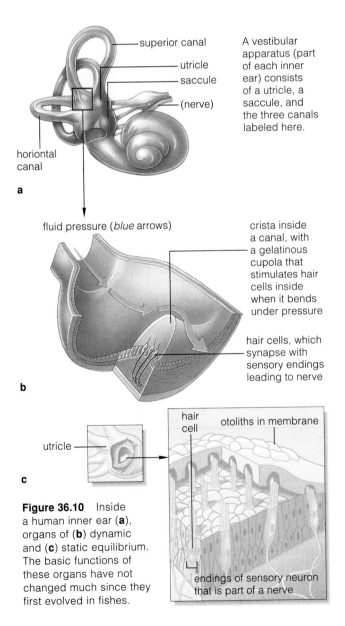

superior canal
utricle
saccule
(nerve)

A vestibular apparatus (part of each inner ear) consists of a utricle, a saccule, and the three canals labeled here.

horiontal canal

a

fluid pressure (*blue* arrows)

crista inside a canal, with a gelatinous cupola that stimulates hair cells inside when it bends under pressure

hair cells, which synapse with sensory endings leading to nerve

b

hair cell

otoliths in membrane

utricle

c

endings of sensory neuron that is part of a nerve

Figure 36.10 Inside a human inner ear (**a**), organs of (**b**) dynamic and (**c**) static equilibrium. The basic functions of these organs have not changed much since they first evolved in fishes.

In humans, organs of equilibrium are located in the part of the inner ear called the **vestibular apparatus** (Figures 36.9 and 36.10*a*). Cristae, the organs concerned with *dynamic* equilibrium, detect accelerated, rotational motions of the head. One is present at the swollen base of each of three semicircular canals that are oriented in three directions. When your head rotates horizontally, vertically, or diagonally, fluid in a canal corresponding to that direction is displaced in the opposite direction. The fluid presses against a cupola, a gelatinous mass into which hair cells project (Figure 36.10*b*). **Hair cells** are a type of mechanoreceptor. When they bend under pressure, the deformation to their plasma membrane may trigger action potentials. These hair cells synapse with the endings of nearby sensory neurons, the axons of which help form a vestibular nerve.

An organ of *static* equilibrium, which is activated when the animal starts and stops moving in a straight line, resides on the floor of each utricle and saccule of the vestibular apparatus (Figure 36.10*c*). A thickened membrane lying over the floor is weighted by deposits of calcium carbonate crystals. These crystalline masses are **otoliths**, or "ear stones." When the membrane slides in response to linear acceleration, hair cells projecting into it bend and are thereby activated.

Motion sickness may result when monotonous linear, angular, or vertical motion overstimulates hair cells in the canals of the organs of balance. It also may result from conflicting signals from the ears and eyes about motion or the head's position. People who are carsick, airsick, or seasick suffer nausea and may vomit; action potentials triggered by the sensory input reach a brain center that governs the vomiting reflex.

Organs of equilibrium help keep the body balanced in relation to gravity, velocity, acceleration, and other forces that influence its position and movement. In humans, such organs occur in the vestibular apparatus of the inner ear.

Properties of Sound

Many arthropods and most vertebrates have a sense of **hearing**, which is the perception of sounds. Sounds are traveling vibrations, or wavelike forms of mechanical energy. Think of what happens when you clap your hands and produce waves of compressed air. Each time hands clap together, molecules are forced outward, so a low-pressure state is created in the region they vacated. Such pressure variations are depicted as wave forms, in which *amplitude* corresponds to loudness (Figure 36.11). Typically, amplitude is measured in decibels. Every ten of these units signifies a tenfold increase in intensity above the faintest sound that humans can hear. The *frequency* of a sound is the number of wave cycles per second. Each cycle extends from the start of one wave peak to the start of the next peak. The more cycles per second, the higher are the frequency and perceived pitch.

Evolution of the Vertebrate Ear

After vertebrates invaded the land, the sense of hearing became far more important than it had been in aquatic habitats. Although the parts of the *internal* ear dealing with equilibrium did not change much, other parts slowly expanded into structures that could receive and process the faint sounds traveling through air. Also, the **middle ear** evolved, and structures inside it amplified and transmitted air waves to the inner ear. In reptiles, a shallow depression formed on each side of the head and led to an eardrum (a thin membrane that rapidly

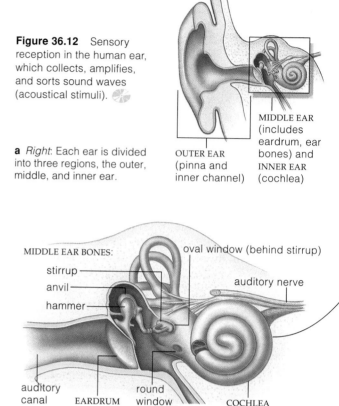

Figure 36.12 Sensory reception in the human ear, which collects, amplifies, and sorts sound waves (acoustical stimuli).

a *Right*: Each ear is divided into three regions, the outer, middle, and inner ear.

OUTER EAR (pinna and inner channel)

MIDDLE EAR (includes eardrum, ear bones) and INNER EAR (cochlea)

MIDDLE EAR BONES:
stirrup
anvil
hammer

oval window (behind stirrup)
auditory nerve

auditory canal   EARDRUM   round window   COCHLEA

b Components of the human ear. External flaps of the outer ear collect sound waves, which move into and auditory canal and arrive at the eardrum (alson called the tympanic membrane).

vibrates in response to air waves). Behind the eardrum of all existing crocodiles, birds, and mammals is an air-filled cavity and small bones, which transmit vibrations to the inner ear. Long ago, among early fishes, the bones structurally supported gill pouches, then they became part of the jaw joint. Thus, among reptiles, birds, and mammals, bones that once functioned in gas exchange became specialized for feeding, then for hearing.

Among mammals, an **external ear** that functions in collecting sound waves also became well developed. The external ear of nearly all species has a pinna (sound-collecting, skin-covered flaps of cartilage, typically with elaborate folds, that project from both sides of the head) and a deep channel leading to the middle ear.

Sensory Reception in the Human Ear

Figure 36.12 illustrates the outer, middle, and inner ear of humans. Like other mammals and also birds, humans have a pea-size but highly developed **cochlea**, the portion of the inner ear that receives and sorts out sound waves. Here we find **acoustical receptors**, or vibration-sensitive mechanoreceptors. The ones in the human ear are hair cells, which bend back and forth in

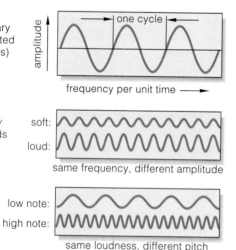

Sounds differ in *amplitude* (they vary in pressure, depicted here as wave forms) and in *frequency* (number of wave cycles/unit time).

amplitude

one cycle

frequency per unit time →

A sound's *intensity* (loudness) depends on its amplitude.

soft:
loud:

same frequency, different amplitude

A sound's *pitch* (tone) depends on its frequency.

low note:
high note:

same loudness, different pitch

Figure 36.11 Properties and examples of sound waves. Compared to the pure tone of a tuning fork, most sounds are combinations of sound waves of different frequencies. Combinations of overtones give the sounds their timbre, or quality, which is one of the properties that help you recognize, say, voices of people you know over the phone.

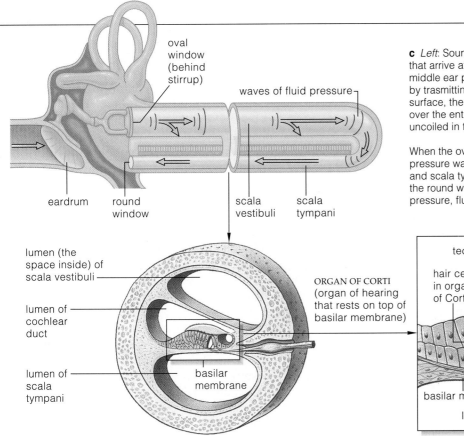

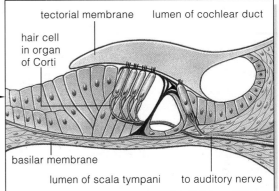

c *Left*: Sound reception in the inner ear. Sound waves that arrive at the eardrum make it vibrate. Bones of the middle ear pick up the vibrations and amplify the stimulus by trasmitting the force of pressure waves to a smaller surface, the oval window. This is and elastic membrane over the entrance to the inner ear, the coclea (shown uncoiled in this diagram for clarity).

When the oval window bows in and out, it produces fluid pressure waves in two ducts, called the scala vestibuli and scala tympani. The waves reach another membrane, the round window. When this membrane bulges under pressure, fluid moves back and forth in the inner ear.

d Pressure waves are sorted out at the cochlear duct, the third duct of the coiled inner ear. One of its membranes (basilar membrane) starts out narrow and stiff. It is broader and flexible deep in the coil. At different points, the membrane vibrates more strongly to sounds of different frequencies.

e At the organ of Corti, 16,000 hair cells project into a jellylike flap (techtorial membrane) over the basilar membrane. When the flap vibrates suitably, hair cells bend, action potentials arise, and messages travel along an auditory nerve to the brain.

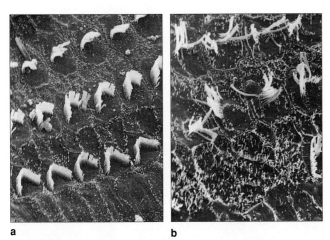

a　　　　　　　**b**

Figure 36.13 Results of an experiment on the effect of intense sound on the inner ear. (**a**) From a guinea pig's ear, three rows of hair cells that project into the tectorial membrane in the organ of Corti. (**b**) Hair cells in the same organ after twenty-four hours of exposure to noise levels comparable to extremely loud music.

To give you a sense of what "loud" is, a ticking watch measures 10 decibels (100 times louder than the hearing threshold for humans). A normal conversation is about 60 decibels (about a million times louder), a food blender operating at high speed is about 90 decibels (a billion times louder), and a raging rock concert, about 120 decibels (a trillion times louder).

response to pressure waves. This is the start of a flow of information along an auditory nerve that leads from the receptors to the brain. Hair cells can be permanently damaged by prolonged exposure to intense sounds. They are not adapted to amplified music, jet planes, and other recent developments (Figure 36.13).

Echolocation

Earlier, you read that a bat brain responds to signals generated by echolocation, or the use of echoes from ultrasounds that the bat itself produces. Ultrasounds are extremely high-frequency waves, in the 100-decibel range, which humans cannot even hear. Dolphins and whales also emit ultrasounds that travel through water. By perceiving frequency variations in the echoes, these mammals also can pinpoint the distance and direction of movement of one another and of predators or prey.

Hearing among land vertebrates involves structures that collect, amplify, and sort out sound waves traveling through air. In the inner ear, sound waves produce fluid pressure variations, which hair cells transduce into action potentials.

Requirements for Vision

All organisms are sensitive to light. Although they have no photoreceptor cells, sunflowers track the sun as it arcs across the sky. Even single-celled amoebas abruptly stop moving when you shine a light on them. We do not find photoreceptor cells until we poke about among the invertebrates. Even then, not all of the species that have photoreceptors "see" as you do. Many are able to detect a change in light intensity, as a mollusk on the seafloor might do when a fish swims above it. However, they are unable to discern the size or shape of such objects.

At the minimum, the sense of **vision** requires two components: eyes, and a capacity for image formation in brain centers that receive and interpret patterns of visual stimulation being sent to them. Such images are pieced together from information about the shape, brightness, position, and movement of visual stimuli. **Eyes** are sensory organs that contain a tissue with a dense array of photoreceptors. All photoreceptors have this in common: *They incorporate pigment molecules that can absorb photon energy, which can be converted into excitation energy in sensory neurons.*

Nearly all photoreceptor cells have some portion of their surface infolded to a small or large extent, which increases the surface area available for photochemical reactions. Their structure can be traced to the kind of epidermal cells that gave rise to them. The *rhabdomeric* photoreceptors are derived from cells having microvilli. They predominate in the flatworms, mollusks, annelids, arthropods, and echinoderms. By contrast, the *ciliary* photoreceptors have a photoreceptive surface derived from the membrane of a cilium. We find these among the cnidarians, some flatworms, and all vertebrates.

A Sampling of Invertebrate Eyes

SIMPLE EYES Earthworms and some other animals have photoreceptors dispersed through their integument, so they can respond to light even though they have no eyes. The simplest eye is an **ocellus** (plural, ocelli), a patch or cuplike depression of the integument in which photoreceptors and pigmented cells are concentrated. Figure 36.14a shows one of these. Such eyes allow the animal to use light for orienting the body, detecting a predator's shadow, or influencing biological clocks.

COMPLEX EYES Vision evolved among fast-moving predators that had to discriminate quickly among prey and other objects that crossed their rapidly changing visual field. Remember, a **visual field** is simply the part of the outside world that an animal sees. At the least, different groups of photoreceptors must sample the

light intensity of different parts of the visual field. The brain interprets the intensities as contrasting parts of a visual image. The quality of the resulting image varies among invertebrates. With the exception of cephalopods and a few other animals, it usually is not good. Why? Good image formation requires many photoreceptors in the eye, and the eyes of most invertebrates are too small to have a lot of them. For example, a planarian's pigment cup only has about 200, compared with the 70,000 per square millimeter in an octopus eye.

Image formation benefits from a **lens**, a transparent body that bends all light rays from a given point in the visual field so that they converge onto photoreceptors (compare Section 4.2). Abalones have a spherical lens that bends incoming light rays but not to the same points, so images that form are blurred (Figure 36.14b). Things improve with a space between the back of the lens and the photoreceptors, as in a snail eye. They get better when the eye has a **cornea**, a transparent cover that directs light rays onto the lens, as happens in a snail or conch (Figure 36.14c,d). Invertebrates on land are better adapted for visually discriminating objects. This is the case for spiders and

epidermis —
photoreceptor —
pigmented cell —

a

epidermis —
transparent body (lens) —
photoreceptor —
sensory cell —

b

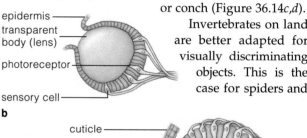

cuticle —
epidermis —
lens —
photoreceptor —
sensory neuron —

c

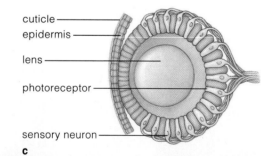

d

Figure 36.14 Organization of invertebrate eyes, longitudinal section. There are far more photoreceptors than can be shown in these diagrams. (**a**) A limpet's ocellus, a shallow depression in the epidermis that incorporates light-sensitive receptors. (**b**) An abalone eye, with its spherical, transparent lens. (**c**) Eye of a land snail. (**d**) Well-developed eyes of a conch, here peering into the waters of the Great Barrier Reef along the east coast of Australia.

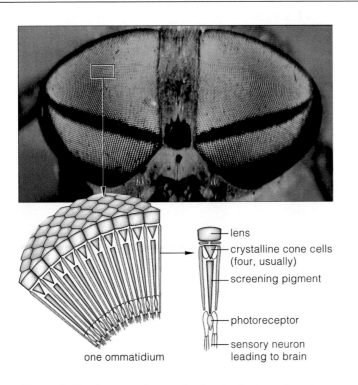

Figure 36.15 Compound eyes of a deerfly. The lens of each photosensitive unit (ommatidium) directs light onto a crystalline cone, which focuses light on photoreceptor cells below it.

one ommatidium

lens
crystalline cone cells (four, usually)
screening pigment
photoreceptor
sensory neuron leading to brain

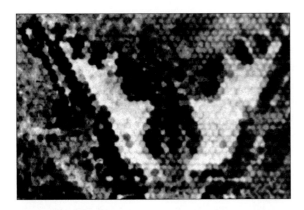

Figure 36.16 Approximation of light reception in an insect eye. This image of a butterfly was formed when a photograph was taken through the outer surface of a compound eye that had been detached from an insect. It may not be what the insect "sees," because integration of signals sent to the brain from photoreceptors may produce a sharper image. This representation is useful insofar as it suggests how the overall visual field may be *sampled* by separate ommatidia.

most other arthropods having simple eyes, in which a single lens services all the photoreceptors. It also is the case with the complex, *compound* eyes of crustaceans and insects. A **compound eye** has many closely packed photosensitive units, each with a bundle of rhabdomeric photoreceptors (Figure 36.15). Some have thousands of units, called ommatidia (singular, ommatidium).

Figure 36.17 Eye of an octopus, one of the mollusks called cephalopods. The pupil can constrict into slits and flare open in response to changing light conditions or to the presence of a potential mate or enemies. From many experiments, we know cephalopods form distinct images. They can sense the size, shape, and vertical/horizontal orientations and possibly color of objects. They also are nearsighted, although they can snag tiny prey three meters away.

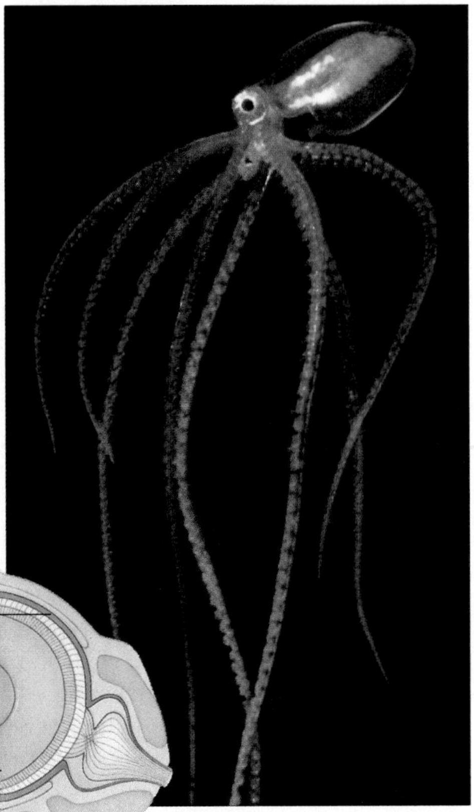

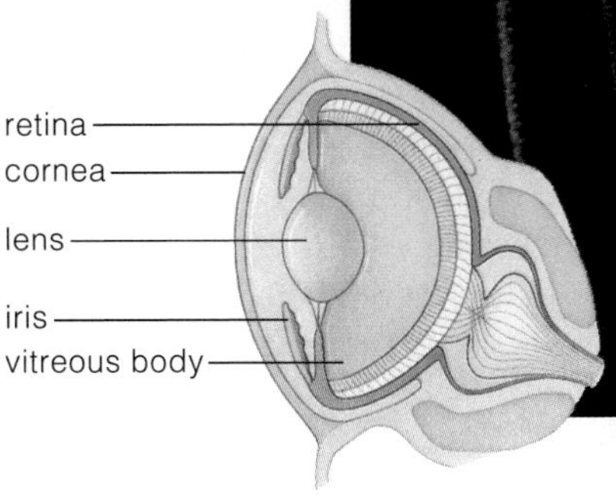

retina
cornea
lens
iris
vitreous body

According to the mosaic theory of image formation, each unit samples only a small part of the visual field. The brain builds images based on signals about differences in light intensities, with each unit contributing a small bit to its formation of a visual mosaic (Figure 36.16).

Of all invertebrates, octopuses and squids have the most complex eyes. Like vertebrates, they have **camera eyes**, so named because the eyeball is structured along the lines of a camera. Its interior is a darkened chamber. Light can enter it only through the pupil, an opening in a ring of contractile tissue called the iris (equivalent to a camera's diaphragm). Behind the pupil is a lens that focuses light onto the **retina**, a tissue with a densely packed array of photoreceptors (in a camera, onto light-sensitive film). Axons of sensory nerves converge as an optic tract that leads to the brain (Figure 36.17).

The cephalopods and vertebrates are only remotely related, so similarities in the structure and functions of their eyes may be a result of convergent evolution. By one theory, a group of genes that originally had roles in the development of the nervous system was coopted for the task of eye formation. Whether this happened more than once in different lineages is not yet known.

Vision requires eyes (sensory organs with a dense array of photoreceptors) and a capacity for image formation in the brain, based on patterns of visual stimulation.

STRUCTURE AND FUNCTION OF VERTEBRATE EYES

The Eye's Structure

Reflect on some points made in the preceding section. *Vision* requires eyes (sensory organs that incorporate a tissue of densely packed photoreceptors) and image formation in brain centers that not only receive but also interpret patterns of stimulation from different parts of the eyeballs. Such images, again, require information about the shape, brightness, position, and movement of visual stimuli. With this in mind, take a look at Figure 36.18, which shows the human eye. Like most vertebrate eyes, the eyeball has a three-layered wall (Table 36.2).

The eyeball's outermost layer consists of the sclera and cornea. The sclera, the dense, fibrous "white" of the eye, protects most of the eyeball. The cornea, made of transparent collagen fibers, covers the rest of it.

The middle layer has a choroid, ciliary body, iris, and pupil. The choroid is a dark-pigmented, vascularized tissue. It absorbs light that photoreceptors have not absorbed and prevents it from scattering inside the eye. The doughnut-shaped, pigmented iris is suspended behind the cornea (the Latin *iris* refers to the rings of a rainbow). The dark "hole" in its center is the pupil, the entrance for light. When rays of bright light strike the eye, circular muscles inside the iris contract and thus shrink the size of the pupil. In dim light, radial muscles contract and so enlarge the pupil.

The inner layer, which includes the retina, is at the back of your eyeball. In birds of prey, it is more toward the eyeball's roof (Figure 36.19).

The eye's interior has a lens. A clear fluid called the aqueous humor bathes the lens, which consists of layers of transparent proteins. The vitreous body, a jellylike substance, fills the chamber behind the lens.

Figure 36.19 Here's looking at you! In owls and some other birds of prey, photoreceptors are concentrated more on top of the inner eyeball, not at back. Such birds look down more than up when they fly and scan the ground for a meal. When they are on the ground, they cannot see something overhead very easily unless they turn their head almost upside-down.

Eyes with a large pupil and iris tell us something about the life-style of their owners. Animals that move about actively at night or in dimly lit habitats are less likely to stumble, bump into objects, or fall off cliffs when they intercept as much of the available light as they can in a given interval. The greater the amount of incoming light, the better is the visual discrimination. A

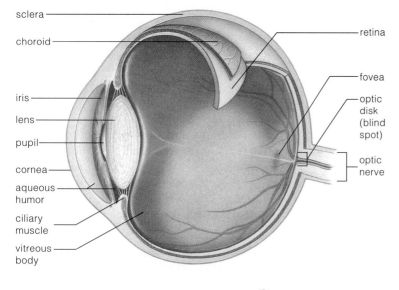

Figure 36.18 Structure of the human eye.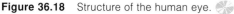

Table 36.2	Components of the Vertebrate Eye	
THREE LAYERS FORMING THE WALL OF EYEBALL:		
Fibrous Tunic:	Sclera. *Protects eyeball*	
	Cornea. *Focuses light*	
Vascular Tunic:	Choroid. *Blood vessels nutritionally support wall cells; pigments prevent light scattering*	
	Cilary body. *Its muscles control lens shape; its fine fibers hold lens in upright position*	
	Iris. *Adjustments here control incoming light*	
	Pupil. *Serves as entrance for light*	
Sensory Tunic:	Retina. *Absorbs and transduces light energy*	
	Fovea. *Increases visual acuity*	
	Start of optic nerve. *Carries signals to brain*	
INTERIOR OF EYEBALL:		
Lens	*Focuses light on photoreceptors*	
Aqueous humor	*Transmits light, maintains pressure*	
Vitreous body	*Transmits light, supports lens and eyeball*	

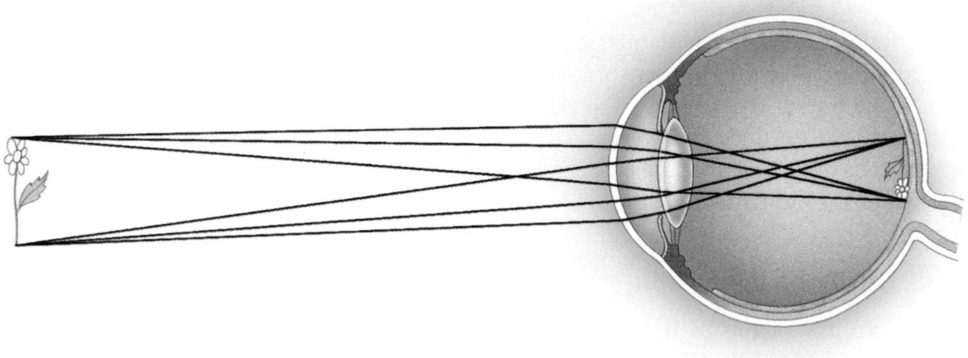

Figure 36.20 Pattern of retinal stimulation in the human eye. Being curved, the transparent cornea that is positioned in front of the pupil changes the trajectories of light rays as they enter the eye. The pattern is upside-down and reversed left to right, compared to the stimulus in the visual field.

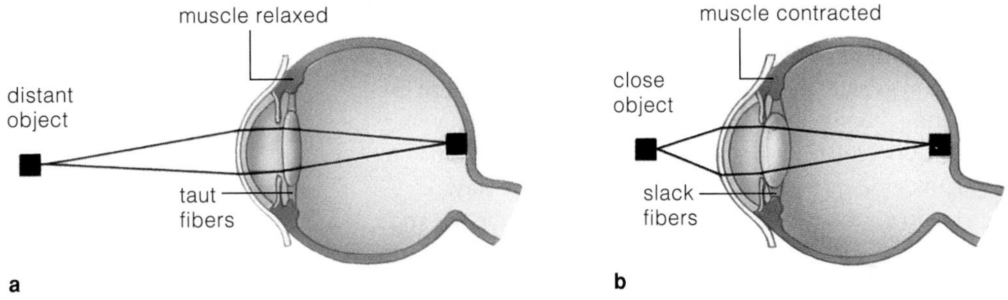

Figure 36.21 Two focusing mechanisms. A ciliary muscle encircles the lens and attaches to it. (**a**) When the muscle relaxes, the lens flattens; the focal point moves farther back and brings distant objects into focus. (**b**) When the muscle contracts, the lens bulges; the focal point moves closer and brings close objects into focus.

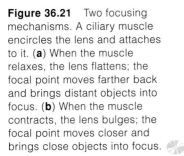

muscle relaxed

distant object

taut fibers

a

muscle contracted

close object

slack fibers

b

large pupil lets in large amounts of light. A pupil that is ringed by a large iris can be dilated or constricted to admit more or less light, which is useful for animals that are active at night as well as during the day.

When light rays do converge at the back of the eye, they stimulate the retina in a distinct pattern. Because the cornea has a curved surface, light rays coming from a given point in the visual field hit it at different angles, so their trajectories change. (As described in Section 4.2, rays of light bend at the boundaries between substances of different densities, and bending sends them in new directions.) Because of their newly angled trajectories, the light rays that converge at the back of the eyeball stimulate the retina in a pattern that is upside-down and reversed left to right, relative to the original source of the light rays. Figure 36.20 is a simplified diagram of this outcome.

Visual Accommodation

Light rays emanating from sources at varying distances from the eye strike the cornea at different angles, which means they could end up being focused at different distances behind it. This would do the brain no good. However, adjustments in the position or shape of the lens normally focus all incoming stimuli onto the retina. We call these lens adjustments **visual accommodation**. Without them, light rays from distant objects would be improperly focused in front of the retina, and rays from close objects would be focused behind it.

In fishes and reptiles, eye muscles move the lens forward or back, like a camera's focusing apparatus. Extending the distance between the lens and the retina moves the focal point forward; shrinking the distance moves it back. By contrast, in birds and mammals, the shape of the lens is adjusted. A ciliary muscle encircles the lens and attaches to it by fiberlike ligaments (Figure 36.18). When this muscle relaxes, the lens flattens and thereby moves the focal point farther back, as in Figure 36.21a. When the muscle contracts, the lens bulges and moves the focal point forward, as in Figure 36.21b.

In some cases, the lens cannot be adjusted enough to make the focal point match up precisely with the retina. For example, in some people, the eyeball is not shaped quite right, and the position of the lens is either too close or too far away from the retina. When this is the case, accommodation alone cannot bring about an exact match. As described in the next section, eyeglasses can correct both visual disorders, which are commonly called nearsightedness and farsightedness.

For most vertebrate eyes, the eyeball has three layers, and it encloses a lens, an aqueous humor, and a vitreous body.

A protective sclera and a light-focusing cornea make up the outer layer. The middle layer has a vascularized, pigmented choroid and other parts that admit and control incoming light. Photoreception occurs at the retina of the inner layer.

Adjustments in the positioning or shape of the lens focus incoming visual stimuli onto the retina.

We conclude this chapter with one of the best examples of neuronal architecture, the sensory pathway from the retina to the brain. This is how raw visual information is received, transmitted, and combined; and it leads to awareness of light and shadows, of colors, of near and distant objects in the outside world.

Organization of the Retina

Information flow begins as light strikes the retina. The retina's basement layer, a pigmented epithelium, covers the choroid. Resting on this layer are densely packed arrays of **rod cells** and **cone cells**, which are two classes of ciliary photoreceptors (Figure 36.22). Rod cells detect very dim light. During the night or in darkened places, they contribute to coarse perception of movement by detecting changes in light intensity across the visual field. Cone cells detect bright light. They contribute to sharp vision and color perception during the day.

Distinct layers of neurons are located above the rods and cones. The first to accept the visual information from the photoreceptors are bipolar sensory neurons, then ganglion cells. Axons of these ganglion cells form the two optic nerves to the brain (Figure 36.23).

Before signals depart from the retina, they converge dramatically. The input from 125 million photoreceptors converges on a mere 1 million ganglion cells. Signals also flow laterally among horizontal cells and amacrine cells. Both kinds of neurons act in concert to dampen or

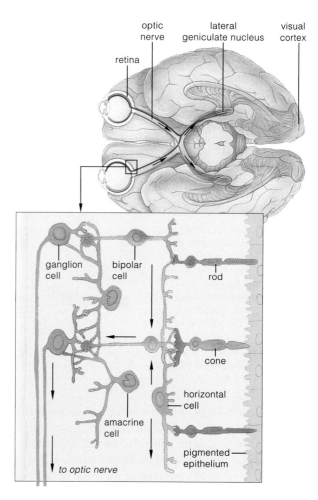

Figure 36.23   The sensory pathway from the retina to the brain.

enhance the signals before ganglion cells get them. Thus, *a great deal of synaptic integration and processing goes on even before visual information is sent to the brain.*

Neuronal Responses to Light

ROD CELLS   A rod cell's outer segment contains several hundred membrane disks, each peppered with about 10^8 molecules of a visual pigment called rhodopsin. The membrane stacking and the extremely high density of pigments greatly increase the chances of intercepting packets of light energy—that is, photons of particular wavelengths. The action potentials that result from the absorption of even a few photons can lead to conscious awareness of objects in dimly lit surroundings, as in dark rooms or late at night.

Rhodopsin effectively absorbs wavelengths in the blue-to-green portion of the visible spectrum (Section 7.2). Absorption changes the shape of the pigment. This triggers a cascade of reactions that alter activity at ion

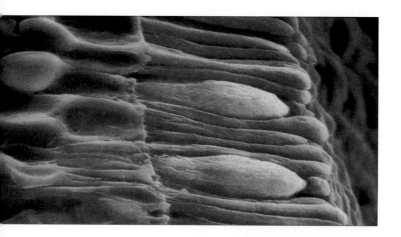

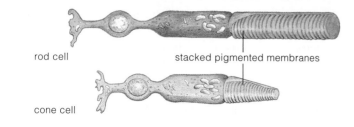

rod cell   stacked pigmented membranes

cone cell

Figure 36.22   Mammalian photoreceptors: rods and cones.

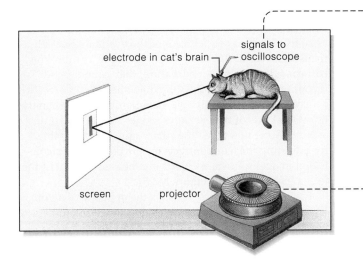

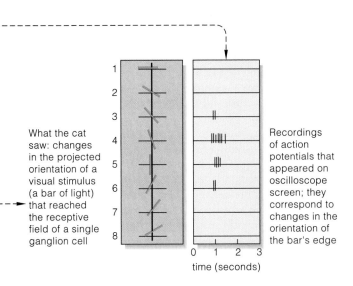

What the cat saw: changes in the projected orientation of a visual stimulus (a bar of light) that reached the receptive field of a single ganglion cell

Recordings of action potentials that appeared on oscilloscope screen; they correspond to changes in the orientation of the bar's edge

time (seconds)

Figure 36.24 Example of experiments into the nature of the receptive fields for visual stimuli. David Hubel and Torsten Wiesel implanted an electrode in an anesthetized cat's brain. They placed the cat in front of a small screen upon which different patterns of light were projected—in this case, a hard-edged bar. Light or shadow falling on part of the screen excited or inhibited signals sent to a single neuron in the visual cortex.

Tilting the bar at different angles produced changes in the neuron's activity. A vertical bar image produced the strongest signal (*numbered 5 in the sketch*). When the bar image tilted slightly, signals were less frequent. When it tilted past a certain angle, signals stopped.

channels and ion pumps across the rod cell's plasma membrane. Gated sodium channels close, the voltage difference changes across the membrane, and a graded potential results. This potential reduces the ongoing release of a neurotransmitter with inhibitory effects on adjacent sensory neurons. Released from inhibition, the sensory neurons start sending signals about the visual stimulus on toward the brain.

CONE CELLS The sense of color and of daytime vision starts with photon absorption by red, green, and blue cone cells, each with a different kind of visual pigment. Here again, photon absorption reduces the release of a neurotransmitter that otherwise inhibits the neurons that are adjacent to the photoreceptors. Near the center of the retina is a funnel-shaped depression called the fovea. This is the retinal area with the greatest density of photoreceptors and the greatest visual acuity. Its cone cells are the ones that discriminate the most precisely between adjacent points in space.

RESPONSES AT RECEPTIVE FIELDS Collectively, the neurons in the eye respond to light in organized ways. The retinal surface is organized into receptive fields, restricted areas that influence the activity of individual

sensory neurons. For example, for each ganglion cell, the field is a tiny circle. Some cells respond best to a tiny spot of light, ringed by dark, in the field's center. Others respond to a rapid change in light intensity, to a spot of one color, or to motion. During one experiment, sensory cells generated signals after they detected a suitably oriented bar (Figure 36.24). Such cells do not respond to diffuse, uniform illumination, which is just as well; doing so would produce a confusing array of signals to the brain.

ON TO THE VISUAL CORTEX The part of the outside world that an individual actually sees is its visual field. The right side of both retinas intercepts light from the *left* half of the visual field; the left side intercepts light from the *right* half. The optic nerve leading out of each eye delivers the signals about a stimulus from the *left* visual field to the right cerebral hemisphere, and signals from the *right* visual field to the left hemisphere (Figure 36.23).

Axons of the optic nerves end in a layered brain region, the lateral geniculate nucleus. Each layer has a map corresponding to receptive fields of the retina. Its interneurons deal with one aspect of the stimulus— form, movement, depth, color, texture, and so on. After initial processing, all the visual signals travel rapidly, at the same time, to different parts of the visual cortex. There, final integration produces organized electrical activity and the sensation of sight.

Each eye is an outpost of the brain, collecting and analyzing information about the distance, shape, brightness, position, and movement of visual stimuli. Its sensory pathway starts at the retina and proceeds along an optic nerve to the brain.

Organization of visual signals begins at receptive fields in the retina, it is further processed in the lateral geniculate nucleus, and it is finally integrated in the visual cortex.

DISORDERS OF THE HUMAN EYE

Of all the sensory receptors that the human brain requires to help keep you alive and functioning independently in the environment, fully two-thirds are located in your pair of eyes. They are photoreceptors, and they do more than detect light. They also allow you to see the world in a rainbow of colors. The eyes are the single most important source of information about the outside world.

Given their essential role, it is no surprise that so much attention is paid to eye disorders, as brought about by injuries, inherited abnormalities, diseases, and advancing age. The consequences range from relatively harmless conditions, such as nearsightedness, to total blindness. Each year, many millions of people must deal with such consequences.

FOCUSING PROBLEMS   Other heritable abnormalities arise from misshapen features of the eye that affect the focusing of light. *Astigmatism*, for example, results from corneas with an uneven curvature; they cannot bend incoming light rays to the same focal point.

Nearsightedness, or myopia, commonly occurs when the horizontal axis of the eyeball is longer than the vertical axis. It also occurs when the ciliary muscle responsible for adjusting the lens contracts too strongly. The outcome is that images of distant objects are focused in front of the retina instead of on it (Figure 36.25).

Farsightedness, or hyperopia, is the opposite of myopia. Here, the vertical axis of the eyeball is longer than the horizontal axis (or the lens is "lazy"). As a result, close images are focused behind the retina (Figure 36.26).

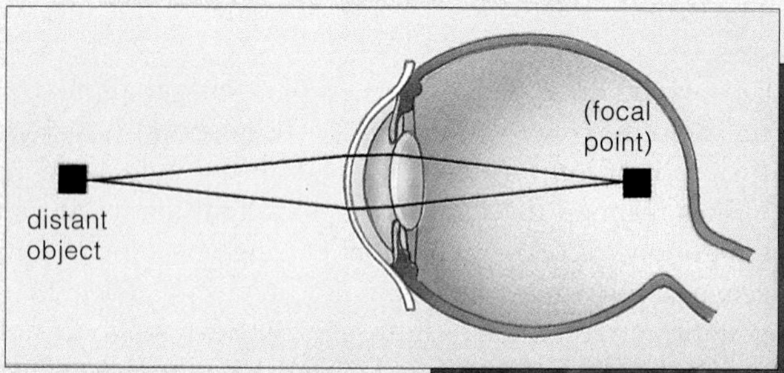

Figure 36.25 Example of the focal point in individuals who have nearsighted vision. The birds are flamingos at a lake in Tanzania, East Africa. To a nearsighted person, the birds in flight in the distance would be out of focus.

COLOR BLINDNESS   Occasionally, some or all of the cone cells that selectively respond to light of red, green, or blue wavelengths fail to develop in individuals. The rare individuals who have only one of three kinds of cones are totally *color-blind*. They can perceive the world only in shades of gray.

Consider a common genetic abnormality, *red-green color blindness*. It is an X-linked, recessive trait that shows up most often in males. The retina of affected individuals lacks some or all of the cone cells with pigments that normally respond to light of red or green wavelengths. Most of the time, people who are affected by red-green color blindness merely have trouble distinguishing red from green in dim light. Some cannot distinguish between the two even in bright light.

Even a normal lens loses some of its natural flexibility as a person grows older. That is why people over forty years old often start wearing eyeglasses.

EYE DISEASES   The structure and functioning of the eyes are vulnerable to infectious diseases. For example, *histoplasmosis*, a fungal infection of the lungs that is described in Section 24.5, can lead to retinal damage, which can cause partial or total loss of vision. As another example, *Herpes simplex*, a virus that causes skin sores, also can infect the cornea and cause it to ulcerate.

Trachoma is a highly contagious disease that has blinded millions, mostly in North Africa and the Middle East. The culprit is a bacterium that also is responsible for the sexually transmitted disease chlamydia (Section 45.14).

The eyeball and the lining of the eyelids (the conjunctiva) become damaged. The damaged tissues are entry points for bacteria that can cause secondary infections. In time the cornea can become so scarred that blindness follows.

AGE-RELATED PROBLEMS   You have probably heard of *cataracts*, a gradual clouding of the lens. It is a problem associated with aging, although it also may arise through eye injury or diabetes. Possibly cataracts form when the transparent proteins that make up the eye's lens undergo structural changes. The clouding may skew the trajectory of incoming light rays. If the lens becomes opaque, light cannot enter the eye at all.

Glaucoma, a different age-related problem, results when excess aqueous humor accumulates inside the eyeball. The

blood vessels that service the retina collapse under the increased fluid pressure. Vision deteriorates as sensory neurons of the retina and optic nerve die off.

Although chronic glaucoma often is associated with advanced age, the conditions that give rise to the disorder actually start to develop in middle age. If detected early enough, the fluid pressure that tends to build up in the eye can be relieved by drugs or surgery before damage becomes severe.

EYE INJURIES   *Retinal detachment* is the eye injury that we read about most often. It may follow a physical blow to the head or illness that causes a tear in the retina. As the jellylike vitreous body oozes through the torn region, the retina is lifted away from the underlying choroid. In

time it may peel away entirely, leaving its blood supply behind.

Early symptoms of a detached retina include blurred vision, flashes of light that occur even in the absence of outside visual stimulation, and loss of peripheral vision. Without medical intervention, the person may become totally blind in the damaged eye.

NEW TECHNOLOGIES   Today a variety of tools can be used to correct some eye disorders. In *corneal transplant surgery*, for example, a defective cornea can be removed, then an artificial cornea (made of clear plastic) or a natural cornea from a donor can be stitched in its place. Within a year, the patient can be fitted with eyeglasses

Figure 36.26   Example of the focal point in farsighted vision. To a farsighted person, the birds standing in water in the foreground would be out of focus.

or contact lenses. Similarly, cataracts can be surgically corrected by removing the lens and replacing it with an artificial one, although the operation is difficult and not always successful.

Severely nearsighted people sometimes opt for *radial keratotomy*, a still-controversial surgical procedure in which tiny, spokelike incisions are made around the edge of the cornea to flatten it more. When all goes well, the adjustment eliminates the need for corrective lenses. Sometimes the result is overcorrected or undercorrected vision, and more surgery is required

As a final example, retinal detachment may be treated with *laser coagulation*, a painless technique in which a laser beam seals off leaky blood vessels and "spot welds" the retina to the underlying choroid.

SUMMARY

1. A stimulus is a specific form of energy that the body detects by means of sensory receptors. A sensation is an awareness that stimulation has occurred. Perception is understanding what the sensation means.

2. Sensory receptors are the endings of sensory neurons or specialized cells adjacent to them. They respond to stimuli, which are specific forms of energy, such as light and mechanical pressure. Animals respond to aspects of the internal or external environment when they have receptors that are sensitive to the energy of stimuli.

   a. Mechanoreceptors, such as free nerve endings, can detect mechanical energy related to touch, pressure, and motion and changes in position.

   b. Thermoreceptors detect the presence of or changes in radiant energy from heat sources.

   c. Pain receptors (nociceptors) detect tissue damage.

   d. Chemoreceptors, such as olfactory receptors and taste receptors, detect chemical substances dissolved in fluids that are bathing them.

   e. Osmoreceptors detect changes in water volume (hence solute concentrations) in the surrounding fluid.

   f. Photoreceptors, which include the rods and cones of the retina in the human eye, detect light.

3. A sensory system has sensory receptors for specific stimuli and nerve pathways from those receptors to receiving and processing centers in the brain. The brain assesses a particular stimulus based on which nerve pathway is delivering the signals, the frequency of signals traveling along each axon of that pathway, and the number of axons that were recruited into action.

4. Somatic sensations include touch, pressure, pain, temperature, and muscle sense. The receptors associated with these sensations are not localized in a single organ or tissue. The special senses include taste, smell, hearing, balance, and vision. The receptors associated with these senses typically reside in sensory organs, such as eyes, or some other particular body region.

5. The senses of taste and smell both involve sensory pathways from chemoreceptors to processing regions in the cerebral cortex and limbic system.

6. Organs of equilibrium, as in the vertebrate inner ear, detect gravity, velocity, acceleration, and other forces that affect the body's positions and movements. The vertebrate sense of hearing (sound perception) requires components of the outer, middle, and inner ear that respectively collect, amplify, and sort out sound waves.

7. Vision requires eyes (sensory organs with a dense array of photoreceptors, as in a retina) and a capacity for image formation in the brain, based upon incoming patterns of visual stimulation. The sense of vision and discrimination among objects evolved first among fast-moving, predatory animals.

Review Questions

1. What is a stimulus? When sensory receptors detect a specific stimulus, what happens to the stimulus energy? *36.1*

2. Name six categories of sensory receptors and the type of stimulus energy that each kind detects. *36.1*

3. How do somatic sensations differ from special senses? *36.1*

4. What is pain? Describe one type of pain receptor. *36.2*

5. Describe the properties of sound. Which evolved first, the sense of balance or sense of hearing? Do both senses require participation of the outer, middle, and inner ear? *36.4, 36.5*

6. How does vision differ from light sensitivity? What sensory organs and structures does vision require? *36.6, 36.7*

7. Label the component parts of the human eye: *36.7*

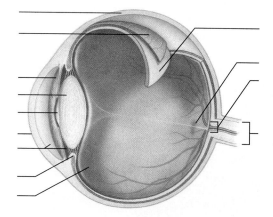

8. How does the vertebrate eye focus light rays from a visual stimulus? What do nearsighted and farsighted mean? *36.7, 36.9*

9. On the bell-shaped rim of the jellyfish shown in Figure 36.27 are tiny saclike structures that hold calcium particles next to a sensory cilium. When the bell tilts, the particles slide over the cilium. Would you assume that these structures contribute to the sense of taste, smell, balance, hearing, or vision? *36.4*

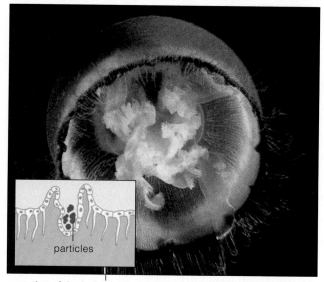

particles

section of the bell margin

Figure 36.27 Sensory structures located on the bell-shaped rim of a medusa (jellyfish).

Self-Quiz *(Answers in Appendix IV)*

1. A _____ is a specific form of energy that is detected by a sensory receptor.

2. Conscious awareness of a stimulus is called a _____ .

3. _____ is understanding what particular sensations mean.

4. Each sensory system consists of _____ .
   a. nerve pathways from specific receptors to the brain
   b. sensory receptors
   c. brain regions that deal with sensory information
   d. all of the above

5. _____ is (are) a decrease in the response to an ongoing stimulus.
   a. Perception   c. Visual accommodation
   b. Sensory adaptation d. both b and c

6. _____ detect mechanical energy associated with changes in pressure, position, or acceleration.
   a. Chemoreceptors   c. Photoreceptors
   b. Mechanoreceptors d. Thermoreceptors

7. _____ are chemoreceptors.
   a. Taste receptors   d. Photoreceptors
   b. Olfactory receptors e. both a and b
   c. Auditory receptors   f. both a and c

8. _____ detect infrared energy (heat).
   a. Chemoreceptors   c. Photoreceptors
   b. Mechanoreceptors d. Thermoreceptors

9. Detecting light energy is the function of _____ .
   a. chemoreceptors   c. photoreceptors
   b. mechanoreceptors d. thermoreceptors

10. Vision requires _____ .
    a. a tissue with dense arrays of photoreceptors
    b. eyes
    c. image-forming centers in brain
    d. all of the above

11. The outer layer of the human eyeball includes the _____ .
    a. lens and choroid   c. retina
    b. sclera and cornea d. both a and c

12. The inner layer of the human eyeball includes the _____ .
    a. lens and choroid   c. retina
    b. sclera and cornea d. both a and c

13. Match each term with the most suitable description.
    _____ somatic senses
    _____ stimulus
    _____ special senses
    _____ variations in stimulus intensity
    _____ sensory receptor

    a. sensory neuron endings or specialized cells next to them
    b. taste, smell, hearing, balance, and vision
    c. form of energy that a specific sensory receptor can detect
    d. encoded in the frequency and number of action potentials
    e. touch, pressure, temperature, pain, and muscle sense

Critical Thinking

1. Wayne, on standby for the last flight from San Francisco to New York, was assigned the last available seat on the plane. It was in the last row, where vibrations and noise from the engines are most pronounced. When Wayne got off the plane in New York, he was speaking very loudly and having trouble hearing what other people were saying. The next day, things were back to normal. Speculate on what happened to his sense of hearing during the flight.

2. Juanita made an appointment with her doctor because she was experiencing recurring episodes of dizziness. Her doctor immediately asked her whether "dizziness" meant she had sensations of lightheadedness, as if she were going to faint, or whether it meant she had sensations of *vertigo*—that is, a feeling that she herself or objects near her were spinning around. Why did her doctor consider this clarification important early in his evaluation of her condition?

3. Michael, now three years old, experiences chronic *middle-ear infection*, which is becoming quite common among youngsters enrolled in crowded day-care centers. This year, despite antibiotic treatment, an infection became so advanced that he had some trouble hearing. Then his left eardrum ruptured and a thickened, jellylike substance dribbled out. His mother was startled and upset, but the pediatrician told her not to worry, that if the eardrum had not ruptured on its own, he would have had to insert a small drainage tube into it. Speculate on why the pediatrician concluded that this would have been a necessary surgical intervention.

Selected Key Terms

acoustical receptor *36.5*	otolith *36.4*
camera eye *36.6*	pain *36.2*
chemoreceptor *36.1*	pain receptor *36.1*
cochlea *36.5*	perception *36.1*
compound eye *36.6*	pheromone *36.3*
cone cell *36.8*	photoreceptor *36.1*
cornea *36.6*	retina *36.6*
echolocation *CI*	rod cell *36.8*
encapsulated receptor *36.2*	sensation *36.1*
external ear *36.5*	sensory adaptation *36.1*
eye *36.6*	sensory system *36.1*
free nerve ending *36.2*	somatic sensation *36.1*
hair cell *36.4*	somatosensory cortex *36.2*
hearing *36.5*	special sense *36.1*
inner ear *36.4*	taste receptor *36.3*
lens *36.6*	thermoreceptor *36.1*
mechanoreceptor *36.1*	vestibular apparatus *36.4*
middle ear *36.5*	vision *36.6*
ocellus *36.6*	visual accommodation *36.7*
olfactory receptor *36.3*	visual field *36.6*
osmoreceptor *36.1*	

Readings

Kandel, E., and J. Schwartz. 1995. *Essentials of Neural Science and Behavior*. Norwalk, Connecticut: Appleton and Lange.

Nicholls, J., A. Martin, and B. Wallace. 1992. *From Neuron to Brain*. Third edition. Sunderland, Massachusetts: Sinauer.

Romer, A., and T. Parsons. 1986. *The Vertebrate Body*. Sixth edition. New York: Saunders.

Sherwood, L. 1997. *Human Physiology*. Third edition. Belmont, California: Wadsworth.

Wright, Karen. April 1994. "The Sniff of Legend." *Discover* 15(4): 60–67. Speculation on the existence of a sensory pathway activated by human pheromones.

Zeki, S. September 1992. "The Visual Image in Mind and Brain." *Scientific American* 267(3): 68–76.

Web Site See *http://www.wadsworth.com/biology* for practice quiz questions, hypercontents, BioUpdates, and critical thinking. The Wadsworth Biology Resource Center provides a wealth of information fully organized and integrated by chapter.

37 ENDOCRINE CONTROL

Hormone Jamboree

In the 1960s, at her camp in a forest by the shore of Lake Tanganyika in Tanzania, the primatologist Jane Goodall let it be known that bananas were available. One of the first chimpanzees attracted to the delicious food was a female—Flo, as she came to be called (Figure 37.1). Flo brought along her two offspring, an infant female and a juvenile male, and exhibited commendable parental behavior toward them.

Three years passed, and Goodall observed that Flo's preoccupation with motherhood gave way to a preoccupation with sex. She also observed that male chimpanzees followed Flo to the camp by the lake and stayed for more than the bananas.

Sex, as Goodall discovered, is *the* premier binding force in the social life of chimpanzees. These primates do not mate for life as eagles do, or wolves. Before the rainy season begins, mature females that are entering their fertile cycle

Figure 37.1 (**a**) Jane Goodall in Gombe National Park, near the shores of Lake Tanganyika, scouting for chimpanzees. (**b**) Flo and three of her offspring, all subjects of long-term field observations that clarified the central role of sex—and of the hormones that orchestrate it—in the social life of these primate relatives of humans.

become sexually active. As one of the more dramatic responses to changing concentrations of hormones in their bloodstream, the external sexual organs of the females become swollen and vivid pink. The swellings are strong visual signals to males. They are the flags of sexual jamborees, of great gatherings of highly stimulated chimps in which any males present may copulate in sequence with the same females.

The gathering of many flag-waving females draws together individuals that forage alone or in small family groups for most of the year. It reestablishes the bonds that unite their rather fluid community. As infants and juveniles play with one another and with the adults, their aggressive and submissive jostlings help map out future dominance hierarchies. Consider Flo, a high-ranking member of the social hierarchy. Through her sexual attractiveness and direct solicitations, she built alliances with many male chimps. Through her status and aggressive behavior, she helped her male offspring win confrontations with other young male chimps.

The hormone-induced swelling during a female's fertile period lasts somewhere between ten and sixteen days. Yet she is fertile for only one to five days. Sex hormones induce the swelling even after a female has become pregnant. Almost certainly, sexual selection has favored the prolonged swelling. Males groom a sexually attractive female more often, protect her, give her more food, and let her tag along to new foraging sites. The more that males accept a female, the higher she rises in the social hierarchy—and the more her individual offspring benefit.

Through their effects, hormones help orchestrate the growth, development, and reproductive cycles of nearly all animals, from the invertebrate worms to chimpanzees and humans. They influence minute-by-minute and day-to-day metabolic functions. Through interplays with one another and with the nervous system, hormones have profound influence over the physical appearance, the well-being, and the behavior of individuals. And even the behavior of individuals affects whether they will survive, either on their own or as part of social groups.

This chapter focuses mainly on hormones—on their sources, targets, and interactions as well as the mechanisms involved in their secretion. If the details start to seem remote, remember that this is the stuff of life. Hormones underwrote Flo's appearance, behavior, and rise through the chimpanzee social hierarchy—and just imagine what they have been doing for you.

KEY CONCEPTS

1. For nearly all animals, hormones and other signaling molecules have central roles in integrating the activities of individual cells in ways that benefit the whole body.

2. Only the cells with molecular receptors for a specific hormone are its targets. Hormones operate by serving as signals for change in the activities of their targets.

3. Many types of hormones influence gene transcription and protein synthesis in target cells. Other types call for alterations in existing molecules and structures in cells. Some exert their effects by binding with and altering membrane characteristics, such as the permeability of the plasma membrane to a particular solute.

4. Some hormones help the body adjust to short-term shifts in diet and in levels of activity. Others help induce long-term adjustments in cell activities that bring about bodily growth, development, and reproduction.

5. Among vertebrates, the hypothalamus and pituitary gland interact in ways that coordinate the activities of a number of endocrine glands. Together, they exert control over many of the body's functions.

6. Besides hormonal signals, neural signals, changes in local chemical conditions, and environmental cues serve as the triggers for hormonal secretion.

THE ENDOCRINE SYSTEM

Hormones and Other Signaling Molecules

Throughout their lives, cells must respond to changing conditions by taking up and releasing various chemical substances. In vertebrates, the responses of millions to many billions of cells must be integrated in ways that benefit the whole body.

Integration of cell activities depends upon signaling molecules. These include hormones, neurotransmitters, local signaling molecules, and pheromones. Each type of signaling molecule acts on target cells, which are any cells that have receptors for the molecule and that may alter their activities in response to it. A target may or may not be adjacent to the cell that sends the signal.

By definition, **hormones** are the secretory products of endocrine glands, endocrine cells, and some neurons, and they travel the bloodstream to nonadjacent target cells. They are this chapter's focus.

By contrast, **neurotransmitters** are released from axon endings of neurons, then act swiftly on target cells by diffusing across the tiny gap that separates them. Section 34.3 describes their sources and their action. Also, **local signaling molecules,** released by many types of body cells, alter conditions within localized regions of tissues. Prostaglandins, for instance, target smooth muscle cells in bronchiole walls, which then constrict or dilate and so alter air flow in lungs.

Pheromones, nearly odorless secretions of particular exocrine glands, diffuse through water or air to targets outside the animal body. These hormone-like secretions act on cells of other individuals of the same species and help integrate social behavior, such as behaviors related to sexual reproduction. Pheromones are common signals among animals. For example, female silk moths secrete bombykol as a sex attractant (Section 36.3); and soldier termites secrete alarm signals when ants attack their colony (Section 51.9). Researchers have also discovered a pheromone detector, a *vomeronasal* organ, in humans. Do pheromones act at the subconscious level to trigger inexplicable impressions, such as spontaneous good or bad "feelings" about someone you just met? We do not know the answer, but studies are under way.

Discovery of Hormones and Their Sources

The word "hormone" dates back to the early 1900s. Then, W. Bayliss and E. Starling were attempting to find out what triggers the secretion of pancreatic juices when food is traveling through the canine gut. As they knew, acids are mixed with food inside the stomach, and the pancreas then secretes an alkaline solution after the acidic mixture has been propelled forward, into the small intestine. Was the nervous system or something else stimulating the pancreatic response?

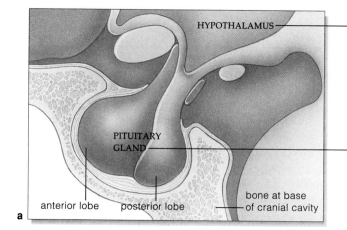

a

Figure 37.2 (**a**) A major neural-endocrine control center. The pituitary gland interacts intimately with the hypothalamus, a brain region that secretes some hormones. (**b**) *Facing page:* Overview of the key components of the human endocrine system and of the primary effects of their main hormonal secretions. The system also includes endocrine cells of many organs, including the liver, kidneys, heart, small intestine, and skin.

To find the answer, Bayliss and Starling blocked nerves—but not blood vessels—leading to a laboratory animal's upper small intestine. Later, when acidic food entered the small intestine, the pancreas still secreted the alkaline solution. Even more telling, extracts of cells from the intestinal lining—a glandular epithelium—also induced the response. Glandular cells had to be the source of the pancreas-stimulating substance.

The substance came to be called secretin. Proof of its existence and mode of action confirmed a centuries-old idea: *The bloodstream picks up internal secretions that can influence the activities of organs inside the body.* Starling coined the word "hormone" for such internal glandular secretions (after *hormon,* meaning to set in motion).

Later on, researchers identified many other kinds of hormones and their sources. Figure 37.2 is a simplified picture of the locations of the following major sources of hormones in the human body. Bear in mind, these sources are typical of most vertebrates:

Pituitary gland

Adrenal glands (*two*)

Pancreatic islets (*numerous cell clusters*)

Thyroid gland

Parathyroid glands (*in humans, four*)

Pineal gland

Thymus gland

Gonads (*two*)

Endocrine cells of the stomach, small intestine, liver, kidneys, heart, placenta, skin, adipose tissue, and other organs

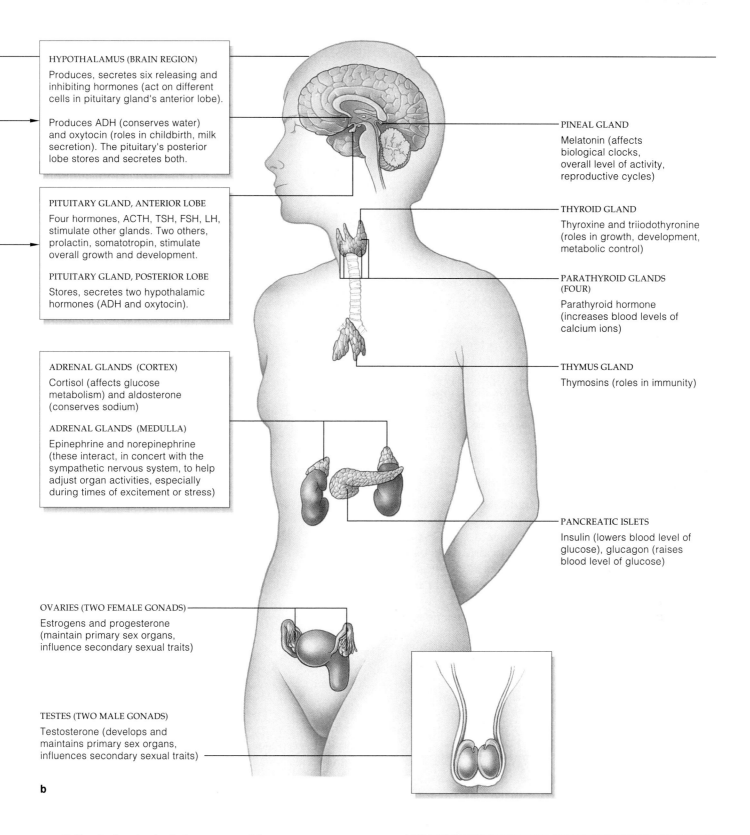

HYPOTHALAMUS (BRAIN REGION)

Produces, secretes six releasing and inhibiting hormones (act on different cells in pituitary gland's anterior lobe).

Produces ADH (conserves water) and oxytocin (roles in childbirth, milk secretion). The pituitary's posterior lobe stores and secretes both.

PITUITARY GLAND, ANTERIOR LOBE

Four hormones, ACTH, TSH, FSH, LH, stimulate other glands. Two others, prolactin, somatotropin, stimulate overall growth and development.

PITUITARY GLAND, POSTERIOR LOBE

Stores, secretes two hypothalamic hormones (ADH and oxytocin).

ADRENAL GLANDS (CORTEX)

Cortisol (affects glucose metabolism) and aldosterone (conserves sodium)

ADRENAL GLANDS (MEDULLA)

Epinephrine and norepinephrine (these interact, in concert with the sympathetic nervous system, to help adjust organ activities, especially during times of excitement or stress)

OVARIES (TWO FEMALE GONADS)

Estrogens and progesterone (maintain primary sex organs, influence secondary sexual traits)

TESTES (TWO MALE GONADS)

Testosterone (develops and maintains primary sex organs, influences secondary sexual traits)

PINEAL GLAND

Melatonin (affects biological clocks, overall level of activity, reproductive cycles)

THYROID GLAND

Thyroxine and triiodothyronine (roles in growth, development, metabolic control)

PARATHYROID GLANDS (FOUR)

Parathyroid hormone (increases blood levels of calcium ions)

THYMUS GLAND

Thymosins (roles in immunity)

PANCREATIC ISLETS

Insulin (lowers blood level of glucose), glucagon (raises blood level of glucose)

b

Collectively, the body's sources of hormones came to be called the **endocrine system**. The name implies that there is a separate control system for the body, apart from the nervous system. (*Endon* means within; *krinein* means separate.) Later, however, biochemical research and electron microscopy studies revealed that endocrine sources and the nervous system function in intricately connected ways, as you will see shortly.

Integration of cell activities depends on hormones and other signaling molecules. Each type of signaling molecule acts on target cells, which are any cells that have receptors for it and that may alter their activities in response to it.

Components of the endocrine system and certain neurons produce and secrete hormones, which the bloodstream takes up and distributes to nonadjacent target cells.

The Nature of Hormonal Action

Hormones and other signaling molecules interact with protein receptors of target cells, and their interactions have diverse effects on physiological processes. Some kinds of hormones induce target cells to increase their uptake of glucose, calcium, or some other substance from the surroundings. Other kinds stimulate or inhibit target cells into altering their rates of protein synthesis, modifying the structure of proteins or other elements in the cytoplasm, or changing cell shape.

Two factors exert considerable influence over the responses to hormonal signals. *First,* different hormones activate different cellular mechanisms. *Second,* not all cell types are equipped to respond to a given signal. For example, many cell types have receptors for cortisol, so this hormone has widespread effects on the body. By contrast, only a few cell types have the receptors for hormones that stimulate a highly directed response.

Let us now briefly consider the effects of two main categories of these signaling molecules: the steroid and peptide hormones (Table 37.1).

Characteristics of Steroid Hormones

Steroid hormones, recall, are lipid-soluble molecules derived from cholesterol (Section 3.5). The cholesterol remodeling proceeds at multiple-enzyme systems in the endoplasmic reticulum and the mitochondria of steroid-secreting cells. We find such cells in adrenal glands and primary reproductive organs. Testosterone, one of the sex hormones, is an example of the final products.

Consider testosterone's effects on the development of the secondary sexual traits associated with maleness. The developmental steps will proceed as normal only if target cells have working receptors for testosterone. In a genetic disorder called *testicular feminization syndrome,* these receptors are defective. Genetically, the affected person is male (XY); he has functional testes that are

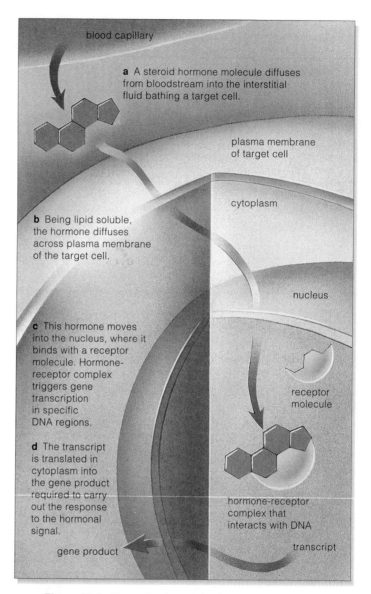

a A steroid hormone molecule diffuses from bloodstream into the interstitial fluid bathing a target cell.

blood capillary

plasma membrane of target cell

cytoplasm

b Being lipid soluble, the hormone diffuses across plasma membrane of the target cell.

nucleus

c This hormone moves into the nucleus, where it binds with a receptor molecule. Hormone-receptor complex triggers gene transcription in specific DNA regions.

receptor molecule

d The transcript is translated in cytoplasm into the gene product required to carry out the response to the hormonal signal.

hormone-receptor complex that interacts with DNA

gene product

transcript

Figure 37.3 Example of a mechanism by which a steroid hormone initiates changes in a target cell's activities.

able to secrete testosterone. Target cells cannot respond to the hormone, however, so the secondary sexual traits that do develop are like those of females.

How does a steroid hormone such as testosterone exert effects on cells? Being lipid soluble, it may diffuse directly across the lipid bilayer of a target cell's plasma membrane (Figure 37.3). Once inside the cytoplasm, the hormone molecule usually moves into the nucleus and binds to some type of protein receptor. Or it binds to a receptor molecule in the cytoplasm in some cases, then the hormone-receptor complex enters the nucleus. The configuration of the complex allows it to interact with a specific region of the cell's DNA. Different complexes inhibit or stimulate transcription of certain gene regions

Table 37.1	Two Main Categories of Hormones
Type	Examples
Steroid and steroid-like hormones	Estrogens (feminizing effects), progestins (related to pregnancy), androgens (such as testosterone; masculinizing effects), cortisol, aldosterone. In addition, thyroid hormones and vitamin D act like steroids.
Peptide hormones:	
Peptides	Glucagon, ADH, oxytocin, TRH
Proteins	Insulin, somatotropin, prolactin
Glycoproteins	FSH, LH, TSH

Characteristics of Peptide Hormones

Traditionally, **peptide hormones** have been defined as water-soluble signaling molecules that may incorporate anywhere from 3 to 180 amino acids. Various peptides, polypeptides, and glycoproteins fall into this category. All peptide hormones issue signals right at a receptor of a target cell's plasma membrane. The signals activate specific membrane-bound enzyme systems that initiate reactions leading to the cellular response.

Consider a liver cell with receptors for glucagon, a peptide hormone. Every receptor spans the plasma membrane, and part of it extends into the cytoplasm. After a receptor binds glucagon, the response requires assistance from a **second messenger**. Such messengers are small molecules in the target cell's cytoplasm, and they relay a signal from a hormone-receptor complex at the plasma membrane into the cell interior. In this case, cAMP (cyclic adenosine monophosphate) is the second messenger. As Figure 37.4 indicates, glucagon binding triggers activity at a membrane-bound enzyme system. First an activated form of the enzyme adenylate cyclase converts ATP to cyclic AMP, which starts a cascade of reactions. Many molecules of cyclic AMP form and act as signals for the conversion of many molecules of an enzyme, a protein kinase, to active form. These act on other enzymes, and so on until a final reaction converts stored glycogen in the cell to glucose. In short order, the number of molecules involved in the final response to the glucagon-receptor complex is enormous.

Or consider a muscle cell with receptors for insulin, a protein hormone. Among other things, a signal from a hormone-receptor complex triggers the movement of molecules of glucose transporter proteins through the cytoplasm and insertion into the plasma membrane, so that the cell can take up glucose faster. In addition, it switches on enzymes of glucose metabolism.

Bear in mind, there are other hormone categories, including the catecholamines such as epinephrine. Like glucagon, epinephrine combines with specific surface receptors and triggers the release of cyclic AMP as a second messenger that assists in the cellular response.

Hormones interact with receptors located either at the plasma membrane or within the cytoplasm of target cells.

Steroid hormone-receptor complexes typically enter a target cell and interact with the cell's DNA. Also, some types interact with membranes and alter membrane functions.

Peptide hormones do not exert their effects by entering a cell. When they bind to a membrane receptor, the binding itself is a signal for enzyme-mediated, intracellular events. Often a second messenger inside the cytoplasm relays the signal into the cell interior.

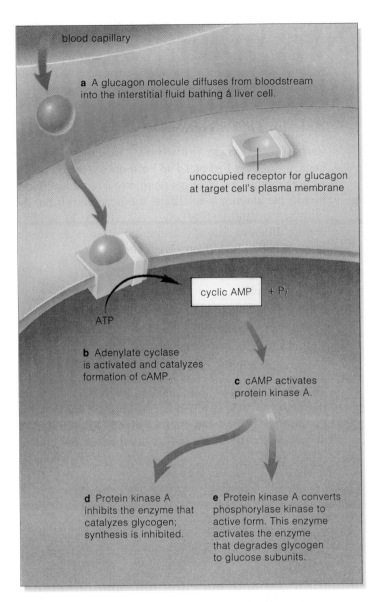

blood capillary

a A glucagon molecule diffuses from bloodstream into the interstitial fluid bathing â liver cell.

unoccupied receptor for glucagon at target cell's plasma membrane

cyclic AMP + P$_i$

ATP

b Adenylate cyclase is activated and catalyzes formation of cAMP.

c cAMP activates protein kinase A.

d Protein kinase A inhibits the enzyme that catalyzes glycogen; synthesis is inhibited.

e Protein kinase A converts phosphorylase kinase to active form. This enzyme activates the enzyme that degrades glycogen to glucose subunits.

Figure 37.4 Example of a mechanism by which a peptide hormone initiates changes in a target cell's activities. When the hormone glucagon binds at a receptor, it initiates reactions inside the cell. In this case cyclic AMP, which is one type of second messenger, relays the signal into the cell interior.

---

into mRNA. The translation of such mRNA transcripts results in enzymes and other proteins that can carry out a response to the hormonal signal.

Steroid hormones also may exert effects in another way. It now appears they can bind to cell membranes and alter the membrane properties in ways that modify the functions of the target cell.

One final point: Thyroid hormones and vitamin D are not steroid hormones, but they do behave like them. Also, the genes coding for their receptors are part of the same group that codes for steroid hormones.

THE HYPOTHALAMUS AND PITUITARY GLAND

Deep in the forebrain is the **hypothalamus**. This brain region monitors internal organs and activities related to their functioning, such as eating and sexual behavior. It also secretes some hormones. Suspended from its base by a slender stalk of tissue is a lobed gland, about the size of a pea. The hypothalamus and this **pituitary gland** interact as a major neural-endocrine control center.

The pituitary's *posterior* lobe secretes two hormones that are synthesized by the hypothalamus. Its *anterior* lobe produces and secretes its own hormones, most of which help control the release of hormones from other endocrine glands (Table 37.2). The pituitary of many vertebrates—*not* humans—has an intermediate lobe as well. In many cases, the third lobe secretes a hormone that governs reversible changes in skin or fur color.

Posterior Lobe Secretions

Figure 37.5*a* shows the cell bodies of certain neurons in the hypothalamus. Their axons extend down into the posterior lobe, then terminate next to a capillary bed. The neurons produce antidiuretic hormone (ADH) and oxytocin, then store them in the axon endings. After either hormone is released into the interstitial fluid, it diffuses into capillaries, then travels the bloodstream to its targets. ADH acts on cells of nephrons and collecting ducts in the kidneys, which filter the blood and rid the body of excess water and salts (in the form of urine). ADH promotes water reabsorption when the body must conserve water. Oxytocin has roles in reproduction. It triggers contractions of the uterus during labor, and it causes milk release when offspring are being nursed.

Anterior Lobe Secretions

ANTERIOR PITUITARY HORMONES Inside the pituitary stalk, a capillary bed picks up hormones secreted by the hypothalamus and delivers them into a *second* capillary bed in the anterior lobe. There, the hormones leave the bloodstream and then act on target cells. As Figure 37.6 shows, different cells of the anterior pituitary secrete six hormones that they themselves produce:

Corticotropin	ACTH
Thyrotropin	TSH
Follicle-stimulating hormone	FSH
Luteinizing hormone	LH
Prolactin	PRL
Somatotropin (or growth hormone)	STH (or GH)

The effects of these hormones are widespread through the body. ACTH and TSH orchestrate secretions from the adrenal glands and thyroid gland, respectively, as described shortly. FSH and LH act through the gonads

---

Table 37.2   Hormones Released From the Mammalian Pituitary Gland

Pituitary Lobe	Secretions	Designation	Main Targets	Primary Actions
POSTERIOR Nervous tissue (extension of hypothalamus)	Antidiuretic hormone	ADH	Kidneys	Induces water conservation as required during control of extracellular fluid volume and solute concentrations
	Oxytocin	OCT	Mammary glands	Induces milk movement into secretory ducts
			Uterus	Induces uterine contractions
ANTERIOR Glandular tissue, mostly	Corticotropin	ACTH	Adrenal cortex	Stimulates release of adrenal steroid hormones
	Thyrotropin	TSH	Thyroid gland	Stimulates release of thyroid hormones
	Follicle-stimulating hormone	FSH	Ovaries, testes	In females, stimulates estrogen secretion, egg maturation; in males, helps stimulate sperm formation
	Luteinizing hormone	LH	Ovaries, testes	In females, stimulates progesterone secretion, ovulation, corpus luteum formation; in males, stimulates testosterone secretion, sperm release
	Prolactin	PRL	Mammary glands	Stimulates and sustains milk production
	Somatotropin (or growth hormone)	STH (GH)	Most cells	Promotes growth in young; induces protein synthesis, cell division; roles in glucose, protein metabolism in adults
INTERMEDIATE* Glandular tissue, mostly	Melanocyte-stimulating hormone	MSH	Pigmented cells in skin and other integuments	Induces color changes in response to external stimuli; affects some behaviors

* Present in most vertebrates (not humans). MSH is associated with the anterior lobe in humans.

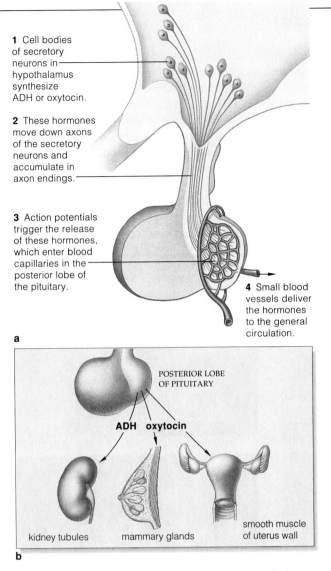

1 Cell bodies of secretory neurons in hypothalamus synthesize ADH or oxytocin.

2 These hormones move down axons of the secretory neurons and accumulate in axon endings.

3 Action potentials trigger the release of these hormones, which enter blood capillaries in the posterior lobe of the pituitary.

4 Small blood vessels deliver the hormones to the general circulation.

a

POSTERIOR LOBE OF PITUITARY

ADH   oxytocin

kidney tubules   mammary glands   smooth muscle of uterus wall

b

Figure 37.5 (**a**) Functional links between the hypothalamus and the posterior lobe of the pituitary gland. (**b**) Main targets of the posterior lobe secretions.

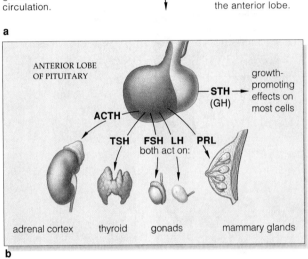
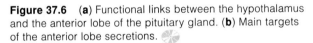

1 Cell bodies of secretory neurons in hypothalamus secrete releasing and inhibiting hormones.

2 First capillary bed, in base of hypothalamus, picks up hormones.

3 Hormones are delivered into second capillary bed, in anterior lobe of pituitary.

5 Hormones secreted from anterior lobe cells enter venules that lead to the general circulation.

4 Releasing or inhibiting hormones diffuse out of the capillaries, act on endocrine cells in the anterior lobe.

a

ANTERIOR LOBE OF PITUITARY

STH (GH)   growth-promoting effects on most cells

ACTH

TSH   FSH LH   PRL
both act on:

adrenal cortex   thyroid   gonads   mammary glands

b

Figure 37.6 (**a**) Functional links between the hypothalamus and the anterior lobe of the pituitary gland. (**b**) Main targets of the anterior lobe secretions.

to influence gamete formation and secretions of the sex hormones required in sexual reproduction, the central topic of Chapter 34. Somatotropin affects metabolism in many tissues and liver secretions that influence growth of bone and soft tissues (Table 33.2 and Figure 37.6*b*). Prolactin affects different cell types but it is best known for stimulating and then sustaining milk production in mammary glands, after other hormones have primed the tissues. In some species, prolactin affects hormone production in the ovaries as well.

REGARDING THE HYPOTHALAMIC TRIGGERS Most of the hypothalamic hormones that act in the anterior lobe of the pituitary are **releasers**, meaning they stimulate the secretion of hormones from target cells. For example, the one called GnRH (gonadotropin-releasing hormone) brings about the secretion of FSH and LH, which are

classified as gonadotropins. Similarly, TRH, a releasing hormone, stimulates the secretion of thyrotropin.

But some hypothalamic hormones are **inhibitors** of secretion from their targets in the anterior pituitary. For instance, the one known as somatostatin brings about a decrease in somatotropin and thyrotropin secretion.

The hypothalamus and pituitary gland produce eight kinds of hormones and interact to control their secretion.

The posterior lobe of the pituitary stores and secretes two hypothalamic hormones, ADH and oxytocin, both of which target specific cell types.

The anterior lobe of the pituitary produces and secretes six hormones, ACTH, TSH, FSH, LH, PRL, and STH. These all trigger the release of other hormones from other endocrine glands, with wide-ranging effects throughout the body.

EXAMPLES OF ABNORMAL PITUITARY OUTPUT

The body does not churn out enormous quantities of hormone molecules. (Two researchers, Roger Guilleman and Andrew Schally, realized this when they isolated the first known releasing hormone. In their four-year attempt to secure TRH, they dissected 500 metric tons of brains and 7 metric tons of hypothalamic tissue from sheep and ended up with only a single milligram of it.) Yet normal body function depends on the tiny amounts.

Endocrine glands release their tiny but significant secretions mainly in short bursts. Elegant controls over the frequency of those abrupt secretory events prevent underproduction or overproduction of a

hormone. When something interferes with the controls, disorders may be the outcome. For example, *gigantism* results when anterior pituitary cells produce too much somatotropin. Proportionally, affected adults are similar to a person of normal size but are much larger (Figure 37.7a). By contrast, *pituitary dwarfism* results when the body does not produce enough somatotropin. Affected adults are proportionally similar to an average person but much smaller (Figure 37.7b). What if somatotropin output becomes excessive in adulthood, when the long bones no longer can lengthen? Then, *acromegaly* results. Bone, cartilage, and other connective tissues in hands, feet, and jaws thicken abnormally. So do epithelia of the skin, nose, eyelids, lips, and tongue (Figure 37.7c).

Age nine

Sixteen

Thirty-three

Fifty-two

Figure 37.7 Examples of the outcome of abnormal secretion of a hormone.

(**a**) Manute Bol, an NBA center, is 7 feet 6-3/4 inches tall owing to excessive secretion of somatotropin (STH) during his childhood.

(**b**) More examples of the effect of STH on body growth. The male at the center of this photograph is affected by gigantism, which resulted from excessive STH production in childhood. The person at the right displays pituitary dwarfism, which resulted from underproduction of STH in childhood. The person at the left is of average height.

(**c**) Acromegaly, which resulted from excessive production of STH during adulthood. Before this female reached maturity, she was symptom-free.

As another example, ADH secretion can diminish or stop when the pituitary's posterior lobe is damaged, as by a blow to the head or by a brain tumor. This is one cause of *diabetes insipidus*. The symptoms include excretion of large volumes of dilute urine, which may cause life-threatening dehydration. Diabetes insipidus responds to hormone replacement therapy based on injections or nasal spray applications of synthetic ADH.

Generally, endocrine glands release very small amounts of hormones in short bursts, the frequency of which depends on control mechanisms. When controls fail, the resulting oversecretion or undersecretion may cause disorders.

SOURCES AND EFFECTS OF OTHER HORMONES

Table 37.3 lists hormones from endocrine sources other than the pituitary. The remainder of this chapter will provide you with a few examples of their effects and of the controls over their output. The examples will make more sense if you keep the following points in mind.

First, hormones often interact with one another. In other words, one or more hormones may oppose, add to, or prime target cells for another hormone's effects. *Second*, negative feedback mechanisms often control the secretions. When a hormone's concentration increases or decreases in some body region, the change triggers events that respectively dampen or stimulate further secretion. *Third*, a target cell may react differently to a hormone at different times. Its response depends on the hormone's concentration and on the functional state of the cell's receptors. *Fourth*, environmental cues may be important mediators of hormonal secretion.

The secretion of a hormone and its effects are influenced by hormone interactions, feedback mechanisms, variations in the state of target cells, and sometimes environmental cues.

Table 37.3   Hormone Sources Other Than the Mammalian Hypothalamus and Pituitary

Source	Secretion(s)	Main Targets	Primary Actions
ADRENAL CORTEX	Glucocorticoids (including cortisol)	Most cells	Promote protein breakdown and conversion to glucose
	Mineralocorticoids (including aldosterone)	Kidney	Promote sodium reabsorption (sodium conservation); help control the body's salt–water balance
ADRENAL MEDULLA	Epinephrine (adrenaline)	Liver, muscle, adipose tissue	Raises blood level of sugar, fatty acids; increases heart rate and force of contraction
	Norepinephrine	Smooth muscle of blood vessels	Promotes constriction or dilation of blood vessel diameter
THYROID	Triiodothyronine, thyroxine	Most cells	Regulate metabolism; have roles in growth, development
	Calcitonin	Bone	Lowers calcium level in blood
PARATHYROIDS	Parathyroid hormone	Bone, kidney	Elevates calcium level in blood
GONADS:			
Testes (in males)	Androgens (including testosterone)	General	Required in sperm formation, development of genitals, maintenance of sexual traits; growth, development
Ovaries (in females)	Estrogens	General	Required for egg maturation and release; preparation of uterine lining for pregnancy and its maintenance in pregnancy; genital development; maintenance of sexual traits; growth, development
	Progesterone	Uterus, breasts	Prepares, maintains uterine lining for pregnancy; stimulates breast development
PANCREATIC ISLETS	Insulin	Liver, muscle, adipose tissue	Lowers sugar level in blood
	Glucagon	Liver	Raises sugar level in blood
	Somatostatin	Insulin-secreting cells	Influences carbohydrate metabolism
THYMUS	Thymosins, etc.	Lymphocytes	Have roles in immune responses
PINEAL	Melatonin	Gonads (indirectly)	Influences daily biorhythms, seasonal sexual activity
STOMACH, SMALL INTESTINE	Gastrin, secretin, etc.	Stomach, pancreas, gallbladder	Stimulate activities of stomach, pancreas, liver, gallbladder required for food digestion, absorption
LIVER	Somatomedins	Most cells	Stimulate cell growth and development
KIDNEYS	Erythropoietin	Bone marrow	Stimulates red blood cell production
	Angiotensin*	Adrenal cortex, arterioles	Helps control blood pressure, secretion of aldosterone (hence sodium reabsorption)
	1,25-hydroxyvitamin D_6* (calcitriol)	Bone, gut	Enhances calcium reabsorption from bone and calcium absorption from gut
HEART	Atrial natriuretic hormone	Kidney, blood vessels	Increases sodium excretion; lowers blood pressure

* Kidneys produce *enzymes* that modify precursors of this substance, which then enters the general circulation as an activated hormone.

FEEDBACK CONTROL OF HORMONAL SECRETIONS

By considering just a few of the endocrine glands listed in Table 37.3, you can sense how feedback mechanisms control hormonal secretions. Briefly, the hypothalamus, pituitary, or both signal these glands to alter secretory activity. The outcome is a change in the concentration of the secreted hormone in blood or elsewhere. With the shift in chemical information, a feedback mechanism trips into action and blocks or promotes further change.

With **negative feedback**, an increase or decrease in the concentration of a secreted hormone triggers events that *inhibit* further secretion. With **positive feedback**, an increase in the concentration of a secreted hormone triggers events that *stimulate* further secretion.

Negative Feedback From the Adrenal Cortex

Humans have a pair of adrenal glands, one above each kidney (Figure 37.2b). Some cells of the **adrenal cortex**, the outer portion of an adrenal gland, secrete hormones such as glucocorticoids. Glucocorticoids help maintain the concentration of glucose in blood and help suppress inflammatory responses. Cortisol, for instance, blocks the uptake and use of glucose by muscle cells. It also stimulates liver cells to form glucose from amino acids.

A negative feedback mechanism operates when the glucose level in blood declines below a set point. This chemical condition is known as *hypoglycemia*. Take a look at Figure 37.8. When the hypothalamus detects the decrease, it secretes CRH in response. This releasing hormone stimulates the anterior pituitary to secrete corticotropin (ACTH). In turn, ACTH stimulates cells of the adrenal cortex to secrete cortisol, which helps raise the glucose level in blood. How? Cortisol stops muscle cells from taking up glucose that blood is delivering through the body. These cells are major glucose users.

During severe stress, painful injury, or prolonged illness, the nervous system overrides feedback control of cortisol secretion. It initiates a *stress response* in which cortisol helps to suppress inflammation. If unchecked, prolonged inflammation damages tissues. That is why doctors often prescribe cortisol-like drugs for asthma and other chronic inflammatory disorders.

Local Feedback in the Adrenal Medulla

The **adrenal medulla** is the inner portion of the adrenal gland (Figure 37.8). It has hormone-secreting neurons that release epinephrine and norepinephrine. (These substances are neurotransmitters in some contexts and hormones in others.) Suppose the axons of sympathetic nerves carry hypothalamic signals that call for secretion of norepinephrine. Molecules of norepinephrine collect in the synaptic cleft between the axon endings and the target cells. In this case, a localized, negative feedback mechanism operates at receptors on the axon endings. The excess norepinephrine binds to the receptors and causes a shutdown of its further release.

In times of excitement or stress, epinephrine and norepinephrine help adjust blood circulation and fat and carbohydrate metabolism. They increase heart rate, trigger vasoconstriction and vasodilation of arterioles in different regions, and dilate airways to the lungs. The controlled activity directs more of the total volume of blood to heart and muscle cells, and more oxygen flows to energy-demanding cells through the body. These are features of the *fight-flight response* (Section 35.3).

Skewed Feedback From the Thyroid

The human **thyroid gland** is located at the base of the neck in front of the trachea, or windpipe (Figures 37.2b and 37.9a,b). Thyroxine and triiodothyronine, its main hormones, have widespread effects. In their absence, many tissues cannot develop normally. Also, the overall metabolic rates of warm-blooded animals, including humans, depend on them. The importance of feedback control over the secretion of these hormones is brought into sharp focus by cases of abnormal thyroid output.

Consider how thyroid hormone synthesis requires iodine, which we obtain from food. Iodine is converted to an iodized form, iodide, when absorbed from the gut. Without iodide, the blood levels of thyroid hormones decline. The anterior pituitary responds by secreting

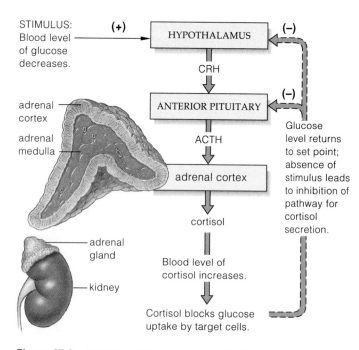

Figure 37.8 Location of the adrenal glands. One rests on top of each kidney. The diagram shows a negative feedback loop that governs secretion of cortisol from the adrenal cortex.

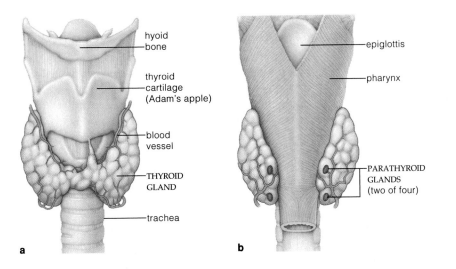

Figure 37.9 Human thyroid gland. (**a**) Anterior and (**b**) posterior views showing the location of four parathyroid glands. (**c**) A mild case of goiter, displayed by Maria de Medici in the year 1625. A rounded neck was considered to be a sign of great beauty during the late Renaissance. It occurred regularly in parts of the world where iodine supplies were insufficient for normal thyroid function.

thyroid-stimulating hormone (TSH). But without iodine, thyroid hormones cannot be made. The feedback signal continues, and so does TSH secretion. Excess TSH in blood overstimulates the thyroid gland, which enlarges in response. The enlargement is a form of *goiter*. Goiter resulting from iodine deficiency is no longer common in countries where people use iodized salt (Figure 37.9c).

When blood levels of thyroid hormones are too low, *hypothyroidism* results. Hypothyroid adults often are overweight, sluggish, dry-skinned, intolerant of cold, confused, and depressed. Affected women commonly show menstrual disturbances.

When blood levels of the thyroid hormones are too high, *hyperthyroidism* results. Affected adults show an increased heart rate, heat intolerance, elevated blood pressure, profuse sweating, and weight loss even when caloric intake increases. Affected individuals typically are nervous and agitated, and have trouble sleeping.

Feedback Control of the Gonads

Gonads are *primary* reproductive organs, which produce gametes and also synthesize and secrete sex hormones. Testes (singular, testis) in males and ovaries in females are examples. Testes secrete testosterone; ovaries secrete estrogens and progesterone. All of these sex hormones influence secondary sexual traits (as they did for the chimps described earlier), and feedback controls govern their secretion. Figure 37.10 is a preview of the feedback loops from ovaries to the hypothalamus and pituitary during the menstrual cycle, a key topic of Chapter 45.

Feedback mechanisms control secretions from endocrine glands. In many cases, feedback loops to the hypothalamus, pituitary, or both govern the secretory activity.

With negative feedback, further secretion of a hormone slows down. With positive feedback, further secretion is enhanced.

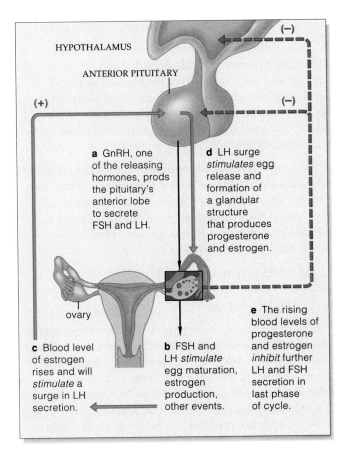

Figure 37.10 Feedback loops to the hypothalamus and pituitary gland from the ovaries during the menstrual cycle, a recurring reproductive event. Positive feedback triggers egg release from an ovary. Negative feedback after its release prevents release of another egg until the cycle is completed.

RESPONSES TO LOCAL CHEMICAL CHANGES

Some endocrine glands or cells don't respond primarily to signals from other hormones or nerves. They respond homeostatically to chemical change in the immediate surroundings, as the following examples illustrate.

Secretions From Parathyroid Glands

Humans have four **parathyroid glands** located on the posterior surface of the thyroid (Figure 37.9b). These glands secrete parathyroid hormone, or PTH, the main regulator of the calcium level in blood. Calcium ions, recall, have roles in muscle contraction, enzyme action, blood clot formation, and other tasks. The parathyroids secrete PTH in response to a low calcium level in blood. Their secretory activity slows when the calcium level rises. PTH acts on cells of the skeleton and kidneys.

PTH induces living bone cells to secrete enzymes that digest bone tissue and thereby release calcium and other minerals to interstitial fluid, then to the blood. It enhances calcium reabsorption from the filtrate flowing through the nephrons of kidneys. PTH also prods some kidney cells to secrete enzymes that act on blood-borne precursors of the active form of vitamin D_3, a hormone (Table 37.3). Vitamin D_3 stimulates intestinal cells to increase the absorption of calcium from the gut lumen. In children who have vitamin D deficiency, insufficient calcium and phosphorus are absorbed, so rapidly growing bones develop improperly. The resulting bone disorder, *rickets*, is characterized by bowed legs, a malformed pelvis, and in many cases a malformed skull and rib cage (Figure 37.11).

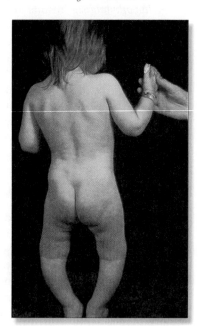

Figure 37.11 A child who is affected by rickets.

Effects of Local Signaling Molecules

Many cells detect changes in the surrounding chemical environment and alter their activity, often in ways that counteract or amplify the change. The cells secrete various local signaling molecules, the effects of which are confined to the immediate vicinity of change. Target cells take up most signaling molecules so rapidly that few enter the general circulation.

Prostaglandins are examples of signaling molecules. Cells in many tissues continually produce and release a variety of prostaglandins. But the rate of synthesis often increases in response to local chemical changes. Section 45.4 includes a fine example of this response.

Growth factors are other examples. They influence growth by regulating the rate at which cells divide. Thus an epidermal growth factor (EGF) discovered by Stanley Cohen influences growth of many cell types. A nerve growth factor (NGF) discovered by Rita Levi-Montalcini helps assure the survival of neurons and influences the direction of their growth in an embryo. One experiment demonstrated that certain immature neurons will survive indefinitely in tissue culture when NGF is present but will die within a few days if it is not.

Secretions From Pancreatic Islets

The pancreas is a gland with exocrine and endocrine functions. Its *exocrine* cells secrete digestive enzymes. Its *endocrine* cells are located in about 2 million clusters within the pancreas. Each small cluster, a **pancreatic islet**, contains three types of hormone-secreting cells:

1. *Alpha* cells in the pancreas secrete the hormone glucagon. In between meals, cells throughout the body take up and use glucose from the blood. The blood level of glucose decreases. At such times, glucagon secretion causes glycogen (a storage polysaccharide) and amino acids to be converted to glucose in the liver. In such ways, *glucagon raises the glucose level*.

2. *Beta* cells secrete the hormone insulin. After meals, when the blood glucose level is high, insulin stimulates glucose uptake by muscle and adipose cells especially. It promotes the synthesis of proteins and fats, and it inhibits protein conversion to glucose. Thus, *insulin lowers the glucose level*.

3. *Delta* cells secrete somatostatin, a hormone that helps control digestion. It also can block secretion of insulin and glucagon.

Figure 37.12 shows how pancreatic hormones interact to maintain the level of glucose in blood even though the times and amounts of food intake vary. Bear in mind, insulin is the only hormone that prods cells to take up and store glucose in forms that can be rapidly tapped when required. Its central role in carbohydrate, protein, and fat metabolism becomes clear when we observe people who cannot produce enough insulin or who lack body cells that can respond to it.

For example, insulin deficiency may lead to *diabetes mellitus*, a disorder in which excess glucose accumulates in blood, then in urine. Urination becomes excessive, so the body's water-solute balance is disrupted. Affected

Figure 37.12 Some of the homeostatic controls over glucose metabolism.

Following a meal, glucose enters the bloodstream faster than cells can use it. The blood glucose level rises, and pancreatic beta cells are stimulated to secrete insulin. Insulin's main targets—liver, fat, and muscle cells—not only use glucose but store excess amounts of it in the form of glycogen.

Between meals, the blood glucose level decreases. Pancreatic alpha cells are stimulated to secrete glucagon. The target cells with receptors for this hormone convert glycogen back to glucose, which enters the blood.

Glucose metabolism also is subject to indirect controls. For example, the hypothalamus commands the adrenal medulla to secrete certain hormones. The hormones speed the conversion of glycogen to glucose in the liver and slow the reverse process, especially in cells of the liver, adipose tissue, and muscle tissue.

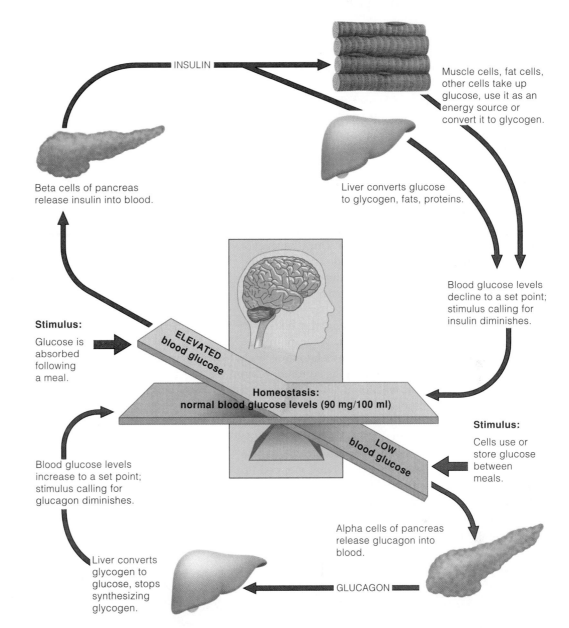

INSULIN

Muscle cells, fat cells, other cells take up glucose, use it as an energy source or convert it to glycogen.

Beta cells of pancreas release insulin into blood.

Liver converts glucose to glycogen, fats, proteins.

Blood glucose levels decline to a set point; stimulus calling for insulin diminishes.

Stimulus:
Glucose is absorbed following a meal.

ELEVATED blood glucose

Homeostasis:
normal blood glucose levels (90 mg/100 ml)

LOW blood glucose

Stimulus:
Cells use or store glucose between meals.

Blood glucose levels increase to a set point; stimulus calling for glucagon diminishes.

Alpha cells of pancreas release glucagon into blood.

Liver converts glycogen to glucose, stops synthesizing glycogen.

GLUCAGON

people become dehydrated and thirsty—abnormally so. Without a steady supply of glucose, their body cells start depleting their own fats and proteins as sources of energy. Weight loss is one outcome. Another is that ketones accumulate in blood and urine. Ketones are normal acidic products of fat breakdown. When they accumulate, they contribute to water losses and alter the body's acid-base balance. Such imbalances disrupt brain function. In extreme cases, death may follow.

In "type 1 diabetes," the body mistakenly mounts an autoimmune response against its insulin-secreting beta cells. Certain lymphocytes identify the beta cells as foreign and destroy them. A combination of genetic susceptibility and environmental triggers produces the disorder, which is less common but more immediately dangerous than other forms of diabetes. Usually, the symptoms first appear in childhood and adolescence (the disorder is also known as juvenile-onset diabetes). Type 1 diabetic patients survive with insulin injections.

In "type 2 diabetes," insulin levels are close to or above normal, but the target cells cannot respond to insulin. As affected persons grow older, their beta cells produce less and less insulin. Type 2 diabetes usually is manifested during middle age. Affected persons lead normal lives by controlling their diet and weight, and sometimes by taking drugs to enhance insulin action or secretion.

The secretions from some endocrine glands and cells are direct homeostatic responses to a change in the localized chemical environment.

HORMONAL RESPONSES TO ENVIRONMENTAL CUES

This last section of the chapter invites you to reflect on a key point. An individual's growth, development, and reproduction begin with genes and hormones, and so does behavior. *But certain environmental factors commonly influence gene expression and hormonal secretion, and they do so in predictable ways.* Chapter 51 invites analysis of the environmental influence on animal behavior. For now, it is enough to consider the following examples.

Daylength and the Pineal Gland

Embedded in the brain is a photosensitive organ, the **pineal gland** (Section 35.5). In the absence of light, the gland secretes the hormone melatonin. Thus the level of melatonin in the blood varies from day to night, and with the seasons. The variations influence the growth and development of gonads, the primary reproductive organs. In a variety of species, they have important roles in reproductive cycles and reproductive behavior.

Think about a hamster. In winter, when nighttime darkness is longest, the blood level of melatonin is high and sexual activity is suppressed. In the summer, when daylength is longest, the melatonin level is low, and hamster sex reaches its peak. Or think about a male white-throated sparrow (Figure 37.13a). In the fall and winter, melatonin indirectly suppresses growth of its gonads by inhibiting gonadotropin secretion. It does so until days start to lengthen in spring. Now, stepped-up gonadal activity leads to production of hormones that influence singing behavior, as described in Section 51.1. With his distinctive song, the male sparrow defines his territory and may hold the interest of a mate.

Does melatonin also influence human behavior? Perhaps. Clinical observations and studies suggest that decreased melatonin secretion may trigger **puberty**, the age during which human reproductive organs and structures start to mature. For example, in cases where disease caused the destruction of an individual's pineal gland, puberty began prematurely.

Melatonin is known to act on certain neurons that lower your body's core temperature and that make you drowsy after sunset, when light is waning. At sunrise, when melatonin secretion slows, your core temperature increases and you wake up and become active.

An internal, biological clock governs the cycle of sleep and arousal. It seems to tick in synchrony with daylength. Think of night workers who try to sleep in the morning but end up staring groggily at sunbeams on the ceiling. Think of travelers from the United States to Paris who go through four days of "jet lag." Two or three hours past midnight they are sitting up in bed, wondering where the coffee and croissants are. Two hours past noon they are ready for bed. They will shift to a new routine when the melatonin signals arrive at their target neurons on Paris time.

In winter, some individuals experience *winter blues*. They get abnormally depressed, go on carbohydrate binges, and have an almost overwhelming desire to sleep (Figure 37.13b). Winter blues might arise when a biological clock gets out of sync with the seasonally shorter daylengths. Intriguingly, clinically administered doses of melatonin make the seasonal symptoms worse. And exposure to intense light, which can shut down pineal activity, may lead to dramatic improvement.

Figure 37.13 (**a**) A male white-throated sparrow, belting out a song that began, indirectly, with an environmentally induced decline in melatonin secretion. (**b**) Annie blanketing her winter blues.

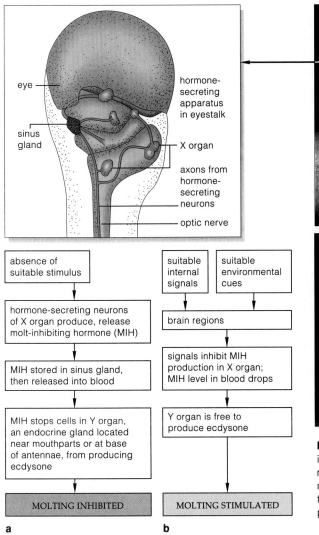

Figure 37.14 (**a,b**) Steps in the hormonal control of molting in crustaceans, including crabs (**c**). The steps differ a bit in insects, which do not use a molt-inhibiting hormone. Rather, stimulation of the insect brain causes certain neurons to secrete ecdysiotropin. This hormone induces different neurons to produce still another hormone that targets ecdysone-producing cells in prothoracic glands. (**d**) This insect, a cicada, is emerging from its old cuticle.

Comparative Look at a Few Invertebrates

Although this chapter's focus has been on vertebrates, do not lose sight of the fact that all organisms produce signaling molecules of one sort or another. Consider the hormonal control of **molting**, a periodic discarding and replacement of a hardened cuticle that otherwise would limit increases in body mass. As described in Section 26.15, molting occurs during the life cycle of all insects, crustaceans, and other invertebrates with thick cuticles.

Although details vary from group to group, molting is largely under the control of ecdysone. This steroid hormone is derived from cholesterol and is chemically related to many important vertebrate hormones. In insects and crustaceans, molting glands produce and store ecdysone, then release it for distribution through the body at molting time. Hormone-secreting neurons in the brain seem to regulate its release. The hormone-secreting neurons apparently respond to a combination of environmental cues, including light and temperature, as well as internal signals.

Figure 37.14 gives examples of the control steps, which differ in crustaceans and insects.

During premolt and molting periods, coordinated interactions among ecdysone and other hormones bring about major structural and physiological changes. The interactions cause the old cuticle to detach from the epidermis and muscles. They induce the dissolving and recycling of the cuticle's inner layers. The interactions also trigger shifts in metabolism and in the composition and volume of the internal environment. They promote cell divisions, secretions, and pigment formation, all of which go into producing a new cuticle. Simultaneously, hormonal interactions control heart rate, muscle action, color changes, and other physiological processes.

Environmental cues, such as changes in light intensity from day to night and seasonal changes in daylength, influence certain hormonal secretions.

SUMMARY

1. The cells of complex animals continually exchange substances with the body's internal environment. Their myriad withdrawals and secretions are integrated in ways that ensure cell survival through the whole body.

2. Integration of cell activities requires the stimulatory or inhibitory effects of signaling molecules.

   a. Signaling molecules are chemical secretions from a cell that adjust the behavior of other, target cells.

   b. Any cell with molecular receptors for a signaling molecule is a target. The target cells may or may not be next to the cell that sends the signal.

   c. There are different kinds of signaling molecules. Hormones as well as neurotransmitters, local signaling molecules, and pheromones are the main kinds.

   d. Certain steroids, steroidlike molecules, amines, peptides, proteins, and glycoproteins are hormones.

3. In target cells, hormones influence gene activation, protein synthesis, and alterations in existing enzymes, membranes, and other cellular components. Hormones exert their physiological effects through interactions with specific protein receptors at the plasma membrane or in the cytoplasm of target cells.

   a. Steroid hormones are lipid soluble. Complexes of steroid hormones and receptors interact with DNA and possibly with membranes of their targets.

   b. Protein hormones are water soluble. Some enter the cytoplasm complexed with receptors. The effect of others is exerted with the help of membrane transport proteins and second messengers in the cytoplasm, some of which trigger the actual response.

4. The posterior lobe of the pituitary stores and secretes two hypothalamic hormones, ADH and oxytocin. ADH targets cells in kidneys and affects extracellular fluid volume. Oxytocin acts on cells in mammary glands and the uterus to influence reproductive events.

5. The hypothalamic hormones called releasing and inhibiting hormones control secretions from different cells of the anterior lobe of the pituitary gland.

6. The anterior lobe makes and secretes six hormones, ACTH, TSH, FSH, LH, PRL, and STH. These trigger secretions from the adrenal cortex, thyroid, gonads, and mammary glands. By doing so, they exert wide-ranging responses throughout the body.

7. The vertebrate body has other sources of hormones, including the adrenal medulla, the parathyroid, thymus, and pineal glands, pancreatic islets, and endocrine cells in the stomach, small intestine, liver, and heart.

8. Hormone interactions, feedback mechanisms, the number and kind of receptors on target cells, variations in the state of target cells, and sometimes cues from the environment influence the secretion of a hormone and its effects.

9. Many cellular responses to hormones help the body adjust to short-term shifts in diet and levels of activity. Other kinds help bring about long-term adjustments for growth, development, and reproduction.

   a. In general, secretion of hormones such as insulin and parathyroid hormone can change rapidly when the extracellular concentration of some substance must be controlled homeostatically.

   b. Hormones such as somatotropin have prolonged, gradual, often irreversible effects, as on development.

Review Questions

1. Name the endocrine glands that occur in most vertebrates and state where each is located in the human body. *37.1*

2. Distinguish among hormones, neurotransmitters, local signaling molecules, and pheromones. *37.1*

3. A hormone molecule binds to a receptor on a cell membrane. It does not enter the cell, however. Rather, binding activates a second messenger inside the cell that triggers an amplified response to the hormonal signal. State whether the signaling molecule is a steroid hormone or a peptide hormone. *37.2*

4. Which secretions of the posterior lobe of the pituitary gland have the targets indicated? (Fill in the blanks.) *37.3*

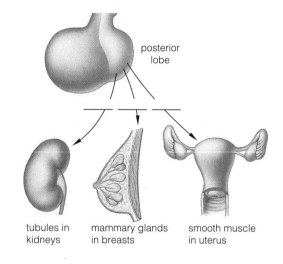

posterior lobe

tubules in kidneys   mammary glands in breasts   smooth muscle in uterus

5. Which secretions of the anterior lobe of the pituitary gland have the targets indicated? (Fill in the blanks.) *37.3*

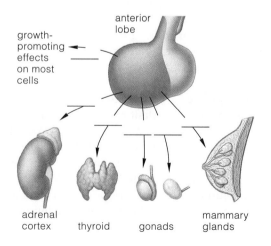

anterior lobe

growth-promoting effects on most cells

adrenal cortex   thyroid   gonads   mammary glands

Self-Quiz (Answers in Appendix IV)

1. _____ are molecules released from a signaling cell that have effects on target cells.
   a. Hormones
   b. Neurotransmitters
   c. Pheromones
   d. Local signaling molecules
   e. both a and b
   f. a through d

2. Hormones are products of _____ .
   a. endocrine glands or cells
   b. some neurons
   c. exocrine cells
   d. a and b
   e. a and c
   f. a, b, and c

3. ADH and oxytocin are hypothalamic hormones secreted from the _____ lobe of the pituitary gland.
   a. anterior
   b. posterior
   c. intermediate
   d. secondary

4. _____ has effects on body tissues in general.
   a. ADH
   b. Oxytocin
   c. Bombykol
   d. Somatotropin

5. Which do *not* stimulate hormone secretions?
   a. neural signals
   b. local chemical changes
   c. hormonal signals
   d. environmental cues
   e. All of the above can stimulate hormone secretion.

6. _____ lowers blood sugar levels; _____ raises it.
   a. Glucagon; insulin
   b. Insulin; glucagon
   c. Gastrin; insulin
   d. Gastrin; glucagon

7. The pituitary detects a rising hormone concentration in blood and inhibits the gland secreting the hormone. This is a _____ feedback loop.
   a. positive
   b. negative
   c. long-term
   d. b and c

8. Second messengers include _____ .
   a. steroid hormones
   b. protein hormones
   c. cyclic AMP
   d. both a and b

9. Match the hormone source with the closest description.
   ____ adrenal gland
   ____ thyroid gland
   ____ parathyroids
   ____ pancreatic islets
   ____ pineal gland
   ____ thymus gland
   a. affected by daylength
   b. key roles in immunity
   c. raise blood calcium level
   d. epinephrine source
   e. insulin, glucagon
   f. hormones require iodine

Critical Thinking

1. The zebra offspring being nursed in Figure 37.15 is too young to nourish itself by eating grasses. Its source of nutrients is its mother's milk. Explain how secretions from the hypothalamus and both lobes of the pituitary gland influence the production and secretion of milk.

2. In winter, with its far fewer daylight hours compared to the summer, Maxine became very depressed, craved carbohydrate-rich foods, and stopped exercising regularly. And she put on a great deal of weight. Her doctor diagnosed her condition as *seasonal affective disorder* (SAD), or the winter blues. Maxine was advised to purchase a cluster of intense, broad-spectrum lights and to sit near them at least an hour every day. The cloud of depression started to lift quickly. Use your understanding of the secretory activity of the pineal gland to explain why Maxine's symptoms appeared and why the prescribed therapy worked.

3. Marianne is affected by *type 1 insulin-dependent diabetes*. One day, after injecting herself with too much insulin, she starts to shake and feels confused. Her doctor recommends a glucagon injection. What caused her symptoms? How would an injection of glucagon help?

Figure 37.15 Female zebra nursing her offspring.

4. Through recombinant DNA technology, somatotropin (growth hormone) is now available commercially for treating pituitary dwarfism. Although it is illegal to do so, some athletes are using somatotropin instead of anabolic steroids. (Here you may wish to compare Section 38.11.) They do so because somatotropin cannot be detected by the drug test procedures employed in sports medicine. Explain how the athletes might believe this hormone can improve their performance.

5. *Osteoporosis* is a condition in which loss of calcium results in thin, brittle bones. Combined with other treatments, vitamin D_3 injections are sometimes recommended. Explain why.

Selected Key Terms

adrenal cortex *37.6*
adrenal medulla *37.6*
endocrine system *37.1*
gonad *37.6*
hormone *37.1*
hypothalamus *37.3*
inhibitor (hypothalamic) *37.3*
local signaling molecule *37.1*
molting *37.8*
negative feedback *37.6*
neurotransmitter *37.1*
pancreatic islet *37.7*
parathyroid gland *37.7*
peptide hormone *37.2*
pheromone *37.1*
pineal gland *37.8*
pituitary gland *37.3*
positive feedback *37.6*
puberty *37.8*
releaser (hypothalamic) *37.3*
second messenger *37.2*
steroid hormone *37.2*
thyroid gland *37.6*

Readings

Goodall, J. 1986. *The Chimpanzees of Gombe*. Cambridge, Massachusetts: Belknap Press of Harvard University Press.

Goodman, H. 1994. *Basic Medical Endocrinology*. Second edition. New York: Raven Press.

Hadley, M. 1995. *Endocrinology*. Fourth edition. Englewood Cliffs, New Jersey: Prentice-Hall.

Snyder, S. October 1985. "Molecular Basis of Communication Between Cells." *Scientific American* 253(4): 132–141.

Web Site See *http://www.wadsworth.com/biology* for practice quiz questions, hypercontents, BioUpdates, and critical thinking. The Wadsworth Biology Resource Center provides a wealth of information fully organized and integrated by chapter.

38

PROTECTION, SUPPORT, AND MOVEMENT

Of Men, Women, and Polar Huskies

In 1989 Will Steger and his dogsled team walked on ice for seven months, endured temperatures of −113°F, and lived through a blizzard that lasted for more than seven weeks. They crossed Antarctica—all 6,023 kilometers (3,741 miles) of it. In 1995 that legendary polar explorer set out with four men, two women, and thirty-three sled dogs to cross 3,220 kilometers of the Arctic Ocean in one season. Ice blankets this northernmost ocean in winter, but the ice becomes treacherously thin during the spring thaw. The sled dogs were with the team for two-thirds of the journey. They were flown out only when the team encountered too much melting ice and had to switch to using canoes.

To Steger's mind, the polar huskies were the heroes of the polar crossings, the members of the team that worked hardest and pulled all the weight (Figure 38.1). They are a mixed breed, the traits of which have been modified through years of artificial selection among Canadian and Greenland huskies (bred for size and strength), Siberian huskies (bred for intelligence), and Alaskan racing dogs (bred for spirit and endurance). Steger's polar huskies show a combination of these traits as well as the loyalty of cared-for pets.

A husky's leg bones are sturdy yet lightweight. Its forelegs move freely, thanks to a rib cage that is deep but not too broad. Its hind legs have massive muscles. These are not the muscles of sprinting greyhounds or cheetahs. They are the muscles of a load-pulling, long-distance runner. The husky also has tough, calloused foot pads—cushions against sharp ice and frozen rock. Like many other mammals, it has a fur coat. The coat's underhair, a dense, soft insulative layer, traps heat. Its coarser, longer, and slightly oily guard hairs protect the insulative layer from wear and tear. On winter nights, the husky settles into a comfortable position and covers its nose with its furry tail, oblivious of drifting snow.

Figure 38.1 In Ely, Minnesota, Will Steger and his polar huskies warming up for their Arctic crossing.

Steger and his teammates, Victor Boyarksy, Julie Hanson, Martin Hignell, Paul Pregont, and Takako Takano, could not even approach the polar husky's stamina and built-in protection against the elements. Long before the polar crossings, they were adhering to a regimen of diet and exercise to put their arm and leg muscles in peak condition for the extraordinary effort that lay ahead. Human legs are not adapted for load-pulling motion, but rather for long-distance walking. Also, human skin cannot withstand bitter cold. Lacking the fur coat of mammals that evolved in polar climates, the team had to depend on special clothing that could insulate and protect them from cold without restricting body movements. From this perspective, it was human ingenuity that allowed humans to keep company with the huskies, which are supremely adapted for the challenges of life on ice.

With this chapter, we turn to the three systems that together are responsible for the superficial features, shape, and movements of most kinds of animals. Figure 38.2 serves as the starting point for our consideration of the structural organization and functions of these systems. Traveling from the outside in, it depicts the integumentary system (skin and its derived structures), muscle system, and skeletal system of one of the more familiar vertebrates.

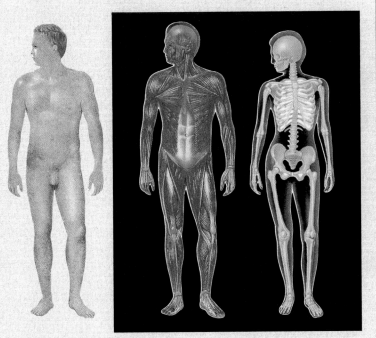

Figure 38.2 From left to right, overview of the integumentary, muscular, and skeletal systems of humans.

KEY CONCEPTS

1. Nearly all animals have an integument, some type of skeleton, and muscles. An integument is the outer covering of the animal body, and a prime example is vertebrate skin.

2. Skin protects the body from abrasion, ultraviolet radiation, bacterial attack, and other environmental insults. It also contributes to overall body functioning, as when it helps control moisture loss.

3. Three categories of skeletal systems are common in the animal kingdom. We call them hydrostatic skeletons, exoskeletons, and endoskeletons. Each has internal body fluids or structural elements, such as bones, against which a contractile force can be applied.

4. Bones are collagen-rich, mineralized organs. They function in movement, protection and support of soft organs, and mineral storage. Blood cells form in some. Ligaments or cartilage bridges joints between bones, and tendons attach them to skeletal muscles.

5. Many responses to changes in external and internal conditions involve muscles that move the animal body or parts of it. In response to stimulation, the cells of muscle tissue can contract, or shorten.

6. Smooth muscle and cardiac muscle are responsible for the motions of internal organs. Skeletal muscle helps move the body's limbs and other structural elements and maintain their spatial positions.

7. Many myofibrils, which are threadlike structures in muscle cells, are divided into sarcomeres. The sarcomere is the basic unit of contraction. Each has parallel arrays of actin and myosin filaments. ATP-driven interactions between actin and myosin shorten the sarcomeres of a muscle and collectively account for its contraction.

INTEGUMENTARY SYSTEM

Animals ranging from invertebrate worms to humans have an outer covering, or **integument** (after the Latin *integere*, meaning to cover). Most coverings are tough, pliable barriers against many environmental insults.

The integument of roundworms and insects, crabs, and other arthropods is a protective **cuticle,** hardened with chitin. Chitin, remember, is a polysaccharide that incorporates nitrogen atoms. Sections 26.8 and 26.15 describe this type of covering.

Vertebrate Skin and Its Derivatives

For vertebrates, the integument consists of a covering called **skin** as well as a variety of structures derived from epidermal cells of the skin's outer tissue layers (Figures 38.3 and 38.4). Beyond the typical assortment of epithelial tissues and glands, great variation exists within vertebrate groups as well as between them. For example, birds have the unique epidermal derivatives

hair

EPIDERMIS

DERMIS

hypodermis (subcutaneous layer)

oil gland

hair follicle

blood vessels

sensory neuron

sweat gland

smooth muscle

a

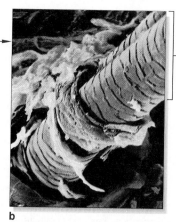

Dead, flattened epidermal cells around a shaft of hair that is projecting from the skin surface. Refer to Section 3.7, which includes a description of the molecular structure of hair.

b

c

Figure 38.3 (**a**) Structure of vertebrate skin. The uppermost portion is the epidermis; the lower portion is the dermis. (**b**) Scanning electron micrograph of a hair. (**c**) This display of skin is provided by naked mole-rats (*Heterocephalus glaber*), which live underground in burrows. Most of their hairs occur on the snout and are modified for sensory functions.

called feathers as well as bills and claws. Many herbivorous mammals have hooves and horns, porcupines have quills, you and your primate relatives have nails, and so on. As you read in Sections 27.4 and 27.5, the hagfishes have bare skin, which they coat with a lathering of slime; and epidermal cells of many fishes give rise to hardened scales of diverse thicknesses, shapes, and colors.

The skin itself has two regions: an outermost **epidermis** and an underlying **dermis.** Below this, a tissue region called the hypodermis anchors the skin to underlying structures yet still allows it to move a bit (Figure 38.3*a*). Fats that become stored in the hypodermis help insulate the body and cushion some of its parts.

Your skin weighs about four kilograms (nine pounds). Stretched out, its surface area would be fifteen to twenty square feet. For the most part, human skin is as thin as a paper towel. It thickens only on the soles of the feet and in other regions that are subjected to pounding or abrasion.

Figure 38.4 A few examples of skin and of structures that are derived from it.

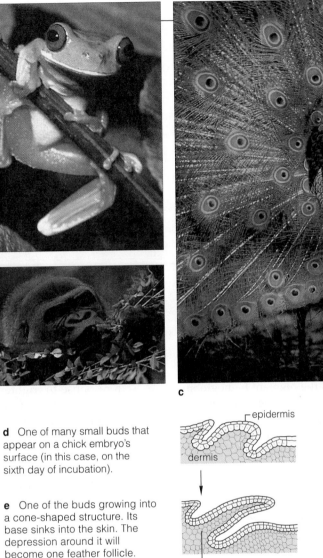

a

b

c

(**a**) Richly pigmented skin of a tree frog. Its pigment-producing cells, called chromatophores, reside primarily in the dermis. The kind deepest in the dermis produces a silvery pigment. Wavelengths reflected and scattered from such pigment molecules appear blue. Other kinds of pigment cells higher in the dermis produce a yellow pigment, which filters the blue wavelengths to produce a rich green coloration. Section 33.1 has a detailed diagram of the skin of a poisonous frog.

(**b**) Richly pigmented skin and hairs of the mountain gorilla. An abundance of melanin-producing cells in the hair follicles supply the pigments. Each hair consists of a cuticle around a central shaft. The cuticle is composed of dead, flattened cells that are packed with the protein keratin.

Very little melanin is distributed among the epidermal layers of naked mole-rats (Figure 38.3c). Blood inside vessels that thread through the dermis contributes to the skin's reddish hue.

(**c**) Peacock feathers, one of the more spectacular examples of structures derived from skin. (**d–f**) This series of diagrams shows how each feather grows from its base, at a region of actively dividing epidermal cells.

d One of many small buds that appear on a chick embryo's surface (in this case, on the sixth day of incubation).

e One of the buds growing into a cone-shaped structure. Its base sinks into the skin. The depression around it will become one feather follicle.

f A layer of horny cells at the cone's surface differentiates into a sheath. An epidermal layer just beneath the sheath will give rise to a feather. The dermis beneath this region of epidermis is richly supplied with blood vessels. It becomes the pulp, which nourishes the growing feather but does not contribute to its structure.

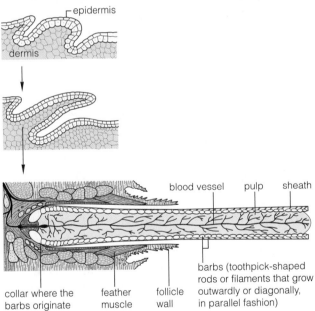

epidermis

dermis

blood vessel   pulp   sheath

barbs (toothpick-shaped rods or filaments that grow outwardly or diagonally, in parallel fashion)

collar where the barbs originate

feather muscle

follicle wall

Functions of Skin

No garment ever made comes close to the qualities of skin. What besides skin holds its shape after repeated stretchings and washings, blocks harmful rays from the sun, kills many bacteria on contact, holds in moisture, fixes small cuts and burns, *and* can last as long as you do? Skin also produces vitamin D, required for calcium metabolism. It plays a passive role in adjusting internal temperature; the nervous system can rapidly adjust the flow of blood (which transports metabolic heat) to and from skin's great numbers of tiny blood vessels. And signals from sensory receptor endings in skin help the brain assess what is going on in the outside world.

The body of most animals has an integument, a protective covering that usually is tough yet pliable. A number of invertebrate species have a cuticle; vertebrates have skin, which consists of epidermis and dermis.

Structure of the Epidermis and Dermis

Like puff pastry, epidermis consists of sheetlike layers; it is a *stratified* epithelium. Its cells are structurally and functionally knit together by an abundance of junctions of the sort described in Section 33.1. Its inner sheets are composed of many living, rapidly dividing cells. The most abundant are **keratinocytes**, which produce the tough, water-insoluble protein keratin. One reason why skin is such a strong, cohesive integument is that the adhesion junctions between keratinocytes are anchored to numerous, crosslinked keratin fibers inside them.

Other cells, the **melanocytes**, produce and donate the brownish-black pigment melanin to keratinocytes. Melanin screens out harmful ultraviolet radiation from the sun. Humans generally have the same number of melanocytes, but skin color varies owing to differences in the distribution and metabolic activity of these cells. For example, melanocytes in albinos cannot produce all of the enzymes required for melanin production. Pale skin contains little melanin, so the pigment hemoglobin inside red blood cells is not masked. The skin appears pink because hemoglobin's red color shows through thin-walled blood vessels and the epidermis itself, both of which are transparent. Carotene, which is a yellow-orange pigment, also contributes to skin color.

Section 38.3 describes two less common cell types in epidermis, the Langerhans and Granstein cells.

The rapid, ongoing mitotic divisions push epidermal cells from deeper layers toward the skin's free surface. Because of pressure from the continually growing cell mass and from normal wear and tear at the surface, older cells are dead and flattened by the time they reach the outer layers (Figure 38.5). There, they are abraded off or flake away on an ongoing basis. Besides replacing the outermost, keratinized layer, the rapid cell divisions also help skin mend quickly after cuts or burns.

Beneath the epidermis is the dermis. This is mostly dense connective tissue with many elastin fibers (which resist daily stretching) and collagen fibers (which impart strength). Blood vessels, lymph vessels, and receptor endings of sensory nerves thread through the dermis. Nutrients from the bloodstream reach living epidermal cells by diffusing through the dermal ground tissue.

Sweat Glands, Oil Glands, and Hairs

Human skin typically has sweat glands, oil glands, and husklike cavities, or follicles, for hairs. These reside mostly in the dermis but epidermal cells give rise to them.

Fluid secreted by **sweat glands** is 99 percent water, with dissolved salts, traces of ammonia, vitamin C, and other substances. You have 2.5 million sweat glands, controlled by sympathetic nerves. One type abounds in

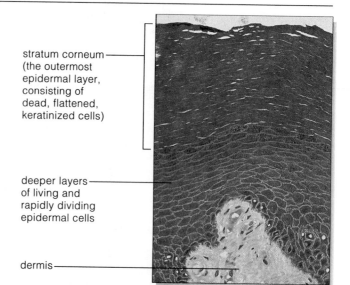

stratum corneum (the outermost epidermal layer, consisting of dead, flattened, keratinized cells)

deeper layers of living and rapidly dividing epidermal cells

dermis

Figure 38.5 Micrograph of a section through human skin.

your palms, soles, forehead, and armpits. They have roles in controlling body temperature and in *cold sweats*, a response to frightening, unsettling situations (Section 35.3). Secretions from a different type of sweat gland increase during stress, pain, and sexual foreplay and prior to menstruation.

Except on the palms of hands and the soles of feet, skin contains **oil glands** (also called sebaceous glands). Oil glands lubricate and soften hair and the skin, and their secretions kill many potentially harmful surface bacteria. *Acne* is an inflammation of the skin that occurs after bacteria have successfully infected oil gland ducts.

Each **hair**, a flexible structure of mostly keratinized cells, has a root embedded in skin and a shaft above its surface (Figure 38.3). Cells divide near the root's base, are pushed upward, then flatten and die. Flattened cells of the shaft's outer layer overlap like roof shingles. When mechanically abused, these frizz out as "split ends." An average human scalp has about 100,000 hairs, although genes, nutrition, and hormones influence hair growth and density. Protein deficiency causes hair to thin (amino acids are required for keratin synthesis). So do high fever, emotional stress, and excess vitamin A intake. When the body produces abnormal amounts of testosterone, *hirsutism*, or excessive hairiness, may be one result. This hormone influences patterns of hair growth and other secondary sexual traits.

The skin's multiple layers of keratinized, melanin-shielded epidermal cells help the body conserve water, avoid damage by ultraviolet radiation, and resist mechanical stress.

Sweat glands, oil glands, hairs, and other structures derived from epidermal cells are embedded largely in the dermis. Blood vessels, lymph vessels, and the receptor endings of sensory neurons also reside in the dermis.

38.3 SUNLIGHT AND SKIN

THE VITAMIN D CONNECTION Even when you do little more than sit outside in the sun, you are giving some of the epidermal cells in your skin the opportunity to make vitamin D, or cholecalciferol. This steroid-like compound helps the body absorb calcium from food. When exposed to sunlight, some type of cell in the skin produces it from a precursor molecule that is related to cholesterol. The cells then release vitamin D to the bloodstream, which transports it to absorptive cells in the intestinal lining. This is a hormone-like action, which means the skin acts like an endocrine gland when exposed to sunlight.

Humans, remember, evolved beneath the intense sun of the African savanna, so skin alone would have provided our early ancestors with enough vitamin D. When humans started moving out of tropical environments and, later, into caves, animal skins, and layers of clothing, they also started to depend more on dietary sources of the essential vitamin D, as described in Section 42.9.

SUNTANS AND SHOE-LEATHER SKIN Do you like to tan your body by rotating beneath the sun's rays or a tanning lamp, like a chicken in an oven broiler? If so, think about what a broiler does to the chicken. The sun's ultraviolet wavelengths stimulate melanin production in skin cells. Continued exposure increases melanin concentrations in light skin and visibly darkens it, thereby producing the "tan" that so many people covet (Figure 38.6). Tanning does protect the body against ultraviolet radiation. Even in naturally dark skin, however, prolonged exposure to sunlight causes the elastin fibers in connective tissue of the dermis to clump together, so skin loses its resiliency. In time it starts to look like old shoe leather.

In this respect, tanning accelerates the *aging* of skin. As any person grows older, epidermal cells divide less often.

Skin gets thinner and more susceptible to injury. Glandular secretions that once kept it soft and moistened dwindle. Collagen and elastin fibers in the dermis break down and become sparser, so skin loses elasticity and its wrinkles deepen. By themselves or in combination with excessive tanning, prolonged exposure to dry wind and tobacco smoke also accelerates the aging process.

SUNLIGHT AND THE FRONT LINE OF DEFENSE Besides the cells that produce melanin and keratin, skin contains two other types of cells, which help defend the surface of the body against invasion by pathogenic cells and protect it against cancer. These components of the epidermis are called Langerhans cells and Granstein cells.

Langerhans cells are phagocytes that develop in bone marrow, then they take up stations in skin. After they engulf virus particles or bacterial cells, they pepper the surface of their plasma membrane with molecular alarm signals that mobilize the body's immune system.

Ultraviolet radiation can damage Langerhans cells. This might be why sunburns can trigger *cold sores*, the small, painful blisters that announce the recurrence of a *Herpes simplex* infection. Nearly everyone harbors the *H. simplex* virus. It remains hidden in the face, inside a ganglion (a cluster of neuron cell bodies). Sunburns and other stress factors can activate the virus. Virus particles move down the neurons to their endings in skin. There they infect epithelial cells and cause skin eruptions.

When ultraviolet radiation damages Langerhans cells, it weakens one of the body's first lines of defense against invasions. It also can activate proto-oncogenes and trigger cancerous transformation of skin cells. As described in Section 13.4, skin cancers grow rapidly and can spread to adjacent lymph nodes unless they are surgically removed.

In some as-yet-undetermined way, **Granstein cells** apparently interact with the white blood cells that can put the brakes on immune responses in the skin. By issuing suppressor signals, they help keep the responses from spiraling out of control. Although the functions of Granstein cells are not completely understood, these cells are known to be less vulnerable than the Langerhans cells to the damaging effects of ultraviolet radiation.

Figure 38.6 Demonstration of how shoe-leather skin forms.

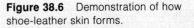

TYPES OF SKELETONS

Operating Principles for Skeletons

Many of the responses an animal makes to external and internal conditions involve the movement of the whole body or parts of it. The activation, contraction, and relaxation of muscle cells bring about the movements. But muscle cells alone cannot produce them. *All muscles require the presence of some medium or structural element against which the force of contraction can be applied.* A skeletal system fulfills this requirement.

Three types of skeletons predominate in the animal world. With a **hydrostatic skeleton**, the muscles work against an internal body fluid and redistribute it within a confined space. Like a filled waterbed, the confined fluid resists compression. By contrast, an **exoskeleton** has rigid, *external* body parts, such as shells, that can receive the applied force of muscle contraction. An **endoskeleton** has rigid, *internal* body parts, such as bones, that receive the applied force of contraction.

Examples From the Invertebrates

Many invertebrates with a soft body have a hydrostatic skeleton. Reflect on the sea anemone, with its soft, vase-shaped body and saclike gut (Figure 38.7). Its body wall incorporates longitudinal and radial muscles. Between meals, when the longitudinal muscles are contracted (shortened) and radial ones are relaxed (lengthened), a sea anemone looks short and squat. When the body lengthens into an upright feeding position, its radial muscles are contracting (and forcing some fluid out of the gut cavity), and its longitudinal ones are relaxing.

Or reflect on the earthworm. This annelid, recall, has a series of coelomic compartments, each with its own muscles, nerves, and bristles. Its body moves forward by contracting and relaxing its fluid-filled segments one after another (Section 26.14). By coordinating contractions on one side or the other of the segments, the worm also thrashes sideways and moves forward and back.

As another example, jumping spiders have a hinged exoskeleton, as other arthropods do. In addition, they use body fluids to transmit force when

a Resting position   **b** Feeding position

Figure 38.7 Outcomes of contractile force applied against the sea anemone's hydrostatic skeleton. This invertebrate's body wall has contractile cells running longitudinally (parallel with the body axis) and radially around the gut cavity. (**a**) In this case, radial cells are relaxed and longitudinal ones are contracted. Anemones typically are in this resting position at low tide, when currents cannot bring food morsels to them. (**b**) Radial cells are contracted, longitudinal ones are relaxed, and the body is extended to its upright feeding position.

they leap at prey. Muscle contractions cause blood in body tissues to surge into the hind leg spines. It is like giving a water-filled rubber glove a quick squeeze to make the glove's skinny fingers rigidly erect. Figure 38.8 shows this resourceful use of hydraulic pressure (*hydraulic* means fluid pressure inside tubes).

The hinged arthropod exoskeleton has advantages. Some of its hard parts can be moved like levers by sets of muscles attached to them. Thus small contractions can bring about large movements of wings or some other body parts. This is especially true of the cuticle of a winged insect. It extends over all body segments and over gaps between segments (Figure 38.9). At these gaps, the cuticle remains pliable. It acts like a hinge when muscles alternately raise and lower either the wing or the body parts to which the wings are attached.

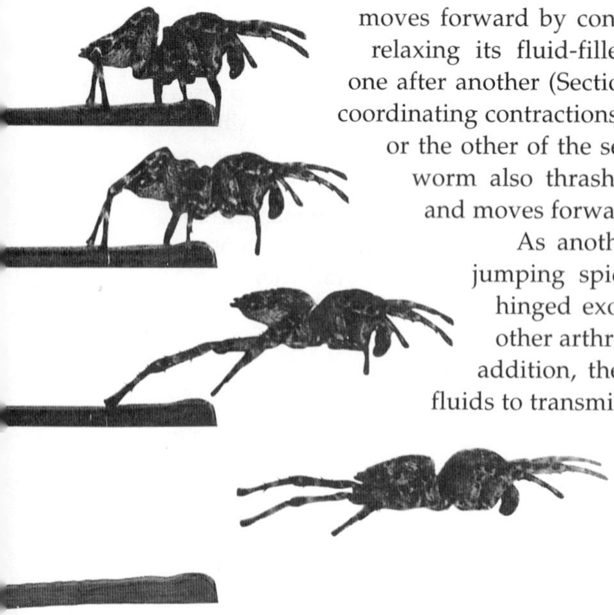

Figure 38.8 Time-lapse photographs of the leap of the jumping spider *Sitticus pubescens*. Its leaping depends partly on the hydraulic extension of its hind legs when blood surges into them under high pressure.

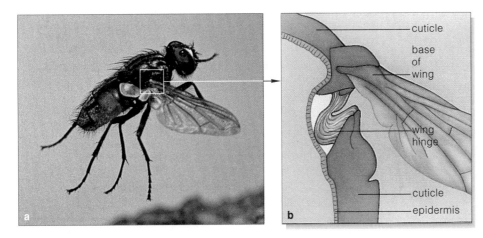

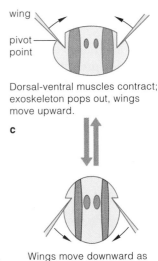

cuticle

base
of
wing

wing
hinge

cuticle

epidermis

wing

pivot
point

Dorsal-ventral muscles contract;
exoskeleton pops out, wings
move upward.

c

Wings move downward as
exoskeleton pops back and
swiftly stretch the muscles.

d

Figure 38.9   Housefly wing movement, an outcome of the contraction of sets of muscles that extend from dorsal to ventral regions of the exoskeleton. The contractile force works against the exoskeleton, near hinge points where wings are attached. When the muscles contract, the exoskeleton pops inward and wings move up. When the exoskeleton pops outward by elastic force, the wings move down. The quick-stretched muscles invite a fast repeat of the events.

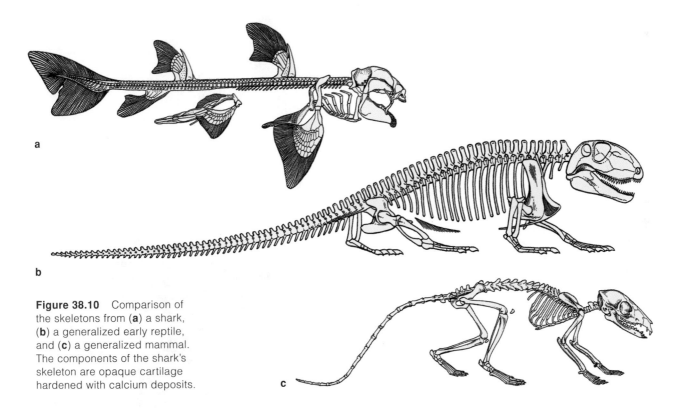

Figure 38.10   Comparison of the skeletons from (**a**) a shark, (**b**) a generalized early reptile, and (**c**) a generalized mammal. The components of the shark's skeleton are opaque cartilage hardened with calcium deposits.

Examples From the Vertebrates

Vertebrates have endoskeletons of one sort or another. For example, sharks have a skeleton of an opaque form of cartilage, hardened with calcium deposits (Figure 38.10*a*). Some other fishes have a flexible skeleton of an elastic, translucent form of cartilage that almost looks like glass. For most other vertebrates, however, the endoskeleton consists mainly of bone (Figure 38.10*b,c*).

We turn next to the functions and the characteristics of bones. Afterward, we will consider how different types of bones are arranged in the human skeletal system.

Animal skeletons have structural elements or body fluids against which the force of contraction can be applied.

Functions of Bone

By definition, **bones** are complex organs that function in movement, protection, support, mineral storage, and formation of blood cells, as listed in Table 38.1. Bones that interact with skeletal muscles maintain or change the positions of body parts. Bones also support and anchor muscles. Certain bones form hard compartments that enclose and protect the brain, the lungs, and other internal organs. Bones are reservoirs for mineral ions, the deposits and withdrawals of which help maintain body fluids and support metabolic activities. Only some bones are   sites of blood cell formation.

Table 38.1 Functions of Bone
1. *Movement.* Bones interact with skeletal muscles to maintain or change the position of body parts.
2. *Support.* Bones support and anchor muscles.
3. *Protection.* Many bones form hard compartments that enclose and protect soft internal organs.
4. *Mineral storage.* Bones are a reservoir for mineral ions, the deposits and withdrawals of which help maintain ion concentrations in body fluids.
5. *Blood cell formation.* Some bones contain regions where blood cells are produced.

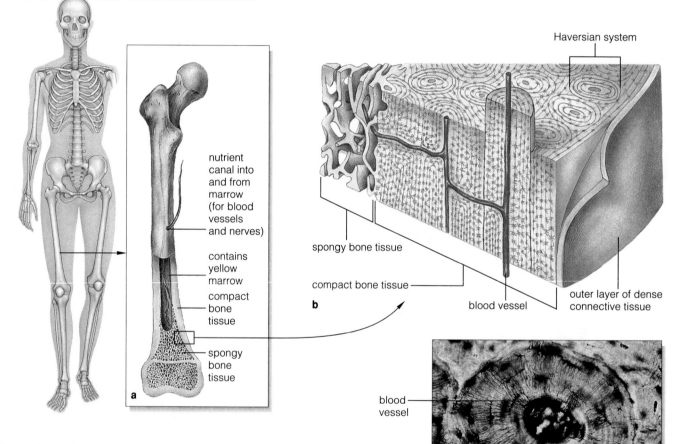

nutrient canal into and from marrow (for blood vessels and nerves)

contains yellow marrow

compact bone tissue

spongy bone tissue

a

b

Haversian system

spongy bone tissue

compact bone tissue

blood vessel

outer layer of dense connective tissue

blood vessel

space occupied by living bone cell

c

75 μm

Figure 38.11 (a) Structure of a femur, one of the long bones of mammals. Femurs also are called thighbones. (b) Appearance of spongy bone tissue and compact bone tissue in a femur. The thin, dense layers of compact bone tissue form cylindrical arrays around interconnecting canals, which contain blood vessels and nerves. Each array is called a Haversian system. (c) Micrograph of one Haversian system. The blood vessel at its center services osteocytes, living bone cells in small spaces in bone tissue. Small tunnels connect neighboring spaces.

Bone Structure

Human bones range in size from tiny middle earbones to the clublike femurs, or thighbones. Bones are long, short (or cubelike), flat, and irregular. All of them have connective, epithelial, and bone tissues. The bone tissues are calcium hardened, with living cells and collagen fibers in a ground substance.

Consider the thighbone in Figure 38.11. *Compact* bone tissue in its shaft and at its ends resists mechanical shock. This type of bone tissue is deposited as many thin, dense, cylindrical layers around interconnected canals that contain blood vessels and nerves. Each cylindrical array is called a Haversian system. The blood

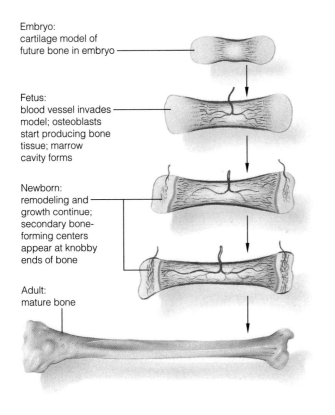

Embryo:
cartilage model of
future bone in embryo

Fetus:
blood vessel invades
model; osteoblasts
start producing bone
tissue; marrow
cavity forms

Newborn:
remodeling and
growth continue;
secondary bone-
forming centers
appear at knobby
ends of bone

Adult:
mature bone

Figure 38.12 Long bone formation, starting with osteoblast activity in a cartilage model (here, already formed in the embryo). Bone-forming cells are active first in the shaft region, then at the knobby ends. In time, the only cartilage left is at the ends.

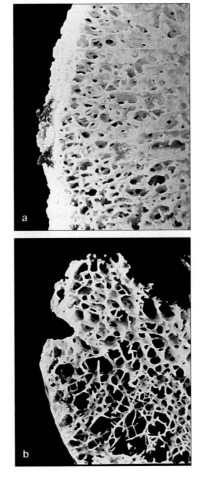

a

b

Figure 38.13 An example of bone tissue affected by osteoporosis.

(**a**) Section through normal bone tissue. In such tissue, the mineral deposits continually replace the withdrawals.

(**b**) After the onset of osteoporosis, replacements of mineral ions lag behind withdrawals. In time the tissue erodes, and bones become hollow and brittle.

vessels and nerves inside it service living bone cells. *Spongy* bone tissue in the bone ends and shaft imparts strength without adding much weight. Its abundant spaces make the tissue appear spongy, but its flattened parts are firm. **Red marrow**, a major site of blood cell formation, fills the spaces in some bones, such as the breastbone. The cavities in most mature bones contain **yellow marrow**. Yellow marrow consists largely of fat. It converts to red marrow and produces new red blood cells when blood loss from the body is severe.

HOW BONES DEVELOP The cartilage models for many bones form in the animal embryo (Figure 38.12). Bone-forming cells, or osteoblasts, secrete material inside the model's shaft and onto its surface. A marrow cavity opens up as cartilage breaks down inside the model. In time, osteoblasts become surrounded by their secretions and thereafter are called **osteocytes**, or living bone cells. Their metabolic activities maintain mature bones.

BONE TISSUE TURNOVER Minerals are continually deposited *and* withdrawn from bone tissue in ways that maintain the adult body's calcium levels. In this **bone tissue turnover**, bone cells secrete enzymes that digest bone tissue. The released ions of calcium and other minerals enter interstitial fluid, then the blood, which distributes them to metabolically active cells.

For instance, a thighbone becomes much thicker and stronger as its bone cells deposit minerals at the shaft's surface. At the same time, it becomes less heavy as other bone cells destroy bone tissue within its shaft.

Exercising stimulates calcium deposition and tends to increase bone density. Stress or injury triggers its withdrawal. Also, as a person ages, the backbone, hip bones, and other bones decrease in mass, especially in women. Figure 38.13 shows the effect of *osteoporosis*. A weakened backbone may collapse, curve abnormally, and lower the rib cage, which puts stress on internal organs. Decreasing osteoblast activity, calcium loss, sex hormone deficiencies, excessive intake of protein, and decreased physical activity contribute to the disorder.

Bones are collagen-rich, mineralized organs that function in movement, protection, support, storage of calcium and other minerals, and blood cell formation.

Appendicular and Axial Portions

The human skeleton has 206 bones, which anatomists have subdivided into appendicular and axial portions. Its *appendicular* portion has pectoral girdles (at the shoulders), pelvic girdle (at the hips), and paired arms, hands, legs, and feet. Its pectoral girdles have slender collarbones, and flat shoulder blades. Fall on an outstretched arm and you might dislocate a shoulder or fracture a collarbone, which is flimsily arranged and the bone most frequently broken.

The human skeleton's *axial* portion includes skull bones, twelve pairs of ribs, and the breastbone. It also has twenty-six **vertebrae** (singular, vertebra). These are bony segments of the vertebral column, or backbone. The curved backbone extends from the skull's base to the pelvic girdle. There the backbone transmits the weight of the torso to the lower limbs. A spinal cord threads through a series of bony canals at the rear of the column. Between are **intervertebral disks**—cartilaginous shock absorbers and flex points that permit movement.

Sometimes a severe or rapid shock forces a disk to slip out of place or rupture. Such painful, *herniated disks* are a less-than-advantageous outcome of bipedalism. Remember, the primate ancestors of the human lineage were quadrupedal. The first hominids started to walk upright about 4 million years ago, and a pronounced S-shaped curve in the backbone was a result. Today, the older we get, the longer we have been fighting gravity in a compromised way, and the more back pain we suffer.

Skeletal Joints

Joints are areas of contact or near-contact between bones, and each has a distinctive bridge of connective tissue. Very short connecting fibers join bones at *fibrous* joints. Straps of cartilage join them at *cartilaginous* joints. Long straps of dense connective tissue, or **ligaments,** bridge the gap between the bones at *synovial* joints.

Fibrous joints hold teeth in their sockets. They also connect the flat skull bones of a fetus. At childbirth, the loose connections allow the bones to slide over each other a bit and thereby prevent skull fractures. The skull of a newborn still has fibrous joints as well as membranous areas known as "soft spots," or fontanels. In childhood, the fibrous tissue hardens and skull bones are fused into a single unit.

Table 38.2   Components of the Human Skeleton

APPENDICULAR PORTION:
   Pectoral girdles: clavicle (collarbone) and
               scapula (shoulder blade)
   Arm bones: humerus, radius, ulna
   Hand bones: carpals, metacarpals, phalanges (of fingers)
   Pelvic girdle (six fused bones at the hip)
   Leg bones: femur (thighbone), patella, tibia, fibula
   Foot bones: tarsals, metatarsals, phalanges (of toes)

AXIAL PORTION:
   Skull: cranial bones and facial bones
   Rib cage: sternum (breastbone) and ribs (12 pairs)
   Vertebral column:vertebrae (26) and intervertebral disks

Cartilaginous joints bridging vertebrae, ribs, and the breastbone permit slight movements. Synovial joints, such as knee joints, move freely. Ligaments stabilize the knee joints, as in Figure 38.14. Where one bone touches another, cartilage cushions them and absorbs shocks. A flexible capsule of dense connective tissue surrounds the region of contact. Cells of a membrane that lines the capsule's interior secrete a fluid that lubricates the joint.

Like many other joints, the knee joint is vulnerable to stress. This is the joint that lets you swing, bend, and turn the long bones below it. When you run, it absorbs the force of your weight each time the foot below it hits the ground. Stretch or twist a knee joint suddenly and too far, and you may *strain* it. Tear its ligaments or tendons and you *sprain* it. Move the wrong way and you may well dislocate the attached bones. During collision sports, such as football, blows to the knees frequently sever a ligament. The severed portion must be reattached surgically before ten days pass. Why? Phagocytes in a lubricating fluid within the joint normally clean up after everyday wear and tear. When presented with torn ligaments, they will indiscriminately turn the tissue to mush.

Joint inflammation as well as degenerative disorders are collectively called "arthritis." In *osteoarthritis*, cartilage at the knees and other freely movable joints wears off as a person ages. Joints in the fingers, knees, hips, and the vertebral column are affected most often. In *rheumatoid arthritis*, synovial membranes in joints become inflamed and thickened, cartilage degenerates, and bone deposits

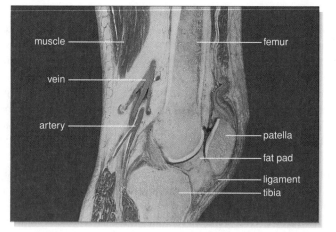

muscle   femur
vein
artery   patella
         fat pad
         ligament
         tibia

Figure 38.14   Human knee joint, longitudinal section.

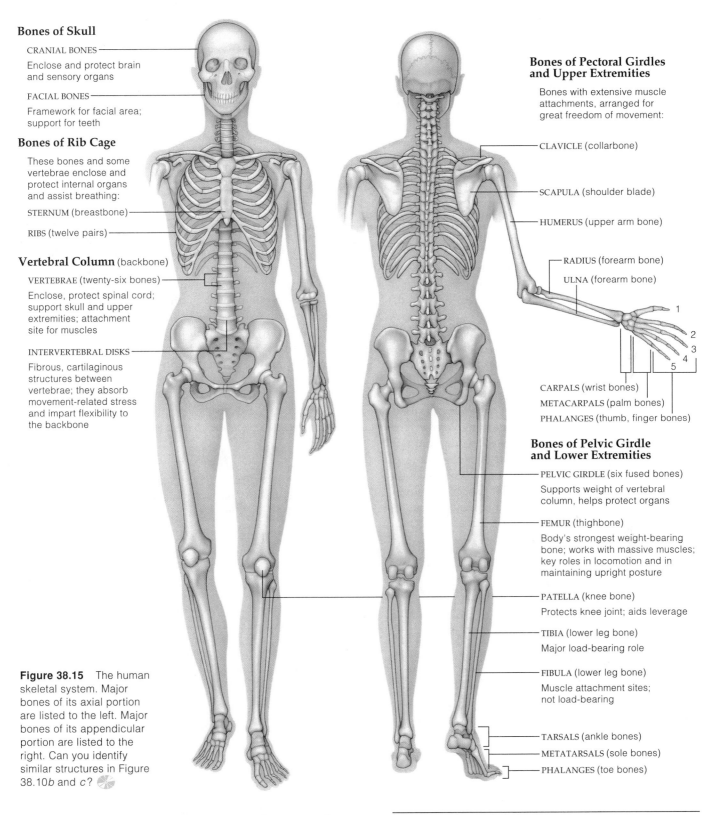

Figure 38.15 The human skeletal system. Major bones of its axial portion are listed to the left. Major bones of its appendicular portion are listed to the right. Can you identify similar structures in Figure 38.10b and c?

build up. This degenerative disorder may be triggered by a bacterial or viral infection, but it also appears to have a genetic component. It can begin at any age, but symptoms usually emerge before age fifty.

A human skeleton has an axial portion (a backbone, skull bones, and rib cage) and an appendicular portion (pelvic girdle, pectoral girdles, and arm, hand, leg, and foot bones).

How Muscles and Bones Interact

Skeletal muscles are the functional partners of bones. Each skeletal muscle contains bundles of hundreds to many thousands of muscle cells, which look like long, striped fibers. In muscle tissue, remember, muscle cells contract (shorten) in response to adequate stimulation. They lengthen in response to gravity and other loads. When you dance, breathe, scribble notes, or tilt your head, contracting muscle cells are helping to move your body or change the positions of some of its parts.

Connective tissue bundles muscle cells together and extends beyond them to form **tendons**. Each tendon, a cord or strap of dense connective tissue, attaches some muscle to bone (Figure 38.16). Most of the attachment sites are like a car's gearshift. *They form a lever system, in which a rigid rod is attached to a fixed point but able to move about at it.* Muscles connect to bones (rigid rods) near a joint (fixed point). As the muscles contract, they transmit force to bones and make them move.

Skeletal muscles interact with one another as well as with bones. Some are arranged in pairs or groups that work together to promote the same movement. Others work in opposition, with the action of one opposing or reversing the action of another. Figure 38.17a shows how opposing muscle groups work to move frog legs. Also look at Figure 38.17b, then extend your right arm forward. Now place your left hand over the biceps in the upper arm and slowly "bend your elbow." Feel the biceps contract? Even when your biceps contracts only a bit, it causes a large movement in the forearm bone that is connected to it. This is true of most leverlike arrangements.

Tendons can rub against bones, but sheaths help reduce the resulting friction.

Figure 38.16 Tendon sheath. This is not the same as a bursa, a small sac also filled with synovial fluid. Bursae are cushions interposed *between* bone and skin or bone and tendons.

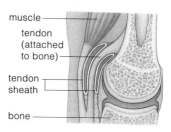

Tendons slide within the sheaths, which are fluid-filled membranous sacs wrapped around them (Figure 38.16). Your knees, wrists, and finger joints have such sheaths.

Bear in mind, only *skeletal* muscle is the functional partner of bone. As mentioned earlier, smooth muscle is mainly a component of the walls of internal organs, such as the stomach (Section 33.3). Cardiac muscle is present only in the wall of the heart. We will consider the structure and functioning of smooth muscle and cardiac muscle in later chapters in this unit.

3 Now the first muscle group in the frog's upper hindlimb contracts again, drawing the limb back toward the body.

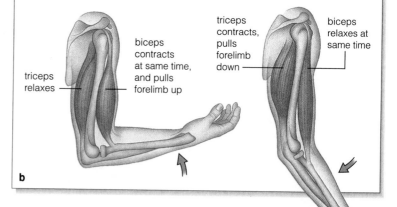

2 An opposing muscle group attached to the same limb contracts forcefully and pulls the limb back. The contractile force, directed against the ground, propels the frog forward.

1 A muscle group attached to each upper hindlimb contracts and pulls the leg forward a bit.

triceps relaxes

biceps contracts at same time, and pulls forelimb up

triceps contracts, pulls forelimb down

biceps relaxes at same time

Figure 38.17 (a) A frog demonstrating how a small decrease in the length of contracting muscles can produce a large movement. Its leap depends on opposing muscle groups attached to the upper limb bone of each hind leg. One muscle group pulls the limb forward and toward the body's midline. Another pulls it back and away from the body. (b) Two opposing muscle groups in a human arm. When a triceps contracts, the forearm extends straight out. When the triceps relaxes and its opposing partner (biceps) contracts, the elbow joint flexes and the forearm bends upward.

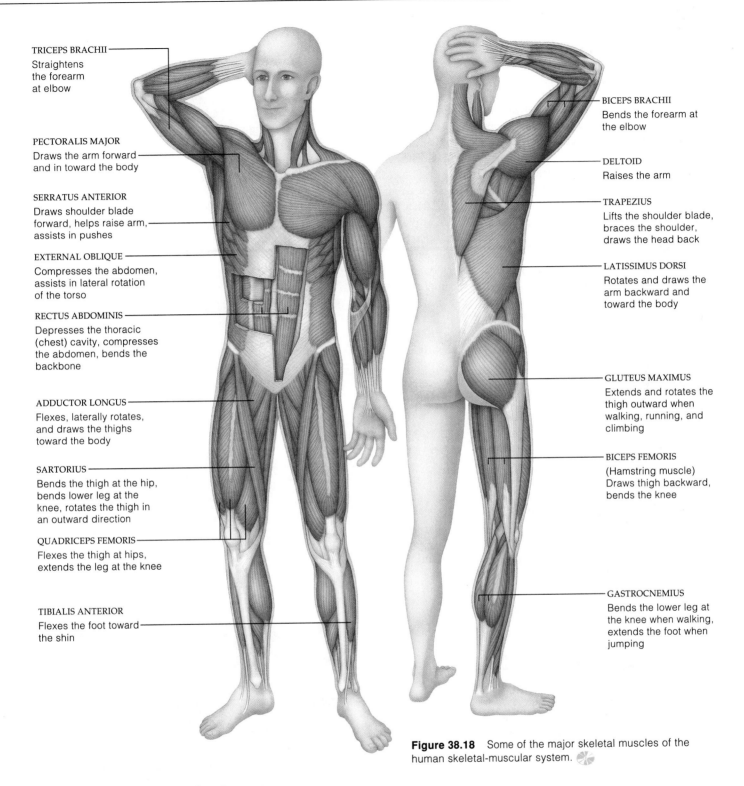

TRICEPS BRACHII
Straightens the forearm at elbow

PECTORALIS MAJOR
Draws the arm forward and in toward the body

SERRATUS ANTERIOR
Draws shoulder blade forward, helps raise arm, assists in pushes

EXTERNAL OBLIQUE
Compresses the abdomen, assists in lateral rotation of the torso

RECTUS ABDOMINIS
Depresses the thoracic (chest) cavity, compresses the abdomen, bends the backbone

ADDUCTOR LONGUS
Flexes, laterally rotates, and draws the thighs toward the body

SARTORIUS
Bends the thigh at the hip, bends lower leg at the knee, rotates the thigh in an outward direction

QUADRICEPS FEMORIS
Flexes the thigh at hips, extends the leg at the knee

TIBIALIS ANTERIOR
Flexes the foot toward the shin

BICEPS BRACHII
Bends the forearm at the elbow

DELTOID
Raises the arm

TRAPEZIUS
Lifts the shoulder blade, braces the shoulder, draws the head back

LATISSIMUS DORSI
Rotates and draws the arm backward and toward the body

GLUTEUS MAXIMUS
Extends and rotates the thigh outward when walking, running, and climbing

BICEPS FEMORIS
(Hamstring muscle) Draws thigh backward, bends the knee

GASTROCNEMIUS
Bends the lower leg at the knee when walking, extends the foot when jumping

Figure 38.18 Some of the major skeletal muscles of the human skeletal-muscular system.

Human Skeletal-Muscular System

The human body has more than 600 skeletal muscles, some superficial, others deep in the body wall. Some, such as facial muscles, attach to the skin. The trunk has muscles of the thorax, backbone, abdominal wall, and pelvic cavity. Other groups of muscles attach to upper and lower limb bones. Figure 38.18 shows a few of the main skeletal muscles and lists their functions. We turn next to the mechanisms underlying their contraction.

Skeletal muscles transmit contractile force to bones and make them move. Tendons strap skeletal muscles to bones.

Functional Organization of Skeletal Muscle

Bones move—they are pulled in some direction—when the skeletal muscles that are attached to them shorten. When a skeletal muscle shortens, its component muscle cells are shortening. When a muscle cell shortens, many units of contraction within that cell are shortening. The basic units of contraction are called **sarcomeres**.

Figure 38.19 gives you an idea of how bundles of cells in a skeletal muscle run parallel with the muscle itself. Inside each muscle cell are myofibrils, threadlike structures all bundled together in parallel array. Each myofibril is functionally divided into many sarcomeres, arranged one after another along its length. Dark bands called Z lines define the two ends of every sarcomere.

As Figure 38.19c shows, a sarcomere contains many filaments, oriented parallel with its long axis. Certain differences in their length and positioning give rise to the striped appearance of skeletal muscle (and cardiac muscle). Some of the filaments are thin; others are thick. Each *thin* filament is like two strands of pearls, twisted together. The "pearls" are molecules of **actin**, a globular protein with contractile functions:

— one actin molecule

portion of one thin filament

Other proteins (coded *green*) are near actin's surface grooves. Each *thick* filament is made of molecules of **myosin**, another protein with contractile functions, in parallel array. A myosin molecule has a long tail and a double head that projects from the filament's surface:

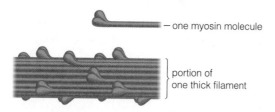

— one myosin molecule

portion of one thick filament

Thus the myofibrils, muscle cells, and muscle bundles of a skeletal muscle all run in the same direction. What function does this consistent, parallel orientation serve? It focuses the force of muscle contraction onto a bone in a particular direction.

Sliding-Filament Model of Contraction

How do sarcomeres shorten and bring about contraction of a skeletal muscle? The answer lies with sliding and pulling interactions among the sarcomere's filaments. A set of actin filaments extends from each Z line partway

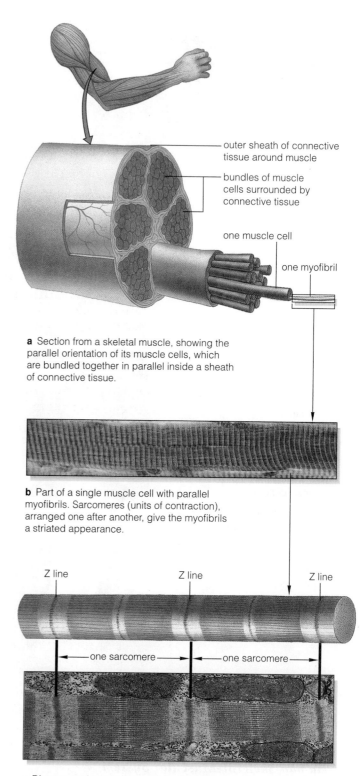

— outer sheath of connective tissue around muscle

— bundles of muscle cells surrounded by connective tissue

— one muscle cell

— one myofibril

a Section from a skeletal muscle, showing the parallel orientation of its muscle cells, which are bundled together in parallel inside a sheath of connective tissue.

b Part of a single muscle cell with parallel myofibrils. Sarcomeres (units of contraction), arranged one after another, give the myofibrils a striated appearance.

Z line   Z line   Z line

one sarcomere   one sarcomere

c Diagram and transmission electron micrograph of two of the sarcomeres from a myofibril. This closer view shows how dark "bands," called Z lines, define the two ends of each sarcomere. Mitochondria (the oval-shaped organelles) adjacent to the sarcomeres provide ATP energy for muscle action.

Figure 38.19   Components of a skeletal muscle.

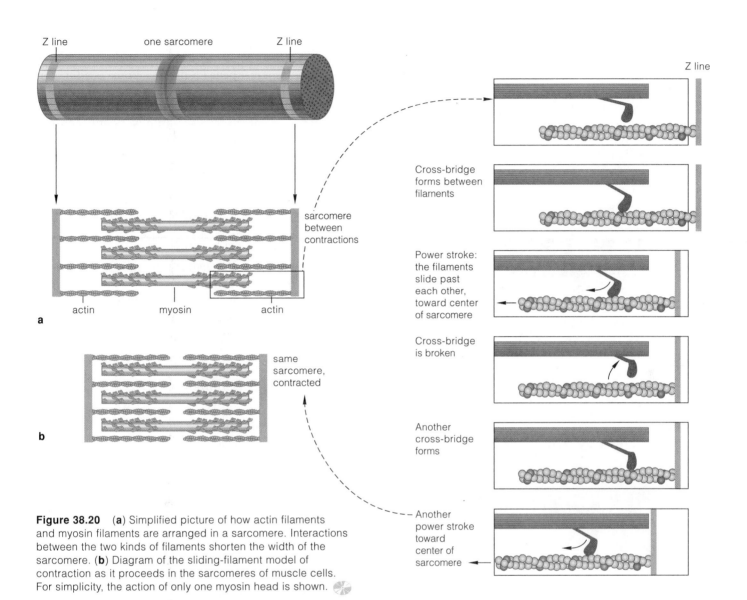

Figure 38.20 (**a**) Simplified picture of how actin filaments and myosin filaments are arranged in a sarcomere. Interactions between the two kinds of filaments shorten the width of the sarcomere. (**b**) Diagram of the sliding-filament model of contraction as it proceeds in the sarcomeres of muscle cells. For simplicity, the action of only one myosin head is shown.

to the center portion of each sarcomere. A set of myosin filaments partially overlaps the sets of actin, but it does not extend all the way to the Z lines. As Figure 38.20 indicates, when a muscle is contracting, these myosin filaments are physically sliding along and pulling each of the two sets of actin filaments toward the center of the sarcomere. The sarcomere shortens as a result. The physical interaction of actin and myosin is a key premise of the **sliding-filament model** of muscle contraction.

The myosin and actin filaments interact by way of **cross-bridge formation**. As indicated in Figure 38.20, these particular cross-bridges are attachments between a myosin head and a binding site on actin. The myosin heads are activated when messages from the nervous system stimulate the muscle cell. They physically attach to an adjacent actin filament and tilt in a short power stroke, driven by ATP energy, toward the sarcomere's

center. During the power stroke, the heads pull the actin filament along with them. Another energy input makes the heads let go, attach to another region of the filament, tilt in another power stroke, and so on down the line. A single contraction of a sarcomere requires a whole series of power strokes.

A skeletal muscle shortens through combined decreases in the length of its numerous sarcomeres. Sarcomeres are the basic units of contraction.

The parallel orientation of a muscle's component parts directs the force of contraction toward a bone that must be pulled in some direction.

By energy-driven interactions between the myosin and actin filaments, the many sarcomeres of a muscle cell shorten and collectively account for its contraction.

The Control Pathway

When skeletal muscles contract, they help move the body and its assorted parts at certain times, in certain ways—and they do so in response to commands from the nervous system. The commands, which reach the muscles by way of motor neurons, can stimulate or inhibit contraction of the sarcomeres in muscle cells.

Like all cells, a muscle cell shows a difference in electric charge across its plasma membrane. That is, the cytoplasm just beneath the membrane is a bit more negative than interstitial fluid outside it. But only in muscle cells, neurons, and other *excitable* cells does the difference in charge reverse abruptly, briefly, and in a predictable way in response to adequate stimulation.

The abrupt reversal in charge, an **action potential**, occurs as charged ions flow across the membrane in an accelerating way. The electrical commotion, remember, spreads without diminishing along the membrane, away from the point of stimulation (Section 34.1).

Suppose action potentials arise in a muscle cell. They spread rapidly away from the stimulation point, then along the small, tubelike extensions of the plasma membrane shown in Figure 38.21. The tubes connect with a system of membranous chambers that thread lacily around the muscle cell's myofibrils. That system, called the **sarcoplasmic reticulum**, takes up, stores, and releases calcium ions in controlled ways.

The Control Mechanism

The arrival of action potentials causes an outward flow of calcium ions from the sarcoplasmic reticulum. The released ions diffuse into the myofibrils and reach actin

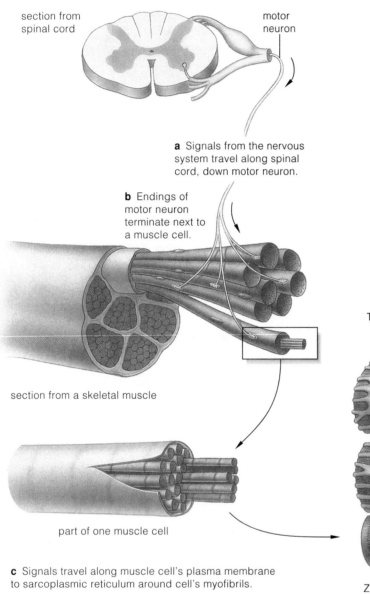

section from spinal cord

motor neuron

a Signals from the nervous system travel along spinal cord, down motor neuron.

b Endings of motor neuron terminate next to a muscle cell.

section from a skeletal muscle

part of one muscle cell

c Signals travel along muscle cell's plasma membrane to sarcoplasmic reticulum around cell's myofibrils.

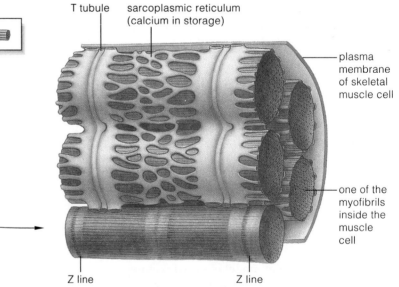

T tubule   sarcoplasmic reticulum (calcium in storage)

plasma membrane of skeletal muscle cell

one of the myofibrils inside the muscle cell

Z line   Z line

d Signals trigger the release of calcium ions from sarcoplasmic reticulum threading among the myofibrils. The arrival of calcium allows actin and myosin filaments in the myofibrils to interact and bring about contraction.

Figure 38.21   Pathway for signals from the nervous system that stimulate or inhibit contraction of skeletal muscle. The plasma membrane of each muscle cell surrounds the cell's myofibrils and connects with inward-threading T tubules. These membranous tubes are close to the sarcoplasmic reticulum, a calcium-storing system that functions in the control of contraction.

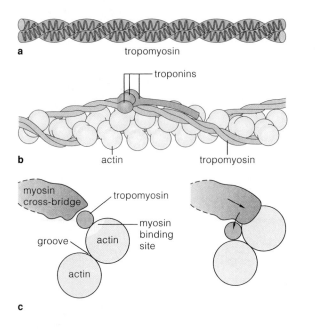

a   tropomyosin

troponins

b   actin   tropomyosin

myosin cross-bridge   tropomyosin

myosin binding site

groove   actin

actin

c

Figure 38.22 Arrangement of troponins, tropomyosin, and actin filaments in skeletal muscle cells. When calcium binds with a troponin, tropomyosin moves away from the actin and thereby exposes the cross-bridge binding sites.

Figure 38.23
Three of the metabolic routes by which ATP forms in muscle cells in response to the demands of physical exercise.

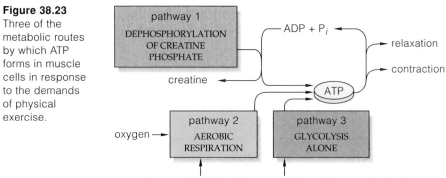

pathway 1

DEPHOSPHORYLATION OF CREATINE PHOSPHATE

ADP + P$_i$

relaxation

contraction

creatine

ATP

oxygen →

pathway 2
AEROBIC RESPIRATION

pathway 3
GLYCOLYSIS ALONE

glucose from bloodstream and from glycogen breakdown

filaments. Before this happened, the muscle was at rest (it was not contracting). Its actin binding sites were blocked and myosin could not form cross-bridges with them. However, the arrival of calcium ions clears the binding sites, so that contraction can proceed. After the contraction, the calcium ions are actively transported back into the membrane storage system.

What blocks cross-bridge binding sites in a muscle at rest? Two proteins, tropomyosin and troponin, are located in or near the surface grooves of actin filaments (Figure 38.22). At a low calcium level, the proteins are joined so tightly together that the tropomyosin is forced slightly outside the groove. In this position, it blocks the cross-bridge binding site. When the calcium level

rises, however, calcium binds with the troponin and thereby alters its shape. When troponin is in its altered shape, it has a different molecular grip on tropomyosin, which is now free to move into the groove and expose the binding site (Figure 38.22c).

Sources of Energy for Contraction

All cells require ATP, but only in muscle cells does the demand skyrocket in so short a time. When a muscle cell at rest is called upon to contract, phosphate donations from ATP must proceed twenty to a hundred times faster. But the cell has only a small supply of ATP at the start of contractile activity. At such times, it forms ATP by a very fast reaction. An enzyme simply transfers phosphate from **creatine phosphate**, an organic compound, to ADP. The cell has about five times as much creatine phosphate as ATP, so this reaction is good for a few contractions. And it is enough to buy time for a relatively slower ATP-forming pathway to kick in (Figure 38.23).

During prolonged, moderate exercise, the oxygen-requiring reactions of aerobic respiration typically can provide most of the ATP required for contraction. For the first five to ten minutes, a muscle cell taps its store of glycogen for glucose, the starting substrate. For the next half hour or so of sustained activity, the muscle cell depends on glucose and fatty acid deliveries from the blood. For contractile activity longer than this, fatty acids are the main fuel source (Section 8.6).

What happens when exercise is so intensive that it exceeds the capacity of the respiratory system and circulatory system to deliver oxygen for the aerobic pathway? At such times, glycolysis alone will contribute more of the total ATP that is being produced. Remember, by this set of anaerobic reactions, a glucose molecule is only partly broken down, so the net ATP yield is small. But muscle cells can use this metabolic route as long as glycogen stores continue to provide glucose.

After intense exercise, deep, rapid breathing helps repay the body's **oxygen debt**, incurred when ATP use by muscles exceeded the aerobic pathway's deliveries.

It takes commands from the nervous system to initiate action potentials in muscle cells. The action potentials are signals for cross-bridge formation, hence for contraction.

During exercise, the availability of ATP inside muscle cells affects whether contraction will proceed, and for how long.

PROPERTIES OF WHOLE MUSCLES

Muscle Tension and Muscle Fatigue

Whether a muscle actually shortens during cross-bridge formation depends on the external forces acting upon it. Collectively, the cross-bridges exert **muscle tension**. By definition, this is a mechanical force that a contracting muscle exerts on an object, such as a bone. Opposing it is a load, either the weight of an object or gravity's pull on the muscle. Only when muscle tension exceeds the load does a stimulated muscle shorten.

An *isometrically* contracting muscle develops tension but does not shorten. It supports a load in a constant position, as when you hold a glass of lemonade in front of you. An *isotonically* contracting muscle shortens and moves a load. With *lengthening* contraction, though, an external load is greater than the muscle tension, so the muscle lengthens during the period of contraction. This happens to leg muscles when you walk down stairs.

A muscle's tension relates to the formation of cross-bridges in its cells and the number of cells recruited into action. Consider a **motor unit**: a motor neuron and all muscle cells that form junctions with its endings. By stimulating the motor unit with an electrical impulse, we can induce an action potential and make a recording of an isometric contraction. It takes a few milliseconds for tension to increase, then it peaks and declines. This response is a **muscle twitch** (Figure 38.24a). Its duration depends on the load and cell type. For example, fast-acting muscle cells rely on glycolysis (not efficient but fast) and use up ATP faster than slow-acting cells do.

If another stimulus is applied before the response is over, the muscle twitches again. **Tetanus** is a large contraction resulting from the repeated stimulation of a motor unit, so that twitches mechanically run together. (In a disease by the same name, toxins interfere with muscle relaxation.) Figure 38.24d shows a recording of tetanic contraction.

Continuous, high-frequency stimulation that keeps a muscle in a state of tetanic contraction leads to *muscle fatigue,* or a decline in tension. After a few minutes of rest, a fatigued muscle will contract again in response to stimulation. The extent of recovery depends largely on how long and how frequently it was stimulated before. Muscles associated with brief, intense exercise (such as weightlifting) fatigue fast but recover fast. The muscles associated with prolonged, moderate exercise fatigue slowly but take longer to recover, often up to twenty-four hours. The molecular mechanisms causing muscle fatigue are unknown, but glycogen depletion is a factor.

Effects of Exercise and Aging

A muscle's properties depend on how often, how long, and how intensely it is put to use. With regular **exercise**

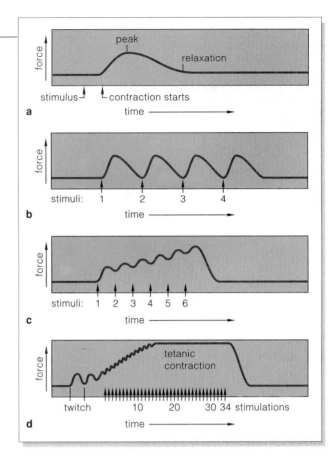

Figure 38.24 Recordings of twitches in muscles artificially stimulated in different ways. (**a**) A single twitch. (**b**) Two stimulations per second cause a series of twitches. (**c**) Six per second cause a summation of twitches, and (**d**) about twenty per second cause tetanic contraction.

(that is, increased levels of contractile activity), muscle cells do not increase in number. But they increase in size and metabolic activity, and they become more resistant to fatigue. Consider *aerobic exercise*, which is not intense but is long in duration. Aerobic exercise increases the number of mitochondria in both fast and slow muscle cells, and it increases the blood capillaries that service them. Such physiological changes improve endurance. By contrast, *strength training* (intense, short-duration exercise such as weightlifting) affects fast-acting muscle cells. These form more myofibrils and more enzymes of glycolysis. Strong, bulging muscles of the sort shown in Figure 38.25 may result, although they don't have much endurance. They fatigue rapidly.

Muscle tension decreases in adult humans thirty to forty years old. These people may exercise just as long and intensely as younger ones, but their muscles cannot adapt (change) in response to the same extent. Even so, some adaptation can be beneficial. Aerobic exercise can improve blood circulation. And, as it turns out, even modest strength training slows the loss of muscle tissue that is an inevitable part of the aging process.

Properties of muscles vary with age and levels of activity.

38.11 PEBBLES, FEATHERS, AND MUSCLE MANIA

A male penguin has a bit of a problem when scouting for a female penguin, because superficially they both look too much alike. The only way he can find one is to drop a pebble at the feet of a likely prospect. If those feet belong to a male, the pebble may be perceived as an insult, and it may start a fight. If female, his stony overture to courtship might be ignored or fancied, depending on her receptivity at that moment.

Peacocks and the males of many other species have no such problem. For them, sexual dimorphism is visible and quite pronounced (Figure 38.4c). Maybe the peacock's eye-stopping feathers evolved through sexual selection, with peahens serving as the deciders of their reproductive success. Charles Darwin certainly thought so. He viewed cases of extreme sexual dimorphism as the outcome of male competition for females and of females choosing among males.

What about extreme sexual dimorphism in body size, which involves increases in muscle mass? This may be one measure of how much mammalian males invest in fighting capacity. Possibly the cost of securing and using resources for growth and maintenance of a massive body is offset by great reproductive rewards.

All of this might make you wonder: Exactly what are the "rewards" for astoundingly muscled human athletes (Figure 38.25)? Is it a misdirected will to win, or a sign of modern athletic competition? Consider this: Each year in the United States alone, about a million athletes use anabolic steroids. And the vast majority of professional football players have used them to increase their "brute power." Even adolescent boys use them as a way to gain the winning edge in wrestling, football, and weightlifting tournaments.

Anabolic steroids are synthetic hormones. They mimic testosterone, a sex hormone that (among other things) governs secondary sexual traits. Testosterone makes boys get a deeper voice; more hair on their face, underarms, and pubic skin; and greater muscle mass in the arms, legs, shoulders, and elsewhere. Besides this, testosterone stimulates heightened aggressive behavior, which is often associated with maleness.

Anabolic steroids stimulate the synthesis of protein molecules, including proteins in muscle cells. Supposedly, they induce very rapid gains in muscle mass and muscle strength when taken during weight-training and exercise programs. This claim is disputed; results of most studies are based on too few subjects. Even so, not everyone believes that these drugs do enough damage to outweigh the edge they presumably give in athletic competition— or the wealth and hero status they accord the "winners."

Yet steroid-using athletes do suffer minor and major side effects. In men, acne, baldness, shrinking testes, and infertility are early signs of toxicity. The symptoms begin when a high level of anabolic steroids in blood triggers a sharp decline in the body's production of testosterone. Anabolic steroids also may trigger early heart disease. Even brief or occasional use may damage the kidneys or set the stage for cancer of the liver, testes, and prostate gland. Among women, anabolic steroids deepen the voice, and they produce pronounced facial hair. Menstrual cycles become irregular. Breasts may shrink, and the clitoris may become grossly enlarged.

Not every steroid user develops severe physical side effects. Far more common are mental difficulties, called 'roid rage or body-builder's psychosis. In such cases, users become irritable and increasingly aggressive. Some men become wildly aggressive, uncontrollably manic, and delusional. For example, one steroid user accelerated his car to high speed and deliberately drove it into a tree; and doesn't that make you wonder just how superior some of us are to a pebble-toting penguin.

Figure 38.25 A human male with pumped-up biceps.

SUMMARY

1. Most animals have an integumentary system, which covers the body's surface. Examples include the cuticle of roundworms and arthropods, as well as vertebrate skin and the structures derived from it.

2. Skin protects against abrasion, ultraviolet radiation, dehydration, and many pathogenic bacteria. It also helps control internal temperature (blood flow to the skin can dissipate heat). Sensory receptors in skin detect stimuli in the external environment. When exposed to sunlight, skin serves an endocrine function; it produces vitamin D, a hormone-like substance required for absorption of calcium from food.

3. Skin consists of two regions: an outermost epidermis and an underlying dermis. The most abundant cells are keratinocytes (keratin producers). Also present are the melanocytes (melanin producers), and Langerhans cells and Granstein cells (which help defend the body against pathogens and cancer cells).

    a. Epidermis consists primarily of multiple layers of dead, keratinized, and melanin-shielded epithelial cells.

    b. The dermis is where rapid cell divisions produce replacements for cells shed on an ongoing basis. Hair, oil glands, sweat glands, and other structures derived from epidermal cells are embedded mainly in the dermis.

4. Movement of the animal body or parts of it requires contractile cells and some medium or structure against which contractile force can be applied.

    a. In hydrostatic skeletons, as in sea anemones, body fluids accept the contractile force and are redistributed inside a confined space.

    b. In exoskeletons, as in insects and other arthropods, rigid external body parts accept the contractile force.

    c. In endoskeletons, rigid internal body parts (mainly bones) receive the applied force of contraction.

5. Bones are complex organs that have osteocytes (living bone cells) in a mineralized, collagen-containing ground substance. Bones function in the movement, protection, and support of body parts; in mineral storage; and, in bones with red and yellow marrow, blood cell formation.

6. The human skeleton is divided into two portions.

    a. The appendicular region includes the pelvic girdle, pectoral girdles (collarbone and shoulderblade), arm and hand bones, and leg and foot bones.

    b. The axial region includes the skull bones, rib cage, and vertebral column, or backbone. The backbone is composed of vertebrae (bony segments) cushioned by cartilaginous, intervertebral disks.

7. Skeletal joints are areas of contact or near-contact where connective tissue bridges adjacent bones. Fibrous joints (short fibers), cartilaginous joints (of cartilage), or synovial joints (with straplike ligaments) bridge the gap between bones at different joints.

8. Cells of smooth, cardiac, and skeletal muscle contract (shorten) in response to adequate stimulation.

9. Tendons attach skeletal muscles to bones. The skeletal muscles and bones interact as a system of levers, with rigid rods (bones) moving at fixed points (joints). Many muscles work together or in opposition to bring about movement or positional changes in body parts.

10. Inside each skeletal muscle cell are many myofibrils arranged parallel with its long axis. These threadlike structures contain actin and myosin filaments organized in parallel. Each myofibril is functionally divided into sarcomeres, the basic units of contraction. The parallel orientation of the muscle's components directs the force of contraction at a bone to be pulled in some direction.

11. In response to stimulation, skeletal muscles shorten by decreases in the length of all of their sarcomeres. The nervous system stimulates muscle cells. Its commands are delivered by motor neuron endings that terminate on muscle cells, and they can trigger action potentials at the muscle cell's plasma membrane.

12. Here are the main points of the sliding-filament model of muscle contraction:

    a. Action potentials cause the release of calcium ions from a membrane system (sarcoplasmic reticulum) that threads around the cell's myofibrils. Calcium diffuses inside sarcomeres, binds to actin filaments, and makes the binding sites change shape. Then myosin filament heads can form cross-bridges with actin filaments.

    b. Each cross-bridge is a *brief* attachment between a myosin head and an actin binding site. Cross-bridges form during repeated, ATP-driven power strokes. The repeated strokes make actin filaments slide past myosin filaments, and collectively they shorten the sarcomere.

13. Muscle cells obtain the ATP required for contraction by three metabolic pathways:

    a. Dephosphorylation of creatine phosphate. This is direct, fast, and good for a few seconds of contraction.

    b. Aerobic respiration. This pathway predominates during prolonged, moderate exercise.

    c. Glycolysis. This pathway takes over when intense exercise exceeds the body's capacity to deliver oxygen to muscle cells.

14. Muscle tension refers to a mechanical force created by cross-bridge formation. The load (gravity or weight of objects) is an opposing force. A stimulated muscle shortens when tension exceeds the load and lengthens when it is less than the load. Levels of exercise and aging affect the properties of muscles.

    a. A motor unit is a motor neuron and all muscle cells that form junctions with its endings.

    b. A muscle twitch is a brief, weak contraction in response to a single action potential at a motor unit. Tetanus is a large contraction resulting from repeated stimulation at a motor unit.

Review Questions

1. List the functions of skin, then distinguish between its regions. Is the hypodermis part of skin? *38.1*

2. Name four cell types in skin and their functions. *38.2, 38.3*

3. Distinguish between:
   a. sweat gland and oil gland *38.2*
   b. hydrostatic skeleton, exoskeleton, and endoskeleton *38.4*
   c. red marrow and yellow marrow *38.5*
   d. ligament and tendon *38.6, 38.7*

4. What are the functions of bones? *38.5*

5. Name the three types of muscle, then state the function of each and where they are located in the body. *38.7*

6. Look at Figures 38.19 and 38.20. Then, on your own, sketch and label the fine structure of a muscle, down to one of its individual myofibrils. Identify the basic unit of contraction in the myofibrils. *38.8*

7. What role does calcium play in the control of contraction? What role does ATP play, and by what routes does it form? *38.9*

Self-Quiz *(Answers in Appendix IV)*

1. Nearly all animals have a(n) _____ system that protects the body from abrasion, ultraviolet radiation, bacterial attack, and other environmental stresses.

2. _____ and _____ systems work together to move the body and specific body parts.

3. The three categories of muscle tissue are _____ , _____ , and _____ .

4. Which is *not* a function of skin?
   a. resist abrasion   c. initiate movement
   b. restrict dehydration   d. help control temperature

5. _____ are shock pads and flex points.
   a. Vertebrae   c. Marrow cavities
   b. Femurs   d. Intervertebral disks

6. Blood cells form in _____ .
   a. red marrow   c. certain bones only
   b. all bones   d. a and c

7. In a skeletal muscle cell, the _____ is the basic unit of contraction.
   a. myofibril   c. muscle fiber
   b. sarcomere   d. myosin filament

8. Muscle contraction requires _____ .
   a. calcium ions   c. action potential arrival
   b. ATP   d. all of the above

9. ATP for muscle contraction can be formed by _____ .
   a. aerobic respiration   d. a and b only
   b. glycolysis   e. a and c only
   c. creatine phosphate breakdown   f. a, b, and c

10. Match the M words with their defining feature.
   ____ muscle   a. actin's partner
   ____ muscle twitch   b. all in the hands
   ____ muscle tension   c. blood cell production
   ____ melanin   d. decline in tension
   ____ myosin   e. brownish-black pigment
   ____ marrow   f. motor unit response
   ____ metacarpals   g. force exerted by cross-bridges
   ____ myofibrils   h. muscle cells bundled in
   ____ muscle fatigue   connective tissue
                          i. threadlike parts in muscle cell

Critical Thinking

1. For Rizza and other young women, the recommended daily allowance (RDA) of calcium is 800 milligrams per day. During Rizza's pregnancy, the RDA is 1,200 milligrams per day. Why would a pregnant woman need the larger amount? What might happen to her bones without it?

2. Kate found a malnourished, stray cat that had just given birth to a litter of four kittens. The cat was trying to hunt, but her muscles were twitching so badly she could scarcely walk. Kate recalled what she had learned about muscle contraction. She offered milk as well as solid food to the cat—which, after several days, regained muscle control and was able to care for her kittens. How might milk help restore muscle function?

3. Compared to most people, Joe and other long-distance runners have a greater number of muscle fibers with more mitochondria. Sprinters have a greater number of muscle fibers that have more of the enzymes necessary for glycolysis but fewer mitochondria. Think about how these two forms of exercise differ and explain why the muscle fibers differ between the two kinds of runners.

Selected Key Terms

actin *38.8*	ligament *38.6*
action potential *38.9*	melanocyte *38.2*
anabolic steroid *38.11*	motor unit *38.10*
bone *38.5*	muscle tension *38.10*
bone tissue turnover *38.5*	muscle twitch *38.10*
creatine phosphate *38.9*	myosin *38.8*
cross-bridge formation *38.8*	oil gland *38.2*
cuticle *38.1*	osteocyte *38.5*
dermis *38.1*	oxygen debt *38.9*
endoskeleton *38.4*	red marrow *38.5*
epidermis *38.1*	sarcomere *38.8*
exercise *38.10*	sarcoplasmic reticulum *38.9*
exoskeleton *38.4*	skeletal muscle *38.7*
Granstein cell *38.3*	skin *38.1*
hair *38.2*	sliding-filament model *38.8*
hydrostatic skeleton *38.4*	sweat gland *38.2*
integument *38.1*	tendon *38.7*
intervertebral disk *38.6*	tetanus *38.10*
joint *38.6*	vertebra (vertebrae) *38.6*
keratinocyte *38.2*	yellow marrow *38.5*
Langerhans cell *38.3*	

Readings

Brusca, R., and G. Brusca. 1990. *Invertebrates*. Sunderland, Massachusetts: Sinauer.

Huxley, H. E. December 1965. "The Mechanism of Muscular Contraction." *Scientific American* 213(6): 18–27. Old article, great illustrations.

Sherwood, L. 1997. *Human Physiology*. Third edition. Belmont, California: Wadsworth.

Weeks, O. December 1989. "Vertebrate Skeletal Muscle: Power Source for Locomotion." *BioScience* 39(11): 791–798.

Web Site See *http://www.wadsworth.com/biology* for practice quiz questions, hypercontents, BioUpdates, and critical thinking. The Wadsworth Biology Resource Center provides a wealth of information fully organized and integrated by chapter.

39 CIRCULATION

Heartworks

For Dr. Augustus Waller, Jimmie the bulldog was no ordinary pooch. Connected to wires and soaked to his ankles in buckets of salty water, Jimmie was a four-footed explorer of the workings of the heart (Figure 39.1*a*). Press your fingers to your chest a few inches left of the center, between the fifth and sixth ribs, and feel the repetitive thumpings of your heart. The same rhythms intrigued Waller and other nineteenth-century physiologists. They wondered: Does each beat of the heart produce a pattern of electrical currents? Could they find out by devising a painless way to record such currents at the body surface?

That's where Jimmie and the buckets of salty water came in. Saltwater happens to be an efficient conductor of electricity. In Waller's experiment, it picked up faint signals from Jimmie's beating heart through the skin of his legs and conducted them to a crude monitoring device. With that device, Waller made one of the first recordings of heart activity (Figure 39.1). Today we call such recordings ECGs, an abbreviation for **electrocardiograms**.

A graph of the normal electrical activity of your own heart would look much the same. The pattern emerged a few weeks after you started growing, by way of mitotic cell divisions, from a single fertilized egg inside your mother. Early on, some of the embryonic cells differentiated into cardiac muscle cells, and these started to contract spontaneously. One small patch of the cells took the lead, and it has functioned as your heart's pacemaker ever since.

If all goes well, that patch of cardiac muscle cells will continue to contract as it should until the day you die. It is a natural pacemaker; it sets the baseline rate at which blood is pumped out of the heart, through blood vessels, then back to the heart. The rate is moderate, about seventy beats every minute; but commands from the nervous

BUCKET

JIMMIE   DR. WALLER

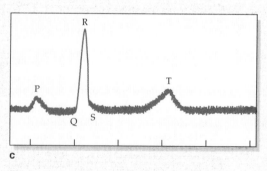

c

Figure 39.1 A bit of history in the making.
(**a**) Jimmie the bulldog, taking part in a painless experiment. (**b**) Augustus Waller and his beloved pet bulldog sharing a quiet moment in Waller's study after the experiment, which yielded one of the world's first electrocardiograms (**c**).

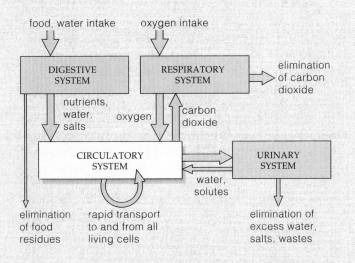

food, water intake   oxygen intake

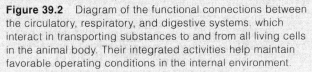

Figure 39.2 Diagram of the functional connections between the circulatory, respiratory, and digestive systems, which interact in transporting substances to and from all living cells in the animal body. Their integrated activities help maintain favorable operating conditions in the internal environment.

system and endocrine system continually adjust it. When you jog, for example, your skeletal muscle cells demand much more blood-borne oxygen and glucose than they do when you sleep. At such times, your heart starts pounding more than twice as fast, and this helps deliver sufficient blood to them.

We have come a long way from Waller and Jimmie in our monitorings of the heart. Sensors can now detect the faint signals characteristic of an impending heart attack. Internists now use computers to analyze a patient's beating heart, and they use ultrasound probes to build images of it on a video screen. Cardiologists routinely substitute battery-powered pacemakers for malfunctioning natural ones.

With this chapter, we turn to the circulatory system, the means by which substances move rapidly to and from the interstitial fluid that bathes living cells in nearly all animals. The circulatory system continually accepts the oxygen, nutrients, and other substances that the animal secures, by way of respiratory and digestive systems, from the outside environment (Figure 39.2). Simultaneously it picks up carbon dioxide and other wastes from cells and delivers them to the respiratory and urinary systems for disposal. Its smooth operation is absolutely central to maintaining operating conditions in the internal environment within a tolerable range— a state we call homeostasis. 🔯

KEY CONCEPTS

1. All cells survive by exchanging substances with their surroundings. In most animals, substances rapidly move to and from cells by way of a closed circulatory system.

2. Blood, a fluid connective tissue, is the transport medium of circulatory systems. It transports oxygen, carbon dioxide, plasma proteins, vitamins, hormones, lipids, and other solutes. It also transports metabolically generated heat.

3. In birds and mammals, a four-chambered, muscular heart pumps blood through two separate circuits of blood vessels, both of which lead back to the heart.

4. In the pulmonary circuit, the heart pumps oxygen-poor blood to the lungs, where it picks up oxygen; then blood flows back to the heart. In the systemic circuit, the oxygenated blood is pumped from the heart to all body regions, where it gives up oxygen and picks up carbon dioxide before flowing back to the heart.

5. Arteries are large-diameter, low-resistance vessels that rapidly transport oxygenated blood from the heart. Veins are large-diameter, low-resistance vessels that transport oxygen-poor blood to the heart and serve as blood volume reservoirs.

6. Arterioles are sites where the flow volume through each organ is controlled. In response to signals, their diameters widen in some regions and narrow in others. Their coordinated responses divert more of the flow volume to organs that are most active at a given time.

7. Numerous small-diameter, thin-walled capillaries interconnect in capillary beds, which are diffusion zones. As a volume of blood spreads out through a bed, it slows owing to the total cross-sectional area of the capillaries, which is greater than that of arterioles. The slowdown allows time for exchanges with interstitial fluid.

General Characteristics

Imagine that an earthquake has closed off the highways around your neighborhood. Grocery trucks can't enter and waste-disposal trucks can't leave, so food supplies dwindle and garbage piles up. Cells would face similar predicaments if your body's highways were disrupted. The highways are part of a **circulatory system**, which functions in the rapid internal transport of substances to and from cells. The system helps maintain favorable neighborhood conditions, so to speak, and this is a vital task. All of your differentiated cells perform specialized tasks and cannot fend for themselves. Different types interact in coordinated ways to maintain the volume, composition, and temperature of **interstitial fluid**, the tissue fluid that bathes them. A circulating connective tissue—**blood**—interacts with interstitial fluid. Blood makes continual deliveries and pickups that help keep conditions tolerable for enzymes and other molecules that carry out cell activities. Together, interstitial fluid and blood are the body's "internal environment."

Blood flows in blood vessels, or tubes that differ in wall thickness and in diameter. A muscular pump, the **heart**, generates pressure that keeps blood flowing. Like many animals, you have a *closed* circulatory system, in which blood is confined within continuously connected walls of the heart and blood vessels. This is not true of *open* circulatory systems of arthropods, such as insects, and most mollusks. In open systems, blood flows out through the vessels and into sinuses, which are small spaces in body tissues. There, blood mingles with tissue fluids, then moves back to the heart through openings in the blood vessels or the heart wall (Figure 39.3a).

Think about the overall "design" of a closed system. Because the heart pumps incessantly, the *volume* of blood flowing through blood vessels has to equal the heart's output at any given time. The flow's *velocity* (speed) is highest in large-diameter transport vessels. It decreases in specific body regions, where there must be enough time for blood to exchange substances with cells. The required slowdown proceeds at **capillary beds**, where blood spreads out through many small-diameter blood vessels called **capillaries**. During any interval, the same volume of blood is moving forward through the beds as elsewhere in the system. But it is doing so at a more leisurely pace, owing to the great total cross-sectional area of the blood capillaries. The simple analogy shown in Figure 39.3e may help you grasp this concept.

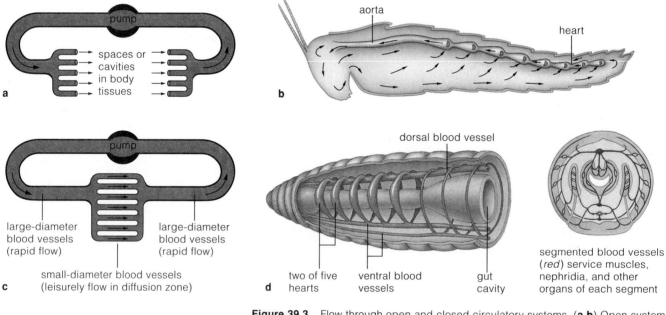

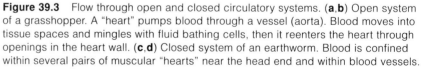

Figure 39.3   Flow through open and closed circulatory systems. (**a,b**) Open system of a grasshopper. A "heart" pumps blood through a vessel (aorta). Blood moves into tissue spaces and mingles with fluid bathing cells, then it reenters the heart through openings in the heart wall. (**c,d**) Closed system of an earthworm. Blood is confined within several pairs of muscular "hearts" near the head end and within blood vessels.

(**e**) Relation between flow velocity and total cross-sectional area in a closed circulatory system. Visualize two fast rivers flowing into and out of a lake. The flow *rate* is the same in all three places; an identical volume of water moves from point *1* to point *3* during the same interval. Flow *velocity* decreases in the lake, for the volume spreads through a larger cross-sectional area and moves forward a shorter distance during the same time.

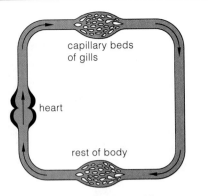 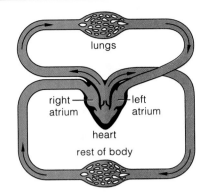 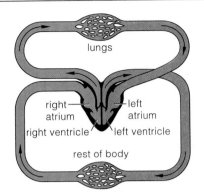

a In fishes, a two-chambered heart (atrium, ventricle) pumps blood in one circuit. Blood picks up oxygen in gills, delivers it to rest of body. Oxygen-poor blood flows back to heart.

b In amphibians, a heart pumps blood through two partially separate circuits. Blood flows to lungs, picks up oxygen, returns to heart. It mixes with oxygen-poor blood still in heart, flows to rest of body, returns to heart.

c In birds and mammals, the heart is fully partitioned into two halves. Blood circulates in two circuits: from the heart's right half to lungs and back, then from the heart's left half to oxygen-requiring tissues and back.

Figure 39.4 Comparison of the closed circulatory systems of vertebrate groups.

Evolution of Vertebrate Circulatory Systems

Humans and other existing vertebrates have a closed circulatory system, although the pump and plumbing of fishes, amphibians, birds, and mammals differ in their details. The differences evolved over hundreds of millions of years and corresponded to the move of some vertebrate lineages onto land. Recall, from Section 27.3, that fishes, the first vertebrates, had gills. Gills, like all respiratory structures, have a thin, moist surface that oxygen and carbon dioxide can diffuse across. Later, in the ancestors of land vertebrates, lungs evolved that supplemented gas exchange. Being *internally moistened* sacs, lungs had advantages for the move onto dry land. Also advantageous were concurrent modifications in circulatory systems, which pick up oxygen from lungs and deliver carbon dioxide wastes to them.

Consider this: In fishes, blood flows in *one* circuit (Figure 39.4*a*). Pressure generated by a two-chambered heart forces it through capillary beds of gills, the largest artery, capillary beds of organs, then back to the heart. The vast gill capillaries offer so much resistance to flow that the pressure drops considerably before the main artery. The blood delivery is fine for the activity level of most fishes. But it would not be sufficient for the more active life-styles of most land vertebrates.

When the amphibians evolved, their heart became partitioned into right and left halves—only partly so, but enough to pump blood through *two* partially separated circuits (Figure 39.4*b*). The separation of flow continued in crocodilians, which are reptiles. It became complete in birds and mammals; their heart acts like two side-by-side pumps (Figure 39.4*c*). The *right* half of their heart pumps oxygen-poor blood to the lungs, where blood picks up oxygen and gives up carbon dioxide. Then the freshly oxygenated blood flows to the heart's left half. This route is the **pulmonary circuit**.

In the **systemic circuit**, the heart's *left* half pumps the freshly oxygenated blood to every tissue and organ where oxygen is used and carbon dioxide forms. Then the oxygen-poor blood flows to the heart's right half.

The double circuit is a rapid and efficient mode of blood delivery. It supports the high levels of activity typical of vertebrates whose ancestors evolved on land.

Links With the Lymphatic System

The heart's pumping action puts pressure on blood flowing through the circulatory system. Partly because of the pressure, small amounts of water and a few of the proteins dissolved in blood are forced out of capillaries and become part of interstitial fluid. However, a rather elaborate network of drainage vessels picks up excess interstitial fluid and reclaimable solutes, then returns them to the circulatory system. This network is part of the **lymphatic system**. Later, you will see how other parts of the lymphatic system help cleanse bacteria and other pathogens from fluid being returned to the blood.

Closed circulatory systems confine blood within one or more hearts and a network of blood vessels. In the open systems, blood also intermingles with tissue fluids.

The closed system of vertebrates transports substances to and from interstitial fluid bathing the body's cells. It is functionally connected with the lymphatic system.

Blood flows rapidly in large-diameter vessels between the heart and capillary beds. There, the flow velocity slows and exchanges are made between blood and interstitial fluid.

In fishes, blood flows in one circuit from and back to the heart. In birds and mammals, blood flows in two circuits, through a heart partitioned as two side-by-side pumps. The double circuit supports the high levels of activity typical of vertebrates that evolved on land.

CHARACTERISTICS OF BLOOD

Functions of Blood

Blood is a connective tissue with multiple functions. It transports oxygen, nutrients, and other solutes to cells. It carries away their metabolic wastes and secretions, including hormones. Blood helps stabilize internal pH. Blood also serves as a highway for phagocytic cells that scavenge tissue debris and fight infections. In birds and mammals, blood helps equalize body temperature. It does this by carrying excess heat from skeletal muscles and other regions of high metabolic activity to the skin, where heat can be dissipated.

Blood Volume and Composition

The volume of blood depends on body size and on the concentrations of water and solutes. Blood volume for average-size adult humans is about 6 to 8 percent of the total body weight. That amounts to about four or five quarts. As is true of all vertebrates, the blood of humans is a sticky fluid, thicker than water and slower flowing. Its components are the **plasma**, **red blood cells**, **white blood cells**, and **platelets**. Plasma normally accounts for 50 to 60 percent of the total blood volume.

PLASMA   Prevent a blood sample in a test tube from clotting, and it separates into a layer of straw-colored liquid, the plasma, which floats over the red-colored cellular portion of blood (Figure 39.5). Plasma is mostly water, and it functions as a transport medium for blood cells and platelets. It also serves as a solvent for ions

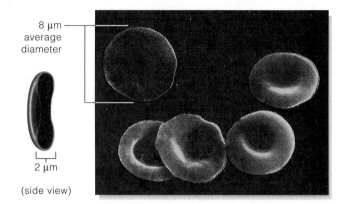

Figure 39.6   Size and shape of red blood cells.

8 μm average diameter

2 μm

(side view)

and molecules, including hundreds of different plasma proteins. Some of the plasma proteins transport lipids and fat-soluble vitamins through the body. Others have roles in blood clotting or in defense against pathogens. Collectively, the concentration of the plasma proteins affects the blood's fluid volume, for it influences the movement of water between blood and interstitial fluid. Glucose and other simple sugars, as well as lipids, amino acids, vitamins, and hormones, are dissolved in plasma. So are oxygen, carbon dioxide, and nitrogen.

RED BLOOD CELLS   Erythrocytes, or red blood cells, are biconcave disks, like doughnuts with a squashed-in center instead of a hole (Figure 39.6). They transport the oxygen used in aerobic respiration and they carry away some carbon dioxide wastes. When oxygen diffuses into

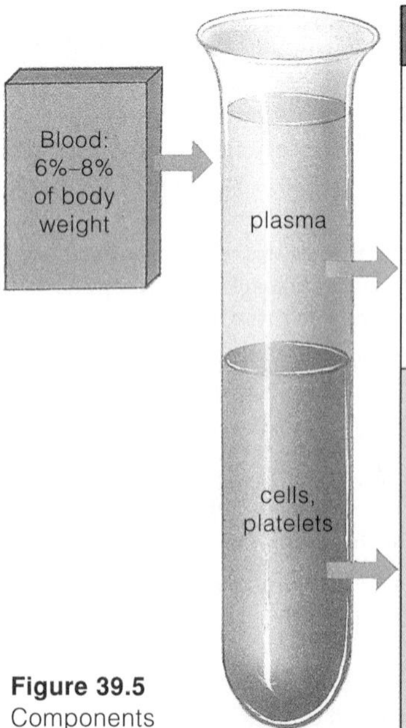

Figure 39.5
Components of blood.

Blood: 6%–8% of body weight

plasma

cells, platelets

Components	Relative Amounts	Functions
Plasma Portion *(50%–60% of total volume):*		
1. Water	91%–92% of plasma volume	Solvent
2. Plasma proteins (albumin, globulins, fibrinogen, etc.)	7%–8%	Defense, clotting, lipid transport, roles in extracellular fluid volume, etc.
3. Ions, sugars, lipids, amino acids, hormones, vitamins, dissolved gases	1%–2%	Roles in extracellular fluid volume, pH, etc.
Cellular Portion *(40%–50% of total volume):*		
1. Red blood cells	4,800,000–5,400,000 per microliter	Oxygen, carbon dioxide transport
2. White blood cells:		
Neutrophils	3,000–6,750	Phagocytosis during inflammation
Lymphocytes	1,000–2,700	Immune responses
Monocytes (macrophages)	150–720	Phagocytosis in all defense responses
Eosinophils	100–360	Defense against parasitic worm
Basophils	25–90	Secretes substances for inflammation, clotting
3. Platelets	250,000–300,000	Roles in clotting

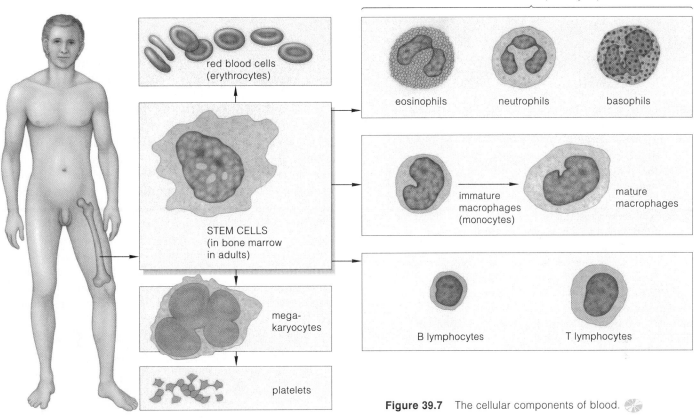

white blood cells (leukocytes)

red blood cells (erythrocytes)

eosinophils   neutrophils   basophils

STEM CELLS (in bone marrow in adults)

immature macrophages (monocytes)   mature macrophages

mega-karyocytes

B lymphocytes   T lymphocytes

platelets

Figure 39.7 The cellular components of blood.

blood, it binds with hemoglobin, the iron-containing pigment that gives red blood cells their color. (Here you may wish to review hemoglobin's molecular structure, as shown in Section 3.7.) Oxygenated blood is bright red. Poorly oxygenated blood is darker red but appears blue inside blood vessel walls near the body surface.

Red blood cells are derived from stem cells in bone marrow (Figure 39.7). Generally speaking, **stem cells** remain unspecialized and retain the capacity for mitotic cell division. Their daughter cells also divide, but only a portion go on to differentiate into specialized types.

Mature red blood cells no longer have their nucleus, nor do they require it. They have enough hemoglobin, enzymes, and other proteins to function for about 120 days. Phagocytes continually engulf the oldest red blood cells or those already dead, but ongoing replacements keep the cell count fairly stable. A **cell count** is a measure of the number of cells of a given type in a microliter of blood. For example, the average number of red blood cells is 5.4 million in males and 4.8 million in females.

WHITE BLOOD CELLS   Leukocytes, or white blood cells, arise from stem cells in bone marrow. They function in daily housekeeping and defense. Many patrol tissues, where they target or engulf damaged or dead cells and anything chemically recognized as foreign to the body. Many others are massed together in the lymph nodes

and spleen, which are part of the lymphatic system. There they divide to produce armies of cells that battle specific viruses, bacteria, and other invaders.

White blood cells differ in size, nuclear shape, and staining traits. There are five categories: neutrophils, eosinophils, basophils, monocytes, and lymphocytes (Figure 39.7). Their numbers can vary, depending on whether an individual is active, healthy, or under siege, as described in the next chapter. The neutrophils and monocytes are search-and-destroy cells. The monocytes follow chemical trails to inflamed tissues. There they develop into macrophages ("big eaters") that can engulf invaders and debris. Two classes of lymphocytes, B cells and T cells, make highly specific defense responses.

PLATELETS   Some stem cells in bone marrow give rise to giant cells (megakaryocytes). These shed fragments of cytoplasm enclosed in a bit of plasma membrane. The membrane-bound fragments are what we call the platelets. Each platelet only lasts five to nine days, but hundreds of thousands are always circulating in blood. They can release substances that initiate blood clotting.

Vertebrate blood has roles in transport, defense, clotting, and maintaining the volume, composition, and temperature of the internal environment.

The body continually replaces blood cells for good reason. Besides aging and dying off regularly, blood cells typically encounter a variety of pathogens that use them as places to complete their life cycle. Besides this, sometimes blood cells malfunction as a result of gene mutations.

RED BLOOD CELL DISORDERS Consider the **anemias**, disorders that result from too few red blood cells or deformed ones. Oxygen levels in blood cannot be kept high enough to support normal metabolism. Shortness of breath, fatigue, and chills follow. *Hemorrhagic* anemias result from sudden blood loss, as from a severe wound; *chronic* anemias result from ongoing but slight blood loss, as from an undiagnosed bleeding ulcer, hemorrhoids, or the monthly blood loss of premenopausal women.

Certain infectious bacteria and parasites replicate inside blood cells, then escape by lysis; they cause some *hemolytic* anemias. Insufficient iron in the diet results in *iron deficiency* anemia, for red blood cells cannot produce enough normal hemoglobin without it. B_{12} *deficiency* anemia is a potential hazard for strict vegetarians and alcoholics. Red blood cells form but they cannot divide without vitamin B_{12}. Normally, meats, poultry, and fish in the diet provide enough of the vitamin.

As described elsewhere in the book, a gene mutation that gives rise to an abnormal form of hemoglobin can result in *sickle-cell* anemia. Another gene mutation, which blocks or lowers the synthesis of the globin chains that make up hemoglobin, causes *thalassemias*. Too few red blood cells can form; those that do are thin and fragile.

Or consider the *polycythemias*—symptoms of far too many red blood cells—which make blood flow sluggish. Some bone marrow cancers can result in this condition. So can "blood doping" by some athletes who compete in strenuous events. They withdraw their own red blood cells, store them, then reinject them a few days prior to competition. The withdrawal triggers red blood cell formation as the body attempts to replace the "lost" cells, so the cell count is bumped up when the withdrawn cells are put back in the body. The idea is to increase the body's oxygen-carrying capacity and endurance. Blood doping leads to temporary high blood pressure and lowers blood's viscosity. Some believe the practice works, but others call it unethical. It is banned from the Olympics.

WHITE BLOOD CELL DISORDERS You may have heard of *infectious mononucleosis*. An Epstein-Barr virus causes this highly contagious disease, which results from too many monocytes and lymphocytes. Following a few weeks of fatigue, aches, low fever, and a chronic sore throat, the patient usually recovers.

Recovery is chancy for *leukemias*, a category of cancers that suppress or impair white blood cell formation in bone marrow. Radiation therapy and chemotherapy can kill the cancer cells. Remissions may last for months or years. (A remission is a symptom-free period of a chronic illness.)

39.4

BLOOD TRANSFUSION AND TYPING

Concerning Agglutination

Whenever blood volume or blood cell counts decline, countermeasures kick in automatically. However, if the volume were to decrease by more than 30 percent, then circulatory shock would follow and could lead to death.

Blood from donors can be transfused into patients affected by a blood disorder or substantial blood loss. Such **blood transfusions** cannot be decided willy-nilly. Why not? *A potential donor and the recipient may not have the same kinds of recognition proteins at the surface of their red blood cells.* Some of the proteins are "self" markers; they identify the cells as belonging to one's own body. During a blood transfusion, if cells from a donor have the "wrong" marker, the recipient's immune system will recognize them as foreign, with serious consequences.

Figure 39.8 shows what happens when blood from incompatible donors and recipients intermingle. In a defensive response called **agglutination**, proteins called

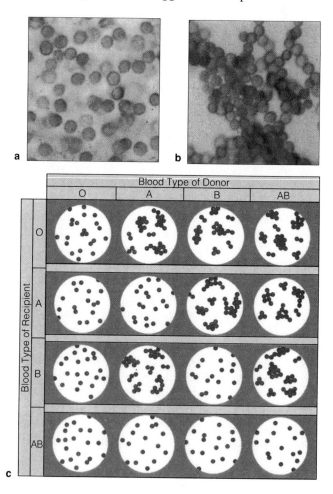

Figure 39.8 Light micrographs showing (**a**) the absence of agglutination in a mixture of two different but compatible blood types and (**b**) agglutination in a mixture of incompatible types. (**c**) Agglutination responses in blood types A, B, AB, and O when mixed with samples of the same and different types.

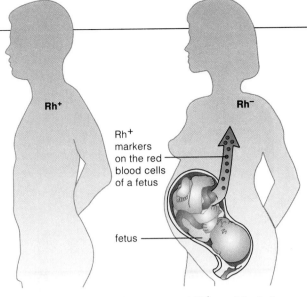

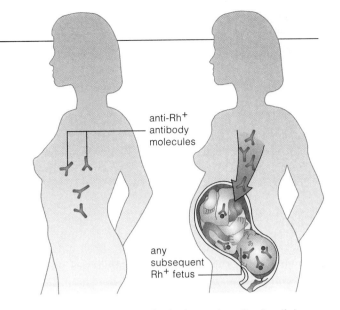

a A forthcoming child of an Rh⁻ woman and Rh⁺ man inherits the gene for the Rh⁺ marker. During pregnancy or childbirth, some of its cells bearing the marker may leak into the maternal bloodstream.

b The foreign marker stimulates her body to make antibodies. If she gets pregnant again and if this second fetus (or any other) inherits the gene for the marker, the circulating anti-Rh⁺ antibodies will act against it.

Figure 39.9 Antibody production in response to Rh⁺ markers on red blood cells of a fetus.

antibodies that are circulating in plasma act against the foreign cells and cause them to clump together. (As described in the next chapter, antibodies bind to specific foreign markers and so target the cell or particle bearing them for destruction by the immune system.) The sheer number of "foreign" cells transfused into the recipient translates into numerous clumps, which can clog small blood vessels and damage tissues. Without treatment, death may follow. The same thing can happen during certain pregnancies if antibodies diffuse from a mother's circulatory system into that of her unborn child.

Based on an understanding of cell surface markers and antibodies, scientists have devised ways to analyze the forms of self markers on a person's red blood cells. Blood typing is based on these analyses.

ABO Blood Typing

Molecular variations in one kind of self-marker on red blood cells are analyzed with **ABO blood typing**. The genetic basis of such variation is described in Section 11.4. People with one form of the marker are said to have type A blood; those with another form have type B blood. Those with both forms of the marker on their red blood cells have type AB blood. Others do not have either form of the marker; they have type O blood.

If you are type A, your antibodies ignore A markers but will act against B markers. If you are type B, your antibodies will ignore B markers but will act against A markers. If you are type AB, your antibodies ignore both forms of the marker, so you can tolerate donations of type A, B, AB, or O blood. If you are type O, you have antibodies against both forms of the marker, so your options are limited to type O donations.

Rh Blood Typing

Rh blood typing is based on the presence or absence of an Rh marker (so named because it was first identified in blood samples of *Rh*esus monkeys). If you are type Rh⁺, your blood cells bear this marker at their surface. If you are type Rh⁻, they don't. Ordinarily, people do not have antibodies against Rh markers. But a recipient of transfused Rh⁺ blood produces antibodies against them, and the antibodies remain in the blood.

If an Rh⁻ woman becomes impregnated by an Rh⁺ man, there is a chance the fetus will be Rh⁺. During pregnancy or childbirth, some of its red blood cells may leak into the mother's bloodstream. If they do, her body will produce antibodies against Rh (Figure 39.9). If she becomes pregnant *again*, Rh antibodies will enter the bloodstream of this new fetus. If its blood is type Rh⁺, then her antibodies will cause its red blood cells to swell, rupture, and release hemoglobin.

Erythroblastosis fetalis is a severe outcome of mixing Rh⁺ and Rh⁻ types. The fetus dies, for too many cells are destroyed. If diagnosed before birth or delivered alive, it can survive if its blood is slowly replaced with transfusions that are free of Rh antibodies. Currently, a known Rh⁻ woman can be treated right after her first pregnancy with an anti-Rh gamma globulin (RhoGam) that will protect her next fetus. The drug will inactivate any Rh⁺ fetal blood cells that are circulating through her bloodstream before she can become sensitized and begin producing potentially dangerous antibodies.

To avoid the symptoms of blood incompatibilities, red blood cells should be typed before transfusions or pregnancies.

HUMAN CARDIOVASCULAR SYSTEM

"Cardiovascular" comes from the Greek *kardia* (heart) and Latin *vasculum* (vessel). In a human cardiovascular system, blood is pumped by a muscular heart into large-diameter **arteries**. It then flows into small, muscular **arterioles**, which branch into the even smaller diameter capillaries introduced earlier. Blood flows continuously from capillaries into small **venules**, and then into large-diameter **veins** that return blood to the heart.

As in most vertebrates, a partition separates the heart into a double pump, which drives blood through two cardiovascular circuits (Figure 39.10). Each circuit has its own set of arteries, arterioles, capillaries, venules, and veins. The *pulmonary* circuit, a short loop, rapidly oxygenates blood. It leads from the heart's right half to capillary beds in both lungs, then returns to the heart's left half. The *systemic* circuit is a longer loop starting at the heart's left half. Its main artery, the **aorta**, accepts the oxygenated blood, which flows on through arterioles, capillary beds in all body regions, and then veins that deliver the oxygen-poor blood to the heart's right half. Figure 39.11 identifies the location of the major blood vessels of both circuits and defines their functions.

For most of the systemic branchings, a given volume of blood flows through one capillary bed. The branch to the intestines is one of the exceptions. First blood picks up glucose and other absorbed substances from one bed, then it moves through another capillary bed, in the liver—an organ with a key role in nutrition. The second bed gives the liver time to process absorbed substances.

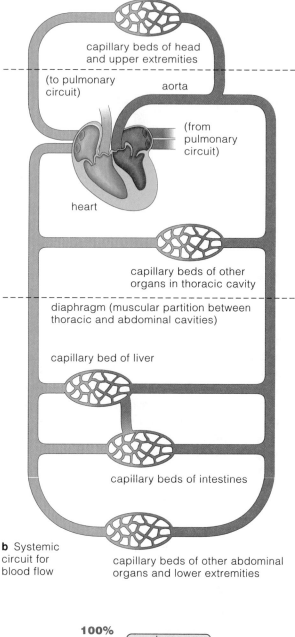

b Systemic circuit for blood flow

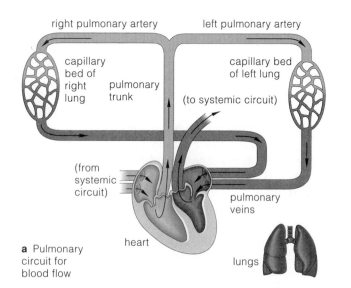

a Pulmonary circuit for blood flow

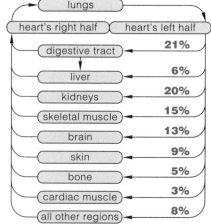

Figure 39.10 (**a**,**b**) Diagrams of the pulmonary and systemic circuits for blood flow through the human cardiovascular system. The blood vessels that are carrying oxygenated blood are color-coded *red*. Those carrying oxygen-poor blood are color-coded *blue*. (**c**) Distribution of the heart's output in a person at rest. The lungs receive all blood pumped out of the heart's right half. The organs serviced by the systemic circuit receive the portions indicated from the heart's left half. Control mechanisms adjust the flow distribution when necessary.

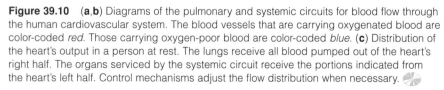

c

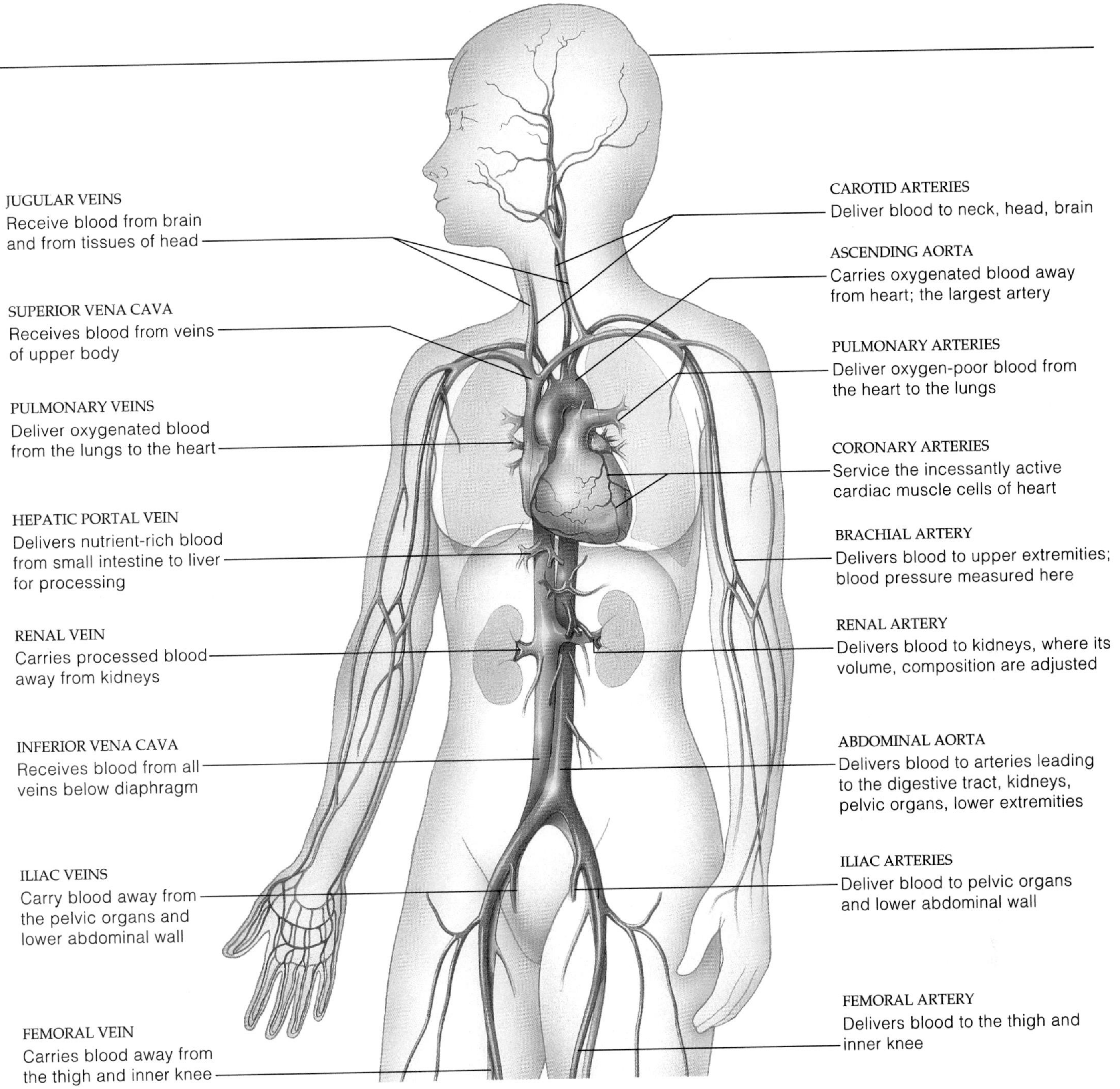

JUGULAR VEINS
Receive blood from brain and from tissues of head

SUPERIOR VENA CAVA
Receives blood from veins of upper body

PULMONARY VEINS
Deliver oxygenated blood from the lungs to the heart

HEPATIC PORTAL VEIN
Delivers nutrient-rich blood from small intestine to liver for processing

RENAL VEIN
Carries processed blood away from kidneys

INFERIOR VENA CAVA
Receives blood from all veins below diaphragm

ILIAC VEINS
Carry blood away from the pelvic organs and lower abdominal wall

FEMORAL VEIN
Carries blood away from the thigh and inner knee

CAROTID ARTERIES
Deliver blood to neck, head, brain

ASCENDING AORTA
Carries oxygenated blood away from heart; the largest artery

PULMONARY ARTERIES
Deliver oxygen-poor blood from the heart to the lungs

CORONARY ARTERIES
Service the incessantly active cardiac muscle cells of heart

BRACHIAL ARTERY
Delivers blood to upper extremities; blood pressure measured here

RENAL ARTERY
Delivers blood to kidneys, where its volume, composition are adjusted

ABDOMINAL AORTA
Delivers blood to arteries leading to the digestive tract, kidneys, pelvic organs, lower extremities

ILIAC ARTERIES
Deliver blood to pelvic organs and lower abdominal wall

FEMORAL ARTERY
Delivers blood to the thigh and inner knee

Figure 39.11 Location and functions of major blood vessels of the human cardiovascular system.

Figure 39.10c shows how the pulmonary and systemic circuits distribute the heart's output to different organs. The percentages listed are for a person at rest. As you will see, the flow distribution along the systemic route is adjusted in response to changes in levels of physical activity and to shifts in external and internal conditions. For instance, the flow to and from a skeletal muscle varies greatly, depending on what it is being called upon to do at a given time. The same is true of skin. When the body gets too cold, the flow of blood (and metabolic heat) is diverted away from skin's vast capillary beds.

When the body is hot, flow to the skin's beds increases and heat radiates from the skin's surface. Only the brain is exempt; it cannot tolerate variations in flow.

The human cardiovascular system consists of two separate circuits, pulmonary and systemic, for blood flow.

The pulmonary circuit is a short loop from the heart's right half, through both lungs, to the heart's left half. Oxygen-poor blood flowing into the circuit rapidly picks up oxygen and gives up carbon dioxide at capillary beds in the lungs.

The longer systemic circuit starts at the heart's left half and aorta. It delivers oxygen and accepts carbon dioxide at capillary beds of all metabolically active regions; then its veins deliver oxygen-poor blood to the heart's right half.

Heart Structure

Think about the fact that the human heart beats about 2.5 billion times during a seventy-year life span, and you know it must be a truly durable pump. Figure 39.12 shows its structure, which speaks of its durability. The pericardium, a double sac of tough connective tissue, protects the heart and anchors it to nearby structures (*peri*, around). A fluid between its layers lubricates the heart during its perpetual twisting motions. The inner layer serves as the outer part of the heart wall. The bulk of that wall, the myocardium, consists of cardiac muscle cells tethered to elastin and collagen fibers. The fibers are so densely crisscrossed, they serve as a "skeleton" against which the force of contraction is applied. The oxygen-demanding cardiac muscle cells have their own, *coronary* circulation; coronary arteries branching off the aorta lead to a capillary bed that services only them. The wall's glistening, innermost layer consists of connective tissue and **endothelium**, a one-cell-thick epithelial sheet. Only the heart and blood vessels have an endothelium.

Each half of the heart has two chambers: an **atrium** (plural, atria) and a **ventricle**. Blood flows into the atria, down into the ventricles, then out through great arteries (the aorta or pulmonary trunk). Between each atrium and ventricle is an AV valve (short for atrioventricular). Between each ventricle and the artery leading out from it is a semilunar valve. Both types are *one-way* valves, with membranes that rhythmically flap open, flap shut, and thereby help keep blood moving in one direction.

Cardiac Cycle

Each time the heart beats, its four chambers go through phases of contraction (systole) and relaxation (diastole). The sequence of contraction and relaxation is a **cardiac cycle**. When relaxed, the atria fill with blood. Increasing fluid pressure forces the AV valves open. Blood flows into the ventricles, which completely fill when the atria contract (Figure 39.13a). As the filled ventricles start to contract, the rising fluid pressure forces the AV valves shut. It rises so sharply above the pressure in the great

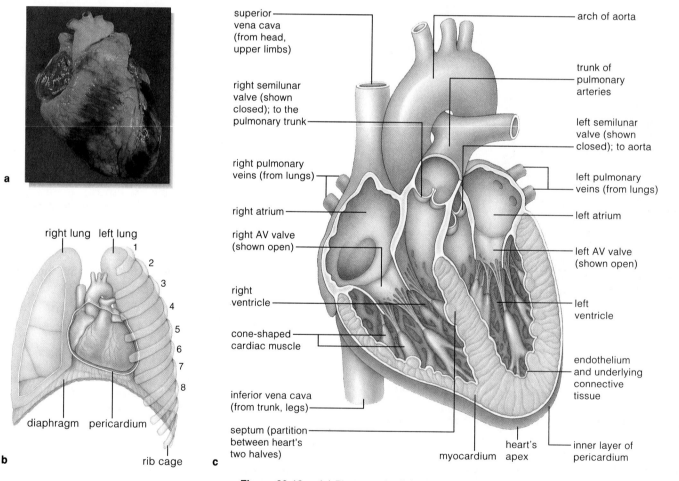

a

b

right lung   left lung
1
2
3
4
5
6
7
8

diaphragm   pericardium

rib cage

c

superior vena cava (from head, upper limbs)

right semilunar valve (shown closed); to the pulmonary trunk

right pulmonary veins (from lungs)

right atrium

right AV valve (shown open)

right ventricle

cone-shaped cardiac muscle

inferior vena cava (from trunk, legs)

septum (partition between heart's two halves)

arch of aorta

trunk of pulmonary arteries

left semilunar valve (shown closed); to aorta

left pulmonary veins (from lungs)

left atrium

left AV valve (shown open)

left ventricle

endothelium and underlying connective tissue

myocardium   heart's apex   inner layer of pericardium

Figure 39.12   (**a**) Photograph of the human heart and (**b**) its location in the thoracic cavity. (**c**) Cutaway view of the human heart, showing its wall and internal organization.

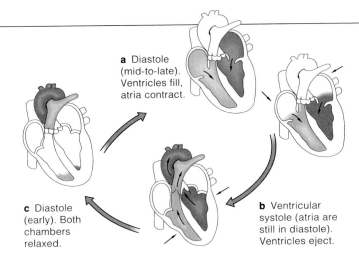

a Diastole (mid-to-late). Ventricles fill, atria contract.

b Ventricular systole (atria are still in diastole). Ventricles eject.

c Diastole (early). Both chambers relaxed.

Figure 39.13 Blood flow during part of a cardiac cycle. Blood and heart movements generate a "lub-dup" sound at the chest wall. At each "lub," AV valves are closing as ventricles contract. At each "dup," semilunar valves are closing as ventricles relax.

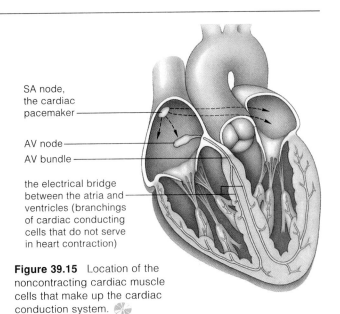

SA node, the cardiac pacemaker

AV node

AV bundle

the electrical bridge between the atria and ventricles (branchings of cardiac conducting cells that do not serve in heart contraction)

Figure 39.15 Location of the noncontracting cardiac muscle cells that make up the cardiac conduction system.

arteries, it forces the semilunar valves open—and blood leaves the heart (Figure 39.13*b*). The ventricles relax, the semilunar valves close, and the already filling atria are ready to repeat the cycle. Thus, in a cardiac cycle, atrial contraction simply helps fill the ventricles. *Contraction of the ventricles is the driving force for blood circulation.*

Mechanisms of Contraction

Like skeletal muscle, cardiac muscle is striated (striped). In response to action potentials, the sarcomeres of its cells also contract in the manner predicted by the sliding-filament model (Section 38.8). This tissue also requires energy for contraction; large numbers of mitochondria in the myocardium provide the ATP. But cardiac muscle cells are structurally unique. They are branching, short, and joined at the end regions. Communication junctions span the plasma membranes of abutting regions and let

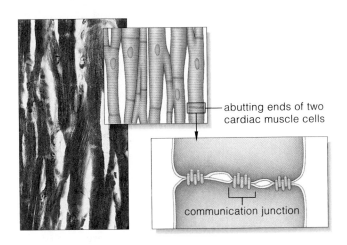

abutting ends of two cardiac muscle cells

communication junction

Figure 39.14 Light micrograph of cardiac muscle cells and sketches showing communication junctions between them.

action potentials spread swiftly among cells, in waves of excitation that wash over the heart (Figure 39.14).

In addition, about 1 percent of the cardiac muscle cells *do not* contract. They function instead as a **cardiac conduction system**. About seventy times each minute, these specialized cells initiate and propagate waves of excitation that travel in rhythmic, orderly sequence from the atria to ventricles. The synchronized excitation underlies the heart's truly efficient pumping. Each wave starts at the SA node, a cluster of cell bodies in the right atrium's wall (Figure 39.15). It passes through the wall to another cell body cluster, the AV node. This is the only electrical bridge between the atria and ventricles, which connective tissue insulates everywhere else. After the AV node, conducting cells are arranged as a bundle in the partition between the heart's two halves. These cells branch, and the branches deliver the excitatory wave up the ventricle walls. The ventricles contract in response with a twisting movement, upward from the heart's apex, that ejects blood into the great arteries.

The SA node fires action potentials faster than the rest of the system and serves as the **cardiac pacemaker**. Its spontaneous, rhythmic signals are the foundation for the normal rate of heartbeat. The nervous system can only *adjust* the rate and strength of contractions dictated by the pacemaker. Even if all nerves leading to a heart are severed, the heart will keep on beating!

The heart's construction reflects its role as a durable pump. Under the spontaneous, rhythmic signals from the cardiac pacemaker, its branching, abutting cardiac muscle cells contract in synchrony, almost as if they were a single unit.

Although the heart has four chambers (each half has one atrium and one ventricle), contraction of the ventricles is the driving force for blood circulation away from the heart.

BLOOD PRESSURE IN THE CARDIOVASCULAR SYSTEM

As you have seen, blood flows to and from the human heart through arteries, arterioles, capillaries, venules, and finally veins. Figure 39.16 shows the structure of these vessels. The arteries are the major transporters of oxygenated blood throughout the body. Arterioles in each region are sites where the volume of blood flow through each organ can be controlled. The capillaries and, to a lesser extent, venules are diffusion zones, and veins are the major transporters of oxygen-poor blood to the heart.

Two key factors influence the rate of flow through each type of blood vessel. First, the flow rate is directly proportional to the pressure gradient between the start and end of the vessel. Second, the flow rate is inversely proportional to the vessel's resistance to flow.

Blood pressure is fluid pressure imparted to blood by heart contractions. The beating heart establishes the higher blood pressure at the beginning of a vessel. As flowing blood rubs against the vessel's inner wall, the friction causes some energy (in the form of pressure) to be lost. The lower pressure at the end of the vessel is due to the frictional losses. Even small changes in blood vessel diameter have major influence over the flow rate. In the pulmonary and systemic circuits, the diameters decrease and present more resistance to flow. And so, because of heart contractions and the subsequent events (mainly frictional losses), blood pressure is highest in the contracting ventricles, still high at the beginning of arteries, then continues to drop until reaching its lowest value—in the relaxed atria (Figure 39.17).

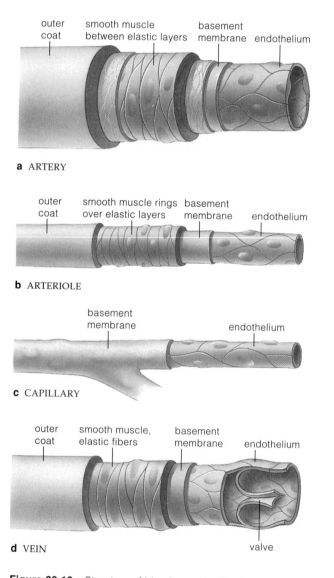

a ARTERY

b ARTERIOLE

c CAPILLARY

d VEIN

Figure 39.16 Structure of blood vessels. The basement membrane around the endothelium of each vessel is a noncellular layer, rich with proteins and polysaccharides.

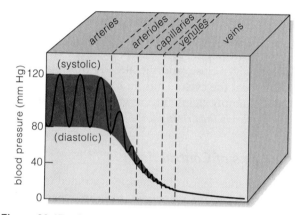

Figure 39.17 Blood pressure. This is a plot of measurements made of the drop in fluid pressure for a given volume of blood moving through the systemic circuit.

Arterial Blood Pressure

From Figure 39.16*a* and the preceding discussion, you see that arteries have a large diameter and present low resistance to flow. Thus they serve as rapid transporters of oxygenated blood. They also are pressure reservoirs that smooth out pulsations in pressure caused by each cardiac cycle. Their thick, muscular, elastic wall bulges as ventricular contraction forces a large volume of blood into them. Then the wall recoils and so forces the blood forward through the circuit while the heart is relaxing.

Suppose you decide to measure your blood pressure daily when you are resting. Typically, you would take a reading from the brachial artery of an upper arm, as in Figure 39.18. *Systolic* pressure is the peak pressure that the contracting ventricles exert against the artery's wall during a cardiac cycle. *Diastolic* pressure is the lowest pressure when blood is draining into the vessels after it. You can estimate the mean arterial pressure as *diastolic pressure + 1/3 pulse pressure* (which is the difference between the highest and lowest pressure readings).

Figure 39.18 Measuring blood pressure.

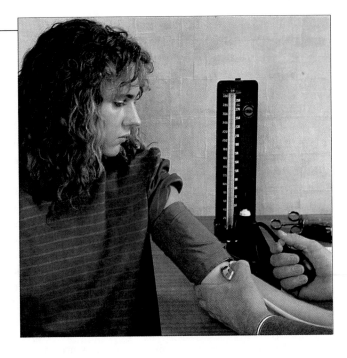

A hollow cuff attached to a pressure gauge is wrapped around a person's upper arm. The cuff is inflated with air to a pressure above the highest pressure of the cardiac cycle (at systole, when the ventricles contract). Above this pressure, no sounds can be heard through a stethoscope positioned below the cuff and over the artery (because no blood is flowing through the vessel). Air in the cuff is slowly released, so that some blood flows into the artery. The turbulent flow causes soft tapping sounds, and when this first occurs, the value on the gauge is the systolic pressure, or about 120 mm mercury (Hg) in young adults at rest. This particular value means that the measured pressure would force mercury to move upward 120 millimeters in a narrow glass column.

More air is released from the cuff. Just after the sounds become dull and muffled, blood flows continuously. So the turbulence and tapping sounds stop. The silence corresponds to the diastolic pressure (at the end of a cardiac cycle, just before the heart pumps out blood). The reading is usually about 80 mm Hg. In this example, pulse pressure (the difference between the highest and lowest pressure readings) is 120 – 80, or 40 mm Hg. Assuming you are an adult in good health, this average resting value stays fairly constant over a few weeks or even months.

Resistance to Flow at Arterioles

Track a volume of blood through the systemic route and you find the greatest pressure drop at arterioles (Figure 39.17), for these offer the greatest resistance to blood flow. The slowdown at arterioles allows time for control mechanisms to divert greater or lesser portions of the total flow volume to different regions. Control is exerted at arteriole walls, which contain rings of smooth muscle. Some control signals cause the smooth muscle cells to relax, which results in **vasodilation**. The word means an enlargement (dilation) of the blood vessel diameter. Other signals cause contraction of the smooth muscle cells, which results in a decrease in the blood vessel diameter, or **vasoconstriction**.

The nervous and endocrine systems exert control over arteriole diameter. For example, sympathetic nerve endings terminate on the smooth muscle cells of many arterioles. Increased activity along the nerves triggers action potentials in the cells to bring about contraction, hence widespread vasoconstriction. Epinephrine and angiotensin are two of the signaling molecules that can trigger changes in arteriole diameter.

Arteriole diameter also is adjusted when changes in metabolic activity shift the localized concentrations of substances in a tissue. Such local chemical changes are "selfish," in that they invite or divert blood flow to meet the tissue's own metabolic needs. Think of what happens as you run. The oxygen level drops in skeletal muscle tissue, and levels of carbon dioxide, hydrogen and potassium ions, and other substances increase. The changed conditions trigger vasodilation of arterioles in the vicinity. Now more blood flows past active muscles, delivering more raw materials and carrying away cell products and metabolic wastes. When muscles relax and demand fewer blood deliveries, the oxygen level increases and triggers vasoconstriction of the arterioles.

Controlling Mean Arterial Blood Pressure

Mean arterial pressure depends on cardiac output and on total resistance through the vascular system. Cardiac output is influenced by controls over the rate and strength of heartbeats, and total resistance mainly by vasoconstriction at the arterioles. Given that organs and tissues vary in their demands for blood, maintaining blood pressure is not easy. The balance between cardiac output and total resistance must be continually juggled.

A **baroreceptor reflex** is the main short-term control over arterial pressure. It starts at baroreceptors, which detect fluctuations in arterial mean pressure and pulse pressure. The most critical of these sensory receptors are located in the carotid arteries (which supply blood to the brain) and aortic arch (which supplies blood to the rest of the body). They continually generate action potentials, and their signals reach a control center in a hindbrain region, the medulla oblongata (Section 35.4). In response, the center adjusts its commands flowing along sympathetic and parasympathetic nerves to the heart and blood vessels. For example, when the mean arterial pressure rises, the center commands the heart to beat more slowly and contract less forcefully.

Long-term control of blood pressure is exerted at kidneys, which adjust the volume and composition of the blood. And that is a topic of another chapter.

Mean arterial pressure is an outcome of controls over the heart's output and over the total resistance to blood flow through the vascular system, as exerted mainly at arterioles.

Capillary Function

THE NATURE OF THE EXCHANGES   Capillary beds, again, are diffusion zones for exchanges between blood and interstitial fluid. When the living cells in any part of the body are deprived of those exchanges, they die quickly; brain cells are dead within four minutes.

By some estimates, the human body has between 10 billion and 40 billion capillaries, which collectively represent an astounding surface area for the exchanges. These components of the circulatory system extend into nearly every tissue, and at least one is as close as 0.001 centimeter to every living cell. Their proximity to cells is crucial, for shorter distances mean faster diffusion rates. (Think about this: It would take years for oxygen to diffuse on its own from your lungs all the way down your legs. By that time your toes, and the rest of you, would long be dead.) Although capillaries are three to seven micrometers across, they deform amazingly; red blood cells (eight micrometers across) squeeze through them single file. Therefore, those oxygen-transporting cells, as well as the substances dissolved in plasma, are in direct contact with the exchange area—the capillary wall—or only a short distance away from it.

And remember, taken as a whole, capillaries present a greater cross-sectional area than the arterioles leading into them, so the flow velocity is not as great. With this slowdown, there is plenty of time for interstitial fluid to exchange substances with the 5 percent or so of the total blood volume that is moving forward through the capillary beds at a given time. Here you may wish to reflect on Figure 39.3e.

Each capillary is a tube of endothelial cells, a single layer thick (Figure 39.16c). The cells abut one another, but the "fit" is different in different parts of the body. In most cases, there are narrow, water-filled clefts between the endothelial cells. In capillaries that service the brain, tight junctions join the cells and clefts are nonexistent; substances cannot "leak" between cells but must move through them. Thus the junctions are a functional part of the blood-brain barrier, as described in Section 35.4.

Oxygen, carbon dioxide, and most other small lipid-soluble substances cross the capillary wall by diffusing through the lipid portion of an endothelial cell's plasma membrane and through its cytoplasm. Certain proteins enter and leave the cells by endocytosis or exocytosis. Small, water-soluble substances such as ions enter and leave at the clefts between endothelial cells. So do white blood cells, as you will see in the chapter to follow.

MECHANISMS OF EXCHANGE   Diffusion is not the only means by which substances are exchanged across the capillary wall. At any time, the fluid pressures acting on the wall may not be balanced, and there may be bulk flow one way or the other across it. Bulk flow, recall, is a movement of water and solutes in the same direction in response to fluid pressure.

As Figure 39.19 shows, the concentrations of water and solutes in blood and in interstitial fluid influence the direction in which the fluid flows. At the beginning of a capillary bed, the outward-directed force of blood pressure is greater than the inward-directed osmotic force of interstitial fluid. A small amount of protein-free plasma is pushed out in bulk through clefts in the wall. This process is called **ultrafiltration**. Farther on in the bed, the balance shifts. Whereas blood pressure has been continually declining, the osmotic force has stayed the same. When the inward-directed force exceeds the outward force of blood pressure, tissue fluid moves through the clefts between endothelial cells and into the capillary. This process is called **reabsorption**.

Normally the outcome is a very small, *net* outward movement of fluid from the capillary bed, which the lymphatic system returns to the blood. Such bulk flow helps maintain the fluid balance between blood and interstitial fluid. This is important, for blood pressure is maintained only when there is adequate blood volume. When the blood volume plummets, as happens during hemorrhage, interstitial fluid can be tapped by way of reabsorption to help counter the loss.

Sometimes blood pressure increases so much that it triggers excessive ultrafiltration. When excess fluid accumulates in interstitial spaces, this condition is called *edema*. It occurs to some extent during physical exercise, when arterioles dilate in many tissue regions. Edema also can result from an obstructed vein or from heart failure. It becomes extreme during *elephantiasis*. As described in Section 26.9, elephantiasis is a disease brought on by roundworm infection and a subsequent obstruction of lymphatic vessels.

Venous Pressure

What happens to the flow velocity after it continues on past the vast cross-sectional area of the capillary beds? Remember, the capillaries merge into venules, or "little veins." These in turn merge into large-diameter veins. The total cross-sectional area is once again reduced and flow velocity increases as blood returns to the heart.

In functional terms, venules are a bit like capillaries. Some solutes diffuse across their wall, which is only a bit thicker than that of a capillary. Some control over capillary pressure also is exerted at these vessels.

Veins are large-diameter, low-resistance transport tubes to the heart (Figure 39.16d). They have valves that prevent backflow. When gravity beckons, venous flow reverses direction and pushes the valves shut. Their wall can bulge considerably under pressure, more so

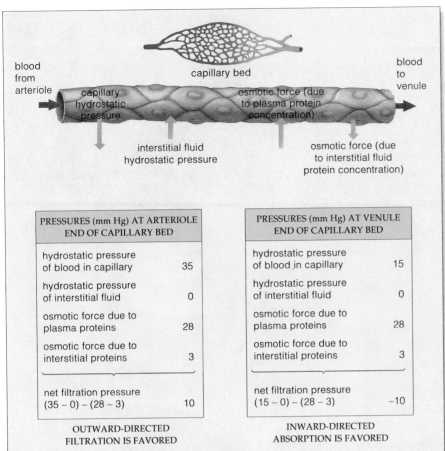

Figure 39.19 Bulk flow in an idealized capillary bed. The fluid movement plays no significant role in diffusion. But it is important in maintaining the distribution of extracellular fluid between the blood and interstitial fluid.

blood from arteriole

capillary bed

blood to venule

capillary hydrostatic pressure

osmotic force (due to plasma protein concentration)

interstitial fluid hydrostatic pressure

osmotic force (due to interstitial fluid protein concentration)

PRESSURES (mm Hg) AT ARTERIOLE END OF CAPILLARY BED	
hydrostatic pressure of blood in capillary	35
hydrostatic pressure of interstitial fluid	0
osmotic force due to plasma proteins	28
osmotic force due to interstitial proteins	3
net filtration pressure (35 – 0) – (28 – 3)	10

OUTWARD-DIRECTED FILTRATION IS FAVORED

PRESSURES (mm Hg) AT VENULE END OF CAPILLARY BED	
hydrostatic pressure of blood in capillary	15
hydrostatic pressure of interstitial fluid	0
osmotic force due to plasma proteins	28
osmotic force due to interstitial proteins	3
net filtration pressure (15 – 0) – (28 – 3)	–10

INWARD-DIRECTED ABSORPTION IS FAVORED

The fluid movements across a capillary wall result from two opposing forces, called ultrafiltration and reabsorption.

At the arteriole end of a capillary, the difference between blood pressure and interstitial fluid pressure leads to filtration. Because of the difference, some plasma (but very few plasma proteins) leaves the capillary by bulk flow through clefts between endothelial cells of the capillary wall. Ultrafiltration is the bulk flow of fluid *out* of the capillary.

Reabsorption is the osmotic movement of some interstitial fluid *into* the capillary. It results from a difference in water concentration between plasma and interstitial fluid. With its dissolved protein components, the plasma has a greater solute concentration, hence a lower water concentration.

Reabsorption near the end of a capillary bed tends to balance ultrafiltration at the beginning. Normally, there is only a small *net* filtration of fluid, which the lymphatic system returns to the blood.

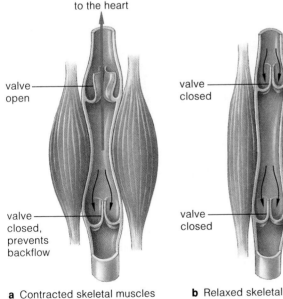

to the heart

valve open

valve closed, prevents backflow

a Contracted skeletal muscles against nearby vein assist blood flow to heart.

valve closed

valve closed

b Relaxed skeletal muscles; venous valves shut, no backflow.

Figure 39.20 How the bulging of contracting skeletal muscles helps increase fluid pressure inside a vein. Notice the structure of each valve (membrane flaps) in the vein's lumen.

than an arterial wall. Thus veins can serve as reservoirs for variable volumes of blood. Collectively, the human body's veins can hold up to 50 to 60 percent of the total blood volume.

A vein wall contains some smooth muscle. When blood must circulate faster, as during physical exercise, the smooth muscle contracts. The wall stiffens and the vein bulges less, so that venous pressure rises and drives more blood to the heart. Also, when limbs are moving, skeletal muscles bulge against veins in their vicinity. They help raise the venous pressure and drive blood back to the heart (Figure 39.20). Rapid breathing also contributes to increased venous pressure. Inhaled air pushes down on internal organs and thereby alters the pressure gradient between the heart and veins.

Capillary beds are diffusion zones for exchanges between blood and interstitial fluid. Here also, a slight amount of bulk flow helps maintain the fluid balance between blood and interstitial fluid.

Venules overlap somewhat with capillaries in function.

Veins are highly distensible blood volume reservoirs and help adjust flow volume back to the heart.

39.9 CARDIOVASCULAR DISORDERS

Cardiovascular disorders are the leading cause of death in the United States. They affect at least 40 million people and kill about 1 million each year. The most common are *hypertension*, which is sustained high blood pressure; and *atherosclerosis*, a progressive thickening of the arterial wall and narrowing of the arterial lumen. Both affect blood circulation and so cause most *heart attacks* (damaged or destroyed heart muscle) and *strokes* (brain damage).

In most heart attacks, a "crushing" pain behind the breastbone lasts a half hour or more. Mild to severe pain may radiate into the left arm, shoulder, or neck. Sweating, shortness of breath, erratic heartbeat, nausea, vomiting, and dizziness or fainting may accompany an attack.

RISK FACTORS Extensive research has correlated the following risk factors with cardiovascular disorders:

1. Smoking (Section 41.7)
2. Genetic predisposition to heart failure
3. High level of cholesterol in the blood
4. High blood pressure
5. Obesity (Section 42.10)
6. Lack of regular exercise
7. Diabetes mellitus (Section 37.7)
8. Age (the older you get, the greater the risk)
9. Gender (until age fifty, males are at much greater risk than females)

The risk factors can be controlled by exercising regularly, eating properly, and not smoking. With too little exercise and too much food, adipose tissue masses increase in the body, more blood capillaries develop to service them, and the heart has to work harder to pump blood through the increasingly divided vascular circuit. In people who smoke, the nicotine in tobacco makes the adrenal glands secrete epinephrine. This hormone, remember, triggers vasoconstriction, which boosts the heart rate and blood pressure. Also, the carbon monoxide in cigarette smoke inhibits the binding of oxygen to hemoglobin, so the heart has to pump harder to get oxygen to cells. Smoking also predisposes people to atherosclerosis even when their blood level of cholesterol is normal.

The next paragraphs describe some of the damage that results from cardiovascular disorders.

HYPERTENSION Hypertension results from gradual increases in the flow resistance through small arteries. In time, blood pressure remains above 140/90, even when a person is resting. Heredity may be a factor; the disorder tends to run in families. Diet is also a factor. For instance, high salt intake can raise blood pressure in susceptible people and increase the heart's workload. The heart may eventually enlarge and fail to pump blood effectively. High blood pressure also may contribute to a "hardening" of arterial walls that hampers delivery of oxygen to the brain, heart, and other vital organs.

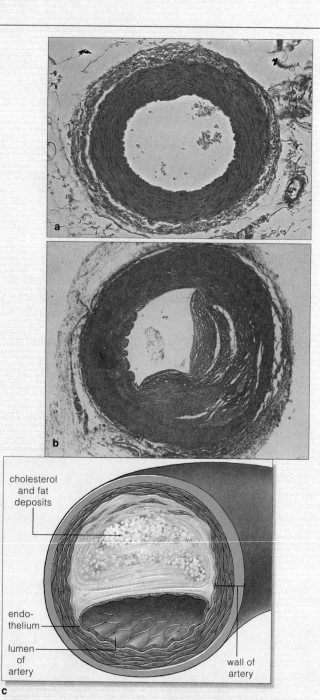

Figure 39.21 Sections from (**a**) a normal artery and (**b**) one with a narrowed lumen. (**c**) Sketch of an atherosclerotic plaque.

Hypertension is called a "silent killer" because affected individuals may show no outward symptoms. Even when they know their blood pressure is high, some people tend to resist helpful medication, changes in diet, and regular exercise. Of 23 million hypertensive Americans, most do not undergo treatment. About 180,000 die each year.

ATHEROSCLEROSIS With *arteriosclerosis*, arteries thicken and lose their elasticity. With atherosclerosis, the condition worsens as cholesterol and other lipids build up in the arterial wall and cause the lumen to narrow (Figure 39.21).

Recall, from the introduction to Chapter 16, that the liver produces enough cholesterol to satisfy the body's needs. Food intake increases the cholesterol level. When circulating in blood, cholesterol is bound to proteins as *low-density* lipoproteins, or **LDLs**. These bind to receptors on cells throughout the body. Cells take up LDLs and their cholesterol cargo for use in cell activities. The excess is attached to proteins as *high-density* lipoproteins, or **HDLs**, and transported back to the liver, where it is metabolized.

In some people, for a variety of reasons, not enough LDL is removed from the blood. As the blood level of LDL increases, so does the risk of atherosclerosis. LDLs, with their bound cholesterol, infiltrate the walls of arteries. In those walls, abnormal smooth muscle cells multiply, connective tissue components increase in mass, and cholesterol accumulates in endothelial cells and the clefts between them. On top of the lipids, calcium deposits actually form microscopic slivers of bone. A fibrous net forms over the entire mass. This *atherosclerotic plaque* sticks out into the arterial lumen (Figure 39.21c).

The bony slivers of the plaque shred the endothelium. Platelets gather at the damaged site, and, in the manner described in Section 39.10, they secrete chemicals that initiate clot formation. The condition worsens as fatty globules in the plaque become oxidized. Many globules take on a form that resembles the surface components of common bacteria—including a type that instigates the formation of bonelike calcium deposits in the lungs. As an awful consequence, the call goes out to bacteria-fighting monocytes. Soon an inflammatory response is under way, and certain chemicals released during the fray activate the genes for bone formation. Normally, this is a good thing; it helps wall off invaders and prevents infection from spreading. It is bad news for arteries.

As plaques and clots grow, they can narrow or block an artery. Blood flow to the tissues serviced by the artery may dwindle to a trickle or stop. A clot that stays in place is a *thrombus*. If it becomes dislodged and then travels the bloodstream, it is an *embolus*. Think of coronary arteries, which have narrow diameters. They are highly vulnerable to clogging by a plaque or clot. When they narrow to one-quarter of their former diameter, the outcome ranges from *angina pectoris*, or mild chest pains, to a heart attack.

Physicians can diagnose atherosclerosis in coronary arteries by *stress electrocardiograms*. These are recordings of the electrical activity of the cardiac cycle while a person is exercising on a treadmill. They also can diagnose the condition by *angiography*. In this procedure, a dye that is opaque to x-rays is injected into the bloodstream.

Severe blockage may require surgery. With *coronary bypass surgery*, a section of an artery from the chest is stitched to the aorta and to the coronary artery below the narrowed or blocked region, as shown in Figure 39.22. With *laser angioplasty*, laser beams directed at the plaques vaporize them. With *balloon angioplasty*, a small balloon inflated within a blocked artery breaks up the plaques.

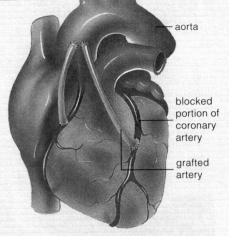

Figure 39.22 Two coronary bypasses (color-coded *green*).

aorta

blocked portion of coronary artery

grafted artery

ARRHYTHMIAS ECGs can be used to detect *arrhythmias*, or irregular heart rhythms (Figure 39.23). Arrhythmias are not always a sign of abnormal conditions. Endurance athletes, for example, may have a below-average resting cardiac rate, or *bradycardia*. As an adaptation to ongoing strenuous exercise, their nervous system has adjusted the cardiac pacemaker's rate of contraction downward. Exercise or stress often causes 100+ heartbeats a minute, or *tachycardia*. *Atrial fibrillation*, an irregular heartbeat, affects more than 10 percent of the elderly and young people with various heart diseases. A coronary occlusion or some other disorder may cause irregular rhythms that rapidly lead to a dangerous condition called *ventricular fibrillation*. In portions of the ventricles, cardiac muscle contracts haphazardly and blood pumping falters. Within seconds, the individual loses consciousness, which might signify impending death. A strong electric shock to the chest may restore normal cardiac function.

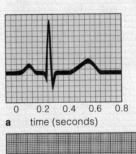

0   0.2   0.4   0.6   0.8

a   time (seconds)

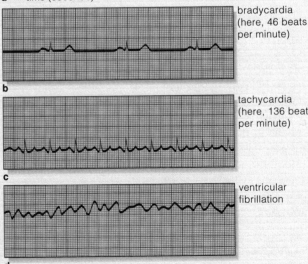

bradycardia (here, 46 beats per minute)

b

tachycardia (here, 136 beats per minute)

c

ventricular fibrillation

d

Figure 39.23 (**a**) An ECG of a single, normal human heartbeat. (**b–d**) Three examples of recordings of arrhythmias.

Small blood vessels are vulnerable to ruptures, cuts, and other damage. **Hemostasis**, a process involving blood vessel spasm, platelet plug formation, and blood coagulation, may repair the damage and stop blood loss.

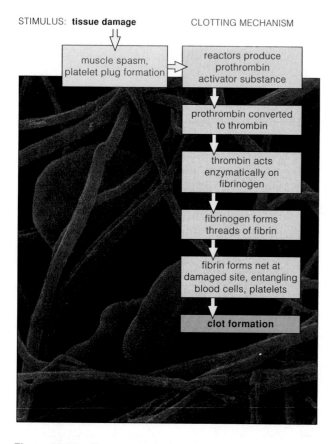

STIMULUS: **tissue damage**   CLOTTING MECHANISM

muscle spasm, platelet plug formation

reactors produce prothrombin activator substance

prothrombin converted to thrombin

thrombin acts enzymatically on fibrinogen

fibrinogen forms threads of fibrin

fibrin forms net at damaged site, entangling blood cells, platelets

clot formation

Figure 39.24 Plasma proteins involved in clot formation. Blood coagulates when damage exposes the collagen fibers in blood vessel walls. Exposure initiates reactions that cause rod-shaped plasma proteins (fibrinogens) to stick together as long, insoluble threads. These adhere to the exposed collagen, forming a net that traps blood cells and platelets. The entire mass is a clot.

As shown in Figure 39.24, smooth muscle in the damaged wall contracts in an automatic response called a spasm. The vasoconstriction temporarily staunches the blood flow. Platelets clump together as a temporary plug in the damaged wall. They also release substances that help prolong the spasm and attract more platelets. Then blood coagulates, or converts to a gel, and forms a clot. Finally, the clot retracts, forming a compact mass, and the breach in the wall is sealed.

The body routinely repairs damage to small blood vessels and prevents blood loss. The repair process includes blood vessel spasm, platelet plug formation, and coagulation.

We conclude this chapter with a brief section on the manner in which the lymphatic system supplements blood circulation. Think of this section as a bridge to the next chapter, on immunity, for the lymphatic system also helps defend the body against injury and attack. The system is composed of drainage vessels, lymphoid organs, and lymphoid tissues. The tissue fluid that has moved into the vessels is called lymph.

Lymph Vascular System

A portion of the lymphatic system called the **lymph vascular system** consists of many tubes that collect and transport water and solutes from interstitial fluid to ducts of the circulatory system. Its main components are **lymph capillaries** and **lymph vessels** (Figure 39.25).

The lymph vascular system serves three functions. First, its vessels are drainage channels for water and plasma proteins that have leaked away from blood at capillary beds and that must be delivered back to the blood circulation. Second, the system also takes up fats that the body has absorbed from the small intestine and delivers them to the blood circulation, in the manner described in Section 42.5. Third, it delivers pathogens, foreign cells and material, and cellular debris from the

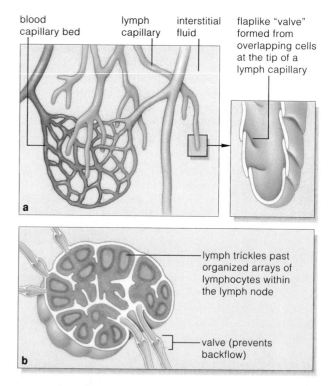

blood capillary bed   lymph capillary   interstitial fluid   flaplike "valve" formed from overlapping cells at the tip of a lymph capillary

a

lymph trickles past organized arrays of lymphocytes within the lymph node

valve (prevents backflow)

b

Figure 39.25 (**a**) Diagram of some of the lymph capillaries at the start of the drainage network called the lymph vascular system. (**b**) Cutaway diagram of a single lymph node. Its inner compartments are packed with organized arrays of infection-fighting white blood cells.

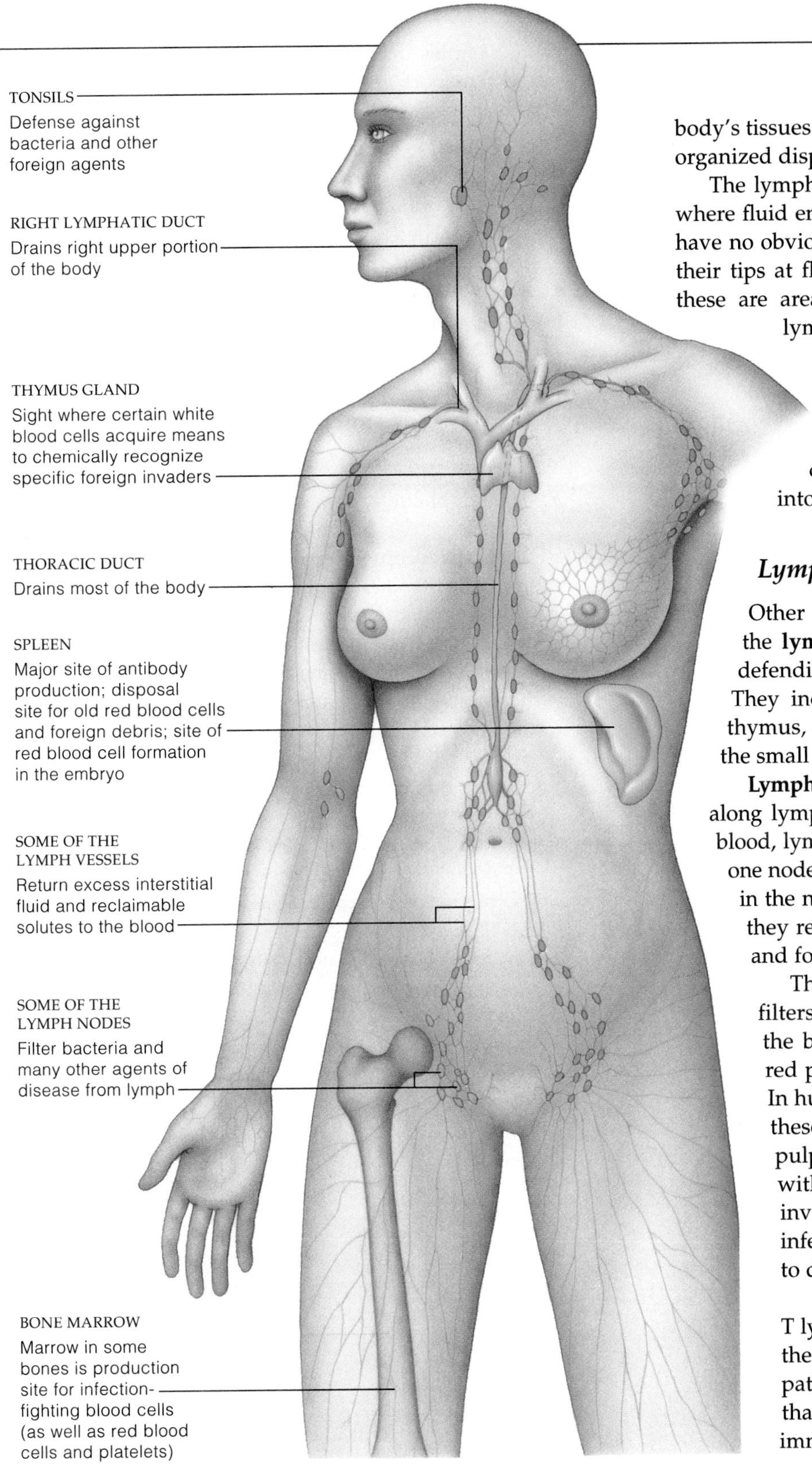

TONSILS
Defense against
bacteria and other
foreign agents

RIGHT LYMPHATIC DUCT
Drains right upper portion
of the body

THYMUS GLAND
Sight where certain white
blood cells acquire means
to chemically recognize
specific foreign invaders

THORACIC DUCT
Drains most of the body

SPLEEN
Major site of antibody
production; disposal
site for old red blood cells
and foreign debris; site of
red blood cell formation
in the embryo

**SOME OF THE
LYMPH VESSELS**
Return excess interstitial
fluid and reclaimable
solutes to the blood

**SOME OF THE
LYMPH NODES**
Filter bacteria and
many other agents of
disease from lymph

BONE MARROW
Marrow in some
bones is production
site for infection-
fighting blood cells
(as well as red blood
cells and platelets)

body's tissues to the lymph vascular system's efficiently organized disposal centers, the lymph nodes.

The lymph vascular system starts at capillary beds, where fluid enters the lymph capillaries. The capillaries have no obvious entrance; water and solutes move into their tips at flaplike "valves." As Figure 39.25a shows, these are areas where endothelial cells overlap. The lymph capillaries merge into lymph vessels, which have a larger diameter. The lymph vessels contain some smooth muscle in their wall as well as valves in their lumen that prevent backflow. They converge into collecting ducts that drain into veins in the lower neck (Figure 39.26).

Lymphoid Organs and Tissues

Other portions of the lymphatic system, called the **lymphoid organs and tissues**, have roles in defending the body against damage and attack. They include lymph nodes, the spleen, and the thymus, as well as tonsils and patches of tissue in the small intestine and appendix.

Lymph nodes are strategically located at intervals along lymph vessels (Figure 39.26). Before entering blood, lymph is filtered as it trickles through at least one node. Masses of lymphocytes take up residence in the nodes after forming in bone marrow. When they recognize an invader, they multiply rapidly and form large armies to destroy it.

The **spleen** is the largest lymphoid organ. It filters pathogens and used-up blood cells from the blood itself. One of its compartments, the red pulp, is a huge reservoir of red blood cells. In human embryos, the red pulp also produces these cells. The other compartment, the white pulp, has masses of lymphocytes associated with blood vessels. If and when a specific invader reaches the spleen during a severe infection, the lymphocytes become mobilized to destroy it, just as they are in lymph nodes.

It is in the **thymus gland** that immature T lymphocytes differentiate in ways that allow them to recognize and respond to specific pathogens. The thymus produces hormones that influence these events. It is central to immunity, the focus of the chapter to follow.

The lymph vascular portion of the lymphatic system returns water and solutes from tissue fluid to blood, and delivers fats and foreign material to lymph nodes. The system's lymph nodes and other lymphoid organs help defend the body against tissue damage and infectious diseases.

Figure 39.26 Components of the human lymphatic system and their functions. The *green* dots represent some of the major lymph nodes. Patches of lymphoid tissue in the small intestine and in the appendix also are part of the lymphatic system.

1. The closed circulatory system of humans and other vertebrates consists of a heart (a muscular pump), blood vessels (arteries, arterioles, capillaries, venules, and veins), and blood. The system functions in rapid internal transport of substances to and from cells.

2. Blood helps maintain favorable conditions for cells. This fluid connective tissue consists of plasma, red and white blood cells, and platelets. It delivers oxygen and other substances to the interstitial fluid around cells. It also picks up cell products and wastes from that fluid.

   a. Plasma, the liquid portion of blood, is a transport medium for blood cells and platelets. Plasma is a solvent for plasma proteins, simple sugars, lipids, amino acids, mineral ions, vitamins, hormones, and several gases.

   b. Red blood cells transport oxygen from the lungs to all body regions. They are packed with hemoglobin, an iron-containing pigment molecule that binds reversibly with oxygen. Red blood cells also transport some carbon dioxide from interstitial fluid to the lungs.

   c. Some phagocytic white blood cells cleanse tissues of dead cells, cellular debris, and anything else detected as not belonging to the body. Other white blood cells (lymphocytes) form great armies that destroy specific bacteria, viruses, and other disease agents.

3. The human heart is an incessantly beating double pump. Each half of the heart has two chambers, called an atrium and a ventricle. Blood flows into the atria, then the ventricles, then on into the great arteries. One-way valves enforce the one-way flow.

4. The heart's partition separates blood flow into two circuits, one pulmonary and the other systemic.

   a. The pulmonary circuit loops between the heart and lungs. Oxygen-poor blood from the systemic veins enters the *right* atrium, is pumped through pulmonary arteries to both lungs, picks up oxygen, then flows through pulmonary veins to the heart's left atrium.

   b. The systemic circuit loops between the heart and all body tissues. Oxygenated blood in the *left* atrium flows into the left ventricle, is pumped into the aorta, then is distributed to capillary beds. There, the blood gives up oxygen and picks up carbon dioxide. Systemic veins return it to the heart's right atrium.

5. Ventricular contraction drives the blood through both circuits. Blood pressure is high in contracting ventricles. It drops successively in arteries, arterioles, capillaries, venules, and veins. It is lowest in relaxed atria.

   a. Arteries are rapid-transport vessels and a pressure reservoir (they smooth out pressure changes resulting from heartbeats and thereby smooth out blood flow).

   b. Arterioles are sites where the volume of flow through organs can be controlled.

   c. Capillary beds are diffusion zones where blood and interstitial fluid exchange substances.

   d. Venules overlap capillaries in function. Veins are rapid-transport vessels and a blood volume reservoir that is tapped to adjust flow volume back to the heart.

6. The cardiac conduction system is the basis for the heart's rhythmic, spontaneous contractions.

   a. Of all the cardiac muscle cells, 1 percent are not contractile. They are specialized to initiate and distribute action potentials, independently of the nervous system. The nervous system only adjusts the rate and strength of the basic contractions; it does not initiate them.

   b. The system's SA node fires action potentials the fastest and is the cardiac pacemaker. Waves of excitation starting here wash over the atria, then down the heart's partition, then up the ventricles. The ventricles contract in a wringing motion that ejects blood from the heart.

7. The lymphatic system has these functions:

   a. Its vascular portion takes up water and plasma proteins that seep out of blood capillaries, then returns them to the blood circulation. It transports absorbed fats and delivers pathogens and foreign material to disposal centers. Lymph capillaries and vessels are components.

   b. Its lymphoid organs and tissues have production centers for lymphocytes. Some are battlegrounds where organized arrays of lymphocytes fight disease agents.

Review Questions

1. Define the functions of the circulatory system and lymphatic system. Distinguish between blood and interstitial fluid. *39.1*

2. Describe the cellular components of blood. Describe the plasma portion of blood. *39.2*

3. Distinguish between systemic and pulmonary circuits. *39.1*

4. State the functions of arteries, arterioles, capillaries, veins, venules, and lymph vessels. *39.1, 39.5, 39.7, 39.8*

5. Distinguish between the functions of the human heart's atria and ventricles. Then label the heart's components: *39.6*

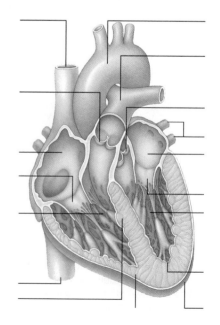

Self-Quiz (Answers in Appendix IV)

1. Cells directly exchange substances with _____.
   a. blood vessels   c. interstitial fluid
   b. lymph vessels   d. both a and b

2. Which are *not* components of blood?
   a. plasma
   b. blood cells and platelets
   c. gases and other dissolved substances
   d. all of the above are components of blood

3. The _____ produces red blood cells, which transport _____ and some _____.
   a. liver; oxygen; mineral ions
   b. liver; oxygen; carbon dioxide
   c. bone marrow; oxygen; hormones
   d. bone marrow; oxygen; carbon dioxide

4. The _____ produces white blood cells, which function in _____ and _____.
   a. liver; oxygen transport; defense
   b. lymph nodes; oxygen transport; pH stabilization
   c. bone marrow; housekeeping; defense
   d. bone marrow; pH stabilization; defense

5. In the pulmonary circuit, the heart's _____ half pumps blood to lungs, then _____ blood flows to the heart.
   a. right; oxygen-poor   c. right; oxygen-rich
   b. left; oxygen-poor   d. left; oxygen-rich

6. In the systemic circuit, the heart's _____ half pumps _____ blood to all body regions.
   a. right; oxygen-poor   c. right; oxygen-rich
   b. left; oxygen-poor   d. left; oxygen-rich

7. Blood pressure is high in _____ and lowest in _____.
   a. arteries; veins   c. arteries; ventricles
   b. arteries; relaxed atria d. arterioles; veins

8. _____ contraction drives blood through the pulmonary circuit and the systemic circuit; blood pressure is highest in contracting _____.
   a. Atrial; ventricles   c. Ventricular; arteries
   b. Atrial; atria   d. Ventricular; ventricles

9. Which is *not* a function of the lymphatic system?
   a. delivers disease agents to disposal centers
   b. produces lymphocytes
   c. delivers oxygen to cells
   d. returns water and plasma proteins to blood

10. Match the type of blood vessel with its major function.
    ____ arteries   a. diffusion
    ____ arterioles   b. control of blood volume distribution
    ____ capillaries   c. transport, blood volume reservoirs
    ____ venules   d. overlap capillary function
    ____ veins   e. transport, pressure reservoirs

11. Match the components with their most suitable description.
    ____ capillary beds   a. two atria, two ventricles
    ____ lymph vascular system b. bathes body's living cells
    ____ heart chambers   c. driving force for blood
    ____ blood   d. zones of diffusion
    ____ heart contractions   e. starts at capillary beds
    ____ interstitial fluid   f. fluid connective tissue

Critical Thinking

1. Shirelle, who is using a light microscope to examine a human tissue specimen, sees red blood cells moving single file through thin-walled tubes. She makes a photomicrograph (Figure 39.27). What type of blood vessel has she captured on film?

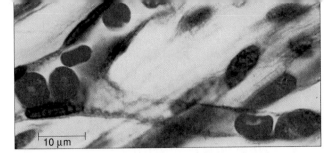

Figure 39.27   Light micrograph of human blood vessels.

2. In individuals who have weak or leaky valves in their veins, fluid pressure associated with the backflow of blood causes venous walls below the valves to bulge outward. In time, the walls become stretched and flabbily distorted, a condition called *varicose veins*. Some people are genetically predisposed to develop the condition, but the cumulative mechanical stress associated with prolonged standing, pregnancy, and aging can contribute to it. With chronic varicosity, the legs themselves become swollen. Explain why this might happen. Also explain why veins close to the leg surface are more susceptible to varicosity than those deeper in the leg tissues.

3. Infection by a hemolytic bacterium (*Streptococcus pyogenes*) may trigger an inflammation that ultimately damages valves in the heart. The disease symptoms of *rheumatic fever* follow. Explain how this disease must affect the heart's functioning and what kinds of symptoms would arise as a consequence.

Selected Key Terms

ABO blood typing 39.4	lymph capillary 39.11
agglutination 39.4	lymph node 39.11
anemia 39.3	lymph vascular system 39.11
aorta 39.5	lymph vessel 39.11
arteriole 39.5	lymphatic system 39.1
artery 39.5	lymphoid organ, tissue 39.11
atrium (heart) 39.6	plasma 39.2
baroreceptor reflex 39.7	platelet 39.2
blood 39.1	pulmonary circuit 39.1
blood pressure 39.7	reabsorption 39.8
capillary (blood) 39.1	red blood cell 39.2
capillary bed 39.1	Rh blood typing 39.4
cardiac conduction system 39.6	spleen 39.11
cardiac cycle 39.6	stem cell 39.2
cardiac pacemaker 39.6	systemic circuit 39.1
cell count 39.2	thymus gland 39.11
circulatory system 39.1	transfusion (blood) 39.4
electrocardiogram CI	ultrafiltration 39.8
endothelium 39.6	vasoconstriction 39.7
heart 39.1	vasodilation 39.7
hemostasis 39.10	vein 39.5
HDL 39.9	ventricle (heart) 39.6
interstitial fluid 39.1	venule 39.5
LDL 39.9	white blood cell 39.2

Readings

Little, R., and W. Little. 1989. *Physiology of the Heart and Circulation*. Fourth edition. Chicago: Year Book Medical.

Robinson, T. F., et al. June 1986. "The Heart as a Suction Pump." *Scientific American* 254(6): 84–91.

Web Site   See *http://www.wadsworth.com/biology* for practice quiz questions, hypercontents, BioUpdates, and critical thinking. The Wadsworth Biology Resource Center provides a wealth of information fully organized and integrated by chapter.

40 IMMUNITY

Russian Roulette, Immunological Style

Until about a century ago, smallpox epidemics swept repeatedly through the world's cities. Some outbreaks were so severe that only half of those who were stricken survived. The survivors were left with permanent scars on the face, neck, shoulders, and arms. But they seldom contracted the same disease again; they were said to be "immune" to smallpox.

No one knew what caused smallpox, but the idea of acquiring immunity was dreadfully appealing. In twelfth-century China, people in good health gambled with deliberate infections. They sought out people with mild cases of smallpox (who were only mildly scarred), removed crusts from the scars, ground them up, and inhaled the powder.

By the seventeenth century, Mary Montagu, wife of the English ambassador to Turkey, was championing inoculation. She went so far as to poke bits of smallpox scabs into her children's skin. Others soaked threads in fluid from sores, then poked the threads into skin.

Some individuals who survived the chancy practices acquired immunity to smallpox, but many developed raging infections. As if the odds were not dangerous enough, the crude inoculation procedures also invited the acquisition of several other infectious diseases.

While this immunological version of Russian roulette was going on, Edward Jenner was growing up in the countryside of England. At the time, it was common knowledge that people who contracted cowpox never got smallpox. (Cowpox is a mild disease that can be transmitted from cattle to humans.) No one thought much about this until 1796, when Jenner, by now a physician, injected material from a cowpox sore into the arm of a healthy boy. Six weeks later, after the reaction subsided, Jenner injected some material from *smallpox* sores into the boy (Figure 40.1a). He had hypothesized that the earlier injection might provoke immunity to

Figure 40.1 (**a**) Statue honoring Edward Jenner's work to develop an immunization procedure against smallpox, one of the most dreaded diseases in human history. (**b**) False-color scanning electron micrograph of a white blood cell being attacked by HIV (*blue particles*), the virus that causes AIDS. Immunologists are working to develop effective weapons against this modern-day scourge.

smallpox. Fortunately he was right; the boy remained free of smallpox. Jenner had developed an effective immunization procedure against a specific pathogen.

The French mocked Jenner's procedure by calling it **vaccination**. The word literally means "encowment." Much later Louis Pasteur, an influential French chemist, developed similar immunization procedures for other diseases. In acknowledgment of Jenner's pioneering work, Pasteur called his procedures vaccinations, also, and only then did the word become respectable.

By Pasteur's time, improved light microscopes were revealing diverse bacteria, fungal spores, and other previously invisible forms of life. As Pasteur himself discovered, microorganisms abound in ordinary air. Did some cause diseases? Probably. Could they settle into food or drink and make it spoil? He demonstrated that they could and did.

Pasteur also found a way to kill most of the suspect disease agents in food or beverages. As he and others knew, boiling killed the agents. He also knew that you cannot boil wine (or beer or milk, for that matter) and end up with the same beverage. He devised a way to heat such beverages at a temperature low enough not to ruin them but high enough to kill most of the resident microorganisms. We still depend on this antimicrobial method, which was named pasteurization in his honor.

In the late 1870s Robert Koch, a German physician, linked a specific microorganism to a specific disease; namely, anthrax. For one experiment, Koch had injected blood from animals with anthrax into healthy ones. The recipients of the injection ended up with blood that teemed with cells of the bacterium *Bacillus anthracis*— and they developed anthrax. Even more convincing, injections of bacterial cells that were cultured outside the animal body also caused the disease!

Thus, by the beginning of the twentieth century, the promise of understanding the basis of infectious disease and immunity loomed large—and the battles against those diseases were about to begin in earnest. Since that time, spectacular advances in microscopy, biochemistry, and molecular biology have increased our understanding of the body's defenses. We now have greater insight into its responses to tissue damage in general and into its immune responses to specific pathogens or tumor cells. The responses are the focus of this chapter.

KEY CONCEPTS

1. The vertebrate body has physical, chemical, and cellular defenses against pathogenic microorganisms, malignant tumor cells, and other agents that can destroy tissues and even the individual itself.

2. During early stages of tissue invasion and damage, white blood cells and certain proteins dissolved in the plasma portion of blood escape from capillaries and execute a rapid, nonspecific counterattack. We call this a nonspecific inflammatory response. Phagocytic white blood cells ingest the invaders. The plasma proteins promote phagocytosis, and some also destroy invaders directly.

3. If the invasion persists, certain white blood cells make immune responses. Those cells can chemically recognize distinct configurations on molecules that are abnormal or foreign to the body, such as those on bacteria and viruses. If the foreign or abnormal molecule triggers an immune response, it is called an antigen.

4. In one type of immune response, some of the white blood cells produce enormous quantities of antibodies. Antibodies are molecules that bind to a specific antigen and tag it for destruction.

5. In another type of immune response, executioner cells directly destroy body cells that have become abnormal, as by infection or by a tumor-producing process.

You continually cross paths with astoundingly diverse **pathogens**—the viruses, bacteria, fungi, protozoans, parasitic worms, and other agents that cause diseases. Having coevolved with most of them, you and other vertebrates have three lines of defense, so you need not lose sleep over this. Most pathogens cannot breach the body surface. If they do, they face chemical weapons and white blood cells, or leukocytes, that attack anything perceived as foreign. Other white blood cells zero in on specific targets. Table 40.1 lists the three lines of defense.

Surface Barriers to Invasion

Most often, pathogens cannot get past skin or the other linings of body surfaces. Think of skin as a habitat of low moisture, low pH, and thick layers of dead cells. Normally harmless bacteria tolerate these conditions. Few pathogens can compete with dense populations of established types unless conditions change. Repeatedly subject your toes to warm, damp shoes, for example, and you may be inviting certain fungi to penetrate the sodden, weakened tissues and cause *athlete's foot*.

Similarly, resident bacteria on the mucous lining of the gut and vagina help protect you—as when lactate, a fermentation product of *Lactobacillus* populations in the vagina, helps maintain a low pH that most bacteria and fungi cannot tolerate. Barriers in branching, tubular

airways leading to your lungs stop airborne pathogens. Air rushing down the tubes flings bacteria against their sticky, mucus-coated walls. In that mucus are protective substances such as **lysozyme**, an enzyme that digests cell walls of bacteria and so invites their death. As a final touch, broomlike cilia in the airways sweep out the trapped and enzymatically whapped pathogens.

The body has additional defenses. Lysozyme and other substances in tears, saliva, and gastric fluid offer protection, as when an outpouring of tears gives the eyes a sterile washing. Urine's low pH and flushing action help bar pathogens from moving into the urinary tract. As another example, diarrhea swiftly flushes irritating pathogens from the gut. Diarrhea must be controlled in children because it causes dehydration, but blocking its action in adults can prolong infection.

Nonspecific and Specific Responses

All animals react to tissue damage. Even simple aquatic invertebrates have phagocytic cells and antimicrobial substances, including lysozymes. But the more complex the animal body, the more complex are the systems that defend it. Think back on the evolution of the circulatory and lymphatic systems in vertebrates that invaded the land (Section 39.1). As circulation of body fluids became more efficient, so did the means for defending the body. Phagocytic cells as well as plasma proteins could move rapidly to tissues under attack, and they could intercept pathogens trickling along the vascular highways.

Vertebrates came to be equipped with sets of plasma proteins. Some proteins promote rapid clot formation after tissue damage; others destroy invading pathogens or target them for phagocytosis. Exquisitely focused responses to specific dangers also evolved.

In short, *internal defenses against a great variety of pathogens are in place even before damage occurs.* Ready and waiting are specialized white blood cells as well as plasma proteins. They take part in a *nonspecific* response to tissue damage in general, not to one pathogen or another. Other white blood cells may recognize unique molecular configurations of a *specific* pathogen. If they do, the resulting "immune" response will run its course whether tissues are damaged or not.

Table 40.1	The Vertebrate Body's Three Lines of Defense Against Pathogens

BARRIERS AT BODY SURFACES (*nonspecific* targets)

Intact skin; mucous membranes at other body surfaces

Infection-fighting chemicals in tears, saliva, etc.

Normally harmless bacterial inhabitants of skin and other body surfaces that can outcompete pathogenic visitors

Flushing effect of tears, saliva, urination, and diarrhea

NONSPECIFIC RESPONSES (*nonspecific* targets)

Inflammation:

1. Fast-acting white blood cells (neutrophils, eosinophils, and basophils)
2. Macrophages (also take part in immune responses)
3. Complement proteins, blood-clotting proteins, and other infection-fighting substances

Organs with pathogen-killing functions (such as lymph nodes)

Some cytotoxic cells (e.g., NK cells) with a range of targets

---

IMMUNE RESPONSES (*specific* targets only)

T cells and B cells; macrophages interact with them

Communication signals (e.g., interleukins) and chemical weapons (e.g., antibodies, perforins)

Intact skin, mucous membranes, antimicrobial secretions, and other barriers at the body surface constitute the first line of defense against invasion and tissue damage.

Inflammation and other internal, nonspecific responses to invasion are the second line of defense.

Immune responses against specific invaders, as executed by armies of white blood cells and their chemical weapons, are the third line of defense.

COMPLEMENT PROTEINS

A set of plasma proteins has roles in both nonspecific and specific defenses. Collectively, these proteins are called the **complement system**. About twenty kinds of complement proteins circulate in the blood in inactive form. If even a few molecules of one kind are activated, they trigger a huge cascade of reactions. They activate many molecules of another complement protein. Each

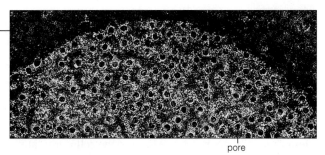

pore

Figure 40.3 Micrograph of a cell surface, showing pores formed by membrane attack complexes.

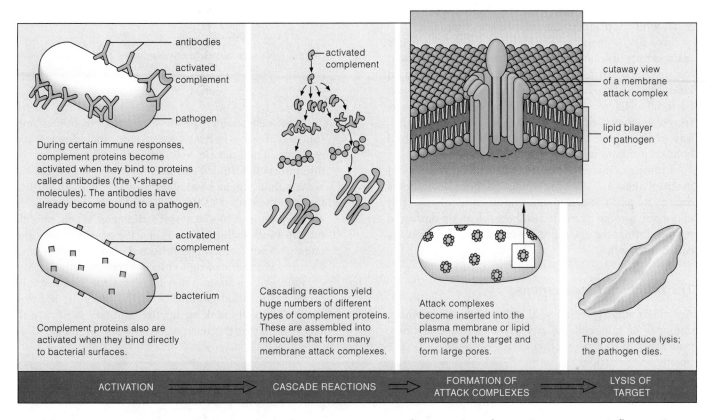

During certain immune responses, complement proteins become activated when they bind to proteins called antibodies (the Y-shaped molecules). The antibodies have already become bound to a pathogen.

Complement proteins also are activated when they bind directly to bacterial surfaces.

Cascading reactions yield huge numbers of different types of complement proteins. These are assembled into molecules that form many membrane attack complexes.

Attack complexes become inserted into the plasma membrane or lipid envelope of the target and form large pores.

The pores induce lysis; the pathogen dies.

ACTIVATION → CASCADE REACTIONS → FORMATION OF ATTACK COMPLEXES → LYSIS OF TARGET

Figure 40.2 Formation of membrane attack complexes. One reaction pathway starts as complement proteins bind to bacterial surfaces; another operates during immune responses to specific invaders. Both pathways produce membrane pore complexes that induce lysis in pathogens.

of these activates many molecules of another kind of protein at the next reaction step, and so on. Deployment of all these molecules has the following effects.

Some of the complement proteins join together to form pore complexes. These molecular structures have an inner channel (Figures 40.2 and 40.3). They become inserted into the plasma membrane of many pathogens, forming pores that induce lysis. **Lysis**, recall, is a gross structural disruption of a plasma membrane or cell wall that leads to cell death. Pore complexes also become inserted into the wall of Gram-negative bacteria. Such cell walls consist of a lipid-rich, outer surface above a peptidoglycan layer. Lysozyme molecules can diffuse through the pores and digest the peptidoglycan.

Some activated proteins promote inflammation, a nonspecific defense response described next. Through their cascades of reactions, they create concentration gradients that attract phagocytic white blood cells to an irritated or damaged tissue. The complement proteins also encourage phagocytes to dine. They can bind to the surface of many types of invaders. When they do so, an invader ends up with a complement "coat." The coat can adhere to phagocytes, and this is rather like putting a basted turkey on a dinner table.

In such ways, the complement proteins target many bacteria, parasitic protistans, and enveloped viruses.

The complement system, a set of twenty or so kinds of plasma proteins circulating in blood, takes part in cascades of reactions that help defend against many bacteria, some parasitic protistans, and enveloped viruses.

The complement system takes part in both nonspecific and specific defenses.

The Roles of Phagocytes and Their Kin

Certain white blood cells take part in an initial response to tissue damage. White blood cells, recall, arise from stem cells in bone marrow. Many of the cells circulate in blood and lymph. A great many take up stations in the lymph nodes as well as the spleen, liver, kidneys, lungs, and brain. Here you may wish to refer to Section 39.2, which introduces the components of blood, and Section 39.11, which introduces the lymphatic system.

Like SWAT teams, three kinds of white blood cells react swiftly to danger in general but are not adapted for sustained battles. **Neutrophils**, the most abundant kind, phagocytize bacteria. They ingest, kill, and digest bacterial cells to simple molecular bits. **Eosinophils** secrete enzymes that make holes in parasitic worms. **Basophils** secrete histamine and other substances that help keep inflammation going after it starts.

Although slower to act, the white blood cells called **macrophages** are the "big eaters." Figure 40.4 shows one of them. A macrophage engulfs and digests just about any foreign agent. It also helps clean up damaged tissues. Immature macrophages circulating in blood are called monocytes.

The Inflammatory Response

An inflammatory response develops when something damages or kills cells of any given tissue. Infections, punctures, burns, and other insults are the triggers. By a mechanism known as **acute inflammation**, the fast-acting phagocytes and complement proteins, as well as other plasma proteins, escape from the bloodstream at capillary beds in the damaged tissue. There they enter interstitial fluid. Localized signs that acute inflammation is under way include redness, heat, swelling, and pain, as listed in Table 40.2.

Table 40.2	Localized Signs of Inflammation and Their Causes
Redness	Arteriolar vasodilation; increase in blood flow to the affected site
Warmth	Arteriolar vasodilation; more blood, carrying more metabolic heat, arrives at site
Swelling	Chemical signals increase capillary permeability; plasma proteins leak out, disrupt fluid balance across wall of capillaries; localized edema
Pain	Nociceptors (pain receptors) stimulated by increased fluid pressure, local chemical signals

Mast cells, which reside in connective tissues and function like basophils, take part in an inflammatory response. They release **histamine** and other substances into interstitial fluid. Their secretions are local chemical signals that trigger vasodilation of arterioles threading through the damaged tissue. Vasodilation, remember, is an increase in a blood vessel's diameter when smooth muscle in its wall relaxes. When the arterioles become engorged with blood, the affected tissue reddens and becomes warmer, owing to blood-borne metabolic heat.

Released histamine also increases the permeability of the thin-walled capillaries in the tissue. It induces endothelial cells making up the capillary wall to pull apart farther at the narrow clefts between them. Thus

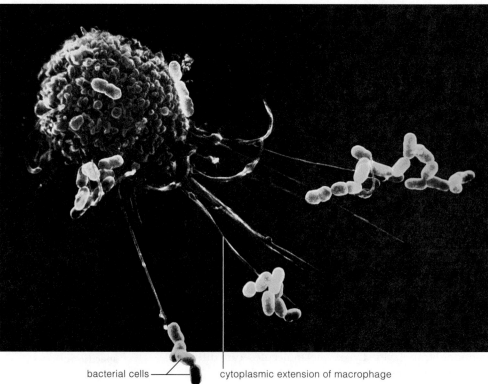

Figure 40.4 False-color scanning electron micrograph of a macrophage (*red*), with some cytoplasmic extensions that made contact with bacterial cells (*green*) in its surroundings. Bacteria are engulfed by this type of phagocytic white blood cell.

bacterial cells —

cytoplasmic extension of macrophage

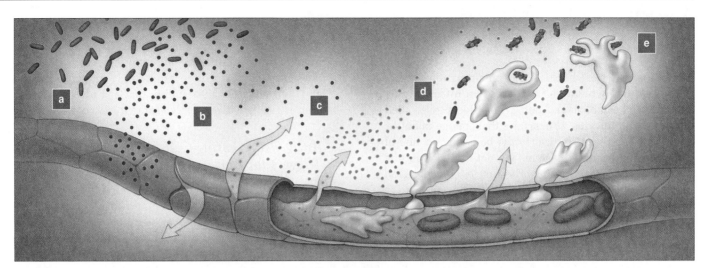

a Bacteria invade a tissue and directly kill cells or release metabolic products that damage tissue.

b Mast cells in tissue release histamine, which then triggers arteriolar vasodilation (hence redness and warmth) as well as increased capillary permeability.

c Fluid and plasma proteins leak out of capillaries; localized edema (tissue swelling) and pain result.

d Complement proteins attack bacteria. Clotting factors wall off inflamed area.

e Neutrophils, macrophages, and other phagocytes engulf invaders and debris. Macrophage secretions attract even more phagocytes, directly kill invaders, and call for fever and for T and B cell proliferation.

Figure 40.5 Acute inflammation in response to a bacterial invasion. The response involves delivering phagocytes and plasma proteins to the tissue. Together, these components of blood inactivate, destroy, or isolate the invaders, remove chemicals and cellular debris, and prepare the tissue for subsequent repair. These are their functions in all inflammatory responses.

the capillaries become "leaky" to plasma proteins that normally do not leave the blood. When some proteins leak out, osmotic pressure increases in the surrounding interstitial fluid. In combination with the higher blood pressure brought about by the increased blood flow to the tissue, ultrafiltration increases and reabsorption decreases across the capillary wall. Localized edema is the outcome of the shift in the fluid balance across the capillary wall. (Here you may wish to review Section 39.8.) The tissue swells with fluid, and its nociceptors give rise to sensations of pain. Typically an individual avoids voluntary movements that might aggravate the pain. This behavior promotes tissue repair.

Within hours of the first physiological responses to the damage, neutrophils are squeezing across capillary walls. They swiftly go to work. Monocytes arrive later, differentiate into macrophages, and engage in sustained action (Figure 40.5). While macrophages are engulfing invaders and debris, they secrete chemical mediators. Mediators called *chemotaxins* attract more phagocytes. An *interleukin* stimulates the formation of B and T cell armies, as described shortly. *Lactoferrin* directly kills bacteria. *Endogenous pyrogen* might trigger the release of prostaglandins, which in turn trigger an increase in the "set point" on the hypothalamic thermostat that controls body temperature. What we call a **fever** is a body temperature that has reached the higher set point.

A fever of about 39°C (100°F) is not a bad thing. It increases body temperature to a level that is "too hot" for the functioning of most pathogens. It also promotes

an increase in a host's defense activities. *Interleukin-1* induces drowsiness, which reduces the body's demands for energy, so more energy can be diverted to the tasks of defense and tissue repair. Macrophages take part in the cleanup and repair operations.

Among the plasma proteins that leak into the tissue are complement proteins and clotting factors of the sort described in Section 39.10. Upon exposure to chemicals secreted by phagocytes and to tissue thromboplastin, fibrin forms and clots develop in the spaces around the inflamed tissue. The clots wall off the inflamed area and typically prevent or delay the spread of invaders and toxic chemicals into the surrounding tissues. After the inflammation subsides, anticlotting factors that had also escaped from the capillaries dissolve the clots.

An inflammatory response develops in a local tissue when cells are damaged or killed, as by infection. It proceeds during both nonspecific and specific defenses of tissues.

Mast cells in damaged or invaded tissues secrete histamine, which causes arterioles to vasodilate and increases capillary permeability to fluid and plasma proteins. The localized vasodilation reddens and warms the tissue. Edema results from the fluid imbalance across the capillary wall. The tissue swelling causes pain.

The response involves phagocytes such as macrophages, which engulf invaders and debris and secrete chemical mediators. It involves plasma proteins, such as complement proteins that target invaders for destruction as well as clotting factors that wall off the inflamed tissue.

THE IMMUNE SYSTEM

Defining Features

Sometimes physical barriers and inflammation are not enough to overwhelm an invader, so an infection may become well established. Then, white blood cells called **B** and **T lymphocytes** form armies that engage in battle.

B and T cells are central to the body's third line of defense—the immune system. We define the **immune system** by two key features. The first is *immunological specificity*, whereby certain kinds of lymphocytes zero in on specific pathogens and eliminate them. The second feature is *immunological memory*, whereby a portion of the T and B cells formed during a first-time confrontation is set aside for a future battle with the same pathogen.

The operating principle for the system is this: *Each kind of cell, virus, or substance bears unique molecular configurations that give it a unique identity.* The unique configurations on an individual's own cells serve as *self* markers. Lymphocytes recognize self markers and will normally ignore them. They also can recognize *nonself* molecular configurations, which are unique to specific foreign agents. When that happens, lymphocytes are stimulated to divide repeatedly, by way of mitosis. And the divisions give rise to huge populations.

As the divisions proceed, subpopulations of the new cells become specialized to respond to the foreign agent in different ways. Some consist of *effector* cells, which are fully differentiated cells that engage and destroy the enemy. Other subpopulations consist of *memory* cells, which enter a resting phase. Instead of engaging in the attack on the specific agent that triggered the initial response, memory cells "remember" it. They will take part in a larger, more rapid response if that same kind of agent shows up again.

Any molecular configuration that triggers formation of lymphocyte armies and is their target is an **antigen**. The most important antigens are certain proteins at the surface of pathogens or tumor cells. As you will see, lymphocytes synthesize receptor molecules that can bind to these configurations. That is how they are able to recognize nonself.

In short, immunological specificity and memory involve three events: *recognition* of antigen, *repeated cell divisions* that form huge populations of lymphocytes, and *differentiation* into subpopulations of effector and memory cells with receptors for one kind of antigen.

Antigen-Presenting Cells—The Triggers for Immune Responses

The plasma membrane of every nucleated cell in the body of every human individual incorporates a variety of proteins. Among these proteins are **MHC markers**, named after the genes that encode the instructions for making them. Certain MHC markers are common at the surface of each nucleated cell in the body. Others are unique to the body's macrophages and lymphocytes.

Think of what happens after a cut allowed bacteria to enter a tissue inside your finger. Lymph vessels pick up some of the inflamed tissue's interstitial fluid and deliver it, along with some invading cells, to the lymph nodes in the vicinity. There, macrophages join the fray. Foreign cells are engulfed and enclosed in vesicles with digestive enzymes that cleave antigen molecules into fragments. The fragments bind to MHC molecules and form **antigen-MHC complexes**. When the vesicles move to the plasma membrane and fuse with it, the complexes are automatically displayed at the macrophage surface.

Any cell displaying processed antigen that is bound with a suitable MHC molecule is an **antigen-presenting cell**. When lymphocytes do encounter such a cell, they take notice (Figure 40.6). *This is the antigen recognition that promotes the cell divisions by which great armies of lymphocytes form.*

Key Players in Immune Responses

The same categories of white blood cells are called into action during each immune response. Figure 40.7 is an overview of how the cells interact. In brief, recognition of antigen-MHC complexes activates **helper T cells**. These produce and secrete substances that induce any responsive T or B lymphocyte to divide and give rise to large populations of effector cells and memory cells. Recognition also activates **cytotoxic T cells**, which can eliminate infected body cells or tumor cells by "touch killing." When they contact a target, they deliver

MHC marker that designates "self" (it occurs only at the surface of body's own cells)

T cells and B cells ignore this

Figure 40.6 Molecular cues that T and B cells either ignore or recognize as a signal to initiate immune responses.

processed antigen, bound to MHC marker, at surface of an antigen-presenting cell

T cells initiate an immune response

antigen (any *unprocessed* foreign or abnormal molecular configuration that lymphocytes recognize as nonself)

B cells initiate an immune response

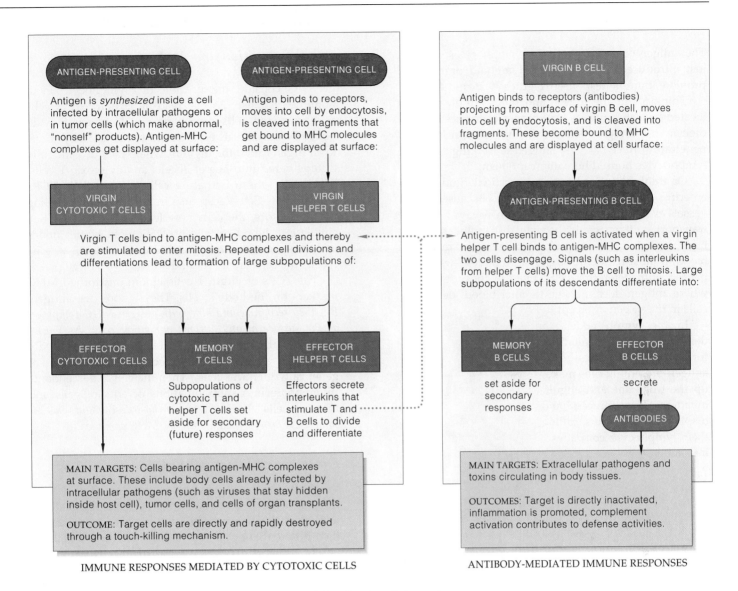

Figure 40.7 Overview of key interactions among B and T lymphocytes during an immune response. Most often, both types of white blood cells are activated when antigen has been detected. An antigen is any large molecule that lymphocytes recognize as not being "self" (normal body molecules). A first-time encounter with antigen elicits a *primary* response. A subsequent encounter with the same type of antigen elicits a *secondary* immune response. A secondary response is larger and more rapid. Memory cells that formed but that were not used during the first battle immediately engage in the second one.

cell-killing chemicals into it. By contrast, B cells make antigen-binding receptor molecules called **antibodies**. When a response is under way, effector B cells secrete staggering numbers of antibody molecules. Only B cells are the basis of *antibody-mediated* responses.

Control of Immune Responses

Antigen provokes an immune response—and removal of antigen stops it. For example, by the time the tide of battle turns, effector cells and their chemical secretions have already killed most of the antigen-bearing agents inside the body. With fewer antigen molecules around

to stimulate the cells, the response declines, then stops. As a final example, inhibitory signals from cells with suppressor roles help shut down immune responses.

Antigens are nonself molecular configurations which, when recognized by certain lymphocytes, trigger immune responses. Helper T cells, cytotoxic T cells, B cells, and their secretions execute these responses.

Following antigen recognition, large T and B cell armies form by repeated mitotic cell divisions. These differentiate into subpopulations of effector cells and memory cells, all of which are sensitized to that one kind of antigen.

The antigen-presenting cells and lymphocytes we have just introduced interact within lymphoid organs that promote immune responses (Figures 39.26 and 40.8).

Think about the tonsils and other lymph nodules located beneath mucous membranes of the respiratory, digestive, and reproductive systems. Just after invaders penetrate surface barriers, antigen-presenting cells and lymphocytes housed here intercept them.

Or think about antigen in interstitial fluid that is entering the lymph vascular system. Because lymph vessels eventually drain into the expressways for blood transport, antigen could become distributed to every body region. However, before antigen can reach the blood, it must trickle through lymph nodes—which are packed with defending cells. Even in those few cases where antigen does manage to enter blood, defending cells in the spleen intercept it.

In the lymph nodes, the defending cells are organized for utmost effectiveness. The antigen-presenting cells make up the front line and engulf invaders. They process and display antigen, thus calling their lymphocyte comrades into action.

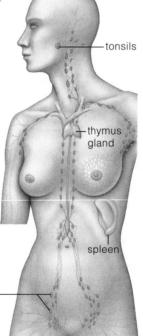

tonsils

thymus gland

spleen

location of antigen-presenting cells and lymphocytes in a lymph node, cross-section

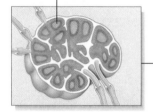

Figure 40.8 Organized arrays of antigen-presenting cells and lymphocytes in lymph nodes.

The cell divisions that produce subpopulations of effector and memory cells proceed in lymph nodes. As lymph drains through, it moves effector activities to the back of the organ and beyond. And all the while, virgin and memory cells circulate through the lymph node, reconnoitering at the front line.

Antigen-presenting cells and lymphocytes intercept and battle pathogens in organized ways within lymphoid organs and tissues, especially the lymph nodes.

T Cell Formation and Activation

Let's first consider the functions of T lymphocytes in an immune response. As you read in Section 39.2, T cells arise from stem cells in bone marrow. However, they do not fully develop in bone marrow. Rather, they travel to an organ called the thymus, where they become fully differentiated into helper T cells and cytotoxic T cells. Specifically, these immature cells acquire their **TCRs** (short for *T-Cell Receptors*) in the thymus. Bristling with receptors, the cells now leave the thymus. They circulate in blood or take up stations in lymph nodes and the spleen, as virgin T cells. In this context, "virgin" means the cells are as yet undisturbed (by antigen).

The TCRs of virgin T cells ignore unadorned MHC markers on the body's cells. They ignore free antigen. *But they recognize and bind with antigen-MHC complexes at the surface of any antigen-presenting cells.* As Figure 40.9 shows, binding induces T cells to divide repeatedly and give rise to large clones. (A clone is a population of genetically identical cells.) Then the clonal descendants differentiate into subpopulations of effector cells and memory cells. *And every one of those descendants has the same TCR for one kind of antigen-MHC complex.*

Functions of Effector T Cells

What actions do subpopulations of effector T cells take? Effector helper T cells secrete interleukins, the chemical mediators that fan repeated mitotic cell divisions and then differentiation of any responsive T and B cells, as described shortly. Effector cytotoxic T cells are killers; they respond to antigen-MHC complexes on body cells that are already infected by intracellular pathogens, such as viruses, and on tumor cells. The complex serves as a "double signal" that tells the killers to attack the cells that bear it (Figure 40.9e).

Effector cytotoxic T cells destroy infected cells with a touch-kill mechanism. They secrete *perforins*, protein molecules that form doughnut-shaped pores in a target cell's plasma membrane. (The pores look similar to the ones shown in Figure 40.3.) These effectors also secrete chemicals that induce cell death by way of **apoptosis**. As described in the introduction to Chapter 15, a target cell is induced to commit suicide. Its cytoplasm dribbles out, its organelles are disrupted, and its DNA becomes fragmented. Having made its lethal hit, the cytotoxic T cell disengages quickly and moves on to new targets.

Cytotoxic T cells also contribute to the rejection of tissue and organ transplants. Parts of MHC markers on donor cells are different enough from the recipient's to be recognized as antigens. But other parts are similar enough to complete the double signal. MHC typing and matching donors to recipients minimize the risk.

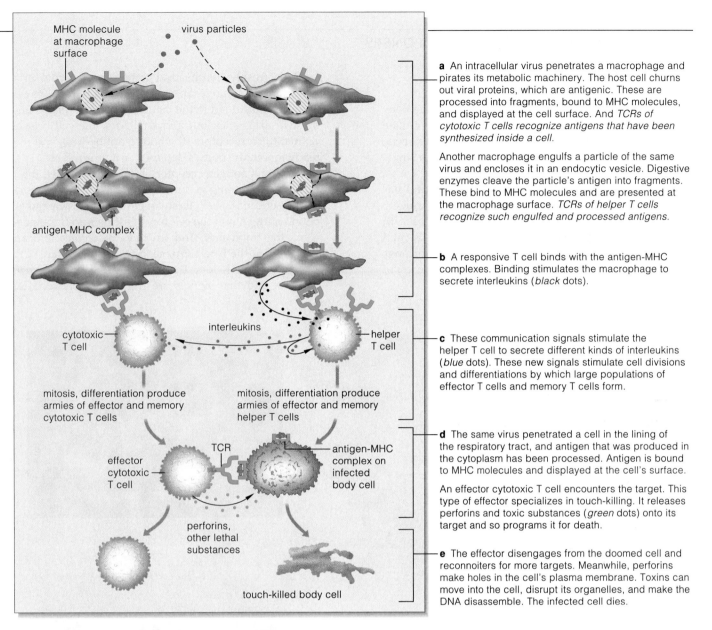

a An intracellular virus penetrates a macrophage and pirates its metabolic machinery. The host cell churns out viral proteins, which are antigenic. These are processed into fragments, bound to MHC molecules, and displayed at the cell surface. And *TCRs of cytotoxic T cells recognize antigens that have been synthesized inside a cell.*

Another macrophage engulfs a particle of the same virus and encloses it in an endocytic vesicle. Digestive enzymes cleave the particle's antigen into fragments. These bind to MHC molecules and are presented at the macrophage surface. *TCRs of helper T cells recognize such engulfed and processed antigens.*

b A responsive T cell binds with the antigen-MHC complexes. Binding stimulates the macrophage to secrete interleukins (*black* dots).

c These communication signals stimulate the helper T cell to secrete different kinds of interleukins (*blue* dots). These new signals stimulate cell divisions and differentiations by which large populations of effector T cells and memory T cells form.

d The same virus penetrated a cell in the lining of the respiratory tract, and antigen that was produced in the cytoplasm has been processed. Antigen is bound to MHC molecules and displayed at the cell's surface.

An effector cytotoxic T cell encounters the target. This type of effector specializes in touch-killing. It releases perforins and toxic substances (*green* dots) onto its target and so programs it for death.

e The effector disengages from the doomed cell and reconnoiters for more targets. Meanwhile, perforins make holes in the cell's plasma membrane. Toxins can move into the cell, disrupt its organelles, and make the DNA disassemble. The infected cell dies.

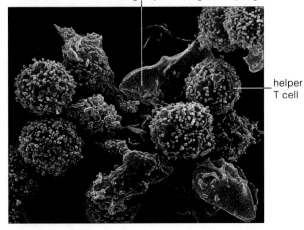

Figure 40.9 Diagram of a T cell-mediated immune response. In this example, the response involves an antigen-MHC complex that activates T cells. The micrograph shows helper T cells associating with an antigen-presenting macrophage.

Regarding the Natural Killer Cells

Other cytotoxic cells, including **natural killer cells** (NK cells), also arise from stem cells in bone marrow. They appear to be lymphocytes, but not T or B cells. Their arousal does not depend on a double signal (that is, an antigen-MHC complex). NK cells reconnoiter for tumor cells and virus-infected cells, then touch-kill them. They possibly recognize odd molecular configurations at the surface of their targets.

T cells arise in bone marrow. Later, in the thymus, they acquire TCRs (receptors for self markers and bound antigen).

Effector helper T cells secrete interleukins that trigger the cell divisions and differentiation into huge armies against specific antigens. Effector cytotoxic cells touch-kill infected cells or tumor cells, even foreign cells of transplants.

ANTIBODY-MEDIATED RESPONSES

B Cells and the Targets of Antibodies

Like T cells, the B cells also arise from stem cells in bone marrow and start down a pathway that will culminate in full differentiation. Along *their* pathway, however, B cells start synthesizing many, many copies of a single kind of antibody molecule.

Although all antibodies are proteins, each kind has binding sites that only match up to a particular antigen. They are more or less Y-shaped, with a tail and with two arms that bear identical antigen receptors. Section 40.9 provides a closer look at these molecules. For now, we can simply think of them as Y-shaped structures, of the sort shown in Figure 40.10.

Each freshly synthesized antibody molecule of a maturing B cell moves to the plasma membrane. Its tail becomes embedded in the membrane's lipid bilayer and the two arms stick out above it. Soon the cell bristles with antigen receptors (the bound antibodies), and it is ready to join the body's defenses as a virgin B cell.

When its antigen receptors lock onto a target, the B cell does something you might not expect. *It becomes an antigen-presenting cell.* First, an endocytic vesicle moves bound antigen into the cell for digestion into fragments. Next, the fragments bind to MHC molecules and are presented at the B cell surface. Now suppose that TCRs of a responsive helper T cell bind to the antigen-MHC complex and that signals are transferred between the

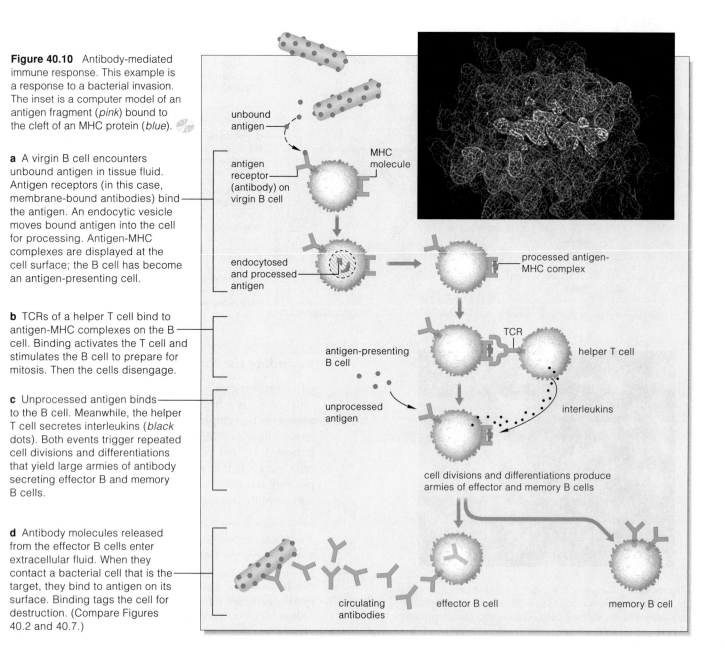

Figure 40.10 Antibody-mediated immune response. This example is a response to a bacterial invasion. The inset is a computer model of an antigen fragment (*pink*) bound to the cleft of an MHC protein (*blue*).

a A virgin B cell encounters unbound antigen in tissue fluid. Antigen receptors (in this case, membrane-bound antibodies) bind the antigen. An endocytic vesicle moves bound antigen into the cell for processing. Antigen-MHC complexes are displayed at the cell surface; the B cell has become an antigen-presenting cell.

b TCRs of a helper T cell bind to antigen-MHC complexes on the B cell. Binding activates the T cell and stimulates the B cell to prepare for mitosis. Then the cells disengage.

c Unprocessed antigen binds to the B cell. Meanwhile, the helper T cell secretes interleukins (*black* dots). Both events trigger repeated cell divisions and differentiations that yield large armies of antibody secreting effector B and memory B cells.

d Antibody molecules released from the effector B cells enter extracellular fluid. When they contact a bacterial cell that is the target, they bind to antigen on its surface. Binding tags the cell for destruction. (Compare Figures 40.2 and 40.7.)

unbound antigen

antigen receptor (antibody) on virgin B cell

MHC molecule

endocytosed and processed antigen

processed antigen-MHC complex

antigen-presenting B cell

TCR

helper T cell

unprocessed antigen

interleukins

cell divisions and differentiations produce armies of effector and memory B cells

circulating antibodies

effector B cell

memory B cell

T and B cell. The cells soon disengage. When the B cell encounters unprocessed antigen, its surface antibodies bind it. The binding, in combination with interleukins secreted from nearby helper T cells, drives the B cell to mitosis. Its clonal descendants differentiate into effector and memory B cells. The effectors (also called plasma cells) produce and secrete huge numbers of antibody molecules. When freely circulating antibody molecules bind antigen, they tag an invader for destruction, as by phagocytes and complement activation.

The main targets of antibody-mediated responses are extracellular pathogens and toxins, which are freely circulating in tissues or body fluids. *Antibodies cannot bind to pathogens or toxins that are hidden in a host cell.*

The Immunoglobulins

During immune responses, B cells produce four classes of antibodies in abundance and lesser quantities of another. Collectively, the five classes of antibodies are called **immunoglobulins**, or **Igs**. They are the protein products of gene shufflings that proceed while B cells mature and while an immune response is under way. The molecules in each class have antigen-binding sites *and* other sites with specialized functions.

IgM antibodies are the first to be secreted during immune responses. They trigger complement cascades, and they bind targets together in clumps—which are more handily eliminated by phagocytes. (Remember the agglutination responses described in Section 39.4?) *IgD* antibodies associate with IgM on virgin B cells, but their function is not yet understood.

IgG antibodies activate complement proteins and neutralize many toxins. These long-lasting antibodies are the only ones to cross the placenta. They can protect the fetus and newborn with the mother's acquired immunities. IgGs also are secreted into the early milk produced by mammary glands, then are absorbed into the suckling newborn's bloodstream.

IgA antibodies enter mucus-coated surfaces of the respiratory, digestive, and reproductive tracts, where they may neutralize infectious agents. Mother's milk delivers them to the mucous lining of a newborn's gut.

IgE triggers inflammation after attacks by parasitic worms and other pathogens. As described later, it also figures in allergies. The tails of IgE antibodies bind to basophils and mast cells, and the antigen receptors face outward. Antigen binding induces basophils and mast cells to release substances that promote inflammation.

Antibodies that are secreted by B cells bind to antigens of extracellular pathogens or toxins and tag them for disposal, as by phagocytes and complement activation.

Carcinomas, sarcomas, leukemia—these chilling words refer to malignant tumors in skin, bone, and other tissues. Such tumors arise when viral attack, irradiation, or chemicals alter genes and cells turn cancerous (Section 15.6). The transformed cells divide repeatedly. Unless they are destroyed or surgically removed, they kill the individual.

Often the surface of a transformed cell bears abnormal proteins and protein fragments bound to MHC. Defense responses to the transformed cells can make the tumor regress but may not be enough to destroy it. Also, some tumors release many copies of the abnormal proteins. If these saturate antigen receptors on the defenders, the tumor may escape detection. If the tumor hides long enough and reaches a critical mass, it may overwhelm the immune system's capacity for an effective response.

Researchers hope to develop procedures to enhance immunological defenses against tumors as well as against certain pathogens. This prospect is called *immunotherapy.*

MONOCLONAL ANTIBODIES   A cancer patient might benefit from injections of mass-produced antibodies against tumor-specific antigens. But mature B cells do not live long enough in culture to mass-produce them. Also, being end cells, B cells cannot reproduce. Some time ago, Cesar Milstein and Georges Kohler showed how to make antibody "factories." They injected an antigen into a mouse, which made antibodies against it. They fused antibody-producing B cells from the mouse with cells extracted from B cell tumors. Some of the descendants of the hybrid cells divided nonstop, and they, too, produced the antibody. Clones of such hybrid cells are now being maintained indefinitely. They make identical copies of antibodies in useful amounts. Their products are known as *monoclonal antibodies.*

Several cancer patients have been inoculated with monoclonal antibodies that are expected to home in on the malignant tumors. In some cases, cell-killing chemicals have been artificially attached to such antibodies. At this writing, the success of clinical trials is limited.

MULTIPLYING THE TUMOR KILLERS   Lymphocytes often infiltrate tumors. Researchers have removed them from a tumor and exposed them to an interleukin (lymphokine). The result of the deliberate exposure is a population of tumor-infiltrating lymphocytes with enhanced killing abilities. The *LAK* cells (short for lymphokine-activated killers) appear to be somewhat effective when injected back into the patient.

THERAPEUTIC VACCINES   *Therapeutic vaccines* might serve as wake-up calls against elusive tumor cells. One idea is to genetically engineer antigen so that it becomes more obvious to killer lymphocytes. For example, Lynn Spitler has been attempting to develop a prostate-specific antigen that is present on all prostate cancers.

Formation of Antigen-Specific Receptors

The variety of antigens in your surroundings is mind-boggling. Collectively, however, antigen receptors of all T and B cell populations in your body show staggering diversity—enough to recognize about a billion different antigens by one estimate. How does the diversity arise?

For the answer, start with the knowledge that all antigen receptors of a single T or B cell are identical—and all are proteins. For example, a Y-shaped antibody molecule consists of four polypeptide chains bonded together (Figure 40.11). Certain parts of each chain, the *variable* regions, fold in ways that produce grooves and bumps with a certain charge distribution. Only antigen that has complementary grooves, bumps, and charge distribution will be able to bind with them.

Receptor diversity begins with rearrangements in the DNA that codes for such variable regions. For example, take a look at Figure 40.12. As a B cell matures, one of many DNA sequences called the V segments becomes joined at random to one of distant DNA sequences called J segments. The DNA intervening between the joined V and J segments loops out and is excised. In this way, the B cell ends up with a unique DNA sequence.

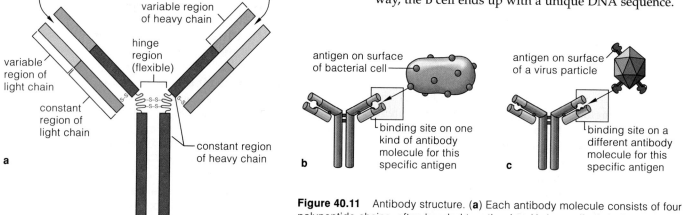

Figure 40.11 Antibody structure. (**a**) Each antibody molecule consists of four polypeptide chains, often bonded together in a Y shape. (**b,c**) At the antigen-binding sites of the molecule, antigen fits into grooves and onto protrusions.

Figure 40.12 Generation of antibody diversity. Antibodies are proteins, and instructions for building proteins are encoded in genes. In the chromosomes with antibody genes, extensive DNA regions contain different versions of segments that code for the variable regions of an antibody molecule.

The different V and J segments shown here are examples. They code for the variable region of a light chain (compare Figure 40.11a). As each B cell matures, a recombination event occurs in this region. In this example, any one of the V segments may be joined to any one of the J segments. Afterward, the DNA intervening between them is excised. The new sequence is attached to a C (*Constant*) segment, and this completes a rearranged antibody gene—which also will be present in all of the descendants of the cell.

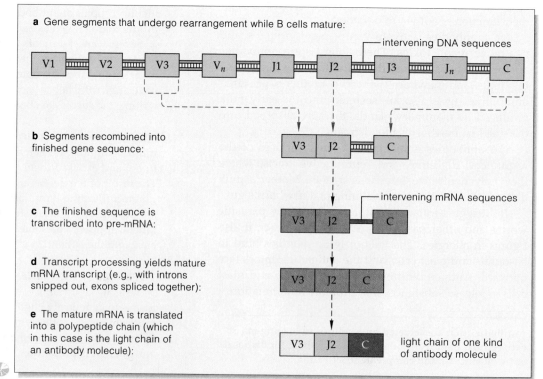

a Gene segments that undergo rearrangement while B cells mature:

intervening DNA sequences

V1 V2 V3 V_n J1 J2 J3 J_n C

b Segments recombined into finished gene sequence:

V3 J2 C

c The finished sequence is transcribed into pre-mRNA:

intervening mRNA sequences

V3 J2 C

d Transcript processing yields mature mRNA transcript (e.g., with introns snipped out, exons spliced together):

V3 J2 C

e The mature mRNA is translated into a polypeptide chain (which in this case is the light chain of an antibody molecule):

V3 J2 C   light chain of one kind of antibody molecule

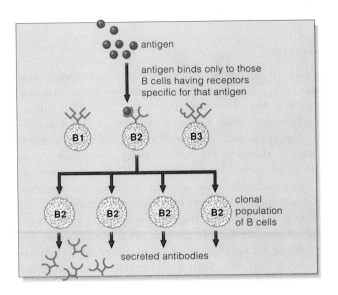

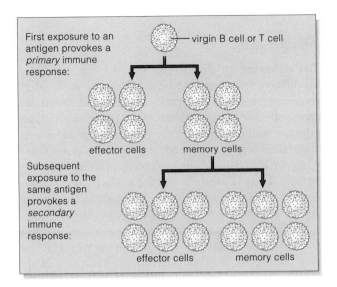

Figure 40.13 Clonal selection of a B cell that produced the specific antibody that can combine with a specific antigen. Only antigen-selected B cells (and T cells) are activated and give rise to a clonal population of immunologically identical cells.

Figure 40.14 Immunological memory. Not all B and T cells are used in a primary immune response to an antigen. A large number continue to circulate as memory cells, which become activated during a secondary immune response.

The same kinds of DNA rearrangements have roles in producing the variable regions of the TCR molecules of maturing T cells.

Some time ago, Mafarlane Burnet developed a *clonal selection* hypothesis that helped point the way to our current view of receptor diversity. He proposed that antigen "chooses" (binds to) one lymphocyte from all the various types in the body, because that lymphocyte has the receptor specific for it. Repeated mitotic cell divisions then give rise to a clone of cells that carry out the response (Figure 40.13).

Immunological Memory

The clonal selection theory explains how an individual can have "immunological memory" of a first encounter with antigen. The term refers to the body's capacity to make a *secondary* immune response to any subsequent encounter with the same type of antigen that provoked the primary response (Figure 40.14).

Memory cells that form during a primary immune response do not engage in that battle. They circulate for years or for decades. Compared to the virgin cells that initiate a primary response, the patrolling battalions consist of far more cells and intercept antigen far sooner. Effector cells form sooner, in greater numbers, so the infection is terminated before the host gets sick. Even greater numbers of memory T and B cells form during a secondary response. Figure 40.15 shows an example of this. In evolutionary terms, the advance preparations against subsequent encounters with a pathogen bestow a survival advantage on the individual.

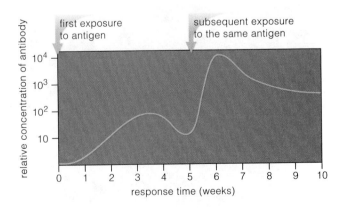

Figure 40.15 Differences in magnitude and duration between a primary and a secondary immune response to the same antigen. In this example, the primary response peaked twenty-four days after it started. The secondary response peaked after only seven days (the span between weeks five and six). Antibody concentration during the secondary response was 100 times greater (10^2 to 10^4).

By recombination of segments, drawn at random from the receptor-encoding regions of DNA, each T cell or B cell receives a gene sequence for one of a billion possible kinds of antigen receptors.

Immunological specificity means the clonal descendants of an antigen-selected cell will react only with the selecting antigen.

Immunological memory refers to the capacity to make a secondary (faster, greater) immune response to a pathogen that caused a primary response in an individual.

Immunization

After reflecting on the chapter introduction, you may already have an understanding that **immunization** refers to various processes that promote increased immunity against specific diseases. With *active* immunization, an antigen-containing preparation known as a **vaccine** is either taken orally or injected into the body, as in Figure 40.16. An initial injection triggers a primary immune response. Later, a subsequent injection (a booster) elicits a secondary response, with the rapid formation of more effector cells and memory cells that can provide long-lasting protection against the disease.

Many vaccines are manufactured from weakened or killed pathogens, as when Sabine polio vaccine is made from weakened polio virus particles. Others are based on inactivated natural toxins, such as the bacterial toxin that causes tetanus. Others are made of harmless viruses, genetically engineered so genes from three or more different viruses are inserted into their DNA or RNA. After vaccination with an engineered virus, the incorporated genes are expressed, antigens are produced, and immunity is established.

Passive immunization helps individuals already infected with pathogens that cause diphtheria, tetanus, measles, hepatitis B, and some other diseases. A person at risk receives injections of purified antibody, the best source of which is some other individual who already has produced a large amount of the required antibody. The effects do not last long, for the patient's B cells are not producing antibodies. Nevertheless, the injection of antibody molecules often can help counter the immediate attack.

Figure 40.16 From the Centers for Disease Control and Prevention, the 1995 immunization guidelines for children living in the United States. Pediatricians routinely immunize infants and children during office visits. Low-cost or free vaccinations are available at many community clinics and health departments.

Allergies

In 8 to 10 percent of the people in the United States alone, normally harmless substances provoke immune responses that cause inflammation, excessive mucus secretion, and other vexing problems. Such substances are **allergens**, and the response to them is an **allergy**. Common allergens are pollen, many drugs and foods, dust mites, fungal spores, insect venom, and cosmetics.

Some individuals are genetically inclined to develop allergies. Infections, emotional stress, or changes in air temperature also may trigger reactions that otherwise might not occur. Upon exposure to certain antigens, IgE antibodies are secreted and bind to mast cells. When the IgE binds antigen, mast cells secrete prostaglandins, histamine, and other substances that fan inflammation. They stimulate mucus secretion and cause airways to constrict. Stuffed sinuses, labored breathing, a drippy nose, and sneezing are key symptoms of the allergic response in *asthma* and *hay fever* (Figure 40.17).

In a few cases, the inflammatory reactions wash through the body and bring about a life-threatening condition called *anaphylactic shock*. For example, a person who is allergic to wasp or bee venom can die within minutes of one sting. Airways to the lungs constrict massively, and fluid escapes rapidly from dilated, grossly permeable capillaries. Blood pressure plummets and may lead to circulatory collapse.

Antihistamines (anti-inflammatory drugs) often relieve the mild, short-term symptoms of allergies. Over time, a patient might try a desensitization program. First, skin tests identify the offending allergens. Inflammatory responses to some can be

Age	Recommended Vaccines
At birth	Hepatitis B
2 months	Hepatitis B, DPT (diphtheria, whooping cough, tetanus), Hib (*Hemophilus influenzae*)
2–4 months	Hepatitis B
4 months	Polio, DPT, Hib
6 months	DPT, Hib
6–18 months	Hepatitis B, polio
12–15 months	Hib, MMR (measles, mumps, rubella)
12–18 months	DPT
4–6 years	Polio, DPT, MMR (final dose of MMR can be at 11–12 years)
11–12 years	DT (diphtheria, tetanus)

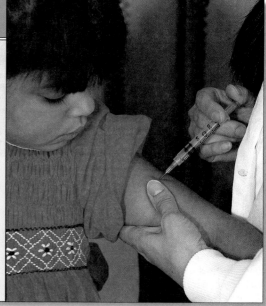

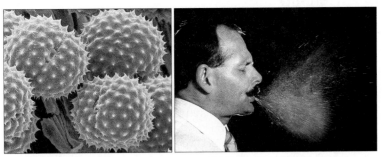

RAGWEED POLLEN . . .   . . . AND SOMETHING IT CAN PROVOKE

Figure 40.17 One of the effects of pollen and other allergens in sensitive people. Allergy sufferers who moved to deserts to escape pollen brought it with them. Half the human population in Tucson, Arizona, is now sensitized to pollen from olive and mulberry trees, planted far and wide in cities and the suburbs.

Figure 40.18 A case of severe combined immunodeficiency, or SCID. Ashanthi DeSilva was born without an immune system. She has a mutated gene for ADA (adenosine deaminase), an enzyme. Without ADA, her cells cannot break down adenosine. As a result, a reaction product accumulates that is toxic to lymphocytes. The disorder's symptoms are caused by infections that cannot be controlled. They include high fever, severe ear and lung infections, diarrhea, and an inability to gain weight.

blocked if the patient can be stimulated to make IgG instead of IgE. Larger doses of specific allergens are administered slowly. Each time, the body makes more circulating IgG and memory cells. The IgG binds with allergen that it encounters and blocks its attachment to IgE, and thereby blocks inflammation.

Autoimmune Disorders

In an **autoimmune response**, the immune system acts against self antigens. Consider *Grave's disorder*, in which the body makes too many thyroid hormone molecules. Feedback mechanisms control the production of the hormones, which affect metabolic rates and the growth and development of many tissues. Affected individuals produce antibodies that bind to receptors on cells that make the hormones. The antibodies do not respond to feedback controls, and they trigger overproduction of thyroid hormones. Disease symptoms typically include elevated metabolic rates, heart fibrillations, excessive sweating, nervousness, and weight loss.

Consider *myasthenia gravis*, a progressive weakening of muscles. Here, antibodies that bind to acetylcholine receptors on skeletal muscle cells cause the disorder.

Finally, consider the chronic inflammation of skeletal joints associated with *rheumatoid arthritis*. Patients are genetically predisposed to this disorder. Macrophages, T cells, and B cells are activated by antigens associated with skeletal joints. Immune responses are made against the body's own collagen molecules and against antibody that has become bound to as-yet-unidentified antigen. Complement activation and inflammation cause more damage in tissues of skeletal joints. So do skewed repair mechanisms. Eventually, the joints fill with synovial membrane cells and become immobilized.

Deficient Immune Responses

When the body has inadequate numbers of functioning lymphocytes, immune responses are not effective. Such

Ashanthi's parents consented to the first federally approved gene therapy for humans. Researchers used genetic engineering methods to splice the ADA gene into the genetic material of a harmless virus. They used the modified virus rather like a hypodermic needle. They allowed it to deliver copies of the "good" gene into Ashanthi's bone marrow cells. Some of the cells incorporated the gene in their DNA and started to synthesize the missing enzyme. At this writing, Ashanthi is now in her teens. Like other ADA-deficient patients who have started treatment, she is doing well.

In other cases, researchers enlist bone marrow stem cells from blood in the umbilical cord of affected newborns. (The cord, which connects the fetus to the placenta during pregnancy, is discarded after childbirth.) They expose the cells to viruses that deliver copies of the ADA gene into them and to factors that stimulate mitotic cell division and growth. The cells are reinserted into the newborns.

severe combined immunodeficiencies (SCIDs) result from heritable disorders as well as from various assaults on the body by outside agents. Deficient or nonexistent immune responses make the person highly vulnerable to infections that are not life threatening to the general population. Figure 40.18 describes one of the heritable disorders. Another is the acquired immunodeficiency syndrome (AIDS). The next section describes how HIV, the virus that causes AIDS, replicates inside certain lymphocytes and destroys the body's capacity to fight infections. We also return to this topic in Section 45.14.

Immunization programs boost immunity to specific diseases.

Certain heritable disorders or attacks by certain pathogens and other agents can result in misdirected, compromised, or nonexistent immunity.

40.11 AIDS—THE IMMUNE SYSTEM COMPROMISED

CHARACTERISTICS OF AIDS   AIDS is a constellation of disorders that follow an infection by a pathogen called the human immunodeficiency virus, or HIV. The virus cripples the immune system, so that the body becomes highly susceptible to usually harmless infections and some otherwise rare forms of cancer.

At this writing, researchers have not developed any vaccine that will work against the known forms of the virus (HIV-1 and HIV-2). Also at this writing, *there is no cure for those already infected*.

By current estimates, more than 1 million Americans are infected. Worldwide, an estimated 22.6 million people are infected; 6 million are already dead. By the turn of the century, the number of infected people may be as high as 60–70 million.

At first an infected person might appear to be in good health, suffering no more than a bout of "the flu." Then he or she starts displaying symptoms that foreshadow AIDS. Typically, symptoms include persistent weight loss, fever, fatigue, bed-drenching night sweats, and many enlarged lymph nodes. In time, diseases resulting from certain opportunistic infections are signs of AIDS. The diseases are rare in the population at large. They include yeast infections of the mouth, esophagus, vagina, and elsewhere, as well as a form of pneumonia caused by *Pneumocystis carinii*. Spots resembling bruises may appear, especially on legs and feet. These are signs of Kaposi's sarcoma, a form of cancer that develops from endothelial cells of blood vessels. The immune system cannot control the infections or cancers, which end up killing the person.

HOW HIV REPLICATES   HIV infects antigen-presenting macrophages and helper T cells. (Immunologists also call helper T cells the CD4 lymphocytes.) HIV is a retrovirus.

Each virus particle has an outermost lipid envelope, which is a bit of plasma membrane that surrounded the particle when it departed from an infected cell. Spiking outward from the envelope are HIV proteins that had become inserted into it. Beneath the lipid envelope, two protein coats—the core proteins—surround two strands of RNA and several copies of reverse transcriptase, an enzyme (compare Sections 16.4 and 22.8).

Once inside a host cell, the viral enzyme uses the RNA as a template for making DNA, which then is inserted into a host chromosome (Figure 40.19). Transcription yields copies of viral RNA. Some transcripts are translated into viral proteins. Others become enclosed in the proteins when new virus particles are put together, and they will function as the hereditary material. The particles are released by budding from the host cell's plasma membrane to start a new round of infection (Figure 40.20). During each round, more macrophages, antigen-presenting cells, and helper T cells are impaired or killed.

The viral genes inserted into the DNA of some host cells remain inactive. They may be activated during a later round of infection.

A TITANIC STRUGGLE BEGINS   Infection marks the onset of a titanic battle between the enemy and the host's immune system. B cells synthesize antibodies in response to HIV antigenic proteins. These are the antibodies that are the basis of diagnostic tests for identifying HIV infection. Armies of helper T cells and cytotoxic T cells also form. However, HIV infects an estimated 2 billion helper T cells and produces 100 million to 1 billion virus particles per day during certain phases of the infection. Every two days, the immune system destroys about half of the virus particles and replaces half of the helper T

Figure 40.19 Life cycle of HIV, one of the retroviruses.

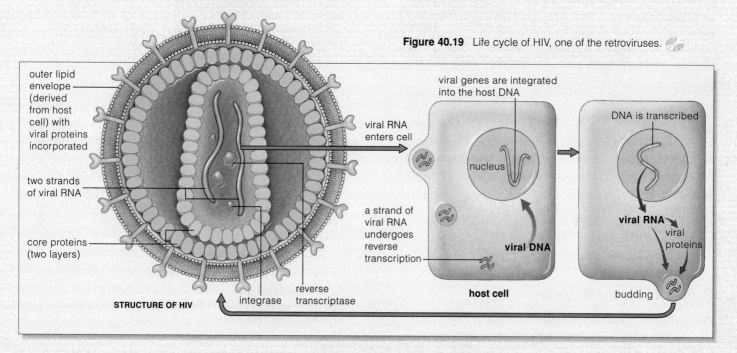

outer lipid envelope (derived from host cell) with viral proteins incorporated

two strands of viral RNA

core proteins (two layers)

STRUCTURE OF HIV

integrase

reverse transcriptase

viral RNA enters cell

a strand of viral RNA undergoes reverse transcription

viral genes are integrated into the host DNA

nucleus

viral DNA

host cell

DNA is transcribed

viral RNA

viral proteins

budding

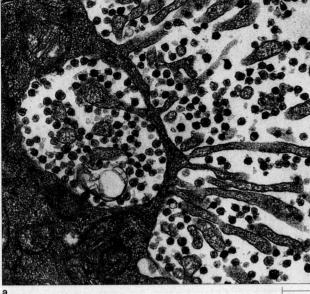

a

490 nm

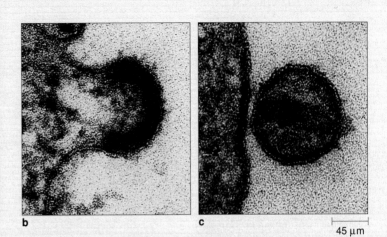

b   c

45 μm

cells lost in the battle. Immense reservoirs of HIV and masses of infected T cells accumulate in lymph nodes. As the battle proceeds, the number of virus particles in the general circulation rises. Gradually, the numbers tilt. The body produces fewer and fewer helper T cells to replace the ones it lost. Although it may take a decade or more, the erosion of the helper T cell count inevitably causes the body to lose its capacity to mount effective immune responses.

Some viruses, including the measles virus, produce far more virus particles in a given day, but the immune system usually wins out. Other viruses, including herpes viruses, can lurk in the body for a lifetime, but the immune system keeps them in check. With HIV, the immune system loses the struggle, then infections and tumors kill the person.

HOW HIV IS TRANSMITTED Like any other virus that infects humans, HIV requires a medium that allows it to leave one host, survive in the environment into which it is released, then enter another host.

HIV is transmitted when some body fluid of an infected person enters the tissues of another person. In the United States, transmission initially occurred most often between males who engaged in homosexual activities, especially anal intercourse. Since then, it has spread among intravenous drug abusers who share blood-contaminated syringes and needles. It also has spread in the heterosexual population, increasingly by vaginal intercourse.

HIV has traveled from infected mothers to offspring during pregnancy, birth, and breast-feeding. Before 1985, contaminated blood supplies accounted for some AIDS cases before health care providers implemented screening. Tissue transplants have caused four infections in 1991. In several developing countries, health care providers have spread HIV by way of contaminated transfusions and reuse of unsterile needles and syringes.

The molecular structure of HIV does not remain stable outside the human body, which is why it must be directly transmitted from one host to another. At this writing, there is no evidence that the virus can be effectively transmitted by way of food, air, water, casual contact, or insect bites. The virus *has* been isolated from human blood, semen,

Figure 40.20 **(a)** Transmission electron micrograph of HIV particles (black specks) escaping from an infected cell. **(b,c)** A virus particle budding from the host cell's plasma membrane.

vaginal secretions, saliva, tears, breast milk, amniotic fluid, urine, and cerebrospinal fluid. It is probably present in other fluids, secretions, and excretions. However, only infected blood, semen, vaginal secretions, and breast milk contain the virus in concentrations that apparently are high enough for successful transmission.

REGARDING PREVENTION AND TREATMENT Developing effective drugs or vaccines against HIV is a formidable challenge. High mutation rates characterize the viral genome, owing to HIV's replication mechanisms and the staggering number of replications in an infected person. Inevitably, natural selection operates in patients who are undergoing drug therapies; it favors drug resistance and fans the evolution of drug-resistant HIV populations. Chemical "cocktails" have had some effect in slowing replication. Among the current drugs of choice are protease inhibitors, AZT (azidothymidine), and ddI (dideoxyinosine). The drugs cannot *cure* infected people, however, for they cannot eliminate the HIV genes that have become incorporated into their DNA.

High mutation rates have another worrisome outcome: They give rise to variations in HIV antigens. The variation makes it difficult for researchers to select effective antigens for vaccines. Another problem is the scarcity of HIV strains with poor cell-killing abilities. Such strains are central to producing antigen that can stimulate the formation of cytotoxic T cell armies—the best protection against HIV. However, over a decade ago, many people in Australia received blood transfusions from a donor whose infection had not been diagnosed. At this writing, neither they nor the donor show any immunodeficiency. And they all carry HIV with a similar defect in the same gene (the nef gene).

In short, *until researchers develop effective vaccines and treatments, checking the spread of HIV depends absolutely on persuading people to avoid or modify social behaviors that put them at risk.* We return to this topic in Section 45.14.

1. Vertebrates fend off many pathogens with physical and chemical barriers at body surfaces. They also are protected by the nonspecific and specific responses of the white blood cells listed in Table 40.3.

a. Nonspecific responses to irritation or damage of tissues include inflammation and involve organs with phagocytic functions, such as the spleen and liver.

b. Immune responses are made against specific pathogens, foreign cells, or abnormal body cells.

2. Intact skin and mucous membranes that line body surfaces are physical barriers to infection. Glandular secretions (as in tears, saliva, and gastric fluid) are examples of chemical barriers. So are the metabolic products of resident bacteria on body surfaces.

3. An inflammatory response develops in body tissues that have become damaged, as by infection.

a. The response starts with arteriole vasoconstriction that increases blood flow to the tissue, which reddens and becomes warmer as a result. Capillary permeability increases, and local edema causes swelling and pain.

b. Pathogens as well as dead or damaged body cells release the substances that trigger increased permeability of capillaries. White blood cells leave the blood and enter the tissue, where they release a number of chemical mediators and engulf invaders. Plasma proteins also enter the tissue. Complement proteins bind pathogens and induce their lysis, and they attract phagocytes. Blood-clotting proteins wall off the damaged tissue.

4. An immune response has these characteristics:

a. It shows specificity, meaning it is directed against antigen. Each antigen is a molecular configuration that lymphocytes recognize as foreign (nonself).

b. Each response also shows memory, meaning that a subsequent encounter with the same antigen triggers a more rapid, secondary response, of greater magnitude.

c. An immune response normally is not made against the body's own self-marker proteins.

5. Macrophages and other antigen-presenting cells process and bind the fragments of antigen to their own MHC markers. Lymphocytes have receptors that can bind to the displayed antigen-MHC complexes. Binding is the start signal for an immune response.

6. After the recognition of antigen, an immune response proceeds through repeated cell divisions that form clones of B and T lymphocytes, which differentiate into subpopulations of effector and memory cells. Chemical mediators such as interleukins secreted by white blood cells drive the responses. Effector helper T cells, cytotoxic T cells, effector B cells, and antibodies act at once. The memory cells are set aside for secondary responses.

7. T cells arise in bone marrow but continue to develop in the thymus, where they acquire TCRs. These T-cell receptors recognize and bind antigen-MHC complexes on antigen-presenting cells. B cells arise in bone marrow. As they mature, they synthesize antigen receptors (that is, antibodies) that become positioned at their surface.

8. Effector cytotoxic T cells can directly destroy virus-infected cells, tumor cells, and cells of tissue or organ transplants. Effector B cells (plasma cells) produce and secrete great numbers of antibodies that freely circulate.

9. Antibodies are protein molecules, often Y-shaped, each with binding sites for one kind of antigen. Only B cells produce them. When antibody binds to antigen, toxins are neutralized, pathogens are tagged for destruction, or attachment of pathogens to body cells is prevented.

10. In active immunization, vaccines provoke immune responses, with production of effector and memory cells. In passive immunization, injections of purified antibodies help the individual through an infection.

11. Allergic reactions are immune responses to some generally harmless substance. Autoimmune responses are misguided attacks triggered by configurations on the body's own cells. Immunodeficiency is a weakened or nonexistent capacity to mount an immune response.

Table 40.3	Summary of Major White Blood Cells and Their Roles in Defense
Cell Type	Main Characteristics
MACROPHAGE	Phagocyte; acts in nonspecific and specific responses; presents antigen to T cells; cleans up and helps repair tissue damage
NEUTROPHIL	Fast-acting phagocyte; takes part in inflammation, not in sustained responses; most effective against bacteria
EOSINOPHIL	Secretes enzymes that attack certain parasitic worms
BASOPHIL AND MAST CELL	Secrete histamines and other substances that act on small blood vessels, thereby producing inflammation; also contribute to allergic reactions
LYMPHOCYTES:	(All take part in most immune responses; following antigen recognition, all form clonal populations of effector cells and memory cells.)
B cell	Effectors secrete four types of antibodies (IgA, IgE, IgG, and IgM) that protect the host in specialized ways
Helper T cell	Effectors secrete interleukins that stimulate rapid divisions and differentiation of both B cells and T cells
Cytotoxic T cell	Effectors kill infected cells, tumor cells, and foreign cells by a touch-kill mechanism
NATURAL KILLER (NK) CELLS	Cytotoxic cell of undetermined affiliation; kills virus-infected cells and tumor cells by a touch-kill mechanism

Review Questions

1. While jogging barefoot along a seashore, some of your toes accidentally land on a jellyfish. Soon the toes are swollen, red, and warm to the touch. Describe the events that result in these signs of inflammation. *40.3*

2. Distinguish between:
   a. neutrophil and macrophage *40.3*
   b. cytotoxic T cell and natural killer cell *40.4, 40.6*
   c. effector cell and memory cell *40.4*
   d. antigen and antibody *40.4*

3. Describe how a macrophage becomes an antigen-presenting cell. *40.4*

4. Why is a vaccine to control AIDS so elusive? *40.11*

Self-Quiz *(Answers in Appendix IV)*

1. _____ are barriers to pathogens at body surfaces.
   a. Intact skin, mucous membranes   d. Urine flow
   b. Tears, saliva, gastric fluid   e. All of the above
   c. Resident bacteria

2. Macrophages are derived from _____ .
   a. basophils   c. monocytes
   b. neutrophils   d. eosinophils

3. Activated complement functions in defense by _____ .
   a. neutralizing toxins   c. promoting inflammation
   b. enhancing resident bacteria   d. forming holes in memory lymphocyte membranes

4. _____ are certain molecules that lymphocytes recognize as foreign and that elicit an immune response.
   a. Interleukins   d. Antigens
   b. Antibodies   e. Histamines
   c. Immunoglobulins

5. The immunoglobulins _____ increase antimicrobial activity in mucus-coated surfaces of some organ systems.
   a. IgA   b. IgE   c. IgG   d. IgM   e. IgD

6. Antibody-mediated responses work best against _____ .
   a. intracellular pathogens   d. both b and c
   b. extracellular pathogens   e. all of the above
   c. extracellular toxins

7. The most important antigens are _____ .
   a. nucleotides   c. steroids
   b. triglycerides   d. proteins

8. _____ would be a target of an effector cytotoxic T cell.
   a. Extracellular virus particles in blood
   b. A virus-infected body cell or tumor cell
   c. Parasitic flukes in the liver
   d. Bacterial cells in pus
   e. Pollen grains in nasal mucus

9. Development of a secondary immune response is based on populations of _____ .
   a. memory cells   d. effector cytotoxic T cells
   b. circulating antibodies   e. mast cells
   c. effector B cells

10. Match the immunity concepts.
    ____ inflammation
    ____ antibody secretion
    ____ a phagocyte
    ____ immunological memory
    ____ vaccination
    ____ allergy
    a. neutrophil
    b. effector B cell
    c. nonspecific response
    d. deliberately provoking an immune response
    e. basis of secondary response
    f. nonprotective immune response

Critical Thinking

1. As described in the chapter introduction, Edward Jenner lucked out. He performed a potentially harmful experiment on a boy who managed to survive it. What would happen if a would-be Jenner tried to do the same thing today?

2. Rob's bumper sticker reads, "Have you thanked your resident bacteria today?" Explain why he appreciates the bacteria that normally reside on the body's skin and mucous membranes.

3. Researchers are attempting to develop a way to get the immune system to accept foreign tissue as "self." Speculate on some of the clinical applications of such a development.

4. Before each flu season, you get an influenza vaccination. This year you come down with "the flu" anyway. What do you suppose happened? (There are at least three explanations.)

5. Infection by *Ebola* virus results in a hemorrhagic fever with a 90 percent mortality rate (Section 22.9). A patient received a blood serum transfusion from another who survived the disease. Explain why the transfusion might increase chances of survival.

6. Ellen developed *chicken pox* when she was in kindergarten. Later in life, when her children developed chicken pox, she remained healthy even though she was exposed to countless virus particles daily. Explain why.

7. Quickly review Section 33.7 on homeostasis. Then write a short essay on how the immune response contributes to stability in the internal environment.

Selected Key Terms

allergen *40.10*	immune system *40.4*
allergy *40.10*	immunization *40.10*
antibody *40.4*	immunoglobulin (Ig) *40.7*
antigen *40.4*	inflammation, acute *40.3*
antigen-MHC complex *40.4*	lysis *40.2*
antigen-presenting cell *40.4*	lysozyme *40.1*
apoptosis *40.6*	macrophage *40.3*
autoimmune response *40.10*	mast cell *40.3*
B lymphocyte (B cell) *40.4*	MHC marker *40.4*
basophil *40.3*	natural killer (NK) cell *40.6*
complement system *40.2*	neutrophil *40.3*
cytotoxic T cell *40.4*	pathogen *40.1*
eosinophil *40.3*	TCR *40.6*
fever *40.3*	T lymphocyte (T cell) *40.4*
helper T cell (CD4 lymphocyte) *40.4*	vaccination *CI*
histamine *40.3*	vaccine *40.10*

Readings

Edelson, R., and J. Fink. June 1985. "The Immunologic Function of Skin." *Scientific American* 252(6): 46–53.

Golub, E., and D. Green. 1991. *Immunology: A Synthesis*. Second edition. Sunderland, Massachusetts: Sinauer.

Janeway, C., Jr. September 1993. "How the Immune System Recognizes Invaders." *Scientific American* 72–79.

Nowak, M., and A. McMichael. August 1995. "How HIV Defeats the Immune System." *Scientific American* 273(2): 58–65.

Tizard, I. 1995. *Immunology: An Introduction*. Fourth edition. Philadelphia: Saunders.

Web Site See *http://www.wadsworth.com/biology* for practice quiz questions, hypercontents, BioUpdates, and critical thinking. The Wadsworth Biology Resource Center provides a wealth of information fully organized and integrated by chapter.

41 RESPIRATION

Conquering Chomolungma

To experienced climbers, Chomolungma may be the ultimate challenge (Figure 41.1). The summit of this Himalayan mountain, also known as Everest, is 9,700 meters (29,128 feet) above sea level. It is the highest place on Earth. Iced-over vertical rock, driving winds, blinding blizzards, and heart-stopping avalanches await the challengers. So does the extreme danger that oxygen-poor air poses to the brain.

Most of us live at low elevations. Of the air we breathe, one molecule in five is oxygen. When we travel 2,400 meters (about 8,000 feet) or more above sea level, the Earth's gravitational pull is weaker, gas molecules spread out more, and the breathing game changes. We face *hypoxia*, or cellular oxygen deficiency. Sensing the deficiency, the brain makes us hyperventilate, or breathe much faster and more deeply than usual. Above 3,300 meters (10,000 feet), hyperventilating can be worrisome. It may lead to significant ion imbalances in cerebrospinal fluid, which can trigger heart palpitations, shortness of breath, headaches, nausea, and vomiting. These are strong clues that our cells are, in a manner of speaking, screaming for oxygen.

Living for months at high elevations helps climbers adapt physiologically to the thinner air. For example, mechanisms kick in that increase the red blood cell count, hence the body's oxygen-carrying capacity. For professional climbers, Chomolungma's base camp is 6,300 meters (19,000 feet) above sea level. At 7,000 meters, oxygen and other gases are extremely diffuse. The oxygen scarcity and low air pressure combine to

Figure 41.1 A climber inching toward the summit of Chomolungma, where oxygen is brutally scarce.

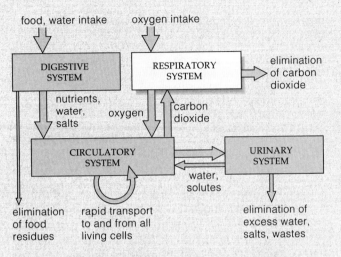

food, water intake   oxygen intake

DIGESTIVE SYSTEM   **RESPIRATORY SYSTEM**   elimination of carbon dioxide

nutrients, water, salts   oxygen   carbon dioxide

CIRCULATORY SYSTEM   **URINARY SYSTEM**

water, solutes

elimination of food residues   rapid transport to and from all living cells   elimination of excess water, salts, wastes

Figure 41.2 Interactions between the respiratory system and other organ systems in complex animals.

make blood capillaries leaky. More plasma escapes through the expanded gaps between endothelial cells of the capillary walls. In the brain and lungs, tissues swell with excess fluid. If the edema is not reversed, climbers will become comatose and die. When stricken, they can inhale bottled oxygen. Rescuers can zip them inside airtight bags, then use a device that pumps in oxygen and removes carbon dioxide until the "air" inside the bag more closely approximates the air at 2,400 meters.

Few of us will ever find ourselves near the peak of Chomolungma, pushing our reliance on oxygen to the limits. Here in the lowlands, disease, smoking, and other environmental insults push it in more ordinary ways, although the risks can be just as great.

The point is this: *Each animal has a body plan that is adapted to the oxygen levels of a particular habitat*. One way or another, by a physiological process known as **respiration**, the body plan allows oxygen to move into the internal environment and carbon dioxide to move out. Why is this important? Animals, remember, have great energy demands. Their cells demand a great deal of oxygen—especially for aerobic respiration, the main metabolic pathway that produces considerable energy and carbon dioxide wastes.

This chapter samples a few **respiratory systems**, which function in the exchange of gases between the body and the environment. Together with other organ systems, they also contribute to homeostasis—that is, to maintaining internal operating conditions for all of the body's living cells (Figure 41.2).

KEY CONCEPTS

1. Of all organisms, multicelled animals require the most energy to drive their metabolic activities. At the cellular level, the energy comes mainly from aerobic respiration, an ATP-producing metabolic pathway that requires oxygen and produces carbon dioxide wastes.

2. By a physiological process called respiration, animals move oxygen into their internal environment and give up carbon dioxide to the external environment.

3. Oxygen diffuses into the animal body as a result of a pressure gradient. The pressure of this gas is higher in air than it is in metabolically active tissues, where cells rapidly use oxygen. Carbon dioxide follows its own gradient, in the opposite direction. Its pressure is higher in tissues, where it is a by-product of metabolism, than it is in the air.

4. In most respiratory systems, oxygen and carbon dioxide diffuse across a respiratory surface, such as the thin, moist epithelium of the human lungs. The blood flowing through the body's circulatory system picks up oxygen and gives up carbon dioxide at this respiratory surface.

5. Gas exchange is most efficient when the rate of air flow matches the rate of blood flow. The nervous system brings the rates into balance by controlling the rhythmic pattern and magnitude of breathing.

The Basis of Gas Exchange

A concentration gradient, recall, is a difference in the number of molecules of a substance between two regions. Like all substances, oxygen or carbon dioxide tends to diffuse down its concentration gradient, or show a net outward movement from the region where its molecules are colliding more frequently (Section 5.3). Respiration is based on the tendency of both gases to diffuse down their respective concentration gradients—or, as we say for gases, down **pressure gradients** that occur between the internal and external environments.

The gases do not exert the *same* pressure. Pump air into a flat tire near a beach in San Diego or Miami, and you fill it with about 78 percent nitrogen, 21 percent oxygen, 0.04 percent carbon dioxide, and 0.96 percent other gases. This is true of dry air anywhere at sea level. At sea level, atmospheric pressure is about 760 mm Hg, as measured by a mercury barometer of the sort shown in Figure 41.3. Oxygen exerts only part of that total pressure on the tire wall, and its "partial" pressure is greater than that of carbon dioxide. Said another way, the **partial pressure** of oxygen—its contribution to the total atmospheric pressure—is $760 \times 21/100$, or about 160 mm Hg. When similarly measured, carbon dioxide's partial pressure is about 0.3 mm Hg.

Gases enter and leave the animal body by crossing a **respiratory surface**. A respiratory surface is a thin layer of epithelium or some other tissue. It must be kept moist at all times, for gaseous molecules cannot diffuse across it unless they are dissolved in fluid. What dictates the number of gas molecules moving across a respiratory surface in a given time? According to **Fick's law**, the more extensive the surface area and the larger the partial pressure gradient, the faster will be the diffusion rate.

Factors That Influence Gas Exchange

SURFACE-TO-VOLUME RATIO All animal body plans promote favorable rates of inward diffusion of oxygen and outward diffusion of carbon dioxide. For example, animals with no respiratory organs are tiny, tubelike, or flattened, and gases diffuse directly across the body surface. These body plans meet a constraint imposed by the surface-to-volume ratio, as described in Section 4.1. To get a sense of the ratio's effects, imagine a flatworm growing in all directions, like an inflating balloon. The worm's surface area does not increase at the same rate as its volume. Once its girth exceeds a single millimeter, the diffusion distance between the body's surface and internal cells will be so great that the flatworm will die.

VENTILATION Large-bodied animals that are highly active have great demands for gas exchange, more than

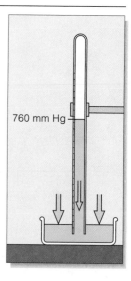

Figure 41.3 Atmospheric pressure as measured with a device called a mercury barometer. Part of the device is a glass tube in which the height of a column of mercury (Hg) can increase or decrease, depending upon air pressure outside the device. At sea level, the mercury rises to about 760 millimeters (29.91 inches) from the base of the tube. At this level, the pressure that the column of mercury exerts inside the tube is equal to the atmospheric pressure on the outside.

760 mm Hg

diffusion alone would be able to satisfy. A variety of adaptations make the exchange rate more efficient. For example, above their gills (respiratory organs), many fishes have tissue flaps that stir the surrounding water by moving back and forth. The stirred water puts more dissolved oxygen closer to the gills, and it carries more carbon dioxide away from them. Among vertebrates, a circulatory system rapidly transports oxygen to cells and carbon dioxide to gills or lungs for disposal. As a final example, you breathe to ventilate your lungs.

TRANSPORT PIGMENTS Rates of gas exchange get a boost with respiratory pigments, mainly **hemoglobin**, that help maintain the steep pressure gradients across a respiratory surface. For example, at the respiratory surfaces inside human lungs, the oxygen concentration is high, and each hemoglobin molecule in blood weakly binds as many as four oxygen molecules. (Here you may wish to refer to Section 3.7.) Then the circulatory system rapidly transports the hemoglobin from the respiratory surface. In oxygen-poor tissues, the oxygen follows its gradient and diffuses out of hemoglobin. By its oxygen-transporting activity, then, hemoglobin helps maintain a pressure gradient that entices oxygen into the lungs.

By a process called respiration, animals take in oxygen for aerobic respiration, an ATP-producing pathway, and remove the pathway's carbon dioxide wastes.

Oxygen and carbon dioxide enter and leave the body by diffusing across a moist respiratory surface. Like other atmospheric gases, they tend to move down their respective pressure gradients. Each gas exerts only part of the total pressure across a respiratory surface.

Gas exchange depends on steep partial pressure gradients between the outside and inside of the animal body. The greater the area of the respiratory surface and the larger the partial pressure gradient, the faster diffusion will proceed.

Flatworms, earthworms, and many other invertebrates are not massive, and their life-styles do not depend on high metabolic rates (Figure 41.4*a*). Demands for gas exchange are simply met by **integumentary exchange**, in which gases diffuse directly across the body's surface covering (integument). This mode of respiration works as long as the surface stays moist; invertebrates that rely solely on it are restricted to aquatic or damp habitats. Integumentary exchange supplements other modes of respiration in amphibians and some other large animals.

Many invertebrates of aquatic habitats have moist, thin-walled respiratory organs called **gills**. Extensively folded gill walls have an increased respiratory surface area that enhances the exchange rates between blood or some other body fluid and the surroundings. Figure 41.4*b* shows the folded gill of a sea hare (*Aplysia*). By supplementing integumentary exchange, the gill helps provide adequate oxygen for this rather large mollusk; some sea hares are 40 centimeters (nearly 16 inches) long.

Invertebrates of dry habitats have small, thick, or hardened integuments that are not well endowed with blood vessels. Although their integuments do conserve

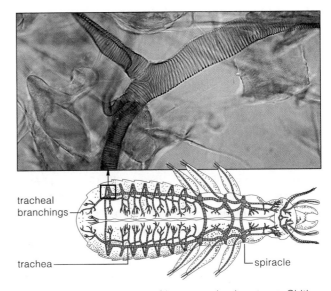

Figure 41.5 General plan of insect tracheal systems. Chitin rings reinforce many of the branching tubes of the system.

precious water, these are not good respiratory surfaces. Such animals have *internal* respiratory surfaces. For example, most spiders have book lungs, or respiratory organs with thin, folded walls that resemble book pages (Section 26.16). Like most insects and the millipedes and centipedes, some spiders rely on a system of internal tubes that function in **tracheal respiration**.

Consider the tracheal system of an insect, as in Figure 41.5. Small openings perforate the insect integument. Each opening, called a spiracle, is the start of a tube that branches inside the body. The last branchings dead-end at a fluid-filled tip, where gases diffuse directly into tissues. The tips of the tubes are especially profuse in muscle and other tissues with high oxygen demands.

We find hemoglobin or other respiratory pigments in many invertebrates, although they are rare in insects.

The respiratory pigments increase the capacity of body fluids to transport oxygen, just like they do in vertebrates. In the species with a well-developed head, oxygenated blood tends to circulate first through the head end, then through the rest of the body.

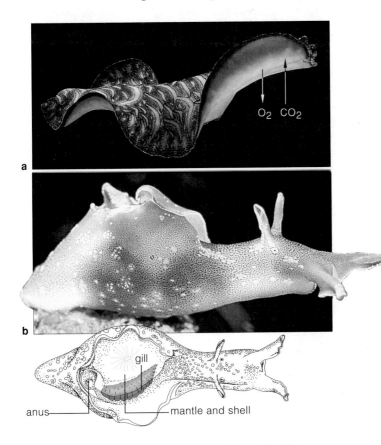

Figure 41.4 Invertebrates of aquatic habitats. (**a**) A flatworm, small enough to get along well without an oxygen-transporting circulatory system. Dissolved oxygen in such habitats reaches individual cells by diffusing across the body surface. (**b**) Gill of a sea hare (*Aplysia*), one of the gastropods.

Flatworms and some other invertebrates that live in aquatic or moist habitats and that do not have a massive body rely on integumentary exchange, in which oxygen and carbon dioxide diffuse directly across the body surface.

Most marine invertebrates and many freshwater types have gills of one sort or another. These respiratory organs have moist, thin, and often highly folded walls.

Most insects, millipedes, centipedes, and some spiders use tracheal respiration. Gases flow through open-ended tubes that start at the body surface and end directly in tissues.

Gills of Fishes and Amphibians

Gills are the respiratory organs of many vertebrates. A few kinds of fish larvae and a few amphibians have *external* gills that project into the surrounding water. Adult fishes have a pair of *internal* gills. These are rows of slits or pockets at the back of the mouth that extend to the body surface, as in Figure 41.6a. Whatever their form, all gills have walls of moist, thin, vascularized epithelium.

In fishes, water flows into the mouth and pharynx, then over arrays of filaments in the gills (Figure 41.6b). Blood vessels thread through the respiratory surfaces in each filament. First the water flows past a vessel that is leading to the rest of the body. The blood inside has less oxygen than the water does, so oxygen diffuses into the blood. The same volume of water flows over a vessel leading into the gills. The water already gave up some oxygen, but it still has more than the blood inside the vessel does, so more oxygen diffuses into the filament. Movement of two fluids in opposing directions is called **countercurrent flow**. By this mechanism, a fish extracts about 80 to 90 percent of the dissolved oxygen flowing past. That is more than the fish would get from a one-way flow mechanism, at far less energy cost.

Lungs

Some fishes and all amphibians, birds, and mammals have a pair of **lungs**, or internal respiratory surfaces in the shape of a cavity or sac. Lungs originated in some lineages of fishes more than 450 million years ago, as pouches off the anterior part of the gut wall (Figure 41.7). They evolved rapidly by way of natural selection, probably because lungs afforded fine advantages. They increased the surface area for gas exchange in oxygen-poor habitats, and they worked better than

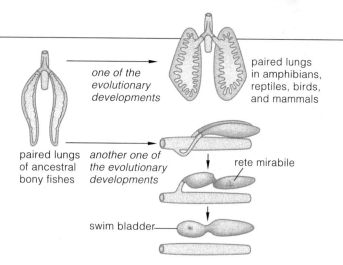

Figure 41.7 Evolution of vertebrate lungs and swim bladders. The esophagus (a tube leading to the stomach) is coded *gold* and respiratory tissues *pink*. Lungs originated as pockets off the anterior part of the gut. They increased the surface area for gas exchange in oxygen-poor habitats. In some fishes, lung sacs became swim bladders. By adjusting the gas volume in these buoyancy devices, a fish can hold its position at different depths.

Trout and other less specialized fishes have a duct between the esophagus and the swim bladder. They surface and gulp air to replenish air in the bladder. Most bony fishes have no such duct. Gases in blood diffuse into their swim bladder, which has a *rete mirabile:* a dense mesh of arteries and veins running in opposing directions. Countercurrent flow through these vessels greatly increases gas concentrations in the bladder. Another region of the bladder promotes reabsorption of gases by body tissues.

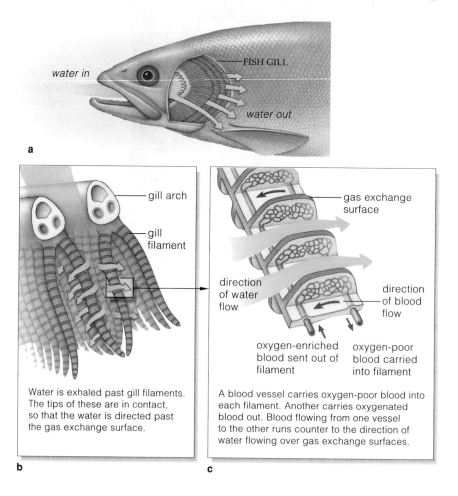

Figure 41.6 Example of a fish gill. (**a**) This is one of a pair of gills. Each gill is located under a bony lid, which has been removed for this sketch. (**b**) Filaments in a fish gill contain vascularized respiratory surfaces. The tips of neighboring filaments touch, so water flowing over them is directed past gas exchange surfaces before being expelled. (**c**) One blood vessel from other body tissues delivers oxygen-poor blood into a filament; another carries oxygenated blood away from it. Blood flowing from one vessel to the other runs counter to the direction of the water flowing over the gas exchange surfaces. Countercurrent flow favors the movement of oxygen (down its partial pressure gradient) into the blood.

Water is exhaled past gill filaments. The tips of these are in contact, so that the water is directed past the gas exchange surface.

A blood vessel carries oxygen-poor blood into each filament. Another carries oxygenated blood out. Blood flowing from one vessel to the other runs counter to the direction of water flowing over gas exchange surfaces.

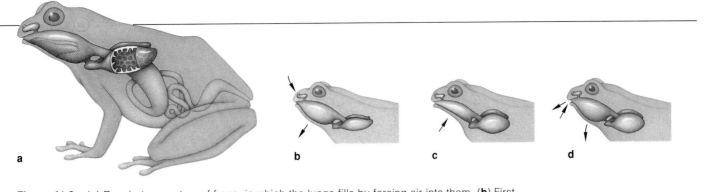

Figure 41.8 (**a**) Respiratory system of frogs, in which the lungs fills by forcing air into them. (**b**) First, the frog lowers the floor of its mouth and inhales air through its two nostrils. (**c**) Then it closes the nostrils, opens the glottis, and elevates the floor of its mouth, so the air has nowhere to go except into the lungs. (**d**) The frog rhythmically ventilates its mouth for a while, which helps move more oxygen into the mouth and more carbon dioxide out of it. (**e**) Finally, the frog contracts muscles in the body wall outside the lungs, the lungs recoil elastically, and air is forced out.

gills could during the move onto dry land. Gills stick together and cannot function at all unless water flows through them and keeps them moist.

Lungfishes of oxygen-poor habitats still have gills. They also use tiny lungs as backups. Amphibians never completed the transition to land. Their skin serves as a respiratory surface for integumentary exchange, in the salamanders especially. Frogs and toads rely more on

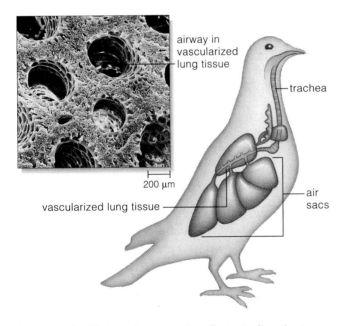

Figure 41.9 Bird respiratory system. Typically, five air sacs are attached to each of two small, inelastic lungs. When a bird inhales, air is drawn into air sacs through tubes, open at both ends, that thread through the vascularized lung tissue. This tissue is the respiratory surface, where gases are exchanged.

When the bird exhales, air is forced out of the sacs, through the small tubes, and out of the trachea. Thus, air is not merely drawn into the bird lungs. Air is continuously drawn through them and across respiratory surfaces. This unique ventilating system supports the high metabolic rates that birds require for flight and other energy-intensive activities.

small lungs for oxygen uptake, but most of the carbon dioxide diffuses across the skin. Frogs also are heavy-duty breathers; they *force* air into the lungs, then empty them by contracting body wall muscles (Figure 41.8).

Frogs can use their lungs for sound production. So can all mammals except whales. Sound originates near the entrance to a larynx, an airway leading to the lungs. Here, part of a mucous membrane folds perpendicularly to the airway. These are **vocal cords**; the gap between them is the **glottis**. Frogs produce sounds by forcing air back and forth through the glottis, between the lungs and paired pouches on the floor of the mouth. Air flow causes the cords to vibrate. Like other vertebrates, frogs produce different sounds by controlling the vibrations.

Paired lungs are the dominant means of respiration in reptiles, birds, and mammals. Breathing moves air by bulk flow into and out of the lungs, where blood capillaries wrap lacily around the respiratory surface. Oxygen and carbon dioxide have steep concentration gradients in the lungs, so they diffuse rapidly across the respiratory surface. Oxygen enters the capillaries and is circulated quickly through the body. In regions where its concentration is low, oxygen diffuses into interstitial fluid, then into cells. Carbon dioxide moves rapidly in the opposite direction and is expelled from the lungs.

This mode of gas exchange is embellished a bit only in the lungs of birds. As described in Figure 41.9, birds are unique in that air not only flows into and out of the lungs, it also flows *through* them. With this exception in mind, we turn next to the human respiratory system, for its operating principles apply to most vertebrates.

A countercurrent flow mechanism in fish gills compensates for low oxygen levels in aquatic habitats. Internal air sacs— lungs—are more efficient in dry land habitats. Amphibians use integumentary exchange, and they also force air into and out of small lungs. Ventilation of paired lungs is the major mode of respiration in reptiles, birds, and mammals.

HUMAN RESPIRATORY SYSTEM

Functions of the Respiratory System

Obtaining oxygen from the air and removing carbon dioxide from the body are the overriding functions of the human respiratory system. A form of ventilation called breathing alternately moves air into and out of a pair of lungs, each of which contains about 300 million outpouchings. Each outpouching is a tiny air sac called an **alveolus** (plural, alveoli). Controls adjust the rate of breathing so that the inflow and outflow of air match the metabolic demands for gas exchange.

The respiratory system's role in respiration ends at alveoli. From that point on, the circulatory system takes over. As you will see, oxygen and carbon dioxide move by diffusion between the alveoli and the pulmonary capillaries that weave around them. (The Latin *pulmo* means lung.)

However, the respiratory system performs other functions. Breathing also is necessary for speech and other forms of vocalization. As you saw in Section 39.8, it enhances the venous return of blood to the heart. It functions to some extent in eliminating excess heat and water. In addition, control over breathing is vital for adjusting the body's acid-base balance. (Carbon dioxide can be used as a building block for carbonic acid, or H_2CO_3. As you read in Section 2.6, H_2CO_3 accepts or gives up hydrogen ions, H^+, depending on the pH. Fast, deep breathing expels more carbon dioxide, so less carbonic acid forms; hence the body loses acid. Slow, shallow breathing has the opposite effect; carbon dioxide accumulates, more H_2CO_3 forms, and the body gains acid.)

ORAL CAVITY

Supplemental airway when breathing is labored

EPIGLOTTIS

Closes off larynx during swallowing

PLEURAL MEMBRANES

Membranes that separate lungs from other organs; also form a thin, fluid-filled cavity that facilitates breathing

LUNG (ONE OF A PAIR)

Lobed, elastic organ of breathing that enhances gas exchange between the body and the outside air

INTERCOSTAL MUSCLES

Rib cage muscles with roles in breathing

DIAPHRAGM

Muscle sheet between the chest cavity and abdominal cavity with roles in breathing

NASAL CAVITY

Chamber in which air is warmed, moistened, and initially filtered; and in which sounds resonate

PHARYNX (THROAT)

Airway connecting the nasal cavity and mouth with the larynx; enhances speech sounds; connects also with the esophagus that leads to the stomach

LARYNX (VOICE BOX)

Airway where sound is produced and where breathing is blocked during swallowing movements

TRACHEA (WINDPIPE)

Airway connecting the larynx with two bronchi that lead into the lungs

BRONCHIAL TREE

Increasingly branched airways starting with the bronchi and ending at air sacs (alveoli)

ALVEOLI

Thin-walled air sacs where oxygen diffuses into the internal environment and carbon dioxide diffuses out

a

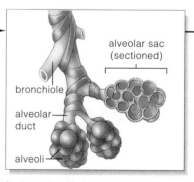

b Location of alveoli relative to the terminal end of a bronchiole

Figure 41.10 **(a)** Components of the human respiratory system and their functions. Muscles, including the diaphragm, and parts of the axial skeleton have secondary roles in respiration. **(b,c)** Location of the alveoli relative to the lung capillaries.

The respiratory system also has built-in mechanisms for dealing with airborne foreign cells and substances that have been inhaled with air. Finally, the system removes, inactivates, or otherwise modifies a number of blood-borne substances before they become circulated through the rest of the body.

From Airways to the Lungs

It will take at least 300 million breaths to get you to age seventy-five. You may find yourself going without food for a few hours or days. But stop breathing even for five minutes and normal brain function is over.

Take a deep breath, then look at Figure 41.10 to get an idea of where the air will travel in your respiratory system. Unless you are out of breath and panting, the air has just entered two nasal cavities, not your mouth. There, mucus secretions warm and moisten it. Ciliated epithelium and hairs in the nasal cavities filter dust and particles from air. Also in the nasal cavities are olfactory receptors that function in the sense of smell (Section 36.3). Now the air is poised at the **pharynx**, or throat. The pharynx is the entrance to the **larynx**, an airway with vocal cords. Right now the **epiglottis**, a tissue flap at the start of the larynx, is pointing upward, so the air moves into the **trachea**, or windpipe. When you swallow, the epiglottis points downward and so closes off the entrance to the trachea. At such times, food or fluid being swallowed enters the esophagus, a tube connecting the pharynx with the stomach.

The trachea branches into two airways, one leading into the tissue of each lung. Each airway is a **bronchus** (plural, bronchi). Its epithelial lining has a profusion of cilia and mucus-secreting cells (Figure 41.11). The lining is a barrier to infection. Bacteria and airborne particles stick to the mucus, then cilia sweep debris-laden mucus toward the mouth. Where it goes from there really is up to you, but possibly the sidewalk is not a suitable destination.

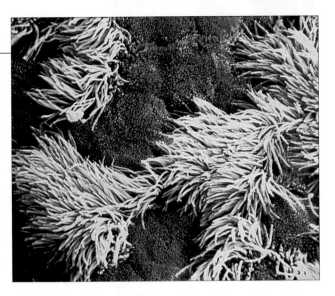

Figure 41.11 False-color scanning electron micrograph of cilia (*gold*) and mucus-secreting cells (*brown*) in a bronchus.

Sites of Gas Exchange in the Lungs

In humans, the lungs are elastic, cone-shaped organs of gas exchange. They are located within the rib cage, to the left and right of the heart and above the **diaphragm**, a muscular partition between the thoracic and abdominal cavity. A thin, pleural membrane lines the outer surface of the lungs and inner surface of the thoracic cavity wall. Visualize the lungs as two baseballs pushed into a partly inflated balloon. The baseballs take up so much space, they press the balloon's opposing sides together. Similarly, the pleural membrane is saclike. The thoracic wall and the lungs press its opposing surfaces together. A thin film of lubricating fluid separates the membrane surfaces and decreases friction between them. During a *pleurisy*, a respiratory ailment, the pleural membrane becomes so inflamed and swollen that friction follows, and breathing can be painful.

Inside each lung, air moves through finer and finer branchings of a "bronchial tree" (Figure 41.10a). These airways are **bronchioles**. Their endings, the *respiratory* bronchioles, bear the cup-shaped alveoli. Most often, alveoli are clustered as larger pouches called alveolar sacs (Figure 41.10b and c). Collectively, alveolar sacs offer a tremendous surface area for gas exchanges with blood. If all the alveolar sacs were stretched out in one layer, they would cover the floor of a racquetball court!

Oxygen uptake and carbon dioxide removal are the major functions of the human respiratory system. In its paired lungs, the circulatory system takes over the remaining tasks of respiration.

The respiratory system also has roles in moving venous blood to the heart, in vocalization, in adjusting the body's acid-base balance, in defense against harmful airborne cells or particles, in removing or modifying a number of blood-borne substances, and in the sense of smell.

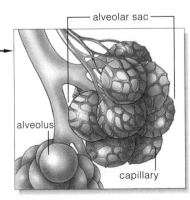

c Location of pulmonary capillaries relative to the alveoli

The Respiratory Cycle

There is a cyclic pattern to breathing, which ventilates the lungs. Each **respiratory cycle** consists of two actions: *inhalation* (a single breath of air drawn into the airways) and *exhalation* (a single breath out). Inhalation always is an active, energy-requiring action. When someone is breathing quietly, it is brought about by the contraction of the diaphragm and, to a lesser extent, the external intercostal muscles. The outcome is an increase in the thoracic cavity volume. Breathe hard, and the volume increases further because neck muscles contract and so elevate the sternum and first two ribs attached to them.

During each respiratory cycle, the thoracic cavity's volume increases, then decreases. *And pressure gradients between air inside and outside the respiratory tract change.* Let's think about the different pressures exerted during a respiratory cycle. The *atmospheric* pressure, 760 mm Hg at sea level, is exerted by the combined weight of all atmospheric gases on all the airways. Before inhalation, *intrapulmonary* pressure (the pressure inside all alveoli) is also 760 mm Hg (Figure 41.12c).

Another pressure gradient helps keep the lungs close to the thoracic cavity wall through the respiratory cycle, even during exhalation, when lungs have a far smaller volume than the thoracic cavity (Figure 41.12b). When the cavity expands, so do the lungs, owing to a pressure gradient that exists across the lung wall. In a person at rest, the *intrapleural* pressure (inside the pleural sac) averages 756 mm Hg, which is less than atmospheric pressure. This pressure is exerted outside the lungs—that is, within the thoracic cavity. While intrapleural pressure is pushing in on the lung's wall, the intrapulmonary pressure is pushing out. The difference in pressure between them (4 mm Hg) is great enough to make the lungs stretch and fill the thoracic cavity.

The cohesiveness of water molecules in the intrapleural fluid (the fluid inside the pleural sac) also helps keep lungs close to the thoracic wall. By analogy, wet two panes of glass and press them together. The panes easily slide back and forth, but they resist being pulled apart. Similarly, intrapleural fluid "glues" the lungs to the wall. Thus, *when the thoracic cavity expands at inhalation, the lungs must expand, also.*

Figure 41.12a shows what happens as you start to inhale. The dome-shaped diaphragm flattens down, and the rib cage is lifted upward and outward. As the thoracic cavity expands, the lungs expand with it. At that time, the air pressure in all alveolar sacs combined is lower than the atmospheric pressure. Fresh air follows the gradient and flows down into the airways, almost to the respiratory bronchioles.

When someone breathes quietly, the second action of the respiratory cycle is passive. The muscles that brought about inhalation relax, and the lungs passively recoil, without further expenditure of energy. The resultant decrease in lung volume compresses the air in the alveolar sacs. Now pressure in the sacs is greater than atmospheric pressure. Air follows the gradient, out from the lungs (Figure 41.12b). Exhalation becomes an active, energy-requiring action only when more air must be expelled rapidly, as

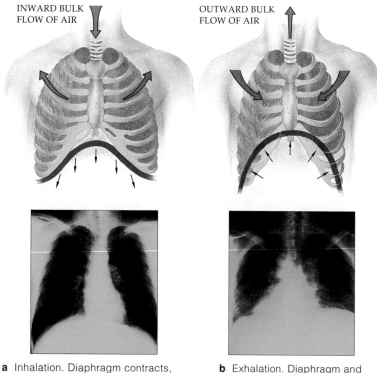

INWARD BULK FLOW OF AIR

OUTWARD BULK FLOW OF AIR

a Inhalation. Diaphragm contracts, moves down. External intercostal muscles contract, lift rib cage up and out. Lung volume expands.

b Exhalation. Diaphragm and external intercostal muscles return to their resting positions. Lungs recoil passively.

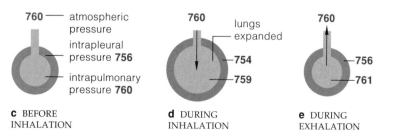

760 — atmospheric pressure
— intrapleural pressure **756**
— intrapulmonary pressure **760**

760 lungs expanded
—754
—759

760
—756
—761

c BEFORE INHALATION

d DURING INHALATION

e DURING EXHALATION

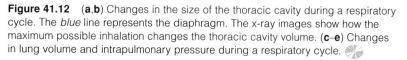

Figure 41.12 (**a,b**) Changes in the size of the thoracic cavity during a respiratory cycle. The *blue* line represents the diaphragm. The x-ray images show how the maximum possible inhalation changes the thoracic cavity volume. (**c–e**) Changes in lung volume and intrapulmonary pressure during a respiratory cycle.

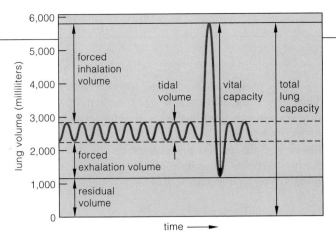

Figure 41.13 Lung volume. During quiet breathing, a tidal volume of air enters and leaves the lungs. Forced inhalation delivers more air to them; forced exhalation releases some air that normally stays in them. A residual volume is trapped in partially filled alveoli even during the strongest exhalation.

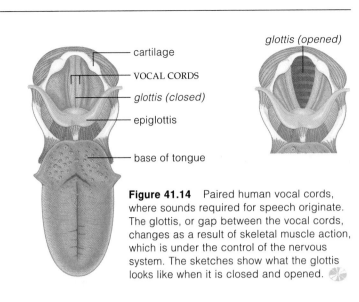

Figure 41.14 Paired human vocal cords, where sounds required for speech originate. The glottis, or gap between the vocal cords, changes as a result of skeletal muscle action, which is under the control of the nervous system. The sketches show what the glottis looks like when it is closed and opened.

during exercise. With active expiration, muscles in the abdominal wall contract, so pressure in the abdomen increases and exerts an upward force on the diaphragm. As the diaphragm is pushed upward, the thoracic cavity volume decreases. Internal intercostal muscles contract and pull the thoracic wall down and inward. The chest wall flattens, and so the thoracic cavity's dimensions decrease further. Lung volume decreases as well when the elastic tissue of the lungs recoils.

Lung Volumes

The lungs can hold up to 5.7 liters or so of air in adult males and 4.2 liters in adult females who are young and in good health. These are average values; a person's age, build, and respiratory health influence the total lung capacity. During quiet breathing, the lungs are far from being fully inflated. In general, they hold 2.7 liters at the end of one inhalation and 2.2 liters at the end of one exhalation. They never do deflate completely. When air flows out and the lung volume is low, the walls of the smallest airways collapse and prevent further air loss.

The volume of air that can move out of the lungs in one breath after maximal inhalation is called the **vital capacity**. Humans rarely use more than half of the total vital capacity, even when they take very deep breaths during strenuous exercise (Figure 41.13). Because the lungs are never empty, gas exchange between alveolar air and blood proceeds even at the end of maximum exhalation. *Hence the concentrations of gases in the blood remain fairly constant throughout the respiratory cycle.*

The volume of air flowing into or out of the lungs in the respiratory cycle, the **tidal volume**, averages 0.5 liter. How much is available for gas exchange? In between breaths, about 0.15 liter stays in the airways; only 0.35 liter of fresh air reaches alveoli. When you breathe, say, ten times a minute, you are supplying your alveoli with (0.35 × 10) or 3.5 liters of fresh air per minute.

Breathing and Sound Production

Near the entrance to the larynx are two paired folds of mucous membrane. The lower pair are the vocal cords (Figure 41.14). During the respiratory cycle, air is forced in and out through the glottis, the gap between them. Air flow makes the cords vibrate, and the vibrations can be controlled in ways that produce different sounds.

Within the folds are thick bands of elastic ligaments connected to various cartilage tissues. When muscles of the larynx contract and relax, the ligaments tighten or slacken, which changes the extent to which the folds are stretched. Under commands from the nervous system, the coordinated action of muscles narrows or widens the glottis. For example, by increasing muscle tension in the vocal cords, a person can decrease the gap between them and make high-pitched sounds or squeaks. The lips, teeth, tongue, and the soft roof over the tongue are enlisted to modify the different sounds into patterns of vocalization, such as speech and song.

When vocal cords are inflamed as an outcome of an infection or irritation, swelling of their mucous lining interferes with their capacity to vibrate. If hoarseness follows, this condition is called *laryngitis*.

Breathing, which ventilates the lungs, has a cyclic pattern. The respiratory cycle consists of inhalation (one breath of air in) and exhalation (one breath of air out).

Inhalation is always an active, energy-requiring process involving contractions mainly of the diaphragm and the external intercostal muscles.

During quiet breathing, exhalation is a passive process. Muscles relax, the thoracic cavity volume decreases, and the lungs recoil elastically. Forceful exhalation is an active process that requires abdominal muscle contraction.

Breathing reverses pressure gradients between the lungs and the air outside the body.

Gas Exchange

At each cup-shaped alveolus in the lungs, oxygen and carbon dioxide passively diffuse across the respiratory surface in response to their partial pressure gradients. An inward-directed gradient for oxygen is maintained because inhalations continually replenish oxygen and cells continually use it. An outward-directed gradient for carbon dioxide is maintained because cells continually produce carbon dioxide and exhalation removes it.

Each alveolus in the lungs consists of a single layer of epithelial cells that is surrounded by a thin basement membrane. And each pulmonary capillary consists of a single layer of endothelial cells. Only a thin film of interstitial fluid separates alveoli from the capillaries. The diffusion distances are so small that gases can flow rapidly across the respiratory surface (Figure 41.15).

Oxygen Transport

Oxygen, like carbon dioxide, does not dissolve well in blood, so it cannot be efficiently transported on its own. In all large animals, demands for oxygen are met with the assistance of hemoglobin molecules. These are the respiratory pigments that are packed inside red blood cells. Each hemoglobin molecule, recall, has quaternary structure. As shown earlier in Section 3.7, it actually is an organized, compact array of four polypeptide chains and four iron-containing, nitrogenous groups called **heme groups**. The iron atom of each heme group binds reversibly with oxygen. *Of all the oxygen inhaled into the human body, 98.5 percent of it is bound to the heme groups of hemoglobin.*

Normally, inhaled air that reaches alveoli has plenty of oxygen. The opposite is true of blood flowing past in the pulmonary capillaries. Thus, in the lungs, oxygen tends to diffuse into the plasma portion of blood. Then it diffuses into red blood cells and rapidly binds with hemoglobin. Hemoglobin that has oxygen bound to it is known as **oxyhemoglobin**, or HbO_2.

At any time, the amount of HbO_2 that forms depends on oxygen's partial pressure. The higher the pressure, the greater will be the oxygen concentration. When plenty of oxygen molecules are around, they tend to randomly collide with the heme binding sites at a faster rate. The encounters continue until all four of the binding sites in hemoglobin are saturated.

HbO_2 molecules hold onto oxygen rather weakly. They give it up in tissues where the partial pressure of oxygen is lower than in the lungs. They give it up even faster in tissues where the blood is warmer, the pH lower, and the partial pressure of carbon dioxide high. Such conditions exist in contracting muscle and other metabolically whipped-up tissues.

Carbon Dioxide Transport

The systemic portion of the circulatory system delivers carbon dioxide to the lungs. It enters blood capillaries in any tissue where its partial pressure is higher than it is in the blood flowing past.

Three mechanisms transport carbon dioxide from the capillary beds of the systemic circuit to the lungs. About 10 percent of the carbon dioxide simply remains dissolved in the blood. Another 30 percent binds with hemoglobin, forming **carbamino hemoglobin** ($HbCO_2$). Most of it—60 percent—is transported in the form of bicarbonate (HCO_3^-). These bicarbonate molecules form after carbon dioxide combines with water in blood plasma. The resulting carbonic acid dissociates (that is, separates) into bicarbonate and hydrogen ions (H^+):

$$CO_2 + H_2O \rightleftharpoons \underset{\text{CARBONIC ACID}}{H_2CO_3} \rightleftharpoons \underset{\text{BICARBONATE}}{HCO_3^-} + H^+$$

In blood plasma, that reaction converts only 1 of every 1,000 carbon dioxide molecules, which doesn't amount to much. It is a different story in red blood cells, which contain the enzyme **carbonic anhydrase**. The action of that enzyme enhances the reaction rate by 250 times! Most of the carbon dioxide not bound to hemoglobin becomes converted to carbonic acid this way. The enzyme-mediated conversion makes the blood level of carbon dioxide drop rapidly. Thus it helps maintain the gradient that promotes the diffusion of carbon dioxide from interstitial fluid into the bloodstream.

What about the bicarbonate ions that form in the reactions? They tend to move out of the red blood cells and into blood plasma. And what about the hydrogen ions? Hemoglobin acts as a buffer for them and keeps the blood from becoming too acidic. (A buffer, recall, is a molecule that combines with or releases H^+ in response to a shift in cellular pH.)

The reactions are reversed when blood reaches the alveoli, where the partial pressure of

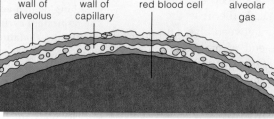

Figure 41.15 What a section through an alveolus and an adjacent lung capillary would look like. Compared to the red blood cell's diameter, the diffusion distance across the capillary wall, the interstitial fluid, and the alveolar wall is extremely small.

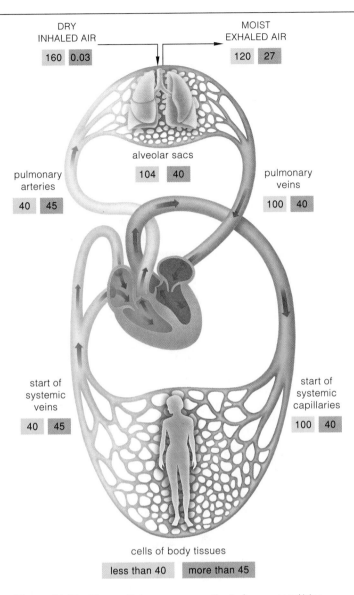

DRY INHALED AIR | 160 | 0.03

MOIST EXHALED AIR | 120 | 27

alveolar sacs | 104 | 40

pulmonary arteries | 40 | 45

pulmonary veins | 100 | 40

start of systemic veins | 40 | 45

start of systemic capillaries | 100 | 40

cells of body tissues | less than 40 | more than 45

Figure 41.16 The partial pressure gradients for oxygen (*blue* boxes) and carbon dioxide (*pink* boxes) in the respiratory tract.

carbon dioxide is lower than it is in nearby capillaries. Carbon dioxide and water form here and diffuse into alveolar sacs, then the gas is exhaled from the body.

The partial pressure gradients for oxygen and for carbon dioxide through the human respiratory system are summarized in Figure 41.16.

Matching Air Flow With Blood Flow

Gas exchange is most efficient when the rate of air flow matches the rate of blood flow. The nervous system acts to balance them by controlling the breathing *rhythm* and its *magnitude* (the rate and depth of the tidal volume).

One respiratory center, in the medulla oblongata (part of the brain stem), controls the rhythmic pattern of breathing. There, a group of neurons shows pacemaker activity; it sets the pace for certain neurons that issue

commands to the muscles involved in inhalation. The signals ultimately stimulate the muscles to contract; in their absence, the muscles relax and exhalation occurs. When exhalation must be active and strong, different neurons in the respiratory center issue commands for the activation of the muscles involved in exhalation.

Higher up in the brain stem, in the pons, different centers smooth out the rhythm set by the respiratory pacemakers. The apneustic center prolongs inhalation; the pneumotaxic center curtails it.

Control of the magnitude of breathing depends on the partial pressures of oxygen and carbon dioxide as well as the H+ concentration in arterial blood. Carbon dioxide levels are the most critical for ongoing adjustments. Chemoreceptors in the brain are highly sensitive to a rise in H+ in the cerebrospinal fluid that bathes them. (Remember, H+ can be a by-product of reactions that proceed when the blood has too much carbon dioxide.) The brain responds by commanding the diaphragm and other muscles to alter their activity; it adjusts the rate and depth of breathing. By contrast, aortic bodies (in the aortic arch) and carotid bodies (where the carotid artery branches) notify the brain when the partial pressure of oxygen in arterial blood plummets below 60 mm Hg. Such life-threatening decreases occur at extremely high altitudes and during severe lung diseases.

In some situations, a person can "forget" to breathe. Breathing that is briefly interrupted and then resumes spontaneously is called *apnea*. Especially during REM sleep (Section 35.8), breathing may stop for one or two seconds or minutes—in a few cases as often as 500 times a night. *Sudden infant death syndrome* (SIDS) occurs when a sleeping infant cannot awaken from an apneic episode, perhaps because its respiratory control centers or chemoreceptors (especially carotid bodies) are not yet fully developed. Infants who sleep on their back or sides are at less risk of SIDS than those positioned on their abdomen. The risk is three times as high among the infants of women who smoked cigarettes or were exposed to secondhand smoke during pregnancy.

Driven by its partial pressure gradient, oxygen diffuses from alveolar air spaces, through interstitial fluid, and into lung capillaries. Carbon dioxide, driven by its partial pressure gradient, diffuses in the opposite direction.

Hemoglobin in red blood cells enormously enhances the oxygen-carrying capacity of blood. Most carbon dioxide is transported in blood in the form of bicarbonate, nearly all of which forms by an enzyme action in red blood cells.

Respiratory centers in the brain stem control the rhythmic pattern of breathing and the magnitude of breathing, in ways that maintain appropriate levels of carbon dioxide, oxygen, and hydrogen ions in arterial blood.

41.7 WHEN THE LUNGS BREAK DOWN

In large cities, in certain workplaces, even near a cigarette smoker, airborne particles and certain gases are present in abnormally high concentrations. And they put extra workloads on the respiratory system.

BRONCHITIS Ciliated, mucous epithelium lines your bronchioles, as Figure 41.11 shows. It is one of the built-in defenses that protect you from respiratory infections. Toxins in cigarette smoke and other airborne pollutants irritate this lining and may lead to *bronchitis*. With this respiratory ailment, excess mucus is secreted when epithelial cells that line the airways become irritated. As mucus accumulates, so do bacteria and other particles stuck in it. Coughing brings up some of the gunk. If the irritation persists, so does coughing. Initial attacks of bronchitis are treatable. However, when the aggravation is allowed to continue, bronchioles become inflamed. Bacteria, chemical agents, or both attack cells of the bronchiole walls. Ciliated cells in the walls are destroyed, and mucus-secreting cells multiply. Fibrous scar tissue forms and in time may narrow or obstruct the airways.

EMPHYSEMA When bronchitis persists, thick mucus clogs the airways. Inside the lungs, tissue-destroying bacterial enzymes attack the stretchable, thin walls of alveoli. The walls break down, and inelastic fibrous tissue forms around them. Gas exchange proceeds at fewer alveoli, which become enlarged. In time, the fine balance between air flow and blood flow is permanently compromised, for the lungs remain distended and inelastic.

Compare Figure 41.17a with b to get a sense of what happens. It becomes difficult to run, walk, even exhale. These are symptoms of *emphysema*, a respiratory ailment that affects about 1.3 million people in the United States alone.

A few people are genetically predisposed to the development of emphysema. They don't have a workable gene for the enzyme antitrypsin, which can inhibit bacterial attacks on alveoli. Poor diet and persistent or recurring colds and other respiratory infections also make a person susceptible to emphysema later in life. *But smoking is the major cause of the disease*. Emphysema can develop slowly, over twenty or thirty years. When emphysema is detected too late, the serious damage to lung tissue cannot be repaired.

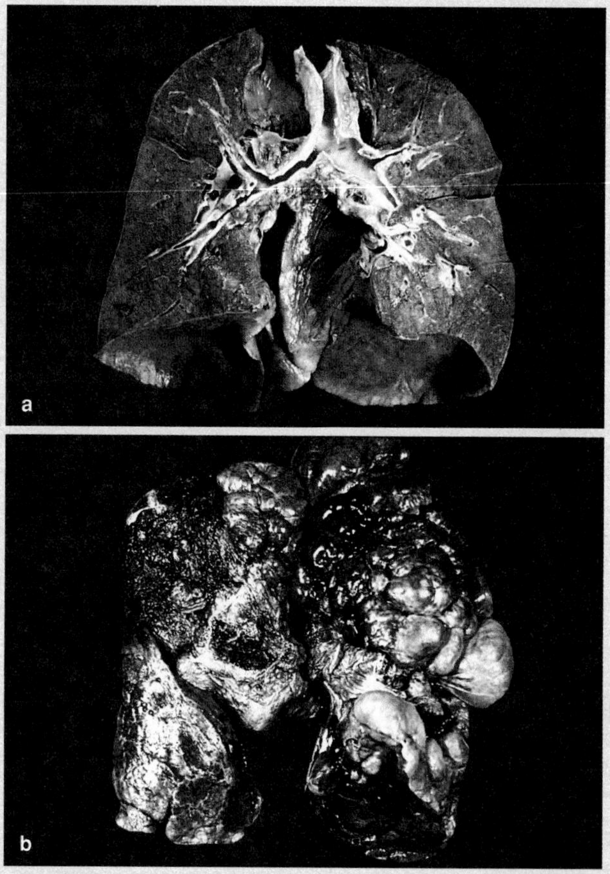

EFFECTS OF CIGARETTE SMOKE Worldwide, smoking now kills 3 million people every year. If current patterns do not change, then by the time today's young smokers reach middle age, 10 million may be dying annually because of the habit. That is one person every three seconds. Even now in the United States, one of 50 million cigarette-puffing people dies of emphysema, chronic bronchitis, or heart disease every thirteen seconds. For every eight of them, one nonsmoker dies of ailments brought on by *secondhand smoke*—of prolonged exposure to tobacco smoke in the surrounding air. And how much thought is given to children who breathe secondhand smoke? Parents and others who otherwise care about children are indirectly increasing their vulnerability to allergies and lung ailments. Smoking is the major cause of lung cancer, which is now the leading cause of death among women. Yet every day, 3,000 to 5,000 Americans light a cigarette for the first time. Even children spend a billion dollars a year on cigarettes. The economy shrinks by 22 billion a year because of the direct medical costs of treating smoke-induced respiratory disorders.

How does cigarette smoke lead to lung damage? Noxious particles in the smoke from just one cigarette immobilize cilia in the bronchioles for several hours. The particles also trigger mucus secretions, which in time clog the airways. They can kill infection-fighting macrophages residing in the respiratory tract. What starts as "smoker's cough" can end in bronchitis and emphysema.

Or consider how cigarette smoke contributes to lung cancer. Inside the body, certain compounds in coal tar and in cigarette smoke become converted to highly reactive intermediates. These are the carcinogens; they provoke uncontrolled cell divisions in lung tissues. On the average, 90 of every 100 smokers who develop lung cancer will die from it. If you now smoke, or are thinking about starting or quitting, you may wish to give serious thought to the information in Figure 41.18.

Figure 41.17 (a) Normal appearance of tissues of human lungs. (b) Lungs from someone affected by emphysema.

Risks Associated With Smoking:	Reducing the Risks by Quitting:
SHORTENED LIFE EXPECTANCY: Nonsmokers live 8.3 years longer on average than those who smoke two packs daily from the midtwenties on.	Cumulative risk reduction; after 10 to 15 years, life expectancy of ex-smokers approaches that of nonsmokers.
CHRONIC BRONCHITIS, EMPHYSEMA: Smokers have 4–25 times more risk of dying from these diseases than do nonsmokers.	Greater chance of improving lung function and slowing down rate of deterioration.
LUNG CANCER: Cigarette smoking is the major cause of lung cancer.	After 10 to 15 years, risk approaches that of nonsmokers.
CANCER OF MOUTH: 3–10 times greater risk among smokers.	After 10 to 15 years, risk is reduced to that of nonsmokers.
CANCER OF LARYNX: 2.9–17.7 times more frequent among smokers.	After 10 years, risk is reduced to that of nonsmokers.
CANCER OF ESOPHAGUS: 2–9 times greater risk of dying from this.	Risk proportional to amount smoked; quitting should reduce it.
CANCER OF PANCREAS: 2–5 times greater risk of dying from this.	Risk proportional to amount smoked; quitting should reduce it.
CANCER OF BLADDER: 7–10 times greater risk for smokers.	Risk decreases gradually over 7 years to that of nonsmokers.
CORONARY HEART DISEASE: Cigarette smoking is a major contributing factor.	Risk drops sharply after a year; after 10 years, risk reduced to that of nonsmokers.
EFFECTS ON OFFSPRING: Women who smoke during pregnancy have more stillbirths, and weight of liveborns averages less (hence, babies are more vulnerable to disease, death).	When smoking stops before fourth month of pregnancy, risk of stillbirth and lower birthweight eliminated.
IMPAIRED IMMUNE SYSTEM FUNCTION: Increase in allergic responses, destruction of defensive cells (macrophages) in respiratory tract.	Avoidable by not smoking.
BONE HEALING: Evidence suggests that surgically cut or broken bones require up to 30 percent longer to heal in smokers, possibly because smoking depletes the body of vitamin C and reduces the amount of oxygen reaching body tissues. Reduced vitamin C and reduced oxygen interfere with production of collagen fibers, a key component of bone. Research in this area is continuing.	Avoidable by not smoking.

Figure 41.18 From the American Cancer Society, a list of the risks incurred by smoking and the benefits of quitting. The photograph shows a few swirls of cigarette smoke poised at the entrance to the two bronchi that lead into the lungs.

EFFECTS OF MARIJUANA SMOKE In 1994, 17 million Americans smoked *pot*—marijuana (*Cannabis*)—at least once to induce light-headed euphoria. The number is rising, especially in junior high and high schools. About 1.5 million are chronic users. They become enamored of the euphoria, but the sensation does not last long. It fades to apathy, depression, and fatigue. Users keep smoking to keep from feeling "wasted." Besides psychological dependency, chronic use can result in throat irritation, persistent coughing, bronchitis, and emphysema.

Breathing at High Altitudes

From Section 41.6 as well as the introduction to this chapter, you know that the partial pressure of oxygen decreases with increasing altitude. The consequences can be dreadful indeed for the occasional climbers of Chomolungma and other cloud-piercing peaks. Yet, like llamas and other species that evolved at high elevations, millions of people comfortably live out their lives 4,800 meters (16,000 feet) or more above sea level.

Compared to people living at lower elevations, the lungs of permanent residents of high mountains have far more alveoli and blood vessels, which formed while they were growing up. Also, the ventricles of their heart enlarged more, so they pump larger volumes of blood. And their muscle tissue has more mitochondria.

Llamas have an additional advantage. Compared to human hemoglobin, llama hemoglobin contains factors that give it a greater affinity for oxygen (Figure 41.19). It picks up oxygen more efficiently at the lower pressures characteristic of high altitudes.

Does this mean that a healthy person who grew up by the seashore cannot ever pick up roots and move to the mountains? No. By mechanisms of **acclimatization**, an individual often can make long-lasting physiological and behavioral adaptations to a new environment that is markedly different from the one left behind. At high altitudes, gradual changes in the pattern and magnitude of breathing and in cardiac output can supplant the initially acute compensations that are made to cellular oxygen deficiency (hypoxia).

Within a few days, the reduction in oxygen delivery stimulates kidney cells to secrete more **erythropoietin**, a hormone that induces stem cells in bone marrow to proliferate and mature into red blood cells. Each second in an adult human, between 2 million and 3 million red blood cells are being churned out as replacements for the ones continually dying. Stepped-up erythropoietin secretion can increase that astounding pace by six times during extreme stress. Increased numbers of circulating red blood cells increase the oxygen-delivery capacity of blood. When the oxygen level increases sufficiently, the kidneys slow down erythropoietin secretion.

Erythropoietin is the major hormonal mediator of red blood cell production. However, human males have a larger muscle mass and greater demands for oxygen. They depend on testosterone, also. Among other things, this sex hormone stimulates increases in the basic rate of red blood cell production. Normally, a sample of blood from males has a larger percentage of red blood cells than an equivalent sample from females.

The compensatory increase in red blood cells comes at a cost. Having many more cells in the bloodstream increases resistance to blood flow, because the blood is now more viscous, or "thicker." Hence the heart must work harder to pump blood.

Carbon Monoxide Poisoning

High elevations are not the exclusive culprits behind cellular oxygen deficiency. Hypoxia also occurs when the partial pressure of oxygen in arterial blood falls owing to *carbon monoxide poisoning*.

Carbon monoxide, a colorless, odorless gas, is a component of exhaust fumes from cars, trucks, and other vehicles. It is also present in the smoke from tobacco and from burning coal and wood. This gas competes with oxygen for binding sites in hemoglobin. And its binding capacity is at least 200 times greater than that of oxygen.

Even small amounts of carbon monoxide can tie up half of the body's hemoglobin. Thus it can impair oxygen delivery to tissues, sometimes to dangerous degrees.

Figure 41.19 (a) Comparison of the binding and releasing capacity of human hemoglobin and llama hemoglobin. (b) Llamas high up in the Peruvian Andes.

(graph) percent saturation of heme binding sites vs. partial pressure of oxygen (mm Hg)

a

b

Respiration in Diving Mammals

THE PHYSIOLOGY OF DIVING   How do whales, seals, and other air-breathing mammals that routinely dive survive underwater, where they cannot inhale oxygen and exhale carbon dioxide? Apparently such mammals (as well as diving birds) share a common, highly specialized pattern of metabolism during their underwater forays.

Here we are not talking about a quick bob under the water's surface. Weddell seals (*Leptonychotes weddelli*) of Antarctic waters typically remain submerged for twenty to twenty-five minutes, and some do so for more than an hour. They usually limit their dives to 400 meters (1,312 feet) or less, but a few have been observed at depths approaching 600 meters—more than a third of a mile down. As two more examples, one sperm whale (Figure 41.20) submerged itself for eighty-two minutes; another, tracked by sonar, reached a depth of 2,250 meters.

By comparison, humans who are trained to dive without oxygen tanks can fully submerge themselves for about three minutes, at most. Career divers, including the ama of Korea and Japan, do only a bit better.

When diving mammals leave the air behind, they take oxygen with them. Hemoglobin binds much of it. So does **myoglobin**, an oxygen-binding protein that is abundant in skeletal muscles. And the lungs serve as a suitcase filled with oxygen. During short dives, aerobic respiration proceeds as usual. During prolonged dives, however, the oxygen is preferentially distributed to the heart and central nervous system, both of which cannot function without the high energy yields of the aerobic pathway. Basically, blood is directed away from most organs. Blood pumped from the heart travels mainly to the lungs, brain, and back to the heart. Skeletal muscles use their own myoglobin-bound oxygen as well as the hemoglobin-bound oxygen in the dense capillary beds that service them. Later on, they switch to anaerobic pathways, and lactate accumulates in tissues. When the animal surfaces, circulation to the muscles is restored.

Diving mammals also have specialized respiratory adaptations. For example, a sperm whale's respiratory system is adapted to collect and store oxygen. When the whale surfaces, it rapidly blows stale air from its lungs, through a tubelike epiglottis and then a blowhole at the body surface. Before the whale dives again, it breathes rapidly, so that its elastic, extensible lungs fill quickly with large volumes of air. During a dive, special valves close its nostrils; and rings of muscle and of cartilage clamp its bronchioles. The whale's respiratory surface is not that large. However, strategically located valves and local networks of blood vessels (plexuses) store and distribute the blood volume and gases in economical fashion. The heart rate slows, metabolism decreases, and so does the rate of oxygen use and carbon dioxide

Figure 41.20   A sperm whale, adapted for prolonged diving. This aquatic mammal is more than twenty-five meters long, on average. It can dive 2,250 meters below the ocean surface and stay there for well over an hour before coming up for oxygen.

formation. In addition, compared to land mammals, the whale's respiratory center is less sensitive to carbon dioxide levels.

HUMAN DEEP-SEA DIVERS   Professional divers know that water pressure increases greatly with depth. They do not dive in deep water without tanks of compressed air (air under pressure). They also make their ascents from the depths with utmost caution.

Why the caution? Because of the increased pressure at depths, more gaseous nitrogen (N_2) than usual has become dissolved in their body tissues. The outcome is *nitrogen narcosis*, or "raptures of the deep." At depths of 45 meters (150 feet), divers become euphoric and drowsy, as if they were tipsy with alcohol. Some have been known to offer the mouthpiece of their airtank to fishes. At lower depths, the divers become clumsy and weak. Below 105 meters, they can slip into a coma.

Gaseous nitrogen is a lipid-soluble gas, so it readily dissolves in the lipid bilayer of cell membranes. By one hypothesis, when it does this to neurons, it suppresses their capacity to respond to action potentials.

Deep-sea divers also are cautious when they return to the surface. When the pressure decreases, N_2 moves back out of the tissues and into the bloodstream. If an ascent is too rapid, N_2 enters the blood faster than the lungs can dispose of it. Then, bubbles of nitrogen may form in the blood and tissues. Too many bubbles cause pain, especially at joints. Hence the common name, "the bends," for what is otherwise known as *decompression sickness*. If bubbles obstruct the blood flow to the brain, deafness, impaired vision, and paralysis may result.

During brief stays at high elevations, the body makes acute compensatory responses to the thinner air. Over time, it makes adaptive adjustments in breathing and cardiac output.

Diving mammals take along oxygen bound to hemoglobin and myoglobin and in their lungs. They also depend on respiratory adaptations for storing and distributing oxygen.

41.9 RESPIRATION IN LEATHERBACK SEA TURTLES

Late at night, near the shoreline of a Caribbean island, biologist Molly Lutcavage and several colleagues move out on a turtle patrol. Finally a large, dark shape, with flippers flailing, emerges from the pale surf. A female Atlantic leatherback sea turtle (*Dermochelys coriacea*) is returning from the sea to nest in the sand (Figure 41.21).

Sea turtles are members of the reptilian lineage, which extends 300 million years back in time. Green turtles, ridleys, loggerheads, leatherbacks—all are endangered or threatened species. The race is on to gather information about their life histories and physiology that may help pull them back from the brink of extinction.

Leatherbacks are the largest and least understood sea turtles. The adult females may weigh more than 400 kilograms (880 pounds). The adult males may weigh nearly twice that amount. Leatherbacks normally leave the water only to breed and lay eggs. During the rest of their lives, they migrate across vast stretches of the open ocean.

Leatherbacks do something no other reptile on Earth can do. They have the capacity to dive 1,000 meters (3,000 feet) below sea level! Before biologists conducted tracking experiments during the mid-1980s, Weddell seals, fin whales, and some other marine mammals were the only nonhuman deep-sea divers known.

To reach such depths, a turtle would have to swim for nearly forty minutes without taking a breath. How does it dive so deeply, for so long, and still have enough oxygen for aerobic metabolism?

Oxygen reserves in the lungs are not enough for the leatherback's prolonged underwater excursions. Even at depths between 80 and 160 meters, pressure exerted by the surrounding water causes air-filled lungs to collapse. However, like the diving marine mammals described in Section 41.8, the leatherback's skeletal muscle cells have an abundance of the oxygen-binding protein myoglobin. In addition, compared to other diving reptiles, they have more blood per unit of body weight, and their blood contains more red blood cells—hence more hemoglobin (Table 41.1).

Imagine yourself as an observer on Lutcavage's team. Their plan is to draw blood samples from a leatherback so that they can study the oxygen-binding capacity of its hemoglobin. Samples of respiratory gases dissolved in the blood may provide insight into how leatherbacks use oxygen. As the researchers know, a nesting female enters a trancelike state during the twenty or so minutes it takes her to deposit eggs in the sand. During that time, she will not resist being gently handled.

The team gets to work. They fit the head of a nesting female with a helmet that is equipped with an expandable plastic sleeve. The fit is snug but not restrictive, so the turtle is able to breathe normally. With each exhalation, the expired gases flow through a one-way valve into a collecting sac. At the same time, the researchers count the number of breaths required to completely fill the sac. (As a baseline, they also tallied the breathing frequency before placing the helmet over her head.) They store the gas samples according to established procedures and set them aside for laboratory analysis.

Working quickly, a member of the team draws blood from a superficial vein in the turtle's neck, and then packs the sample in ice to preserve it. Once they are back at the laboratory, the sample will be subjected to hemoglobin analysis and a red blood cell count. The oxygen level, carbon dioxide level, and pH also will be measured.

Lutcavage and her coworkers did gather samples of blood and respiratory gases from several turtles. They also analyzed skeletal muscle tissue, taken earlier from a drowned turtle. The data gave insight into how leatherbacks manage their spectacularly deep dives.

As expected, calculations of the oxygen uptake and total tidal volume during breathing revealed that the leatherback cannot inhale and hold enough air in the lungs to sustain aerobic metabolism during a prolonged dive. Other oxygen-supplying mechanisms had to be at

Table 41.1	Physiological Comparison of a Few Diving Reptiles and Mammals				
Characteristic	LEATHERBACK TURTLE	GREEN TURTLE	CROCODILE	KILLER WHALE	WEDDELL SEAL
Maximum diving depth (meters)	Greater than 1,000	Less than 100	Less than 30	260	600
Red blood cell count (percentage of total volume)	39	30	28	44	58
Hemoglobin level (grams/deciliter of blood)	15.6	8.8	8.7	16.0	17–22
Myoglobin level (milligram/gram of muscle tissue)	4.9	—*	—*	—*	44.6
Oxygen-carrying capacity (volume percent)	21	7.5–11.9	12.4	23.7	31.6

* Not known.

work. As further analysis revealed, myoglobin levels are so high in leatherbacks, oxygen remains available for skeletal muscle activity during a dive after the air in the lungs gives out (or when their lungs collapse).

Also, leatherbacks have a notable abundance of red blood cells and a type of hemoglobin with a high affinity for binding oxygen. A leatherback's blood may contain a remarkable 21 percent oxygen by volume. That amount is typical of a human, not a "plodding" turtle.

The oxygen-carrying capacity turns out to be the highest ever recorded for a reptile—and it is close to the capacity of deep-diving mammals. Add to this a streamlined body and massive front flippers, and you have a reptile uniquely adapted for diving and maneuvering at great depths.

Studies of leatherbacks are only now under way. Researchers are starting to investigate newer tracking methods to help them monitor a leatherback's metabolism at sea and during dives. So far, it has proved extremely difficult and expensive to track individual turtles in the open ocean.

Figure 41.21 Two female leatherback sea turtles (*Dermochelys coriacea*) returning to water after laying eggs in the sands of an island in the Caribbean Sea.

1. Aerobic respiration is the main metabolic pathway that provides enough energy for active life-styles. It uses oxygen and produces carbon dioxide. The process by which the animal body as a whole acquires oxygen and disposes of carbon dioxide is called respiration.

2. Air is a mixture of oxygen, carbon dioxide, and other gases, each exerting a partial pressure. Each gas tends to move from areas of higher to lower partial pressure. Respiratory systems make use of this tendency.

3. In respiratory systems, oxygen and carbon dioxide diffuse across a respiratory surface: a moist, thin layer of epithelium that is interposed between the external and internal environments.

4. Modes of respiration differ among animal groups.

   a. Invertebrates with a small body mass depend only on integumentary exchange; oxygen and carbon dioxide simply diffuse across the body surface. This mode also persists in some large animals, including amphibians.

   b. Many marine and some freshwater invertebrates have gills, respiratory organs with moist, thin, and often highly folded walls. Most insects and certain spiders use tracheal respiration. Gases flow through open-ended tubes leading from the body surface directly to tissues. Most spiders have book lungs, with leaflike folds.

   c. Fishes have greater energy demands than small invertebrates. They have gills in which a countercurrent flow mechanism compensates for low oxygen levels in aquatic habitats. Paired lungs are the dominant means of respiration in reptiles, birds, and mammals.

5. The airways of the human respiratory system are the nasal cavities, pharynx, larynx, trachea, bronchi, and bronchioles. Many millions of alveoli at the end of the terminal bronchioles are the main sites of gas exchange.

6. Breathing ventilates the lungs. Each respiratory cycle consists of inhalation (one breath in) and exhalation (one breath out). During inhalation, the chest cavity expands, lung pressure decreases below atmospheric pressure, and air flows into the lungs. The events are reversed during normal exhalation.

7. Driven by its partial pressure gradient, oxygen in the lungs diffuses from alveolar air spaces into pulmonary capillaries. It diffuses into red blood cells and binds weakly with hemoglobin. Hemoglobin gives up oxygen at capillary beds in metabolically active tissues. Oxygen diffuses across the interstitial fluid, then into cells.

8. Driven by its partial pressure gradient, carbon dioxide in tissues diffuses from cells, across interstitial fluid, and into the blood. Most of it reacts with water to form bicarbonate. The reactions are reversed inside the lungs; carbon dioxide diffuses from pulmonary capillaries into the air spaces of alveoli, then it is exhaled.

Figure 41.22 Human female at play in the domain of aquatic animals.

Review Questions

1. Define respiratory surface. Why must the oxygen and carbon dioxide partial pressure gradients across it be steep? *41.1*

2. What is the name of the respiratory pigment in red blood cells? What is the name of a major respiratory pigment in skeletal muscle cells? *41.1, 41.8*

3. A few of your friends who have not taken a biology course ask you what insect lungs look like. What is your answer (assuming your instructor is listening)? *41.2*

4. Briefly describe how countercurrent flow through a fish gill is so efficient at taking up dissolved oxygen from water. *41.3*

5. Does the respiratory system of fishes, amphibians, reptiles, birds, or mammals have air sacs that ventilate the lungs? *41.3*

6. Distinguish between:
   a. aerobic respiration and respiration *CI, 41.1*
   b. pharynx and larynx *41.4*
   c. bronchiole and bronchus *41.4*
   d. pleural sac and alveolar sac *41.4, 41.5*

7. Explain why humans (Figure 41.22) cannot survive on their own, for very long, underwater. *41.8*

8. Define the functions of the human respiratory system. In the diagram below, label its components and the major bones and muscles with which it interacts during breathing. *41.4*

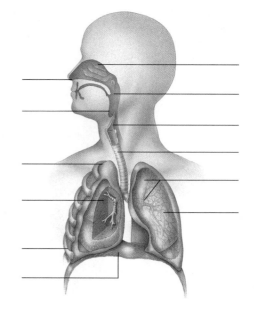

Self-Quiz (*Answers in Appendix IV*)

1. A partial pressure gradient of oxygen exists between _____ .
   a. the atmosphere and the lungs
   b. lungs and metabolically active tissues
   c. air at sea level and air at high altitudes
   d. all of the above

2. The _____ is an airway that connects the nose and mouth with the _____ .
   a. oral cavity; larynx   c. trachea; pharynx
   b. pharynx; trachea   d. pharynx; larynx

3. Oxygen in the air diffuses across _____ as it follows its partial pressure gradient into the human body.
   a. pleural sacs   c. a moist respiratory surface
   b. alveolar sacs   d. both b and c

4. Each human lung encloses a _____ .
   a. diaphragm   c. pleural sac
   b. bronchial tree   d. both b and c

5. In human lungs, gas exchange occurs at the _____ .
   a. two bronchi   c. alveolar sacs
   b. pleural sacs   d. both b and c

6. Breathing _____ .
   a. ventilates the lungs   d. reverses pressure gradients
   b. draws air into airways e. all of the above
   c. expels air from airways

7. During quiet breathing, inhalation is _____ and exhalation is _____ .
   a. passive; passive   c. passive; active
   b. active; active   d. active; passive

8. After oxygen diffuses into pulmonary capillaries, it diffuses into _____ and binds with _____ .
   a. interstitial fluid; red blood cells
   b. interstitial fluid; carbon dioxide
   c. red blood cells; hemoglobin
   d. red blood cells; carbon dioxide

9. Most carbon dioxide in blood is in the form of _____ .
   a. carbon dioxide   c. carbonic acid
   b. carbon monoxide   d. bicarbonate

10. Match the components with their descriptions.
   ____ trachea   a. airway leading into a lung
   ____ pharynx   b. throat
   ____ alveolus   c. fine branch of bronchial tree
   ____ hemoglobin   d. windpipe
   ____ bronchus   e. respiratory pigment
   ____ bronchiole   f. site of gas exchange

Critical Thinking

1. Some cigarette manufacturers conducted public relations campaigns urging their customers to "smoke responsibly." What are some social and biological issues in this controversy? How do these issues apply to the nonsmoking spouse or children of a smoker? To nonsmoking patrons in a restaurant? To the unborn child of a pregnant smoker? In your opinion, what behavior would constitute "responsible smoking"?

2. People sometimes poison themselves with carbon monoxide by building a charcoal fire in an enclosed area. Assuming help arrives in time, what would be the best treatment: (1) placing the victim outdoors in fresh air or (2) rapidly administering pure oxygen? Explain how you arrived at your answer.

3. When you swallow food, muscle contractions force the epiglottis down, to its closed position. This prevents food from going down the trachea and blocking gas flow. Yet each year,

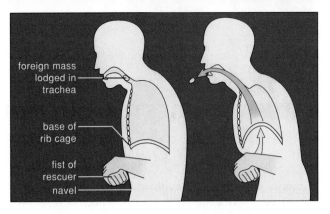

Figure 41.23 Heimlich maneuver to dislodge food stuck in the trachea. Stand behind the victim, make a fist with one hand, then position the fist, thumb-side in, against the victim's abdomen. The fist must be slightly above the navel and well below the rib cage. Now press the fist into the abdomen with a sudden upward thrust. Repeat the thrust several times if needed. The maneuver can be performed on someone who is standing, sitting, or lying down.

several thousand choke to death after food enters the trachea and blocks air flow for as little as four or five minutes. The *Heimlich maneuver*—which is an emergency procedure only— often dislodges food from the esophagus (Figure 41.23). When correctly performed, it elevates the diaphragm forcibly, causing a sharp decrease in the volume of the thoracic cavity and a sudden increase in alveolar pressure. Air forced up the trachea as a result of the increased pressure may be enough to dislodge the obstruction. Once the obstacle is dislodged, a physician must see the person at once. An inexperienced rescuer can inadvertently cause internal injuries or crack a rib.

Reflect on the current social climate in which lawsuits abound. Would you risk performing the Heimlich maneuver to save a relative's life? A stranger's life? Why or why not?

Selected Key Terms

acclimatization *41.8*
alveolus (alveoli) *41.4*
bronchiole *41.4*
bronchus *41.4*
carbamino hemoglobin *41.6*
carbonic anhydrase *41.6*
countercurrent flow *41.3*
diaphragm *41.4*
epiglottis *41.4*
erythropoietin *41.8*
Fick's law *41.1*
gill (fish) *41.2*
glottis *41.3*
heme group *41.6*
hemoglobin *41.1*
integumentary exchange *41.2*

larynx *41.4*
lung *41.3*
myoglobin *41.8*
oxyhemoglobin *41.6*
partial pressure *41.1*
pharynx *41.4*
pressure gradient *41.1*
respiration *CI*
respiratory cycle *41.5*
respiratory surface *41.1*
respiratory system *CI*
tidal volume *41.5*
trachea *41.4*
tracheal respiration *41.2*
vital capacity *41.5*
vocal cord *41.3*

Readings

Hill, R., and G. Wyse. 1992. *Animal Physiology*. Second edition. New York: Harper Collins.

Sherwood, L. 1997. *Human Physiology*. Second edition. Belmont, California: Wadsworth.

Web Site See *http://www.wadsworth.com/biology* for practice quiz questions, hypercontents, BioUpdates, and critical thinking. The Wadsworth Biology Resource Center provides a wealth of information fully organized and integrated by chapter.

DIGESTION AND HUMAN NUTRITION

Lose It — And It Finds Its Way Back

America's fixation on Beautiful People who have nary an ounce of extra fat on their utterly perfect selves is bad enough. After all, we can always rationalize our own extra ounce with the truism that beauty is only skin deep. But the American Medical Association's dire warnings of the Fat Connection to atherosclerosis, heart attacks, strokes, colon cancer, and other ailments is beyond rationalization.

By current standards, the proportion of body fat relative to total tissue mass should be 18 to 24 percent for human females who are less than thirty years old. For males, it should be no more than 12 to 18 percent. An estimated 34 million Americans do not even come close to the standards.

No matter how much we lose by dieting, the lost weight seems to find its way back. This physiological dilemma is an outcome of our evolutionary heritage. Like most other mammals, we have adipose tissue with an abundance of fat-storing cells. Collectively, the cells are an adaptation for survival. They serve as an energy warehouse that can be opened up during times when food is not available. Variations in food intake only influence how empty or full a fat-storing cell will get. Once that cell has formed, it is in the body to stay.

Dieting starts to empty the fat warehouse. The brain interprets this as "starvation" and triggers a metabolic slowdown. The body uses energy far more efficiently—even for the basics, such as breathing and digesting food. That is, it takes less food to do the same things!

As you have probably heard, dieting does no good without *long-term* commitment to physical exercise. Why? When you are dieting, skeletal muscles are also adapting to "starvation," and they burn less energy than before. Jog for four hours or play tennis for eight hours straight, and you may well lose a pound of fat. Meanwhile, your appetite surges, so if and when the dieting stops, "starved" fat cells quickly refill. In 1995, after a decade of careful study, researchers determined that lost weight will be regained unless a person eats less *and* exercises moderately throughout his or her lifetime.

What about gaining weight? Intriguingly, a weight gain triggers an *increase* in metabolism. The body uses 15 to 20 percent more energy than before—until weight drops back to what it was before.

Emotional factors also influence weight gains and losses. Consider *anorexia nervosa*, a potentially fatal eating disorder that is founded on a seriously flawed assessment of body weight. Typically, anorexics have an overwhelming dread of being fat *and* hungry. They starve themselves and often overexercise. Often there are fears about growing up and being sexually mature. Irrational expectations of personal performance are common.

Or consider *bulimia*, an out-of-control, "oxlike appetite." During an hour-long eating binge, a bulimic might ingest more than 50,000 kilocalories' worth of food, then vomit or use laxatives to purge the body

Figure 42.1 Mirror, mirror, on the wall, who is fairest of them all? (And by whose standards?)

of it. Some get into a binge-purge routine as an "easy" way to lose weight. Others suffer severe emotional stress. They may not even like to eat, but at some level the purging relieves them of anger and frustration.

Some bulimics binge-purge once a month. Others do so several times a day. Do they know that repeated purgings can damage the gut? That chronic vomiting brings gastric fluid into the mouth and erodes teeth to stubs? That, at its extreme, bulimia causes the stomach to rupture, and the heart or kidneys to fail?

Are severe eating disorders rare? No. In the United States, an estimated 7 million females and 1 million males are anorexic or bulimic. Most are in their teens and early twenties, but the number also is increasing among preadolescents. Each year, 5 to 6 percent die as a result of complications arising from the disorders.

And with these sobering thoughts in mind, we start a tour of **nutrition**. The word encompasses processes by which an animal ingests and digests food, then absorbs the released nutrients for later conversion to the body's own carbohydrates, lipids, proteins, and nucleic acids.

The nutritional processes proceed at the **digestive system**. This body tube or cavity mechanically and chemically reduces food to particles, then to molecules that are small enough to be absorbed into the internal environment. It also eliminates unabsorbed residues. Other organ systems, especially those shown in Figure 42.2, contribute to the nutritional processes.

KEY CONCEPTS

1. In complex animals, interactions among digestive, circulatory, respiratory, and urinary systems supply the body's cells with raw materials, dispose of wastes, and maintain the volume and composition of extracellular fluid.

2. Most digestive systems have specialized regions for food transport, processing, and storage. Different regions mechanically break apart and chemically break down food, absorb breakdown products, and eliminate the unabsorbed residues.

3. To maintain an acceptable body weight and overall health, energy intake must balance energy output by way of metabolic activity, physical exertion, and so on. Complex carbohydrates provide the most dietary glucose, which typically is the body's main source of immediately usable energy.

4. Nutrition also involves the intake of foods that are good sources of vitamins, minerals, and a number of amino acids and fatty acids that the body itself cannot produce.

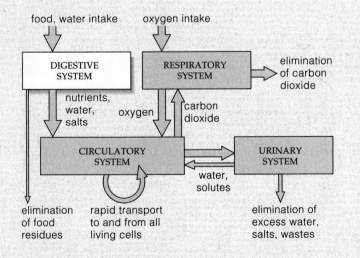

Figure 42.2 Functional links between the digestive, respiratory, circulatory, and urinary systems. These organ systems and others work together to supply cells with raw materials and eliminate wastes.

THE NATURE OF DIGESTIVE SYSTEMS

Incomplete and Complete Systems

Recall, from Chapter 26, that we don't see many organ systems until we get to the flatworms. These are among the invertebrates with an **incomplete digestive system**, a saclike, branching gut cavity that has a single opening at the start of a pharynx (Figure 42.3a). After food enters the sac, it is partly digested and circulated to cells even as residues are being sent back out.

Flatworms, cnidarians, some rotifers, and a few other small invertebrates require nothing more complicated than two-way traffic through a saclike gut that has a single opening. But such a system would not work well enough for the other animals. Instead, they depend on a **complete digestive system**. Basically, this is a tube with a mouth (an opening at one end for food intake) and an anus (an opening at the opposite end for eliminating unabsorbed residues). In between its openings, the tube is divided into regions that are specialized for one-way transport, processing, and storage of raw materials.

As an example, think about the complete digestive system of a frog (Figure 42.3b). In between the frog's mouth and anus are a pharynx, stomach, and small and large intestines. Functionally connected with the small intestine are a liver, gallbladder, and pancreas, which are organs with accessory roles in digestion. Birds, too, have a complete digestive system, with a few unique regional specializations (Figure 42.3c).

Regardless of its complexity, a complete digestive system carries out five overall tasks:

1. **Mechanical processing and motility**. Movements that break up, mix, and propel food material.

2. **Secretion**. Release of digestive enzymes and other substances into the space inside the tube.

3. **Digestion**. Breakdown of food into particles, then into nutrient molecules small enough to be absorbed.

4. **Absorption**. Passage of digested nutrients and fluid across the tube wall and into body fluids.

5. **Elimination**. Expulsion of the undigested and unabsorbed residues from the end of the gut.

In most animals, remember, a digestive system does not act alone. A circulatory system distributes the absorbed nutrients to cells throughout the body. A respiratory system helps cells use the nutrients by supplying them with oxygen for aerobic respiration and removing carbon dioxide wastes. A urinary system counters variations in the kinds and amounts of absorbed nutrients. It helps maintain the composition and volume of the internal environment (Figure 42.2).

Correlations With Feeding Behavior

We can correlate the specializations of any digestive system with feeding behavior. Consider the pigeon in Figure 42.3c. A long tube (esophagus) leads from its bill, the *food-gathering* region, to a *food-processing* region. Being compactly centered in the body mass, the food-processing region is an adaptation of an animal that must balance its body during flight. As for other seed-eating birds, a large *crop*, a stretchable storage organ, balloons from the esophagus. It allows the bird to zoom down on food, to eat a lot, and possibly to make a quick getaway before a predator becomes aware of it. Like other birds active during the day, it fills the crop before

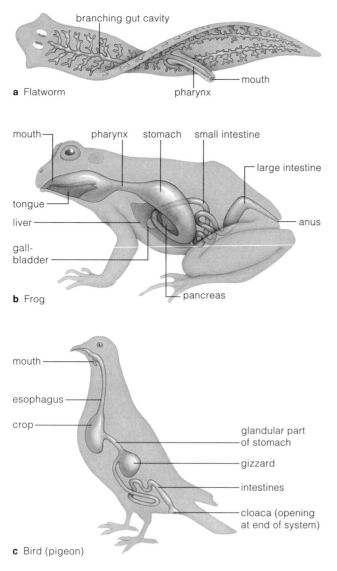

a Flatworm

b Frog

c Bird (pigeon)

Figure 42.3 (**a**) Incomplete digestive system of a flatworm, with two-way traffic of food and undigested material through one opening. (**b,c**) Examples of a complete digestive system, a tube with specialized regions and an opening at each end.

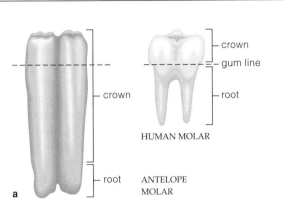

HUMAN MOLAR

ANTELOPE MOLAR

Figure 42.4 Some regional specializations of the complete digestive system of pronghorn antelope (*Antilocapra americana*).

(**a**) Comparison of antelope and human molars. (**b**) Multiple-chambered antelope stomach. The first chamber is a large pouch. The second is smaller, with a honeycombed inner surface (what chefs call tripe). In both chambers, food is mixed with fluid, kneaded, and exposed to the fermentation activities of bacterial and protozoan symbionts. Some of the symbionts degrade cellulose; others synthesize organic compounds, fatty acids, and vitamins. The host uses a portion of these substances. Kneaded food is regurgitated into the mouth, rechewed, then swallowed again. It enters the third chamber, where it is pummeled once more before entering the last stomach chamber.

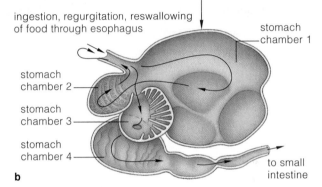

the sun goes down. It releases food during the first few hours after dark, thus decreasing the time of overnight fasting. The glandular lining of the first portion of its stomach secretes enzymes and other substances that aid in digestion. The stomach's second portion, a *gizzard*, is a muscular organ that grinds food much as teeth and jaws do. The intestines are proportionally shorter than they are in ducks, ostriches, and other birds that feed on plant parts rich in tough, fibrous cellulose—which requires a much longer processing time.

We also can correlate feeding behavior with the digestive system of pronghorn antelope (*Antilocapra americana*). Fall through winter, on the mountain ridges from central Canada down into northern Mexico, these animals browse on wild sage. Come spring, they move to open grasslands and deserts, there to browse on new growth (Figure 42.4). While they browse, they keep an eye out for coyotes, bobcats, and golden eagles. And can they eat and run! They reach speeds of 95 kilometers per hour and typically leave predators in the dust.

Now think of your own cheek teeth, or molars, each with a flattened crown that acts as a grinding platform. The crown of an antelope's molars dwarfs yours (Figure 42.4*a*). Why the difference? You probably do not rub your mouth against dirt while eating. An antelope does, and abrasive bits of soil enter its mouth along with tough plant material. Its teeth wear down rapidly, and natural selection has favored more crown to wear down.

Antelopes are **ruminants**, a type of hoofed mammal with multiple stomach chambers in which cellulose

is slowly broken down. The breakdown gets under way in the first two of four stomach sacs (Figure 42.4*b*). There, bacterial symbionts produce cellulose-digesting enzymes, to the antelope's benefit as well as their own. As enzymes act, the antelope regurgitates and rechews the contents of the first two sacs, then swallows again. (That's what "chewing cud" means.) Pummeling the plant material more than once exposes more surface area to the enzymes and gives them more time to act.

Thus the elaborate stomach of ruminants accepts a steady flow of plant material during lengthy feeding times, then slowly liberates nutrients during times of rest. Predatory or scavenging mammals have different specializations. They gorge on food when they can get it, then may not eat again for some time. Part of their digestive system stores food being gulped down too fast to be digested and absorbed. Other, accessory parts of the digestive system help ensure that there will be an adequate distribution of nutrients between meals.

An incomplete digestive system is a saclike body cavity. Both food and residues enter and leave through the same opening. A complete digestive system is a tube with two openings (mouth and anus) and regional specializations.

Complete digestive systems carry out the following tasks in controlled ways: Movements that break up, mix, and propel food material; secretion of digestive enzymes and other substances; breakdown of food; absorption of nutrients; and expulsion of undigested and unabsorbed residues.

OVERVIEW OF THE HUMAN DIGESTIVE SYSTEM

Let's look at your own body as an example of the kinds of organs in a complete digestive system. The human digestive system is a tube with two openings and many specialized organs (Figure 42.5). If the tube of an adult were fully stretched, it would extend 6.5 to 9 meters (21 to 30 feet). From beginning to end, a mucus-coated epithelium lines all the surfaces facing the lumen. (A *lumen* is the space in a tube.) The thick, moist mucus protects the wall of the tube and enhances diffusion across its inner lining. Substances advance in one direction, from the mouth through the pharynx, esophagus, and gastrointestinal tract (gut). The **gut** starts at the stomach and extends through the small intestine, large intestine (colon), and rectum, to the anus. Accessory organs—the salivary glands, gallbladder, liver, and pancreas—secrete substances into different regions of the tube.

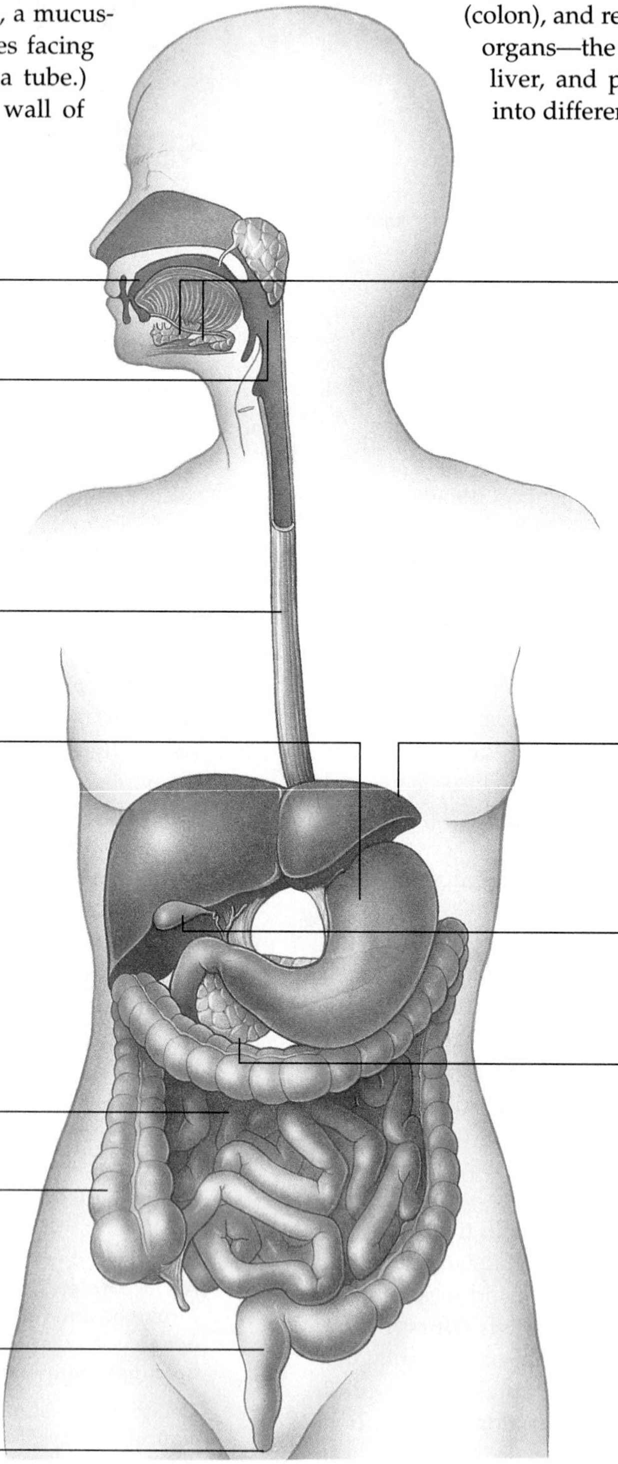

Major Components:

MOUTH (ORAL CAVITY)

Entrance to system; food is moistened and chewed; polysaccharide digestion starts.

PHARYNX

Entrance to tubular part of system (and to respiratory system); moves food forward by contracting sequentially.

ESOPHAGUS

Muscular, saliva-moistened tube that moves food from pharynx to stomach.

STOMACH

Muscular sac; stretches to store food taken in faster than can be processed; gastric fluid mixes with food and kills many pathogens; protein digestion starts.

SMALL INTESTINE

First part (duodenum, C-shaped, about 10 inches long) receives secretions from liver, gallbladder, and pancreas.

In second part (jejunum, about 3 feet long), most nutrients are digested and absorbed.

Third part (ileum, 6–7 feet long) absorbs some nutrients; delivers unabsorbed material to large intestine.

LARGE INTESTINE (COLON)

Concentrates and stores undigested matter by absorbing mineral ions, water; about 5 feet long; divided into ascending, transverse, and descending portions.

RECTUM

Distension stimulates expulsion of feces.

ANUS

End of system; terminal opening through which feces are expelled.

Accessory Organs:

SALIVARY GLANDS

Glands (three main pairs, many minor ones) that secrete saliva, a fluid with polysaccharide-digesting enzymes, buffers, and mucus (which moistens and lubricates food).

LIVER

Secretes bile (for emulsifying fat); roles in carbohydrate, fat, and protein metabolism.

GALLBLADDER

Stores and concentrates bile that the liver secretes.

PANCREAS

Secretes enzymes that break down all major food molecules; secretes buffers against HCl from the stomach.

Figure 42.5 Overview of the component parts of the human digestive system and their specialized functions. Organs with accessory roles in digestion are also listed.

INTO THE MOUTH, DOWN THE TUBE

Food is chewed and polysaccharide breakdown begins in the **oral cavity**, or **mouth**. Thirty-two teeth typically project into an adult's mouth (Figure 42.6). Each **tooth** has an enamel coat (hardened calcium deposits), dentin (a thick, bonelike layer), and an inner pulp (with nerves and blood vessels). It is an engineering marvel, able to withstand years of chemical insults and mechanical stress. Recall, from Section 27.10, that the chisel-shaped incisors shear off chunks of food. Cone-shaped canines tear it. Premolars and molars, with broad crowns and rounded cusps, are good at grinding and crushing food.

Also in the mouth is a **tongue**, an organ consisting of membrane-covered skeletal muscles that function in positioning food in the mouth, swallowing, and speech. On the tongue's surface are circular structures with taste buds embedded in their tissues (Figure 42.7). Each bud contains sensory receptors that can detect chemical

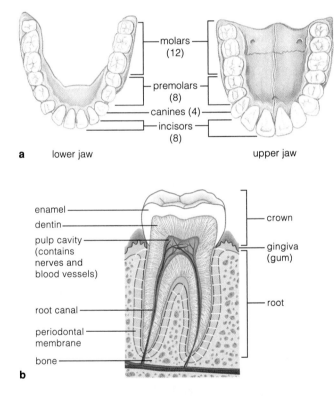

a lower jaw   upper jaw

Figure 42.6 (**a**) Number and arrangement of human teeth. (**b**) Closer look at a molar's two main regions, the crown and root. Enamel caps the crown. It consists of calcium deposits and is the hardest substance in the body.

Normally harmless bacteria live on and between teeth. Daily flossing, gentle brushing, and avoidance of too many sweets help keep their populations in check. In the absence of such preventive measures, conditions favor bacterial infections. These may result in *caries* (tooth decay), *gingivitis* (inflamed gums), or both. In addition, infections can spread to the periodontal membrane, which anchors teeth to the jawbone. In *periodontal disease*, infection slowly destroys the bone tissue around a tooth.

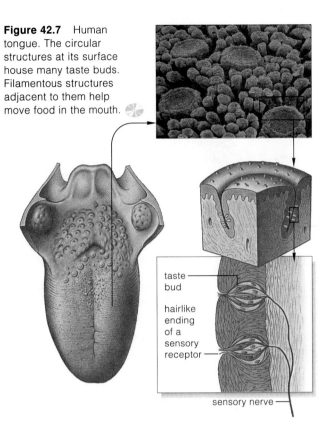

Figure 42.7 Human tongue. The circular structures at its surface house many taste buds. Filamentous structures adjacent to them help move food in the mouth.

differences in dissolved substances. The brain uses this information to give rise to our sense of taste.

Chewing mixes food with **saliva**. This fluid contains an enzyme (salivary amylase), a buffer (bicarbonate, or HCO_3^-), mucins, and water. Salivary glands, beneath and in back of the tongue, produce and secrete saliva through ducts to the free surface of the mouth's lining. Salivary amylase breaks down starch. The HCO_3^- helps maintain the mouth's pH when you eat acidic foods. Modified proteins called mucins help form the mucus that binds food into a softened, lubricated ball, or bolus.

When you swallow, contractions of tongue muscles force boluses into the **pharynx**, the tubular entrance to the esophagus *and* the trachea, an airway to the lungs. As food leaves the pharynx, the epiglottis (a flaplike valve) closes off the trachea and prevents breathing. That is why you normally do not choke on food (Figure 41.23). The **esophagus** connects the pharynx with the stomach. Contractions of its muscular wall propel food past a sphincter into the stomach. A **sphincter** is a ring of smooth muscles, the contractions of which can close off a passageway or an opening to the body surface.

By the action of the mouth's teeth and tongue, food gets chewed, mixed with saliva, and bound into soft, lubricated balls that will be propelled through tubes to the stomach. Enzymes in saliva start the digestion of polysaccharides.

DIGESTION IN THE STOMACH AND SMALL INTESTINE

We arrive now at the premier food-processing organs, the stomach and small intestine. Both have layers of smooth muscles, the contractions of which break apart, mix, and directionally move food (Figure 42.8). The lumen of both organs receives digestive enzymes and other secretions with roles in the chemical breakdown of nutrients into fragments, then into molecules small enough to be absorbed. Table 42.1 lists the names and sources of the major digestive enzymes. Carbohydrate breakdown *starts* in the mouth, and protein breakdown *starts* in the stomach. But digestion of nearly all of the carbohydrates, lipids, proteins, and nucleic acids in food is *completed* in the small intestine.

The Stomach

The **stomach**, a muscular, stretchable sac (Figure 42.8a), serves three major functions. First, it mixes and stores ingested food. Second, its secretions help dissolve and degrade the food, proteins especially. Third, it helps control passage of food into the small intestine.

Glandular epithelium lines the stomach wall that is exposed to the lumen. Each day, the lining's glandular cells secrete about two liters of hydrochloric acid (HCl), mucus, pepsinogens, and other substances that make up

gastric fluid (stomach fluid). Stomach acidity increases when the HCl dissociates into H^+ and Cl^-, and this helps dissolve food to form a liquid mixture called **chyme**. The acidity can kill many pathogens that are ingested with food. It also can cause *heartburn* when gastric fluid backs up into the esophagus.

Protein digestion starts when the high acidity alters the structure of proteins and exposes the peptide bonds. It also converts pepsinogens to active enzymes (pepsins) that cleave the bonds. Protein fragments accumulate in the stomach's lumen. Meanwhile, glandular cells of the stomach lining secrete gastrin. This hormone stimulates the cells that secrete HCl and pepsinogen.

A *peptic ulcer*, an eroded wall region in the stomach or small intestine, results from insufficient secretion of mucus and buffers (or excess pepsin). Heredity, chronic emotional stress, smoking, and excessive use of alcohol or aspirin are contributing factors. Also, nearly everyone who has intestinal ulcers and 70 percent of those who have stomach ulcers are infected by *Helicobacter pylori*. A full course of antibiotic therapy can cure peptic ulcers in patients who test positive for this bacterium.

The stomach empties by waves of contraction and relaxation, of a type called peristalsis. The waves mix chyme and gain force as they approach a sphincter at the base of the stomach (Figure 42.8b). With the arrival of a strong contraction, the sphincter closes, so most of the chyme is squeezed back. A small amount moves into the small intestine. In time, most moves out of the stomach.

The Small Intestine

Figure 42.5 shows three regions of the small intestine: the duodenum, jejunum, and ileum. Each day, the stomach as well as the ducts from three organs, the **pancreas**, **liver**, and **gallbladder**, delivers about 9 liters of fluid to the duodenum. At least 95 percent of the fluid is absorbed across the epithelium that lines the small intestine.

Cells of the intestinal lining and the pancreas secrete digestive enzymes that break down food to monosaccharides (such as glucose), free fatty acids, monoglycerides, amino acids, nucleotides, and nucleotide bases. For instance, like pepsin secreted from cells in the stomach's lining, two pancreatic

Figure labels (a):
esophagus — serosa
pyloric sphincter
longitudinal muscle
circular muscle
oblique muscle
submucosa
duodenum — mucosa

a

b

1
2
3

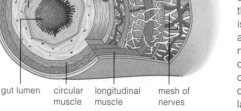

mucosa submucosa serosa blood vessels

gut lumen circular muscle longitudinal muscle mesh of nerves

c

Figure 42.8 (**a**) Stomach structure. (**b**) Peristaltic wave down the stomach. (**c**) Structure of the small intestine of humans. Generally, the gut wall starts with the mucosa (an epithelium and underlying layer of connective tissue) that faces the lumen. The submucosa is a connective tissue layer with blood and lymph vessels and a mesh of nerves that function in local control over digestion. The wall has layers of smooth muscle that differ in orientation. The serosa is an outer layer of connective tissue.

Table 42.1 Major Digestive Enzymes and Their Breakdown Products

Enzyme	Source	Where Active	Substrate	Main Breakdown Products
CARBOHYDRATE DIGESTION:				
Salivary amylase	Salivary glands	Mouth	Polysaccharides	Disaccharides
Pancreatic amylase	Pancreas	Small intestine	Polysaccharides	Disaccharides
Disaccharidases	Intestinal lining	Small intestine	Disaccharides	MONOSACCHARIDES* (e.g., glucose)
PROTEIN DIGESTION:				
Pepsins	Stomach lining	Stomach	Proteins	Protein fragments
Trypsin and chymotrypsin	Pancreas	Small intestine	Proteins	Protein fragments
Carboxypeptidase	Pancreas	Small intestine	Protein fragments	AMINO ACIDS*
Aminopeptidase	Intestinal lining	Small intestine	Protein fragments	AMINO ACIDS*
FAT DIGESTION:				
Lipase	Pancreas	Small intestine	Triglycerides	FREE FATTY ACIDS, MONOGLYCERIDES*
NUCLEIC ACID DIGESTION:				
Pancreatic nucleases	Pancreas	Small intestine	DNA, RNA	NUCLEOTIDES*
Intestinal nucleases	Intestinal lining	Small intestine	Nucleotides	NUCLEOTIDE BASES, MONOSACCHARIDES*

* Breakdown products small enough to be absorbed into the internal environment.

enzymes (trypsin and chymotrypsin) digest proteins to peptides; another cleaves the peptides to free amino acids. The pancreas also secretes bicarbonate, a buffer that helps neutralize HCl arriving from the stomach.

The Role of Bile in Fat Digestion

Fat digestion requires enzyme action. It also requires **bile**, a fluid that the liver secretes continually. The fluid contains bile salts and pigments, cholesterol, and lecithin (one of the phospholipids). When the stomach is empty, a sphincter closes off the main bile duct from the liver. At such times, bile backs up into the saclike gallbladder, where it is stored and concentrated.

By a process called **emulsification**, bile salts speed up fat digestion. Most fats in our diet are triglycerides. Triglyceride molecules are insoluble in water, and they tend to aggregate into large fat globules in the chyme. As movements of the intestinal wall agitate chyme, fat globules break up into small droplets, which become coated with bile salts. Because bile salts carry negative charges, the coated droplets repel each other and remain separated. This suspension of fat droplets, formed by mechanical and chemical action, is the "emulsion."

Compared to fat globules, emulsion droplets give fat-digesting enzymes a much greater surface area to act upon. Thus triglycerides can be broken down much more rapidly to fatty acids and monoglycerides.

Controls Over Digestion

Homeostatic controls, recall, work to counter changes in the internal environment. By contrast, controls over digestion act *before* food is absorbed into the internal environment. The nervous and endocrine systems, and a mesh of nerves in the gut wall, exert the control.

For example, incoming food distends the stomach and stimulates mechanoreceptors in the stomach wall. The resulting signals travel along short reflex pathways to smooth muscles and glands; they also travel longer reflex pathways to the brain. Either way, they stimulate muscles in the gut wall to contract or glandular cells to secrete enzymes, hormones, and other substances into the stomach lumen. Stomach emptying depends on the chyme's volume and composition. For example, a large meal activates more receptors in the stomach wall, so contractions get more forceful and emptying proceeds faster. High acidity or a high fat content in the small intestine calls for hormone secretions that trigger a slowdown in stomach emptying, so food is not moved faster than it can be processed. Fear, depression, and other emotional upsets also trigger such slowdowns.

Gastrointestinal hormones are part of the controls. If chyme contains amino acids and peptides, cells of the stomach lining will secrete the gastrin that stimulates secretion of acid into the stomach. Secretin prods the pancreas to secrete bicarbonate. CCK (cholecystokinin) enhances secretin's action and causes the gallbladder to contract. When the small intestine contains glucose and fat, GIP (glucose insulinotropic peptide) calls for insulin secretion, which stimulates cells to absorb glucose.

Carbohydrate breakdown starts in the mouth, and protein breakdown starts in the stomach.

In the small intestine, most large organic compounds are digested to molecules small enough to be absorbed into the internal environment.

ABSORPTION IN THE SMALL INTESTINE

Structure Speaks Volumes About Function

Unlike the stomach, which can only absorb alcohol and a few other substances, the small intestine is the main site of absorption of the vast majority of nutrients. What makes it so?

Consider Figure 42.9, which shows the structure of the intestinal wall. Focus first on the exuberant folds of the mucosa, which project into the lumen. Look closer and you see amazing numbers of even tinier projections from each one of the large folds. Look closer still and you see that the epithelial cells at the surface of these tiny projections have a brushlike crown of even tinier projections, all exposed to the intestinal lumen.

What is the significance of so much convolution? It has an extremely favorable surface-to-volume ratio, which you read about earlier in Section 4.1. Collectively, the projections from the intestinal mucosa ENORMOUSLY increase the surface area that is available for interacting with chyme and absorbing nutrients from it. Without that immense surface area, absorption would proceed hundreds of times more slowly, which of course would not be enough to sustain human life.

As you can see from Figure 42.9a and b, the absorptive structures on each fold of the intestinal mucosa are **villi** (singular, villus). Although each villus is only about a millimeter long, the mucosa has millions of them. Their very density gives the mucosa a velvety appearance. Inside each villus is an arteriole, a venule, and a lymph vessel, which function in moving substances to and from the general circulation (Figure 42.9e).

The cells making up the epithelial lining of each villus bear **microvilli** (singular, microvillus), which are ultrafine, threadlike projections from their free surface. Each cell has about 1,700 microvilli. Hence its name, "brush border" cell. The cells near the tip of the villus produce some digestive enzymes, as listed in Table 42.1.

Mechanisms of Absorption

Absorption, recall, is the passage of nutrients, water, salts, and vitamins into the internal environment. The vast absorptive surface of the intestinal wall facilitates the process, but so does the action of smooth muscle in the intestinal wall. In **segmentation**, rings of circular muscle contract and relax repeatedly. This creates an oscillating (back and forth) movement that constantly

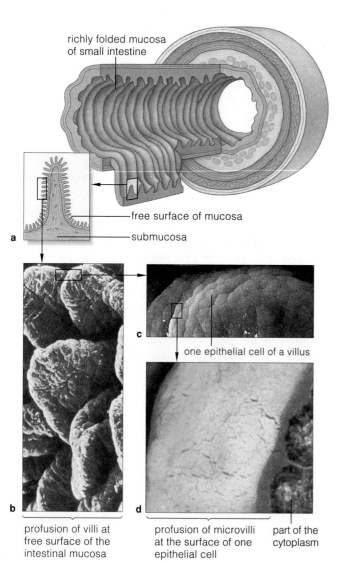

richly folded mucosa of small intestine

a

free surface of mucosa

submucosa

b profusion of villi at free surface of the intestinal mucosa

c

one epithelial cell of a villus

d profusion of microvilli at the surface of one epithelial cell

part of the cytoplasm

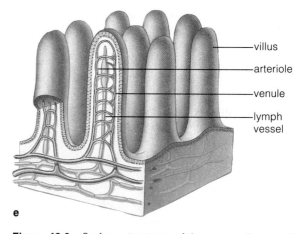

e

villus
arteriole
venule
lymph vessel

Figure 42.9 Surface structures of the mammalian small intestine at increasing magnifications.

(**a,b**) Surface view of the deep, permanent folds of the intestinal mucosa. Each fold bears a profusion of fingerlike absorptive structures called villi. (**c,d**) Individual epithelial cells at the free surface of a villus. Each has a dense crown of microvilli facing the intestinal lumen. (**e**) Blood and lymph vessels in intestinal villi. Monosaccharides and most amino acids that cross the intestinal lining enter blood vessels. Fats enter lymph vessels.

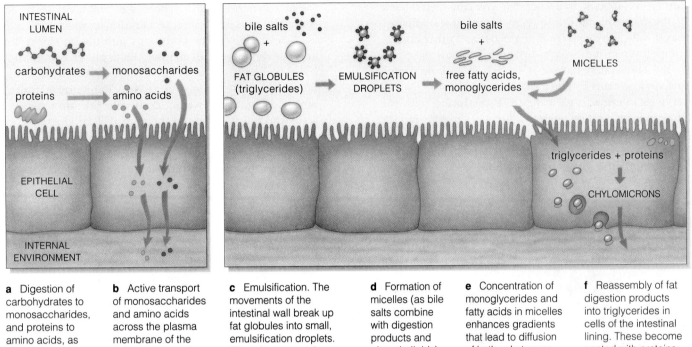

a Digestion of carbohydrates to monosaccharides, and proteins to amino acids, as completed by enzymes secreted from the pancreas and from brush border cells.

b Active transport of monosaccharides and amino acids across the plasma membrane of the cells making up the intestinal lining, then out of the same cells and into the internal environment.

c Emulsification. The movements of the intestinal wall break up fat globules into small, emulsification droplets. Bile salts prevent the globules from re-forming. Pancreatic enzymes digest the droplets to fatty acids and monoglycerides.

d Formation of micelles (as bile salts combine with digestion products and phospholipids). Products can readily slip into and out of the micelles.

e Concentration of monoglycerides and fatty acids in micelles enhances gradients that lead to diffusion of both substances across the lipid bilayer of the plasma membrane of cells making up the lining.

f Reassembly of fat digestion products into triglycerides in cells of the intestinal lining. These become coated with proteins; then they are expelled (by way of exocytosis) into the internal environment.

Figure 42.10 Digestion and absorption in the small intestine.

mixes and forces the contents of the lumen against the wall's absorptive surface:

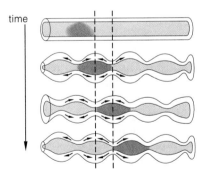

By the time that food is halfway through the small intestine, most of it has been broken apart and digested. Some breakdown products, including monosaccharides and amino acids, are then absorbed in straightforward fashion. So are water and mineral ions, which move across by way of osmosis. Transport proteins in the plasma membrane of brush border cells actively shunt these digestion products across the intestinal lining. By contrast, bile salts assist the absorption of fatty acids and monoglycerides. These products of fat digestion

diffuse across the lipid bilayer of the plasma membrane (Figure 42.10).

By a process called **micelle formation**, bile salts combine with products of fat digestion, forming tiny droplets (micelles). Product molecules in the micelles continuously exchange places with product molecules that are dissolved in the chyme. When concentration gradients favor it, the molecules diffuse out of micelles, into epithelial cells. Fatty acids and monoglycerides recombine in the cells, thus forming triglycerides. Then the triglycerides combine with proteins into particles (chylomicrons) that leave the cells, by exocytosis, and enter the internal environment.

Once they have been absorbed, glucose and amino acids enter blood vessels. The triglycerides enter lymph vessels, which eventually drain into blood vessels.

With its richly folded intestinal mucosa, millions of villi, and hundreds of millions of microvilli, the small intestine offers a vast surface area for absorbing nutrients.

Substances pass through the brush border cells that line the free surface of each villus by active transport, osmosis, and diffusion across the lipid bilayer of plasma membranes.

DISPOSITION OF ABSORBED ORGANIC COMPOUNDS

Earlier in the book, in Section 8.6, you considered some of the mechanisms that govern organic metabolism—specifically, the disposition of glucose and other organic compounds in the body as a whole. You saw examples of the conversion pathways by which carbohydrates, fats, and proteins can be broken apart to molecules that serve as intermediates in the ATP-producing pathway of aerobic respiration. Here, Figure 42.11 rounds out the picture by showing all of the major routes by which organic compounds obtained from the diet are shuffled and reshuffled in the body as a whole.

the main energy source. There is no net breakdown of protein in muscle or other tissues during this period.

Between meals, the body taps into the fat stores. In adipose tissue, cells dismantle fats to glycerol and fatty acids and release these into blood. In the liver, cells break apart glycogen and release the glucose units, which also enter the blood. Body cells take up the fatty acids and glucose and use them for ATP production.

Bear in mind, the liver does more than store and interconvert organic compounds. For example, it helps maintain their concentrations in blood. It inactivates

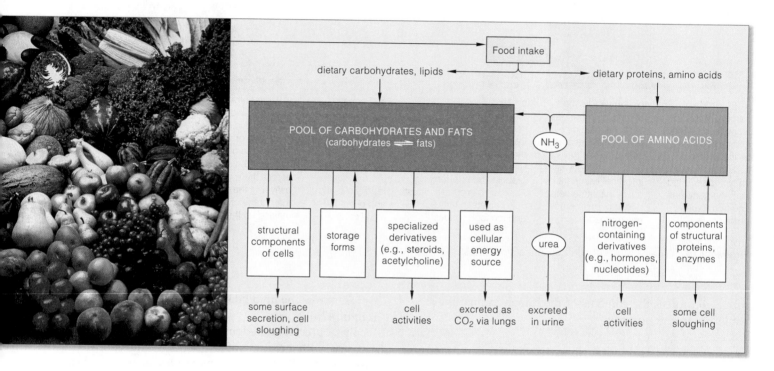

Figure 42.11 Summary of major pathways of organic metabolism. Cells continually synthesize and tear down carbohydrates, fats, and proteins. Urea forms mainly in the liver. See Sections 8.6 and 37.7.

At any time, all living cells in the body are breaking apart most of their own supplies of carbohydrates, lipids, and proteins, then using the breakdown products as energy sources or as building blocks. The nervous system and endocrine system work together to integrate this massive molecular turnover, although a treatment of how they do this would be beyond the scope of this book.

For now, simply consider a few key points about organic metabolism. When you eat, your body builds up pools of materials. Excess amounts of carbohydrates and other organic compounds absorbed from the gut are transformed mostly into fats, which become stored in adipose tissue. Some are converted to glycogen in the liver and in muscle tissue. *While* the compounds are being absorbed and stored, most cells use glucose as

most hormone molecules, which are sent to the kidneys for excretion, in urine. The liver also removes worn-out blood cells from the circulation and inactivates many potentially toxic compounds. For example, ammonia (NH_3), produced during amino acid breakdown, can be toxic in high concentrations. The liver converts NH_3 to urea. This is a less toxic waste product, and it leaves the body by way of the kidneys, in urine.

Just after a meal, the body's cells take up the glucose being absorbed from the gut and use it as a quick energy source. Excess amounts of glucose and other organic compounds become converted mainly to fats that are stored in adipose tissue. Some of the excess is converted to glycogen, which gets stored mainly in the liver and in muscle tissue.

Between meals, the body taps the fat reservoirs as the main energy source. Fats get converted to glycerol and fatty acids, both of which can enter ATP-producing pathways.

THE LARGE INTESTINE

What happens to the material *not* absorbed in the small intestine? It moves into the large intestine, or **colon**. The colon concentrates and stores feces: a mixture of water, undigested and unabsorbed matter, and bacteria. The colon starts at a cup-shaped pouch, the cecum (Figures 42.5 and 42.12). It ascends on the right side of the abdominal cavity, continues across to the other side, then descends and connects with a short tube, the rectum.

Colon Functioning

Material becomes concentrated as water moves across the colon's lining. Cells of the lining actively transport sodium ions out of the lumen. As ion concentrations in the lumen fall, the water concentration increases. Thus water also moves out of the lumen, by way of osmosis. Cells of the lining also secrete mucus, which lubricates the chyme and helps keep it from mechanically damaging the colon wall. Additionally, they secrete bicarbonate, which helps buffer the acidic fermentation products of ingested bacteria that managed to colonize the colon.

Short, longitudinal bands of smooth muscle in the colon wall are gathered at their ends, like a series of full skirts nipped in at elastic waistbands. As they contract and relax, the contents of the lumen move back and forth against the absorptive surface of the colon wall. Nerve plexuses largely control the motion, which is like segmentation in the small intestine but much slower. Whereas the fast transit time through the small intestine does not favor the rapid population growth of ingested bacteria, the colon's slower motion favors it. In general, the colonizers do no harm unless they breach the colon wall and enter the abdominal cavity.

With each new meal, hormonal signals (gastrin) and commands from autonomic nerves induce large portions of the ascending and transverse colon to contract at the same time. Within a few seconds, the lumen's existing contents move as much as three-fourths of the length of the colon and thereby make way for incoming food. The contents are stored in the last portion of the colon until defecation occurs; they distend the rectal wall enough to trigger a reflex action that leads to their expulsion. The nervous system controls the expulsion by stimulating or inhibiting contraction of a muscle sphincter at the anus, the terminal opening of the gut.

The volume of cellulose fiber and other undigested material that cannot be decreased by absorption in the colon is called **bulk**. Bulk contributes to the volume and normal transit time of material through the colon.

Colon Malfunctioning

The frequency of defecation normally ranges from three times a day to once a week. Aging, emotional stress, a

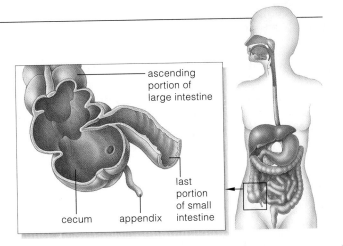

Figure 42.12 Cecum and appendix of the large intestine (colon).

low-bulk diet, injury, or disease can result in delayed defecation, or *constipation*. And the longer the delay, the more water is absorbed, so feces become hard and dry. Besides abdominal discomfort, dull headaches, loss of appetite, and often nausea and depression result.

Also, hard feces may become lodged in the **appendix**, a narrow projection from the cecum (Figure 42.12). The appendix has no known digestive functions. But it is colonized by lymphocytes that provide a line of defense against ingested bacteria. When feces obstruct normal blood flow and mucus secretion to this projection, the stage is set for *appendicitis*, or an inflamed appendix. Unless surgically removed, an inflamed appendix may rupture. Such a rupture will allow bacteria to enter the abdominal cavity and cause life-threatening infections.

The colon also is vulnerable to cancer, which occurs most often among the world's wealthiest and best-fed populations. Too many people skip meals, eat too much and too fast when they do sit down at the table, and generally give their gut erratic workouts. Their diet tends to be rich in sugar, cholesterol, and salt, and low in bulk. Too little bulk extends the transit time of feces through the colon. The longer that irritating, potentially carcinogenic material is in contact with the colon wall, the more damage it can do. Symptoms of *colon cancer* are a change in bowel functioning, rectal bleeding, and blood in feces. Some people appear to be genetically predisposed to develop colon cancer, but a low-fiber diet puts anyone at risk. Intriguingly, colon cancer is almost nonexistent in rural India and Africa, where people cannot afford to eat much more than fiber-rich whole grains. When the same people move to cities in the affluent nations and change their eating habits, the incidence of colon cancer increases.

The large intestine, or colon, functions in the absorption of water and mineral ions from the gut lumen. It also functions in the compaction of undigested residues into feces, for expulsion at the terminal opening of the gut.

HUMAN NUTRITIONAL REQUIREMENTS

You grow and maintain yourself by eating certain foods that are suitable sources of energy and raw materials. Nutritionists measure the energy in **kilocalories**. Each kilocalorie is 1,000 calories of heat energy, or the amount needed to raise the temperature of 1 kilogram of water by 1°C. The raw materials will now be described.

New, Improved Food Pyramids

A few million years ago, our hominid ancestors dined mostly on fresh fruits and other fibrous plant material. From that nutritional beginning, humans in many parts of the world came to prefer low-fiber, high-fat foods. They still do, even when they study **food pyramids**, or charts of well-balanced diets, from elementary school onward. Nutritionists revise the charts on an ongoing basis to reflect additional research. Figure 42.13 shows a recent version. It shows the daily portions of foods that will supply an average-size adult male with 58–60 percent complex carbohydrates, 15-20 percent proteins (less for females), and 20–25 percent fats. Not everyone agrees with the proportions. For example, advocates of *the zone diet* believe that a daily intake of 40 percent complex carbohydrates, 30 percent proteins, and 30 percent fats maintains the body at peak performance levels.

Carbohydrates

Starch and, to a lesser extent, glycogen should be the main carbohydrates in the human diet. These complex carbohydrates are readily broken apart into glucose units, which are the body's main energy source. Starch is abundant in fleshy fruits, cereal grains, and legumes, including beans and peas.

The foods that are rich in complex carbohydrates have an additional advantage—they typically are high in fiber. Section 42.7 highlighted the rather sobering consequences of a low-fiber diet. Epidemiologist Denis Burkitt may have been only half-joking when he put it this way: If you pass a small volume of feces, you have large hospitals.

The carbohydrates called simple sugars do not have the fiber of complex carbohydrates. Neither do they have the vitamins and minerals found in whole foods. Each week, the average American eats as much as 2 pounds of refined sugar (sucrose). You may think this a far-fetched statement, but start taking a closer look at the ingredients listed on the packages of cereal, frozen dinners, soft drinks, and other prepared foods. Many of the sugars are "hidden" as corn syrup, corn sweeteners, dextrose, and so on.

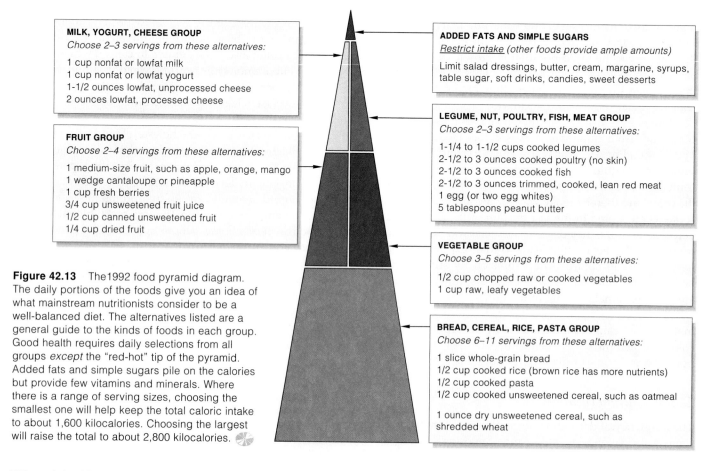

MILK, YOGURT, CHEESE GROUP

Choose 2–3 servings from these alternatives:

1 cup nonfat or lowfat milk
1 cup nonfat or lowfat yogurt
1-1/2 ounces lowfat, unprocessed cheese
2 ounces lowfat, processed cheese

FRUIT GROUP

Choose 2–4 servings from these alternatives:

1 medium-size fruit, such as apple, orange, mango
1 wedge cantaloupe or pineapple
1 cup fresh berries
3/4 cup unsweetened fruit juice
1/2 cup canned unsweetened fruit
1/4 cup dried fruit

ADDED FATS AND SIMPLE SUGARS

Restrict intake (other foods provide ample amounts)

Limit salad dressings, butter, cream, margarine, syrups, table sugar, soft drinks, candies, sweet desserts

LEGUME, NUT, POULTRY, FISH, MEAT GROUP

Choose 2–3 servings from these alternatives:

1-1/4 to 1-1/2 cups cooked legumes
2-1/2 to 3 ounces cooked poultry (no skin)
2-1/2 to 3 ounces cooked fish
2-1/2 to 3 ounces trimmed, cooked, lean red meat
1 egg (or two egg whites)
5 tablespoons peanut butter

VEGETABLE GROUP

Choose 3–5 servings from these alternatives:

1/2 cup chopped raw or cooked vegetables
1 cup raw, leafy vegetables

BREAD, CEREAL, RICE, PASTA GROUP

Choose 6–11 servings from these alternatives:

1 slice whole-grain bread
1/2 cup cooked rice (brown rice has more nutrients)
1/2 cup cooked pasta
1/2 cup cooked unsweetened cereal, such as oatmeal

1 ounce dry unsweetened cereal, such as shredded wheat

Figure 42.13 The 1992 food pyramid diagram. The daily portions of the foods give you an idea of what mainstream nutritionists consider to be a well-balanced diet. The alternatives listed are a general guide to the kinds of foods in each group. Good health requires daily selections from all groups *except* the "red-hot" tip of the pyramid. Added fats and simple sugars pile on the calories but provide few vitamins and minerals. Where there is a range of serving sizes, choosing the smallest one will help keep the total caloric intake to about 1,600 kilocalories. Choosing the largest will raise the total to about 2,800 kilocalories.

No Limiting Amino Acid	Low in Lysine	Low in Methionine, Other Sulfur-Containing Amino Acids	Low in Tryptophan

legumes:
  soybean
  tofu
  soy milk
 	
cereal grains:
  wheat germ
 	
nuts:
 	
milk
cheeses (except
  cream cheese)
yogurt
eggs
meats
	legumes:
	  peanuts
	
	cereal grains:
	  barley
	  buckwheat
	  cornmeal
	  oats
	  rice
	  rye
	  wheat
	
	nuts, seeds:
	  almonds
	  cashews
	  coconut
	  English walnuts
	  hazelnuts
	  pecans
	  pumpkin seeds
	  sunflower seeds
		legumes:
		  beans (dried)
		  black-eyed peas
		  garbanzos
		  lentils
		  lima beans
		  mung beans
		  peanuts
		
		nuts:
		  hazelnuts
		
		fresh vegetables:
		  asparagus
		  broccoli
		  green peas
		  mushrooms
		  parsley
		  potatoes
		  soybeans
		  Swiss chard
			legumes:
			  beans (dried)
			  garbanzos
			  lima beans
			  mung beans
			  peanuts
			
			cereal grains:
			  cornmeal
			
			nuts:
			  almonds
			  English walnuts
			
			fresh vegetables:
			  corn
			  green peas
			  mushrooms
			  Swiss chard

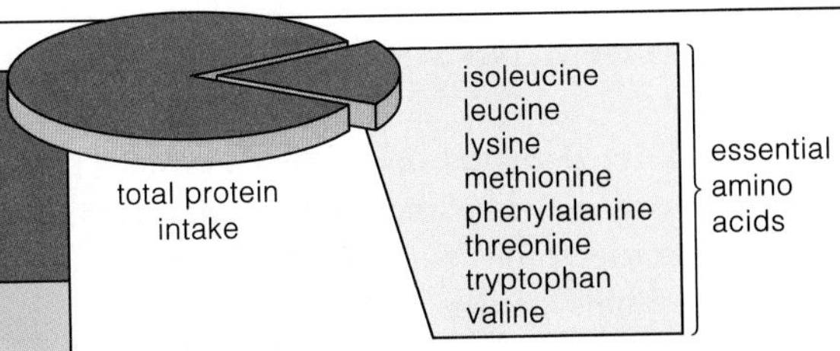

essential amino acids:
isoleucine
leucine
lysine
methionine
phenylalanine
threonine
tryptophan
valine

total protein intake

Figure 42.14 The eight essential amino acids, a small portion of the total protein intake. All must be available at the same time, in certain amounts, if cells are to build their own proteins. Milk and eggs have high amounts of all eight in proportions that humans require; they are among the *complete* proteins.

Nearly all plant proteins are *incomplete*, so vegetarians must be very careful to avoid protein deficiency. For instance, they can combine different foods from the three columns of incomplete proteins at the left. Also, strict vegetarians (who also avoid dairy products and eggs) should be sure to take vitamin B$_{12}$ and vitamin B$_2$ (riboflavin) supplements. Animal protein is a luxury in most traditional societies. Their cuisines have good combinations of plant proteins, including rice/beans, chili/cornbread, tofu/rice, and lentils/wheat bread.

Lipids

The body cannot function without fats and other lipids. For example, the phospholipid lecithin is a required component of cell membranes; and fats serve as energy reserves, cushion many internal organs, and provide insulation beneath the skin. Dietary fats also help the body store fat-soluble vitamins. However, the body can synthesize most of its own fats, including cholesterol, from proteins and carbohydrates. The ones it cannot make are called **essential fatty acids**, but whole foods can provide plenty of them. Linoleic acid is an example. One teaspoon a day of corn oil, olive oil, or some other polyunsaturated fat in food provides enough of it.

Currently, butter and other fats make up 40 percent of the average diet in the United States. Most of the medical community agrees that the proportion should be 30 percent or less.

Like other animal fats, butter is a saturated fat that tends to raise the level of cholesterol in the blood. As described in earlier chapters, cholesterol is used in the synthesis of bile acids and steroid hormones, and it is a required component of our cell membranes. However, too much cholesterol may interfere seriously with blood circulation through the body (Section 39.9). New on the market are edible but nondigestible oils, such as *sucrose polyester*, for individuals who crave fats enough to put up with the digestive upsets associated with the products.

Proteins

Amino acid components of dietary proteins are needed for the body's own protein-building programs. Of the twenty common types, eight are known as the **essential amino acids**. That is, our body cells cannot synthesize them, so we must obtain them from food. The eight amino acids are methionine (or its metabolic equivalent, cysteine), isoleucine, leucine, lysine, phenylalanine (or tyrosine), threonine, tryptophan, and valine.

Most animal proteins are *complete*, meaning their ratios of amino acids match human nutritional needs. Plant proteins are *incomplete*, meaning they lack one or more essential amino acids. Therefore, to get suitable amounts of amino acids, vegetarians must eat certain combinations of different plants (Figure 42.14).

A measure called **net protein utilization** (NPU) is used to compare proteins from different sources. NPU values range from 100 (all essential amino acids occur in ideal proportions) to 0 (one or more amino acids are absent; when eaten alone, the protein is not complete).

Because enzymes and other proteins are vital for the body's structure and function, protein-deficient diets may have severe consequences. Protein deficiency is most damaging to fetuses and infants, for the brain grows and develops rapidly early in life. Unless the mother-to-be eats adequate amounts of protein just before and just after birth, her newborn will be affected by irreversible mental retardation. Even mild protein starvation can retard growth, and it can affect mental and physical performance.

Individuals grow and maintain themselves in good health by eating certain amounts of complex carbohydrates, lipids, and proteins that provide them with adequate raw materials and energy, as measured in kilocalories.

VITAMINS AND MINERALS

Metabolic activity depends on small amounts of more than a dozen organic substances called **vitamins**. Most plants are able to synthesize all of these substances on their own. Animals generally have lost the ability to do so and must obtain vitamins from food.

At the minimum, human cells require the thirteen different vitamins listed in Table 42.2. Each vitamin has specific metabolic roles. Many reactions require several vitamins, so the absence of one affects the functions of others. Metabolism also depends on certain inorganic substances called **minerals**. For example, most of your cells use calcium and magnesium in many reactions. All cells require iron in electron transport chains. Red blood cells cannot function without iron in hemoglobin, the oxygen-carrying pigment in blood. Neurons simply cannot function without sodium and potassium.

People who are in good health get all the vitamins and minerals they need from a balanced diet of whole

Table 42.2 Vitamins: Sources, Functions, and Effects of Deficiencies or Excesses*

Vitamin	Common Sources	Main Functions	Effects of Chronic Deficiency	Effects of Extreme Excess
FAT-SOLUBLE VITAMINS:				
A	Its precursor comes from beta-carotene in yellow fruits, yellow or green leafy vegetables; also in fortified milk, egg yolk, fish liver	Used in synthesis of visual pigments, bone, teeth; maintains epithelia	Dry, scaly skin; lowered resistance to infections; night blindness; permanent blindness	Malformed fetuses; hair loss; changes in skin; liver and bone damage; bone pain
D	D_3 formed in skin and in fish liver oils, egg yolk, fortified milk; converted to active form elsewhere	Promotes bone growth and mineralization; enhances calcium absorption	Bone deformities (rickets) in children; bone softening in adults	Retarded growth; kidney damage; calcium deposits in soft tissues
E	Whole grains, dark green vegetables, vegetable oils	Counters effects of free radicals; helps maintain cell membranes; blocks breakdown of vitamins A and C in gut	Lysis of red blood cells; nerve damage	Muscle weakness, fatigue, headaches, nausea
K	Enterobacteria form most of it; also in green leafy vegetables, cabbage	Blood clotting; ATP formation via electron transport	Abnormal blood clotting; severe bleeding (hemorrhaging)	Anemia; liver damage and jaundice
WATER-SOLUBLE VITAMINS:				
B_1 (thiamin)	Whole grains, green leafy vegetables, legumes, lean meats, eggs	Connective tissue formation; folate utilization; coenzyme action	Water retention in tissues; tingling sensations; heart changes; poor coordination	None reported from food; possible shock reaction from repeated injections
B_2 (riboflavin)	Whole grains, poultry, fish, egg white, milk	Coenzyme action	Skin lesions	None reported
Niacin	Green leafy vegetables, potatoes, peanuts, poultry, fish, pork, beef	Coenzyme action	Contributes to pellagra (damage to skin, gut, nervous system, etc.)	Skin flushing; possible liver damage
B_6	Spinach, tomatoes, potatoes, meats	Coenzyme in amino acid metabolism	Skin, muscle, and nerve damage; anemia	Impaired coordination; numbness in feet
Pantothenic acid	In many foods (meats, yeast, egg yolk especially)	Coenzyme in glucose metabolism, fatty acid and steroid synthesis	Fatigue, tingling in hands, headaches, nausea	None reported; may cause diarrhea occasionally
Folate (folic acid)	Dark green vegetables, whole grains, yeast, lean meats; enterobacteria produce some folate	Coenzyme in nucleic acid and amino acid metabolism	A type of anemia; inflamed tongue; diarrhea; impaired growth; mental disorders	Masks vitamin B_{12} deficiency
B_{12}	Poultry, fish, red meat, dairy foods (not butter)	Coenzyme in nucleic acid metabolism	A type of anemia; impaired nerve function	None reported
Biotin	Legumes, egg yolk; colon bacteria produce some	Coenzyme in fat, glycogen formation and in amino acid metabolism	Scaly skin (dermatitis), sore tongue, depression, anemia	None reported
C (ascorbic acid)	Fruits and vegetables, especially citrus, berries, cantaloupe, cabbage, broccoli, green pepper	Collagen synthesis; possibly inhibits effects of free radicals; structural role in bone, cartilage, and teeth; role in carbohydrate metabolism	Scurvy, poor wound healing, impaired immunity	Diarrhea, other digestive upsets; may alter results of some diagnostic tests

* The guidelines for appropriate daily intakes are being worked out by the Food and Drug Administration.

Table 42.3   Major Minerals: Sources, Functions, and Effects of Deficiencies or Excesses*

Mineral	Common Sources	Main Functions	Signs of Severe Long-Term Deficiency	Signs of Extreme Excess
Calcium	Dairy products, dark green vegetables, dried legumes	Bone, tooth formation; blood clotting; neural and muscle action	Stunted growth; possibly diminished bone mass (osteoporosis)	Impaired absorption of other minerals; kidney stones in susceptible people
Chloride	Table salt (usually too much in diet)	HCl formation in stomach; contributes to body's acid-base balance; neural action	Muscle cramps; impaired growth; poor appetite	Contributes to high blood pressure in susceptible people
Copper	Nuts, legumes, seafood, drinking water	Used in synthesis of melanin, hemoglobin, and some transport chain components	Anemia, changes in bone and blood vessels	Nausea, liver damage
Fluorine	Fluoridated water, tea, seafood	Bone, tooth maintenance	Tooth decay	Digestive upsets; mottled teeth and deformed skeleton in chronic cases
Iodine	Marine fish, shellfish, iodized salt, dairy products	Thyroid hormone formation	Enlarged thyroid (goiter), with metabolic disorders	Goiter
Iron	Whole grains, green leafy vegetables, legumes, nuts, eggs, lean meat, molasses, dried fruit, shellfish	Formation of hemoglobin and cytochrome (transport chain component)	Iron-deficiency anemia, impaired immune function	Liver damage, shock, heart failure
Magnesium	Whole grains, legumes, nuts, dairy products	Coenzyme role in ATP-ADP cycle; roles in muscle, nerve function	Weak, sore muscles; impaired neural function	Impaired neural function
Phosphorus	Whole grains, poultry, red meat	Component of bone, teeth, nucleic acids, ATP, phospholipids	Muscular weakness; loss of minerals from bone	Impaired absorption of minerals into bone
Potassium	Diet provides ample amounts	Muscle and neural function; roles in protein synthesis and body's acid-base balance	Muscular weakness	Muscular weakness, paralysis, heart failure
Sodium	Table salt; diet provides ample to excessive amounts	Key role in body's acid-base balance; roles in muscle and neural function	Muscle cramps	High blood pressure in susceptible people
Sulfur	Proteins in diet	Component of body proteins	None reported	None likely
Zinc	Whole grains, legumes, nuts, meats, seafood	Component of digestive enzymes; roles in normal growth, wound healing, sperm formation, and taste and smell	Impaired growth, scaly skin, impaired immune function	Nausea, vomiting, diarrhea; impaired immune function and anemia

* The guidelines for appropriate daily intakes are being worked out by the Food and Drug Administration.

foods. Generally speaking, specific vitamin and mineral supplements are necessary only for strict vegetarians, the elderly, and people suffering from a chronic illness or taking medication that affects the use of specific nutrients. For example, in one comprehensive study, supplements of vitamin K helped older women retain calcium and so diminish bone loss (osteoporosis) by 30 percent. Preliminary research suggests two vitamins (C and E) and beta-carotene (the precursor for vitamin A) may inactivate free radicals—those rogue metabolic fragments that may contribute to aging and may impair immune function.

No one should take massive doses of any vitamin or mineral supplement except under medical supervision. As Tables 42.2 and 42.3 indicate, excess amounts of many vitamins and minerals are harmful. For example, large doses of at least two vitamins (A and D) actually can damage the body. Like all fat-soluble vitamins, excess amounts can accumulate in tissues and interfere with normal metabolic function. Similarly, sodium is present in plant and animal tissues and is a component of table salt. Sodium has roles in the body's salt–water balance, muscle activity, and nerve function. However, prolonged, excessive intake of sodium may contribute to high blood pressure.

Severe shortages or self-prescribed, massive excesses of vitamins and minerals can disturb the delicate balances in body function that promote health.

42.10 TANTALIZING ANSWERS TO WEIGHTY QUESTIONS

IN PURSUIT OF THE "IDEAL" WEIGHT   To keep your body functioning normally over the long term while simultaneously maintaining an acceptable weight, *caloric intake must be balanced with energy output.*

How many kilocalories should you take in each day to maintain a desired weight? First, multiply that weight (in pounds) by 10 if you are not active physically, by 15 if you are moderately active, and by 20 if you are highly active. Then, depending upon your age, subtract the following amount from the value obtained from the first step:

Age:	25–34	Subtract:	0
	35–44		100
	45–54		200
	55–64		300
	Over 65		400

For example, if you wish to weigh 120 pounds and are highly active, 120 × 20 = 2,400 kilocalories. If you are thirty-five years old, you would take in (2,400 − 100) or 2,300 kilocalories a day. Such calculations provide a rough estimate of caloric intake. Other factors, including height, must also be taken into consideration. An active person 5 feet, 2 inches tall does not require as much energy as an active person who weighs the same but is 6 feet tall.

The question becomes this: *Why* do you wish to weigh a given amount? Are you merely fearful of "being fat"— that is, obese? By definition, **obesity** is an excess of fat in adipose tissues, most often caused by imbalances between caloric intake and energy output. One standard of what constitutes obesity is simply cultural, and it varies from one culture to the next. For example, one female student despaired over her plumpness until she took part in a graduate studies program in Africa. There, great numbers of males considered her to be one of the most desirable females on the planet.

Traditionally, insurance companies have relied on a different standard. They developed charts mainly so that they could identify overweight individuals who might be insurance risks. The medical profession also has used the same charts.

By 1995, researchers were questioning the insurance charts. For example, after a fourteen-year study of nearly 116,000 women, researchers at Harvard University found a strong link between the risk of heart attack and gaining even 11 to 18 pounds in adult life. Women who gained weight after age 18 were by far the group at greatest risk.

In making their correlation, the Harvard researchers took into account age, smoking habits, family history of heart disorders, the use of postmenopausal hormones, and other factors that could influence the risk. *They also found conclusive evidence that exercise is the most crucial factor in controlling weight gain and avoiding heart attacks.*

Generally, thinner people simply live longer. You can use the chart in Figure 42.15 to get an idea of how a 1995 definition of "thin" applies to you.

MY GENES MADE ME DO IT   As you have probably noticed, some people have far more trouble keeping off excess weight than others do. Actually, about one of every three Americans is like this. To some extent, age and emotional state have something to do with it. More importantly, *genes* have a lot to do with it.

Figure 42.15 How to estimate the "ideal" weight for an adult. The values given are consistent with a long-term Harvard study into the link between excessive weight and increased risk of heart disorders. Depending upon certain factors (such as having a small, medium, or large frame), the "ideal" may vary by plus or minus 10 percent.

Weight Guidelines for Women:

Starting with an ideal weight of 100 pounds for a woman who is 5 feet tall, add five additional pounds for each additional inch of height. Examples:

Height (feet)	Weight (pounds)
5' 2"	110
5' 3"	115
5' 4"	120
5' 5"	125
5' 6"	130
5' 7"	135
5' 8"	140
5' 9"	146
5' 10"	150
5' 11"	155
6'	160

Weight Guidelines for Men:

Starting with an ideal weight of 106 pounds for a man who is 5 feet tall, add six additional pounds for each additional inch of height. Examples:

Height (feet)	Weight (pounds)
5' 2"	118
5' 3"	124
5' 4"	130
5' 5"	136
5' 6"	142
5' 7"	148
5' 8"	154
5' 9"	160
5' 10"	166
5' 11"	172
6'	178

Researchers have suspected as much for a long time. Consider their studies of identical twins. (*Identical twins* are born with identical genes.) As an outcome of family problems, the twins who were investigated had been separated at birth and had been raised apart, in different households. Even so, at adulthood, the body weights of those separated twins were similar! People, it seemed, were born with a "set point" for body fat. Could it be that whatever set point an individual inherits is the one he or she will be stuck with for life?

Through many studies and experiments, researchers gathered strong evidence that ultimately confirmed this suspicion. It all started in 1950 with an exceptionally obese mouse (Figure 42.16). But it was not until 1995 that molecular geneticists identified one of its genes that has profound influence over the set point for body fat. They named it the *ob* gene (guess why). What were the clues that led to its discovery? The nucleotide sequence in the gene region of obese mice differs from that in normal mice. The gene is active in adipose tissue, nowhere else. Its biochemical information translates into a type of protein that cells can release into the bloodstream—that is, the gene might specify a hormonal signal.

Jeffrey Friedman discovered the hormone in 1994 and named it **leptin**. A nearly identical hormone was isolated in humans. Leptin is only one of a number of factors that influence the brain's commands to suppress or whip up appetite, but it's a crucial one. By assessing the blood concentrations of incoming hormonal signals, an appetite control center in the hypothalamus "decides" whether the body has taken in enough food to provide enough fat for the day. If so, commands go out to increase metabolic rates—and to stop eating.

If the *ob* gene is mutated, then its expression alone might be enough to disrupt the control center so that a mouse's appetite skyrockets and metabolic furnaces burn less. By extension, the same disruptions might occur in a human with a similar gene mutation even if he or she fully understands the inherent health risks. During one experiment that supports the hypothesis, researchers injected obese mice with leptin. The mice quickly shed their excess weight.

Will researchers go on to develop therapies for obesity in humans? Maybe, but not for five to ten years. As you saw in Chapter 37, hormonal control in humans is tricky business.

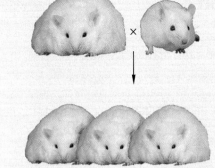

a *1950.* Researchers at the Jackson Laboratories in Maine notice that one of their laboratory mice is extremely obese, with an uncontrollable appetite. Through crossbreeding of this apparent mutant individual with a normal mouse, they produce a strain of obese mice.

b *Late 1960s.* Douglas Coleman of the Jackson Laboratories surgically joins the bloodstreams of an obese mouse and a normal one. The obese mouse now loses weight. Coleman hypothesizes that a factor circulating in blood may be influencing its appetite, but he is not able to isolate it.

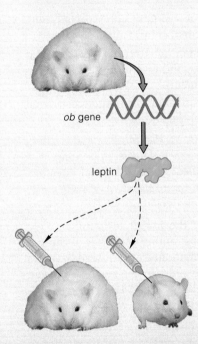

ob gene

leptin

c *1994.* Late in the year, Jeffrey Friedman of Rockefeller University discovers a mutated form of what is now called the *ob* gene in obese mice. Through DNA cloning and gene sequencing, he defines the protein that the mutated gene encodes. The protein, now called leptin, is a hormone that influences the brain's commands to suppress appetite and increase metabolic rates.

d *1995.* Three different research teams develop and use genetically engineered bacteria to produce leptin, which, when injected in obese and normal mice, triggers significant weight loss, apparently without harmful side effects.

Figure 42.16 Chronology of key developments that revealed the identity of a key factor in the genetic basis of body weight.

1. Nutrition refers to all the processes by which the body takes in, digests, absorbs, and uses food.

2. Mammals have a complete digestive system, basically a tube that has two openings (a mouth and an anus), as well as regional specializations. Protective, mucus-coated epithelium lines all exposed surfaces of the tube and facilitates diffusion across the tube wall.

3. As summarized in Table 42.4, the human digestive system includes a mouth, pharynx, esophagus, stomach, small intestine, large intestine (or colon), rectum, and anus. Salivary glands and the liver, gallbladder, and pancreas have accessory roles in the system's functions.

4. These activities proceed in the digestive system:

 a. Mechanical processing and motility (movements that break up, mix, and propel food material).

 b. Secretion (release of digestive enzymes and other substances from the pancreas, liver, and glandular epithelium into the gut lumen).

 c. Digestion (breakdown of food into particles, then into nutrient molecules small enough to be absorbed).

 d. Absorption (diffusion or transport of digested organic compounds, fluid, and ions from the gut lumen into the internal environment).

 e. Elimination (the expulsion of undigested as well as unabsorbed residues at the end of the system).

5. Starch digestion starts in the mouth, and protein digestion starts in the stomach. Digestion is completed and most nutrients are absorbed in the small intestine. The pancreas secretes the main digestive enzymes. Bile from the liver assists in fat digestion.

6. In absorption, cells of the intestinal lining actively transport glucose and most amino acids out of the gut lumen. Fatty acids and monoglycerides diffuse across the lipid bilayer of these cells. In the cells' cytoplasm they are recombined as triglycerides, then are released, by exocytosis, into interstitial fluid.

7. The nervous system, endocrine system, and nerve plexuses in the gut wall interact to govern activities of the digestive system. Many controls operate in response to the volume and composition of food passing through the stomach and intestines. The controls trigger changes in muscle activity and in the rate at which hormones and enzymes are secreted.

8. Nutritionists advise a daily food intake in certain proportions. For example, for an adult male of average body weight: 58–60 percent complex carbohydrates, 12–15 percent protein, and 20–25 percent fats and other lipids. A well-balanced diet of whole foods normally provides all required vitamins and minerals.

9. To maintain a suitable body weight and overall health, caloric intake must balance energy output.

Table 42.4	Summary of the Digestive System
MOUTH (oral cavity)	Start of digestive system, where food is chewed and moistened; polysaccharide digestion begins here
PHARYNX	Entrance to tubular part of digestive and respiratory systems
ESOPHAGUS	Muscular tube, moistened by saliva, that moves food from pharynx to stomach
STOMACH	Sac where food mixes with gastric fluid and protein digestion begins; stretches to store food taken in faster than can be processed; gastric fluid destroys many microbes
SMALL INTESTINE	The first part (duodenum) receives secretions from the liver, gallbladder, and pancreas
	Most nutrients are digested and absorbed in the second part (jejunum)
	Some nutrients are absorbed in the last part (ileum), which delivers unabsorbed material to the colon
COLON (large intestine)	Concentrates and stores undigested matter (by absorbing mineral ions and water)
RECTUM	Distension triggers expulsion of feces
ANUS	Terminal opening of digestive system

Accessory Organs:

SALIVARY GLANDS	Glands (three main pairs, many minor ones) that secrete saliva, a fluid with polysaccharide-digesting enzymes, buffers, and mucus (which moistens and lubricates ingested food)
PANCREAS	Secretes enzymes that digest all major food molecules; secretes buffers against HCl from the stomach
LIVER	Secretes bile (used in fat emulsification); roles in carbohydrate, fat, and protein metabolism
GALLBLADDER	Stores and concentrates bile from the liver

Review Questions

1. Define the five key tasks carried out by a complete digestive system. Then correlate some organs of such a system with the feeding behavior of a particular kind of animal. *42.1*

2. Using the diagram on the next page, list the organs and the accessory organs of the human digestive system. On a separate sheet of paper, list the main functions of each. *42.2; Table 42.4*

3. Name the breakdown products small enough to be absorbed across the intestinal lining, into the internal environment. *42.5*

4. Define segmentation. Does it proceed in the stomach? Does it proceed in the small intestine, colon, or both? *42.5, 42.7*

Self-Quiz (Answers in Appendix IV)

1. The _____ maintains the internal environment, supplies cells with raw materials, and disposes of metabolic wastes.
 a. digestive system d. urinary system
 b. circulatory system e. interaction of all of the
 c. respiratory system systems listed

2. Most digestive systems have regions for _____ food.
 a. transporting c. storing
 b. processing d. all of the above

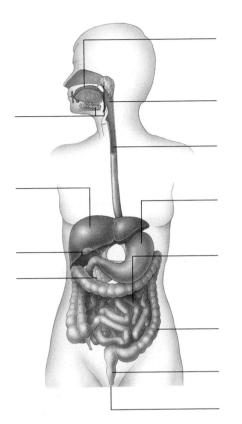

3. Maintaining good health and normal body weight requires that _____ intake be balanced by _____ output.

4. Most of our caloric intake should come from _____ .
   a. complex carbohydrates   c. proteins
   b. simple carbohydrates   d. lipids

5. On its own, the human body cannot produce all of the _____ it requires.
   a. vitamins and minerals   d. a through c
   b. fatty acids   e. a and c
   c. amino acids

6. Secretions from the _____ do *not* assist in digestion and absorption.
   a. salivary glands   c. liver
   b. thymus gland   d. pancreas

7. Digestion is completed and most nutrients are absorbed in the _____ .
   a. mouth   c. small intestine
   b. stomach   d. colon

8. Glucose and most amino acids are absorbed across the gut lining _____ .
   a. by active transport   c. at lymph vessels
   b. by diffusion   d. as fat droplets

9. Bile has roles in _____ digestion and absorption.
   a. carbohydrate   c. protein
   b. fat   d. amino acid

10. Match the organ with its key digestive function(s).
    ____ gallbladder   a. secrete bile and bicarbonate
    ____ stomach   b. digest, absorb most nutrients
    ____ colon   c. store, mix, dissolve food; start protein
    ____ pancreas   breakdown
    ____ salivary   d. store, concentrate bile
         gland   e. concentrate undigested matter
    ____ small   f. secrete substances that moisten food
         intestine   and start polysaccharide breakdown
    ____ liver   g. secrete digestive enzymes, bicarbonate

Critical Thinking

1. A glassful of whole milk contains lactose, proteins, butterfat (mostly triglycerides), vitamins, and minerals. Explain what will happen to each component in your digestive tract.

2. As a person ages, the number of body cells steadily decreases, and energy needs decline. If you were planning an older person's diet, what foods would you emphasize, and why? Which ones would you deemphasize?

3. Using Section 42.10 as a reference, determine your ideal weight and design a well-balanced program of diet and exercise that will help you achieve or maintain that weight.

4. Often, holiday meals are larger than everyday ones and have a high fat content. After stuffing themselves at an early dinner on Thanksgiving Day, Richard and other members of his family feel uncomfortably full for the rest of the afternoon. Based on what you have learned about controls over digestion, propose a biochemical explanation for their discomfort.

5. Your digestive system affords some effective protection against many pathogenic bacteria that can contaminate the kinds of food you eat. Explain some of the ways it can destroy or wash away these microorganisms. (You may wish to refer to Section 40.1, also.)

Selected Key Terms

appendix *42.7*	micelle formation *42.5*
bile *42.4*	microvillus
bulk *42.7*	(microvilli) *42.5*
chyme *42.4*	mineral *42.9*
colon *42.7*	mouth (oral cavity) *42.3*
complete	net protein
digestive system *42.1*	utilization (NPU) *42.8*
digestive system *CI*	nutrition *CI*
emulsification *42.4*	*ob* gene *42.10*
esophagus *42.3*	obesity *42.10*
essential amino acid *42.8*	pancreas *42.4*
essential fatty acid *42.8*	pharynx *42.3*
food pyramid *42.8*	ruminant *42.1*
gallbladder *42.4*	saliva *42.3*
gastric fluid *42.4*	segmentation *42.5*
gut *42.2*	sphincter *42.3*
incomplete	stomach *42.4*
digestive system *42.1*	tongue *42.3*
kilocalorie *42.8*	tooth *42.3*
leptin *42.10*	villus (villi) *42.5*
liver *42.4*	vitamin *42.9*

Readings

Blaser, M. J. February 1996. "The Bacteria Behind Ulcers." *Scientific American* (274).

Sherwood, L. 1997. *Human Physiology*. Second edition. Belmont, California: Wadsworth.

Wardlaw, G., P. Insel, and M. Seyler. 1992. *Contemporary Nutrition: Issues and Insights*. St. Louis: Mosby.

Withers, P. 1992. Chapter 18 of *Comparative Animal Physiology*. New York: Saunders/HBJ.

Web Site See *http://www.wadsworth.com/biology* for practice quiz questions, hypercontents, BioUpdates, and critical thinking. The Wadsworth Biology Resource Center provides a wealth of information fully organized and integrated by chapter.

43 THE INTERNAL ENVIRONMENT

Tale of the Desert Rat

Look closely at a fish or some other marine animal, and you will find that the cells of its body are exquisitely adapted to life in a salty fluid. And yet, some lineages of animals that evolved in the seas moved onto dry land about 375 million years ago. They were able to do so partly because they brought some salty fluid along with them, as an *internal* environment for their cells. Even so, it was not a simple transition. On land, those pioneers and their descendants encountered intense sunlight, dry winds, more pronounced swings in temperature, water of dubious salt content, and sometimes no water at all.

How did the pioneers conserve or replace the water and specific salts they lost as a result of their everyday activities? How did they manage to stay comfortably warm when their surroundings became too cold or too hot? They must have done these things, otherwise the volume, composition, and temperature of their internal environment would have spun out of control. In other words, *how did the land-dwelling descendants of marine animals maintain operating conditions inside the body and thus prevent cellular anarchy?*

Any of their existing descendants provides you with some answers. Think of a kangaroo rat living in an isolated desert of New Mexico (Figure 43.1). After a brief rainy season, the sun bakes the sand for months. The only obvious water is imported, sloshing in the canteens of an occasional

researcher or tourist. Yet with nary a sip of free water, this tiny mammal continually counters threats to its internal environment.

The kangaroo rat waits out the daytime heat inside a burrow, then forages in the cool of night for dry seeds and maybe a succulent. It is not sluggish about this. It hops rapidly and far, searching for seeds and fleeing from coyotes and snakes. All that hopping requires ATP energy and water. Seeds, chockful of energy-rich carbohydrates, provide both. Metabolic reactions that release energy from carbohydrates and other organic compounds also yield water. Each day, such "metabolic water" represents a whopping 90 percent of a kangaroo rat's total water intake. By comparison, it is only about 12 percent of the total intake for your body.

Inside its cool burrow, the kangaroo rat conserves and recycles water. As the animal breathes cool air into

Figure 43.1 Kangaroo rat, a master of water conservation in a New Mexico desert. The chart shows how kangaroo rats and humans gain and lose water. Notice the differences. Also notice that, in both cases, *the losses balance out the gains*—just as they do in all animals. How this happens, and why it absolutely must happen, is the first focus of this chapter.

	KANGAROO RAT	HUMAN
WATER GAIN (milliliters):		
By ingesting solids	6.0	850
By ingesting liquids	0	1,400
By way of metabolism	54.0	350
	60.0	2,600
WATER LOSS (milliliters):		
In urine	13.5	1,500
In feces	2.6	200
By evaporation	43.9	900
	60.0	2,600

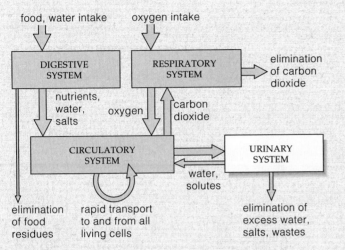

Figure 43.2 Links between the urinary system and other organ systems that contribute to homeostasis, or stability in favorable operating conditions in the animal body.

its warm lungs, water vapor condenses on the epithelial lining inside its nose—so some water can diffuse back inside the body. Also, after a busy night of foraging, the kangaroo rat quickly empties its cheek pouches of seeds, which soak up the small amount of water vapor that escapes by dripping from its nose. When the kangaroo rat eats the dripped-upon seeds, it recycles the water.

A kangaroo rat cannot lose water by perspiring; it has no sweat glands. It can lose water by urinating, but its specialized kidneys do not let it piddle away much. Kidneys filter the blood's water and solutes, including dissolved salts. They adjust *how much water* and *which solutes* return to the blood or leave the body as urine.

Overall, kangaroo rats and all other animals take in enough water and solutes to replace the daily losses (Figure 43.1). How they accomplish their balancing acts will be our initial focus in the chapter. Later, we will take a look at some of the means by which mammals withstand hot, cold, and sometimes unpredictable changes in environmental temperatures on land.

As a starting point, remind yourself of the kinds of fluids inside most animals. **Interstitial fluid** fills the spaces between living cells and other components of tissues. Another fluid, **blood**, transports substances to and from all tissue regions by way of a circulatory system. Taken together, the interstitial fluid and blood are the **extracellular fluid**. In most animals, a well-developed urinary system helps keep the volume and composition of extracellular fluid within tolerable ranges. As you will see, other organ systems, especially those indicated in Figure 43.2, interact with the urinary system in the performance of this homeostatic task.

KEY CONCEPTS

1. Animals are continually gaining and losing water and dissolved substances (solutes). They continually produce metabolic wastes. Even with all the inputs and outputs, the overall volume and composition of the extracellular fluid in the body remain relatively constant.

2. In humans, as in other vertebrates, a urinary system is crucial to balancing the intake and output of water and solutes. This system filters water and solutes from the blood. Then it reclaims both in amounts necessary to maintain extracellular fluid, and it eliminates the rest.

3. Kidneys are blood-filtering organs, and the urinary system of vertebrates has a pair of them. Packed inside each kidney are a great number of tubelike structures called nephrons.

4. At its beginning, each nephron cups around a set of blood capillaries, and it receives water and solutes from them. The nephron returns most of the filtrate to the blood, by giving it up to a second set of capillaries that is intertwined around the nephron's tubular parts.

5. Water and solutes not returned to the blood leave the body as a fluid called urine. During any interval, control mechanisms influence whether the urine is concentrated or dilute. Two hormones, ADH and aldosterone, have key roles in these adjustments.

6. The internal body temperature of animals depends on the balance between heat produced through metabolism, heat absorbed from the environment, and heat lost to the environment.

7. The internal body temperature is maintained within a favorable range through controls over metabolic activity and adaptations in body form and behavior.

The Challenge—Shifts in Extracellular Fluid

Different solid foods and fluids intermittently enter the mammalian gut. Afterward, variable amounts of absorbed water, nutrients, and other substances move into the blood, then into interstitial fluid and on into cells. Such events could easily shift the volume and composition of extracellular fluid beyond tolerable limits. However, the body makes compensatory adjustments that balance out the gains and losses. Within a given time frame, the body takes in as much water and solutes as it gives up.

WATER GAINS AND LOSSES To start, think of a human or some other mammal. It *gains* water mainly by two processes:

Absorption from gut
Metabolism

Considerable water is absorbed from solids and liquids in the gut. As you know, water also forms as a normal by-product of many metabolic reactions. In land mammals, how much water enters the gut in the first place depends on a thirst mechanism. When the body loses too much water, such mammals seek out streams, water holes, and so forth. We will consider the thirst mechanism later in the chapter.

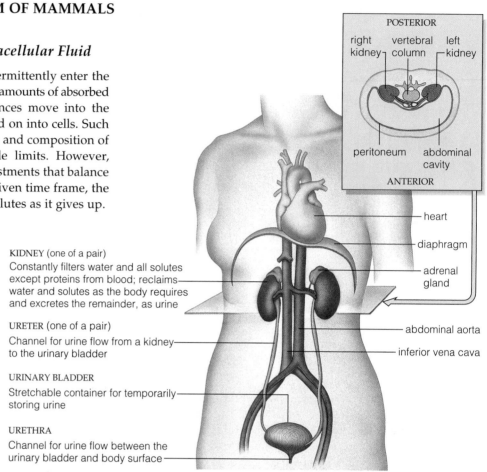

KIDNEY (one of a pair)
Constantly filters water and all solutes except proteins from blood; reclaims water and solutes as the body requires and excretes the remainder, as urine

URETER (one of a pair)
Channel for urine flow from a kidney to the urinary bladder

URINARY BLADDER
Stretchable container for temporarily storing urine

URETHRA
Channel for urine flow between the urinary bladder and body surface

Figure 43.3 Organs of the human urinary system and their functions. The two kidneys, two ureters, and urinary bladder are located *outside* the peritoneum, the membranous lining of the abdominal cavity. Compare Section 26.1.

Normally, the mammalian body *loses* water mainly by way of four physiological processes, as listed here:

Urinary excretion
Evaporation from lungs and skin
Sweating, by mammals that sweat
Elimination, in feces

Urinary excretion affords the most control over water loss. This process eliminates excess water and solutes as urine, a fluid that forms in a urinary system such as that shown in Figure 43.3. Some water also evaporates from respiratory surfaces, and in some species it departs in sweat. A mammal in good health loses very little water from the gut (most is absorbed, not eliminated in feces).

SOLUTE GAINS AND LOSSES There are several ways in which a mammal *gains* solutes, but it does so mainly by these four processes:

Absorption from gut
Secretion from cells
Respiration
Metabolism

For example, nutrients and mineral ions are absorbed from the gut. So are drugs and food additives. Cellular secretions and wastes, including carbon dioxide, enter interstitial fluid, then blood. The respiratory system puts oxygen into blood, and aerobically respiring cells put carbon dioxide into it.

Typically, mammals *lose* solutes in these ways:

Urinary excretion
Respiration
Sweating, by some species

The urine of mammals includes the wastes formed by the breakdown of organic compounds. For example, ammonia forms as amino groups are split from amino acids. Then **urea**, a major waste, forms in the liver when two ammonia molecules join with carbon dioxide. Also in urine are uric acid from nucleic acid breakdown, hemoglobin breakdown products (which give urine much of its color), drugs, and food additives. Besides these solute losses, some mammals also lose mineral ions in sweat. And all mammals lose carbon dioxide, the most abundant waste, by way of respiration.

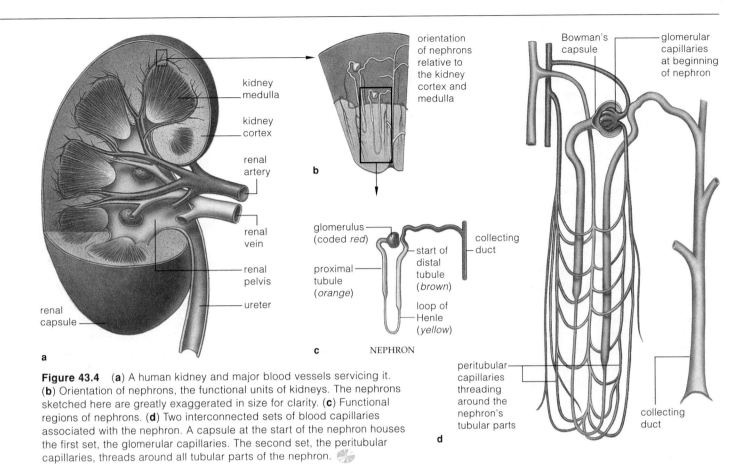

kidney medulla

kidney cortex

renal artery

renal vein

renal pelvis

renal capsule

ureter

a

b orientation of nephrons relative to the kidney cortex and medulla

glomerulus (coded *red*)

proximal tubule (*orange*)

start of distal tubule (*brown*)

collecting duct

loop of Henle (*yellow*)

c NEPHRON

Bowman's capsule

glomerular capillaries at beginning of nephron

peritubular capillaries threading around the nephron's tubular parts

collecting duct

d

Figure 43.4 (**a**) A human kidney and major blood vessels servicing it. (**b**) Orientation of nephrons, the functional units of kidneys. The nephrons sketched here are greatly exaggerated in size for clarity. (**c**) Functional regions of nephrons. (**d**) Two interconnected sets of blood capillaries associated with the nephron. A capsule at the start of the nephron houses the first set, the glomerular capillaries. The second set, the peritubular capillaries, threads around all tubular parts of the nephron.

Components of the Urinary System

Mammals counter shifts in the composition and volume of extracellular fluid mainly by a **urinary system**. This consists of two kidneys, two ureters, a urinary bladder, and a urethra. The **kidneys** are a pair of bean-shaped organs, about as big as an average fist (Figures 43.3 and 43.4). Each has an outer capsule of connective tissue. Blood capillaries thread through its two inner regions, the cortex and medulla.

Kidneys filter water, mineral ions, organic wastes, and other substances from blood, then they adjust the filtrate's composition and return all but about 1 percent to the blood. The tiny portion of unreclaimed water and solutes is urine. By definition, **urine** is a fluid that rids the body of water and solutes that are in excess of the amounts required to maintain extracellular fluid.

Urine flows from a kidney into a **ureter**, one of two tubular channels to the **urinary bladder**. Urine is briefly stored in that muscular sac before flowing on into the **urethra**, a muscular tube that opens at the body surface. Flow from the bladder, or urination, is a reflex action. When the bladder is filled, smooth muscle present in its balloonlike wall contracts as a sphincter around its neck opens. Thus urine is forced out through the urethra. Skeletal muscle surrounds the urethra. Its contraction, which is under voluntary control, prevents urination.

Nephrons—Functional Units of Kidneys

A human kidney has more than a million **nephrons**, slender tubules packed inside lobes that extend from the cortex down through the medulla. At the nephrons, water and solutes are filtered from the blood, and the amount sent back to blood is adjusted.

Each nephron starts as a **Bowman's capsule**. Here its wall cups around *glomerular* capillaries. The cupped wall region and blood vessels are a blood-filtering unit called a **glomerulus** (Figure 43.4c). Next, the nephron has a **proximal tubule** (closest to the capsule), then a hairpin-shaped **loop of Henle** and **distal tubule** (most distant from the capsule). It ends as a **collecting duct**, which is part of a system that leads into the kidney's central cavity (renal pelvis) and the entrance to a ureter.

Blood does not give up all of its water and solutes. The unfiltered part flows into a second set of capillaries around the nephron's tubular parts. In these *peritubular* capillaries, the blood reclaims water and solutes, then flows into veins and back to the general circulation.

A urinary system counters unwanted shifts in the volume and composition of extracellular fluid. In its paired kidneys, water and solutes are filtered from blood. The body reclaims most of this, but the excess leaves the kidneys as urine.

Urine forms in nephrons by three processes: filtration, tubular reabsorption, and tubular secretion. All three depend on properties of cells of the nephron wall. The cells differ in membrane transport mechanisms and permeability from one part of the nephron to the next.

Blood pressure generated by the heart's contractions drives **filtration**, which proceeds at the glomerulus (Figure 43.5). Pressure "filters" blood by forcing water and all solutes except proteins out of the glomerular capillaries. Then the protein-free filtrate moves from the cupped part of the nephron into the proximal tubule.

Tubular reabsorption proceeds along the nephron's tubular regions. Here, most of the filtrate's water and solutes move out of the nephron's lumen (the space enclosed by its wall). Then they move into neighboring peritubular capillaries (Table 43.1 and Figure 43.6).

Tubular secretion occurs at the tubule wall but in the opposite direction of reabsorption. Solutes from the peritubular capillaries enter cells of the wall, which then secrete them into the nephron's lumen. The main solutes are ions of hydrogen and potassium: H^+ and K^+. The process also prevents some metabolites (such as uric acid) and foreign substances (such as drugs) from accumulating in blood.

Factors Influencing Filtration

Each minute, about 1.5 liters (1-1/2 quarts) of blood flow through an adult's kidneys, and 120 milliliters of water and small solutes are filtered into the nephrons. That's 180 liters of filtrate per day! The high filtration rate is possible mainly because glomerular capillaries are 10 to 100 times more permeable to water and small solutes than other capillaries are. Also, blood pressure is high in glomerular capillaries. Why? Compared with most arterioles in the body, the ones delivering blood to the glomerulus have a wider diameter and less resistance to flow. Thus, hydrostatic pressure generated by the heart does not decline as much in kidneys as elsewhere.

At any time, the blood flow to the kidneys affects the filtration rate. Neural, endocrine, and local controls maintain the flow even when blood pressure changes. For example, when you run a race or dance until dawn, the nervous system diverts an above-normal volume of blood away from kidneys, toward the heart and skeletal muscles. It directs coordinated vasoconstriction and vasodilation in different body regions, so less of the blood flow reaches the kidneys. As one more example, cells in the walls of arterioles

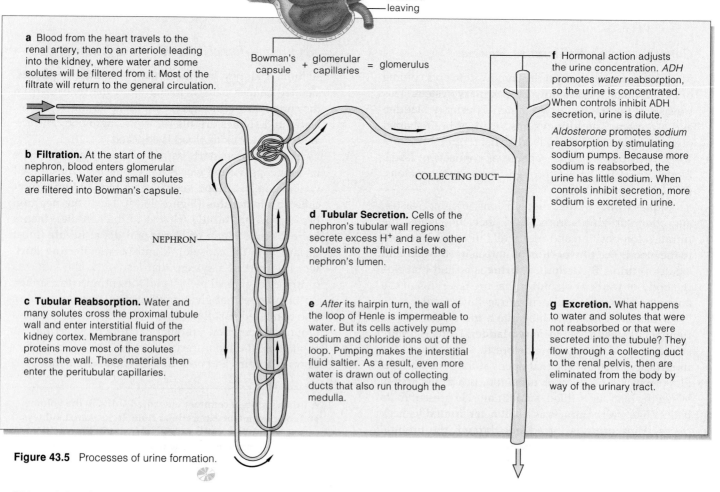

arteriole entering

arteriole leaving

a Blood from the heart travels to the renal artery, then to an arteriole leading into the kidney, where water and some solutes will be filtered from it. Most of the filtrate will return to the general circulation.

Bowman's capsule + glomerular capillaries = glomerulus

f Hormonal action adjusts the urine concentration. *ADH* promotes *water* reabsorption, so the urine is concentrated. When controls inhibit ADH secretion, urine is dilute.

Aldosterone promotes *sodium* reabsorption by stimulating sodium pumps. Because more sodium is reabsorbed, the urine has little sodium. When controls inhibit secretion, more sodium is excreted in urine.

b Filtration. At the start of the nephron, blood enters glomerular capillaries. Water and small solutes are filtered into Bowman's capsule.

COLLECTING DUCT

NEPHRON

d Tubular Secretion. Cells of the nephron's tubular wall regions secrete excess H^+ and a few other solutes into the fluid inside the nephron's lumen.

c Tubular Reabsorption. Water and many solutes cross the proximal tubule wall and enter interstitial fluid of the kidney cortex. Membrane transport proteins move most of the solutes across the wall. These materials then enter the peritubular capillaries.

e *After* its hairpin turn, the wall of the loop of Henle is impermeable to water. But its cells actively pump sodium and chloride ions out of the loop. Pumping makes the interstitial fluid saltier. As a result, even more water is drawn out of collecting ducts that also run through the medulla.

g Excretion. What happens to water and solutes that were not reabsorbed or that were secreted into the tubule? They flow through a collecting duct to the renal pelvis, then are eliminated from the body by way of the urinary tract.

Figure 43.5 Processes of urine formation.

Table 43.1	Average Daily Reabsorption Values for a Few Substances			
	Water (liters)	Glucose (grams)	Sodium (grams)	Urea (grams)
Filtered:	180	180	630	54
Excreted:	1.8	none	3.2	30
Reabsorbed:	99%	100%	99.5%	44%

leading to the glomeruli respond to pressure changes. When blood pressure decreases, they vasodilate, and so the kidneys receive more of the total blood volume. When the pressure rises, they vasoconstrict, so less blood flows in.

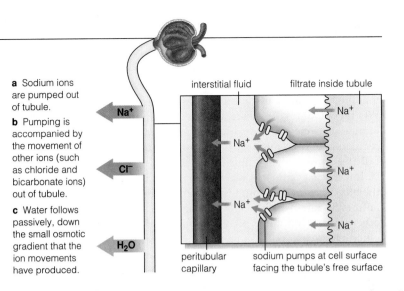

a Sodium ions are pumped out of tubule.

b Pumping is accompanied by the movement of other ions (such as chloride and bicarbonate ions) out of tubule.

c Water follows passively, down the small osmotic gradient that the ion movements have produced.

Figure 43.6 Sodium pumping associated with reabsorption of sodium and water in the kidneys.

Reabsorption of Water and Sodium

REABSORPTION MECHANISM Kidneys precisely adjust how much water and sodium ions the body excretes or conserves. Suppose you drink too much or too little water or wolf down salty potato chips or lose too much sodium in sweat. Responses start promptly as filtrate enters proximal tubules of nephrons. Cells in the tubule wall actively transport some sodium out of the filtrate, and other ions follow sodium into interstitial fluid. Water also leaves the filtrate, by osmosis. The nephron wall is highly permeable in this region, and about two-thirds of the filtrate's water is reabsorbed here (Figure 43.5c,d).

In the medulla, interstitial fluid is saltiest around the hairpin turn of the loop of Henle. Water moves out of the filtrate, by osmosis, *before* the turn. The fluid left behind gets saltier until it matches the interstitial fluid. The loop wall *after* the turn is impermeable to water. But sodium is pumped out by active transport mechanisms (Figure 43.5e). Interstitial fluid gets saltier and attracts more water out of the filtrate just entering the loop.

Fluid arriving at the distal tubule is dilute. The stage is now set for adjustments, which can lead to urine that is highly dilute, concentrated, or anywhere in between.

HORMONE-INDUCED ADJUSTMENTS The cells of distal tubules and collecting ducts have hormone receptors, for ADH and aldosterone. When the extracellular fluid volume falls, the hypothalamus triggers the secretion of **ADH**, or antidiuretic hormone. ADH action makes the tubule walls more permeable to water (Figure 43.5f). More water is reabsorbed, so the urine becomes more concentrated. When the body holds excess water, ADH secretion decreases. The walls become less permeable, less water can be reabsorbed, and urine remains dilute.

Aldosterone promotes sodium reabsorption. The extracellular fluid volume decreases when too much sodium is lost. Sensory receptors in the heart and blood vessels detect the decrease and signal glandular cells in the walls of the arteriole entering the glomerulus. These cells secrete renin, an enzyme that splits off part of a plasma protein. A second reaction converts the protein fragment to the hormone **angiotensin II**, which acts on aldosterone-secreting cells of the adrenal cortex. This is the outer portion of a gland perched on each kidney (Figure 43.3). Aldosterone stimulates cells of the distal tubules and collecting ducts to reabsorb sodium faster, so less sodium is excreted. Conversely, when the body holds excess sodium, aldosterone secretion is inhibited. Less sodium is reabsorbed, so more is excreted.

THIRST BEHAVIOR Near the ADH-secreting cells in the hypothalamus and interacting with them is a **thirst center**. When osmoreceptors signal a decrease in blood volume and a rise in blood solute levels, it induces a sensation of thirst that compels the individual to seek water. Also, besides promoting aldosterone secretion, angiotensin II promotes thirst and ADH secretion. Finally, free nerve endings in the mouth initiate thirst behavior when they detect mouth dryness, an early sign of dehydration.

Concentrated or dilute urine forms in kidneys by filtration, tubular reabsorption, and tubular secretion.

Filtration rates depend mainly on heart contractions, which generate high hydrostatic pressure at the glomerulus of the nephron. They also depend on neural, endocrine, and local control of blood flow being directed to the kidneys.

Reabsorption, which can be adjusted by hormonal controls, helps maintain extracellular fluid. The adjustments rid the body of suitable amounts of water and solutes, in the form of dilute or concentrated urine.

ADH promotes water conservation and concentrated urine. When ADH secretion is inhibited, urine is dilute.

Aldosterone promotes sodium conservation. When its secretion is inhibited, more sodium is excreted in urine.

WHEN KIDNEYS BREAK DOWN

By this point in the chapter, you probably have sensed that the functioning of nephrons is central to good health. Whether by illness or accident, when the nephrons of both kidneys become damaged and no longer perform their regulatory and excretory functions, we call this **renal failure**. Chronic renal failure is irreversible.

Infectious agents that reach the kidneys by way of the bloodstream or through the urethra can cause renal failure. So can ingestion of lead, arsenic, pesticides, and other toxins. Continued high doses of aspirin and some other drugs can do the same thing. Abnormal retention of metabolic wastes, such as the by-products of protein breakdown, can result in *uremic toxicity*. Heart failure, atherosclerosis, hemorrhage, or shock can diminish blood flow and skew the filtration pressure in the kidneys.

Or consider *glomerulonephritis*. In rare instances after a *Streptococcus* throat infection, antibody-antigen complexes become trapped in glomeruli. Unless phagocytes remove them, the complexes continue to activate complement and other agents that bring about widespread inflammation and tissue damage. Or consider how uric acid, calcium salts, and other wastes can settle out of urine and collect in the renal pelvis as *kidney stones*. These hard deposits are usually passed in urine but can become lodged in the ureter or urethra. If they disrupt urine flow, they must be medically or surgically removed to prevent renal failure.

About 13 million people in the United States alone suffer from renal failure. A *kidney dialysis machine* often can restore the proper solute balances. Like the kidney itself, this machine helps maintain the extracellular fluid by selectively removing solutes from blood and adding solutes to it. "Dialysis" refers to an exchange of substances across an artificial membrane that is interposed between two solutions that differ in composition.

In *hemodialysis*, a clinician connects the machine to an artery or to a vein. Then the patient's blood is pumped on through tubes made of a material that is similar to sausage casing or cellophane. The tubes are submerged in a warm saline bath. The mix of salts, glucose, and other substances of the bath sets up the correct concentration gradients with blood. The blood then returns to the body. In *peritoneal dialysis*, fluid of an appropriate composition is introduced into the patient's abdominal cavity, left in place for a specific length of time, then drained out. In this case, the cavity's lining itself, the peritoneum, serves as the membrane for dialysis. For kidney dialysis to have optimum effect, it must be performed three times a week. Each time, the procedure takes about four hours, because the patient's blood must circulate repeatedly in order to improve the solute concentrations in her or his body.

Bear in mind, kidney dialysis is used as a temporary measure in reversible kidney disorders. In chronic cases, the procedure must be used for the rest of the patient's life or until a transplant operation provides her or him with a functional kidney. With treatment and controlled diets, many patients can resume fairly normal activity.

THE ACID-BASE BALANCE

Besides maintaining the volume and composition of extracellular fluid, kidneys help keep it from becoming too acidic or too basic (alkaline). The overall **acid-base balance** is maintained by controlling the concentrations of H^+ and other dissolved ions. To get a sense of the importance of this balance, consider *metabolic acidosis*. This life-threatening condition results when the kidneys cannot secrete enough H^+ to keep pace with all of the H^+ that forms during metabolism.

Buffer systems, respiration, and urinary excretion work in concert to provide control over the acid-base balance. A buffer system, remember, consists of weak acids or bases that help minimize changes in pH by reversibly latching onto and releasing ions (Section 2.6).

Normally, the extracellular pH of the human body should be maintained between 7.37 and 7.43. As you know, acids lower the pH and bases raise it. A variety of acidic and basic substances enter the blood through absorption from the gut and as an outcome of normal metabolism. Typically, cell activities produce an excess of acids, these dissociate into H^+ and other fragments, and pH decreases. The effect is minimized when excess hydrogen ions react with buffer molecules. An example is the *bicarbonate–carbon dioxide* buffer system:

$$H^+ + HCO_3^- \rightleftharpoons H_2CO_3 \rightleftharpoons CO_2 + H_2O$$

$$\underset{\text{BICARBONATE}}{} \qquad \underset{\text{CARBONIC ACID}}{}$$

In this case, the buffer system neutralizes excess H^+, and the carbon dioxide that forms during the reactions is exhaled from the lungs. Like other buffer systems in the body, however, this one only has a temporary effect. It does not *eliminate* excess H^+. Only the urinary system can do so and thereby restore the buffers.

The same reactions proceed in reverse in cells of the nephron's tubular walls. HCO_3^- formed by the reverse reactions moves into interstitial fluid, then peritubular capillaries. Afterward, it enters the general circulation and buffers excess acid. The H^+ formed in the cells is secreted into the nephron and may join with HCO_3^-. The CO_2 that formed can be returned to blood, then exhaled. The H^+ also may combine with phosphate ions or with ammonia (NH_3), then leave the body in urine.

The kidneys work in concert with buffering systems, which neutralize acids, and with the respiratory system to help keep the extracellular fluid from becoming too acidic or too basic (alkaline).

A bicarbonate–carbon dioxide buffer system temporarily neutralizes excess hydrogen ions. The urinary system alone eliminates excess hydrogen ions and restores these buffers.

The bicarbonate–carbon dioxide buffer system is one of the key mechanisms that help maintain the acid-base balance.

ON FISH, FROGS, AND KANGAROO RATS

Now that you have a general sense of how your body maintains water and solute levels, consider what goes on in some other vertebrates, including that kangaroo rat hopping about at the start of the chapter.

Bony fishes and amphibians of freshwater habitats gain water and lose solutes (Figure 43.7a). Water moves into their internal environment by osmosis; they don't gain water by drinking it. The water diffuses across the thin gill membranes in fishes and across the skin in adult amphibians. Excess water leaves as dilute urine, formed inside a pair of kidneys. In both groups of vertebrates, solute losses are balanced by solutes gained from food and by the inward pumping of sodium by certain cells.

Body fluids of herring, snapper, and other marine fishes are about three times less salty than seawater. Such fishes lose water by osmosis, and they replace it by drinking more. They excrete ingested solutes against concentration gradients (Figure 43.7b). Fish kidneys do not have loops of Henle, so urine cannot ever become saltier than body fluids. Cells in the fish gills actively pump out most of the excess solutes in blood.

Figure 43.7c describes the water-solute balancing act in a salmon. This type of fish spends part of its life cycle in freshwater and another part in seawater.

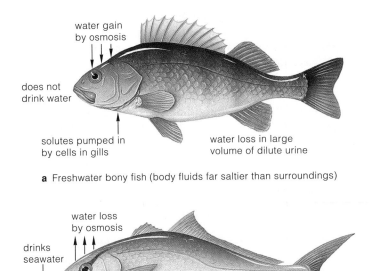

a Freshwater bony fish (body fluids far saltier than surroundings)

b Marine bony fish (body fluids less salty than surroundings)

Figure 43.7 (**a,b**) Water-solute balance in fishes.

(**c**) Water-solute balance by salmon, a type of fish that lives in saltwater and in freshwater. Salmon hatch in streams and later move downstream to the seas, where they feed and mature. They return to home streams to spawn.

For most salmon, salt tolerance is one outcome of changes in the concentrations of certain hormones. The changes seem to be triggered by increasing daylength in spring. Prolactin, a pituitary hormone, plays a key role in sodium retention in freshwater habitats. We know this because a freshwater fish that has its pituitary gland removed will die from sodium loss—but that fish will live if prolactin is administered to it.

Cortisol, a steroid hormone secreted by the adrenal cortex, is crucial to the development of salt tolerance in salmon. Cortisol secretions correlate with an increase in sodium excretion, in sodium-potassium pumping by cells in the salmon's gills, and in absorption of ions and water in the gut. In young salmon, cortisol secretion increases prior to the seaward movement—and so does salt tolerance.

salmon avoiding grizzly while maintaining solute-water balance

c

And about that kangaroo rat! Proportionally, the loops of Henle of its nephrons are astonishingly long, compared to yours. This means a great deal of sodium is pumped out of the nephron. Therefore, the solute concentration in the interstitial fluid around the loops becomes very high. The osmotic gradient between the fluid in the loops of Henle and the urine is *so* steep that nearly all of the water that does reach the equally long collecting ducts gets reabsorbed. In fact, kangaroo rats give up only a tiny volume of urine, which is three to five times more concentrated than the concentrated urine of humans.

The urinary systems of vertebrates differ in their details, such as the length of the nephron's loop of Henle. They are adapted to balance the body's gains in water and solutes with its losses of water and solutes in particular habitats.

We turn now to another major aspect of the internal environment—its temperature. Start by thinking about the temperature around your body. Is the air too hot? Too cold? If you have been sitting for some time, your cells have not been producing much metabolic heat. If you just finished exercising strenuously, metabolic rates have soared, and so has metabolic heat production.

Despite such differences, and assuming you are in good health, the internal temperature of your body remains much the same. A variety of physiological and behavioral responses to change work to maintain that constancy.

Temperatures Suitable for Life

Enzyme-mediated reactions proceed simultaneously in the millions to trillions of cells of a large-bodied animal. Enzymes of most animals typically function within the range of 0°–40°C (32°–104°F) as Table 43.2 shows. Above 41°C, chemical bonds holding an enzyme molecule in its required shape are disrupted. So is enzyme function. When the temperature decreases by 10 degrees, the rate of enzyme activity commonly plummets by 50 percent or more. Therefore metabolism, and life itself, depends on maintaining the **core temperature** within the range of tolerance of the body's enzymes. "Core" refers to the internal temperature of the animal body, as opposed to temperatures of the tissues near its surface.

Heat Gains and Heat Losses

Each metabolic reaction generates heat. If heat were to accumulate internally, core temperature would steadily increase. However, a warm body tends to lose heat to a cooler environment. The core temperature holds steady when the rate of heat loss balances the rate of metabolic heat production. In

Table 43.2	Temperatures Favorable for Metabolism, Compared With Environmental Temperatures
General range of internal temperatures favorable for metabolism:	0°C to 40°C (32°F to 104°F)
Range of air temperatures above land surfaces:	−70°C to +85°C (−94°F to +185°F)
Range of surface temperatures of open ocean:	−2°C to +30°C (+28.4°F to +86°F)

general, the heat content of a large, complex animal depends on the balance between heat gains and losses, which we can express in the following way:

$$\text{CHANGE IN BODY HEAT} = \text{HEAT PRODUCED} + \text{HEAT GAINED} - \text{HEAT LOST}$$

In this summary equation, the heat gains and losses occur by exchanges at body surfaces, such as the skin. Four processes, called radiation, conduction, convection, and evaporation, drive the exchanges.

With **radiation**, an animal gains heat after exposure to radiant energy (as from sunlight) or to any surface that is warmer than the surface temperature of its body.

With **conduction**, an animal contacting a solid object directly gains heat from it or gives up heat to it in response to a thermal gradient between them. Animals lose heat by resting on objects that are cooler than their own temperature, such as a frozen deck chair (Figure 43.8a). They gain heat when they are in direct contact with warmer objects, such as hot sand (Figure 43.8b).

With **convection**, moving air or water transfers heat. Conduction plays a part in this process; heat moves down a thermal gradient between the body and air or water next to it. Mass transfer also plays a part, with currents carrying heat away from or toward the body.

a

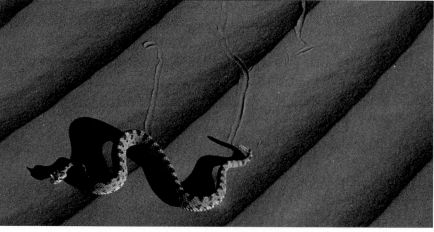

b

Figure 43.8 (a) How might this intrepid tourist sitting on a deck chair of a ship off the coast of Antarctica be gaining and losing heat? (b) A sidewinder making its signature J-shaped track across hot desert sand at dusk. As near as you can tell, how might this rattlesnake be gaining and losing heat?

When heated, air becomes less dense and moves away from the body. Even when there is no breeze, the body loses heat, for its movements create convection.

With **evaporation**, a liquid converts to gaseous form and heat is lost in the process. As described earlier, in Section 2.5, the liquid's heat content provides energy for the conversion. Evaporation from the body surface has a cooling effect, because the water molecules that are escaping carry away some energy with them.

Evaporation rates depend on humidity and on the rate of air movement. If air next to the body is already saturated with water (that is, when the local relative humidity is 100 percent), water will not evaporate. If air next to the body is hot and dry, evaporation may be the only means of countering the metabolic production of heat and the heat gains from radiation and convection.

Ectotherms, Endotherms, and In-Betweens

Animals can adjust the amount of heat lost or gained through changes in behavior and physiology, although some are better equipped than others to do so. Like most animals, snakes, lizards, and other reptiles have low metabolic rates and poor insulation (Figure 43.8b). They rapidly absorb and gain heat, especially the small species. They protect their core temperature primarily by gaining heat from their environment, not by their metabolic activities. Hence we classify these animals as **ectotherms**, which means "heat from outside."

When outside temperatures change, an ectotherm must alter its behavior. This is *behavioral* temperature regulation. For example, the iguana shown at the start of this unit basks on warm rocks, thus gaining heat by conduction. It keeps reorienting its body to expose the most surface area to the sun's infrared radiation. It loses heat after sunset. Before its metabolic rates decrease, it crawls inside crevices or under rocks, where heat loss is not as great and it is not as vulnerable to predators. As another example, meerkats bask in morning sunlight when they emerge from their burrows (Figure 33.1), and they learn to bask beneath heat lamps after being shipped to zoos in cold climates (Figure 43.9).

Most birds and mammals are **endotherms** ("heat from within"). With high metabolic rates, they can stay active under a wide temperature range. (Compared to a foraging lizard of the same weight, a foraging mouse uses up to thirty times *more* energy.) Core temperatures of these animals depend on a balancing of metabolism, controlled heat loss and conservation, and complex behavior. Adaptations in morphology help conserve or dissipate heat associated with the high metabolic rates. Feathers, fur, fat layers, even clothing reduce heat loss (Figure 43.8a). Some mammals in cold habitats have more massive bodies than close relatives in warmer ones.

Figure 43.9 Meerkats keeping warm on a cold winter night in a zoo in Germany.

For example, the snowshoe hare of Canada has a more compact body, far shorter legs, and far shorter ears than the jackrabbit of the American Southwest. Its body has a greater volume of cells for generating metabolic heat relative to the surface area available for heat loss. And heat dissipation from its legs and ears does not begin to match the losses from a jackrabbit's lengthy extremities.

Certain birds and mammals are **heterotherms**. At some times, their core temperature shifts (as it does in ectotherms). At other times, heat exchange is controlled (as in endotherms). For example, given their small size, hummingbirds have high metabolic rates. They locate and sip nectar only in the day. At night, they may shut down almost entirely. Their metabolic rates plummet, and they may get almost as cool as their surroundings. This way, they conserve precious energy.

In general, ectotherms have the advantage in warm, humid regions, such as the tropics. They need not spend much energy to maintain core temperatures, and more energy can be devoted to reproduction and other tasks. In terms of numbers and diversity, reptiles far exceed mammals in tropical regions. Endotherms have the edge in moderate to cold regions. For example, with their high metabolic rates, snowshoe hares, arctic foxes, and some other endotherms can occupy polar habitats, where you would never find a lizard.

The internal, core temperature of an animal's body is being maintained when heat gains and heat losses are in balance.

Metabolic reactions generate heat inside the body. Radiation, conduction, and convection can move heat down thermal gradients that exist between the body and its surroundings. Evaporative heat loss carries heat away from the body.

Besides being morphologically adapted to their habitats, animals can make behavioral and physiological adjustments to environmental temperatures.

Control centers maintain the core temperature of the mammal's body, and they reside in the hypothalamus (Figure 43.10). The centers continually receive input from peripheral thermoreceptors in the skin and from central thermoreceptors in the body. When the temperature deviates from a set point, the centers integrate complex responses involving skeletal muscles, smooth muscle in the arterioles that service skin, and often sweat glands. (Here you may wish to review Section 33.7.) Negative feedback loops back to the hypothalamus shut off the responses when a suitable temperature is reinstated.

Responses to Cold Stress

Table 43.3 lists the major responses that mammals make to cold stress. (Birds make the same responses.) They are called peripheral vasoconstriction, the pilomotor response, shivering, and nonshivering heat production.

Suppose peripheral thermoreceptors detect a decrease in outside temperature. They notify the hypothalamus, which commands smooth muscle in arterioles in the skin

Table 43.3	Responses to Cold Stress
Core Temperature	Responses
36°–34°C	Shivering response, increase in respiration. Increase in metabolic heat output (about 95°F). Peripheral vasoconstriction routes blood deeper in body. Dizziness and nausea set in.
33°–32°C (about 91°F)	Shivering response stops. Metabolic heat output drops.
31°–30°C (about 86°F)	Capacity for voluntary motion is lost. Eye and tendon reflexes inhibited. Consciousness is lost. Cardiac muscle action becomes irregular.
26°–24°C	Ventricular fibrillation sets in (Section 39.9). Death follows (about 77°F).

to contract. The response is **peripheral vasoconstriction**. The diameters of those arterioles constrict, which limits the blood's convective delivery of heat to body surfaces. When your fingers or toes are chilled, all but 1 percent of blood that otherwise would flow to skin is diverted to other regions of the body.

Also, muscle contractions can make hairs (or feathers) "stand up." This **pilomotor response** creates a layer of still air next to the skin and thereby reduces convective and radiative heat loss. Behavioral changes can further minimize exposed surface areas and reduce heat loss, as when cats curl up or when you hold both arms tightly against the body.

A **shivering response** can be made to prolonged cold. Rhythmic tremors begin as skeletal muscles contract, ten to twenty times a second, in response to commands from the hypothalamus. Shivering increases heat production by several times. It has a high energy cost and is not effective for long.

Prolonged or severe cold exposure also leads to a hormonal response that elevates the rate of metabolism. This **nonshivering heat production** is most notable in *brown* adipose tissue. Animals that hibernate or have become acclimatized to cold have this connective tissue. So do human infants. Adults have little unless they are adapted to cold. The tissue is present in Japanese and Korean women who spend six hours a day diving for shellfish in very cold water.

Failure to defend against cold results in *hypothermia*, a condition in which the core temperature falls below normal. In humans, a drop of only a few degrees affects brain

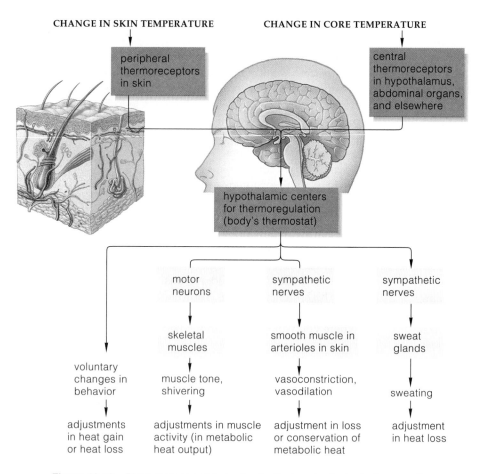

Figure 43.10 Physiological and behavioral adjustments of a typical mammal to changes in outside temperature. This example shows the main pathways for thermoregulation in humans.

Table 43.4 Summary of Mammalian Responses to Changes in Core Temperature

Stimulus	Main Responses	Outcome
Cold stress	Widespread vasoconstriction in skin; behavioral adjustments (e.g, minimizing surface parts exposed)	Conservation of body heat
	Increased muscle action; shivering; nonshivering heat production	Heat production increases
Heat stress	Widespread vasodilation in skin; behavioral adjustments; in some species, sweating, panting	Dissipation of heat from body
	Decreased muscle action	Heat production decreases

Figure 43.11 Evaporative water loss. Besides losing water at the moist respiratory surface of their lungs, horses, humans, and some other mammals have a large number of sweat glands that move water and solutes through pores to the skin surface.

function and leads to confusion; further cooling leads to coma and death. Many mammals can recover from profound hypothermia. But frozen cells may die unless tissues thaw under close medical supervision. Tissue destruction through localized freezing is called *frostbite*.

Responses to Heat Stress

Peripheral vasodilation and evaporative heat loss are the main responses to heat stress (Table 43.4). With **peripheral vasodilation**, hypothalamic signals cause blood vessels in skin to dilate. More blood flows from deeper body regions to skin, where the excess heat it carries is dissipated. **Evaporative heat loss** occurs at moist respiratory surfaces and across skin. Animals that sweat lose more water by this process (Figure 43.11).

For example, humans and some other mammals have sweat glands, which release water and specific solutes through pores at the skin surface. An average-size human has 2-1/2 million or more sweat glands that can produce 1 to 2 liters of sweat in an hour. For every liter of sweat that evaporates, 600 kilocalories of heat energy are lost. Bear in mind, sweat dripping from skin does not dissipate heat; water in sweat must evaporate for that to happen. On hot, humid days, the evaporation rates typically cannot match the rate of sweat secretion; the air's high water content slows it down.

During strenuous exercise, sweating can balance the high rates of heat production in skeletal muscle. With extreme sweating, as might occur in a marathon race, the body loses an important salt—sodium chloride—as well as water. Such losses disrupt the composition and volume of extracellular fluid, and when great enough, the runner collapses and faints.

What about mammals that sweat little or not at all? Some kinds make behavioral responses, such as licking fur, panting, or resting in shade. "Panting" refers to shallow, very rapid breathing that increases evaporative

water loss from the respiratory tract (compare Figure 33.14). Cooling takes place when the water evaporates from the nasal cavity, mouth, and tongue.

Sometimes peripheral blood flow and evaporative heat loss are not enough to counter heat stress in the body, and *hyperthermia* results. With this condition, the core temperature increases above normal. For humans and other endotherms, increases of only a few degrees above normal can be dangerous.

Fever

A *fever*, recall, is part of the inflammatory response to tissue damage. The hypothalamus resets the body's "thermostat," which dictates what the core temperature is supposed to be (Sections 33.7 and 40.3). Mechanisms that increase metabolic heat production and decrease heat loss are brought into play, but they are carried out to maintain a higher temperature. When they do so at the onset of fever, the person feels chilled. When the fever "breaks," peripheral vasodilation and sweating increase as the body attempts to restore the normal core temperature. Then, the person feels warm. By bringing down a fever, aspirin and other anti-inflammatory drugs may prolong healing time. But, without question, they are necessary when fevers approach dangerous levels.

Mammals counter cold stress by widespread vasoconstriction in skin, behavioral adjustments, increased muscle activity, and shivering and nonshivering heat production. They counter heat stress by widespread peripheral vasodilation in skin and by evaporative heat loss.

Control of Extracellular Fluid

1. For cells inside the animal body, the environment consists of certain types and amounts of substances that are dissolved in water. The extracellular fluid fills tissue spaces and blood vessels. Its volume and composition are maintained only when the animal's daily intake and output of water and solutes are in balance. In mammals, the following processes maintain the balance:

   a. Water is gained by absorption from the gut and by metabolism. It is lost by urinary excretion, evaporation from lungs and skin, sweating, and elimination of feces.

   b. Solutes are gained by absorption from the gut, secretion, respiration, and metabolism. They are lost by excretion, respiration, and sweating.

   c. Losses of water and solutes are controlled mainly by adjusting the volume and composition of urine.

2. The urinary system of vertebrates consists of two kidneys, two ureters, a urinary bladder, and a urethra.

3. Kidneys have many nephrons that filter blood and form urine. Each nephron interacts intimately with two sets of blood capillaries: glomerular and peritubular.

   a. The start of a nephron is cup-shaped (Bowman's capsule). It continues as three tubular regions (proximal tubule, loop of Henle, and distal tubule, which empties into a collecting duct).

   b. Together, the Bowman's capsule and the set of highly permeable, glomerular capillaries within it are a blood-filtering unit (glomerulus).

   c. Blood pressure forces water and small solutes out of the capillaries, into the cup. Most of the filtrate is reabsorbed by the tubules and returned to the blood. A portion is excreted as urine.

4. Urine forms in the nephron by three processes:

   a. Filtration of blood at the glomerulus, which puts water and small solutes into the nephron.

   b. Reabsorption. Water and solutes to be retained leave the nephron's tubular parts and enter capillaries that thread around them. A small volume of water and solutes remains in the nephron.

   c. Secretion. A few substances can leave peritubular capillaries and enter the nephron, for disposal in urine.

5. Urine is made more concentrated or less so by two hormones that act on cells of the wall of distal tubules and collecting ducts. ADH conserves water by enhancing reabsorption across the wall. In its absence, more water is excreted. Aldosterone enhances sodium reabsorption. In its absence, sodium is excreted. Angiotensin II induces aldosterone secretion (sodium conservation) and thirst; it also promotes ADH secretion (water conservation).

6. The urinary system acts in concert with buffers and with the respiratory system to maintain the acid-base balance of extracellular fluid.

Control of Body Temperature

1. Maintaining an animal's core (internal) temperature depends on balancing metabolically produced heat and the heat absorbed from and lost to the environment.

2. Animals exchange heat with their environment by four processes:

   a. Radiation. Emission from the body of infrared and other wavelengths. Radiant energy can be absorbed at the body surface, then converted to heat energy.

   b. Conduction. Direct transfer of heat energy from one object to another object in contact with it.

   c. Convection. Heat transfer by air or water currents; involves conduction and mass transfer of heat-bearing currents away from or toward the animal body.

   d. Evaporation. Conversion of liquid to a gas, driven by energy inherent in the heat content of the liquid. Some animals lose heat by evaporative water loss.

3. Core temperatures depend on metabolic rates and on anatomical, behavioral, and physiological adaptations.

   a. For ectotherms, the core temperature depends more on heat exchange with the environment than on heat generated by metabolism.

   b. For endotherms, the core temperature depends largely on high metabolic rates and precise controls over heat produced and heat lost.

   c. For heterotherms, the core temperature fluctuates some of the time, and controls over heat balance come into play at other times.

Review Questions

1. State the function of the urinary system in terms of gains and losses for the internal environment. Name the components of the mammalian urinary system and state their functions. *43.1*

2. Label the component parts of this kidney and nephron. *43.1*

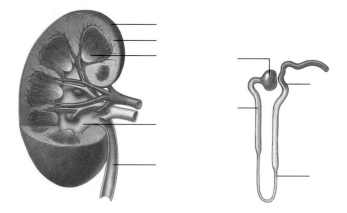

3. Define filtration, tubular reabsorption, and secretion. How does urine formation help maintain the internal environment? *43.2*

4. Which hormones promote (a) water conservation, (b) sodium conservation, and (c) thirst behavior? *43.2*

5. Name and define the physical processes by which animals gain and lose heat. What are the main physiological responses to cold stress and to heat stress in mammals? *43.6, 43.7*

Self-Quiz (Answers in Appendix IV)

1. In mammals, water intake depends on _____ .
   a. absorption from gut   c. a thirst mechanism
   b. metabolism   d. all of the above

2. In mammals, water is lost by way of the _____ .
   a. skin   d. urinary system
   b. respiratory system   e. c and d
   c. digestive system   f. a through d

3. Water and small solutes enter kidneys during _____ .
   a. filtration   c. tubular secretion
   b. tubular reabsorption   d. both a and c

4. Kidneys return water and small solutes to blood by _____ .
   a. filtration   c. tubular secretion
   b. tubular reabsorption   d. both a and b

5. A few substances move out of the peritubular capillaries that thread around tubular parts of the nephron. The substances are moved into the nephron during _____ .
   a. filtration   c. tubular secretion
   b. tubular reabsorption   d. both a and c

6. A nephron's reabsorption mechanism depends on _____ .
   a. osmosis across nephron wall
   b. active transport of sodium across nephron wall
   c. a steep solute concentration gradient
   d. all of the above

7. _____ promotes water conservation.
   a. ADH   c. Low extracellular fluid volume
   b. Aldosterone   d. Both a and c

8. _____ enhances sodium reabsorption.
   a. ADH   c. Low extracellular fluid volume
   b. Aldosterone   d. Both b and c

9. Match the term with the most suitable description.
   ____ glomerulus   a. surrounded by saltiest fluid
   ____ distal tubule   b. extra-long loops of Henle
   ____ loop of Henle   c. involves buffer systems
   ____ acid-base balance   d. blood-filtering unit
   ____ kangaroo rat   e. ADH, aldosterone act here

10. Match the term with the most suitable description.
    ____ ectotherm   a. heat transfer by air or water currents
    ____ endotherm   b. body temperature fluctuates some of
    ____ evaporation   the time, is controlled other times
    ____ heterotherm   c. emission of wavelength energy
    ____ radiation   d. direct heat transfer between one
    ____ conduction   object and another in contact with it
    ____ convection   e. metabolism dictates core temperature
       f. environment dictates core temperature
       g. conversion of liquid to gas

Critical Thinking

1. Fatty tissue holds kidneys in place. Extremely rapid weight loss may cause the tissue to shrink and the kidneys to slip from their normal position. If slippage puts a kink in one or both ureters and blocks urine flow, what may happen to the kidneys?

2. Drink one quart of water in an hour. What changes can you expect in your kidney function and in urine composition?

3. In 1912, the ocean liner *Titanic* left Europe on her maiden voyage across the Atlantic to America. In that same year, a chunk of the leading edge of a Greenland glacier broke off and floated out to sea. Late at night, off the Newfoundland coast, the iceberg and the *Titanic* made an ill-fated rendezvous (Figure 43.12). The *Titanic* was said to be unsinkable. Survival drills had been neglected. There were not enough lifeboats to hold even half the 2,200 passengers. The *Titanic* sank in about 2-1/2 hours.

Figure 43.12 Sinking of the *Titanic*, based on eyewitness accounts.

In less than two hours, rescue ships were on the scene, but 1,517 bodies were recovered from a calm sea. All the dead had on life jackets. None had drowned. Probably they died from _____ . If so, how did their blood flow, metabolism, and skeletal muscle action change prior to death?

4. When iguanas have an infection, they rest for a prolonged period in the sun. Propose a hypothesis to explain why.

5. Out on a first date, Jon takes Geraldine's hand in a darkened theater. *"Aha!"* he thinks. *"Cold hands, warm heart!"* What does this tell us about the regulation of core temperature, let alone Jon?

Selected Key Terms

Salt-Water Balance:
acid-base balance 43.4
ADH 43.2
aldosterone 43.2
angiotensin II 43.2
blood CI
Bowman's capsule 43.1
collecting duct 43.1
distal tubule 43.1
extracellular fluid CI
filtration 43.2
glomerulus 43.1
interstitial fluid CI
kidney 43.1
loop of Henle 43.1
nephron 43.1
proximal tubule 43.1

renal failure 43.3
shivering response 43.7
thirst center 43.2
tubular reabsorption 43.2
tubular secretion 43.2
urea 43.1
ureter 43.1
urethra 43.1
urinary bladder 43.1
urinary excretion 43.1
urinary system 43.1
urine 43.1

Body Temperature:
conduction 43.6
convection 43.6
core temperature 43.6
ectotherm 43.6
endotherm 43.6
evaporation 43.6
evaporative heat loss 43.7
heterotherm 43.6
nonshivering heat production 43.7
peripheral vasoconstriction 43.7
peripheral vasodilation 43.7
pilomotor response 43.7
radiation 43.6

Readings

Flieger, K. March 1990. "Kidney Disease: When Those Fabulous Filters Are Foiled." *FDA Consumer* 24: 26–29.

Sherwood, L. 1997. *Human Physiology*. Second edition. Belmont, California: Wadsworth.

Smith, H. 1961. *From Fish to Philosopher*. New York: Doubleday.

Web Site See *http://www.wadsworth.com/biology* for practice quiz questions, hypercontents, BioUpdates, and critical thinking. The Wadsworth Biology Resource Center provides a wealth of information fully organized and integrated by chapter.

PRINCIPLES OF REPRODUCTION AND DEVELOPMENT

From Frog to Frog and Other Mysteries

With a quavering, low-pitched call that only a female of its kind could find seductive, a male frog proclaims the onset of warm spring rains, of ponds, of sex in the night. By August the summer sun will have parched the earth, and his pond dominion will be gone. But tonight is the hour of the frog!

Through the dark, a hormone-primed female moves toward the vocal male. They meet; they dally in the behaviorally prescribed ways of their species. Then he clamps his forelegs above her swollen abdomen and gives her a prolonged squeeze (Figure 44.1*a*). Out into the water streams a ribbon of hundreds of eggs, which the male blankets with a milky cloud of sperm. Soon afterward, fertilized eggs—**zygotes**—are suspended in the water.

For the leopard frog, *Rana pipiens*, a drama begins to unfold that has been reenacted each spring, with only minor variations, for many millions of years. Within a few hours after fertilization, each zygote divides into two cells, the two divide into four, then the four into eight. In less than twenty hours after fertilization, the

mitotic cell divisions have produced a ball of cells no larger than the zygote. It is an early embryonic stage, of a type known as a blastocyst.

The cells continue to divide, but now they start to interact by way of their surface structures and chemical secretions. At prescribed times, many change shape and migrate to prescribed positions in the embryo. All of those cells inherited the same genetic instructions from the zygote—yet now they are becoming different from one another in appearance and function!

Through their associations, the cells form layers of embryonic tissues and then embryonic organs. A pair of tissue regions at the embryo's surface interact with the tissues beneath them. Together they give rise to a pair of eyes. Within the embryo a heart is forming, and soon it starts an incessant, rhythmic beating. In less than a week, events have transformed the frog embryo into a swimming, algae-eating larva called a tadpole.

Several months pass. Legs form; the tail shortens and disappears. The mouth develops jaws that snap shut on insects and worms. Eventually the transformations lead

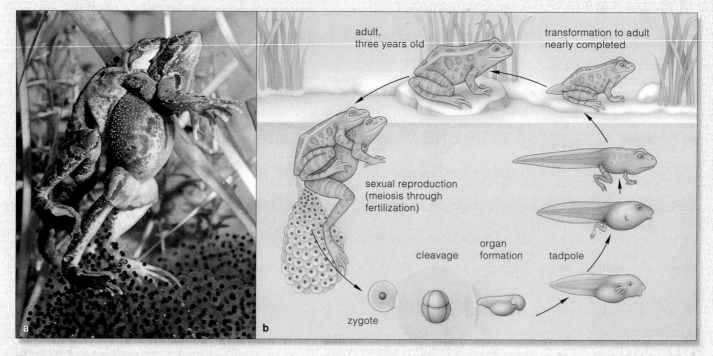

Figure 44.1 Reproduction and development of *Rana pipiens*, the leopard frog. (**a**,**b**) We zoom in on the life cycle as a male leopard frog clasps a female in a reproductive behavior called amplexus. The female releases her eggs into the water, then the male releases sperm over the eggs. A zygote forms when an egg nucleus and a sperm nucleus fuse at fertilization. (**c**) Frog embryos suspended in the water. (**d**) A tadpole. (**e**) A transitional form between a tadpole and the young adult frog (**f**).

to an adult frog. With luck the frog will avoid predators, disease, and other threats in the months ahead. In time it may even call out quaveringly across a moonlit pond, and the life cycle will turn again.

Many years ago you, too, started a developmental journey when a zygote carved itself up. Three weeks into the journey, your embryonic body had the stamp of "vertebrate" on it. A mere five weeks after that, it was a recognizable human in the making!

With this chapter we turn to one of life's greatest dramas—the development of offspring in the image of sexually reproducing parents. The guiding question is this: *How does the single-celled zygote of a frog, human, or any other complex animal become transformed into all of the specialized cells and structures of the adult form?* Some answers will become apparent as we move through a survey of basic developmental principles.

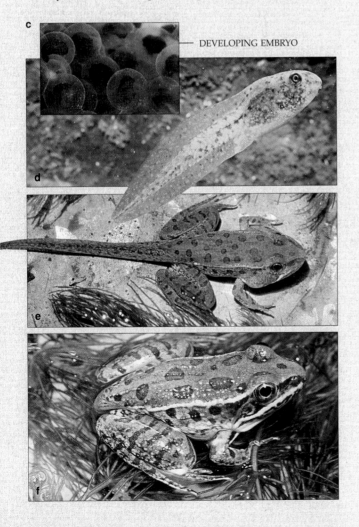

c

DEVELOPING EMBRYO

d

e

f

KEY CONCEPTS

1. Sexual reproduction dominates the life cycle of nearly all animals, but separation into sexes has biological costs. It requires the construction and maintenance of specialized reproductive structures. It also requires hormonal control mechanisms and complex forms of behavior attuned to the environment and to potential mates and rivals.

2. Separation into sexes affords a selective advantage. The offspring show variation in traits, which improves the odds that at least some will survive and reproduce despite unexpected challenges from the environment. This reproductive advantage offsets the biological cost of the separation.

3. The life cycle of humans and many other animals proceeds through six stages of embryonic development. The stages are called gamete formation, fertilization, cleavage, gastrulation, organ formation, and growth and tissue specialization.

4. Each stage of embryonic development builds on the tissues and structures that formed during the stage that preceded it.

5. In a developing embryo, the fate of each type of cell depends partly on cleavage, which distributes different regions of the fertilized egg's cytoplasm to different daughter cells. It also depends on interactions among cells of the embryo. These activities are the foundation for cell differentiation and morphogenesis.

6. With cell differentiation, each type of cell selectively uses certain genes and synthesizes proteins not found in other types, and so becomes unique in structure and function.

7. With morphogenesis, tissues and organs change in size, shape, and proportion. They also become organized relative to one another in prescribed patterns, and they do so at prescribed times.

THE BEGINNING: REPRODUCTIVE MODES

Sexual Versus Asexual Reproduction

In earlier chapters, you read about the cellular basis of **sexual reproduction**. Briefly, by this reproductive mode, meiosis and gamete formation typically occur in two prospective parents. Then, at fertilization, a gamete from one of them fuses with a gamete from the other, thereby forming the zygote, the first cell of a new individual. You also read about **asexual reproduction**, by which a single parent organism produces offspring by various means (but not by gamete formation). Consider now a few structural, behavioral, and ecological aspects of the two reproductive modes.

Think of a scuba diver accidentally kicking a sponge. A tissue fragment breaks away from the sponge body, then grows and develops by mitotic cell divisions and cell differentiation into a new sponge. Or think of one of the flatworms that can undergo transverse fission as it glides through the water. Its body constricts below the midsection. The part behind the constriction grips a substrate and starts a tug-of-war with the part in front of it. After a few hours, it splits off. Both parts go their separate ways, regenerate the missing part, and become a whole worm. Only certain flatworms can do this.

In such cases of *asexual* reproduction, all offspring are genetically the same as the individual parent, or nearly so. Phenotypically they are much the same, also. We can speculate that phenotypic uniformity is useful when each individual's gene-encoded traits are highly adaptive to a limited and more or less consistent set of environmental conditions. Variation introduced into the finely tuned gene package would not do much good, and often it would do harm.

However, most animals live where opportunities, resources, and danger are highly variable. For the most part, such animals reproduce sexually, with female and male parents bestowing different mixes of alleles on the offspring (Section 10.1). The resulting variation in traits improves the odds that some of the offspring, at least, will survive and reproduce even if conditions change in the environment.

Costs and Benefits of Sexual Reproduction

Among animals, separation into sexes is not without cost. Some cells that can serve as gametes must be set aside and nurtured. Housing and delivering gametes near or inside of a prospective mate requires specialized reproductive structures. Often, mating requires special forms of behavior, such as courtship, that can promote fertilization. Mating also requires built-in controls that can synchronize the timing of gamete formation, sexual readiness, even parental behavior in two individuals.

Figure 44.2 *Facing page:* A few examples of where invertebrate and vertebrate embryos develop, how they are nourished, and how (if at all) parents protect them.

(**a**) Snails are *oviparous*, which means that they are egg producers (*ovi-*, egg; *parous*, produce). Parents release the fertilized eggs, which develop on their own, unprotected.

(**b**) Birds also are oviparous. Their fertilized eggs have large reserves of yolk, and they develop and hatch outside the mother's body. Unlike snails, one or both parents expend considerable energy feeding and caring for the young.

Some fishes, lizards, and many snakes are *ovoviviparous*. Their fertilized eggs develop inside the mother; then offspring are born live. Yolk reserves, not the mother's own tissues, sustain the eggs. The copperhead shown in (**c**) is an example. Her offspring are born live inside the relics of egg sacs.

Most mammals are *viviparous*; they produce young that are born live (*vivi-*, alive). (**d,e**) In kangaroos and some other species, embryos are born in unfinished form. The young of this marsupial complete development in a pouch on the mother's ventral surface. Juvenile stages (joeys) continue to draw nourishment from mammary glands inside the mother's pouch. (**f**) By contrast, a human female retains the fertilized egg inside her body. Maternal tissues nourish the developing individual until the time of birth, in the manner described in the next chapter.

Consider the question of *reproductive timing*. How do the mature sperm in one individual become available exactly when the eggs mature in a different individual? Timing depends upon energy outlays for constructing, maintaining, and operating neural as well as hormonal control mechanisms in each parent. Also, parents must produce mature gametes in response to the same cues, such as a seasonal change in daylength, that mark the onset of the most suitable time of reproduction for their species. For example, male and female moose become sexually active only in late summer and early fall. The coordinated timing ensures that their offspring will be born the following spring—when the weather improves and food will be plentiful for many months.

Or consider the challenge of finding and recognizing a potential mate of the same species. Different species invest energy when they synthesize chemical signaling molecules, mating signals such as feathers of certain colors and patterns, and complex sensory receptors that can detect the specific signals being sent. Besides this, males often expend astonishing amounts of energy on courtship routines, as you will read in Chapter 51.

Assuring the survival of offspring is also costly. For example, many invertebrates and bony fishes simply release eggs and motile sperm into the surroundings (Figure 44.1a). The odds for fertilization would not be good if adults produced only *one* sperm or *one* egg each season. Such species invest energy in many gametes, often thousands of them. As another example, nearly all land animals depend on **internal fertilization**, the union

a   **b**

liveborn snake inside egg sac

c

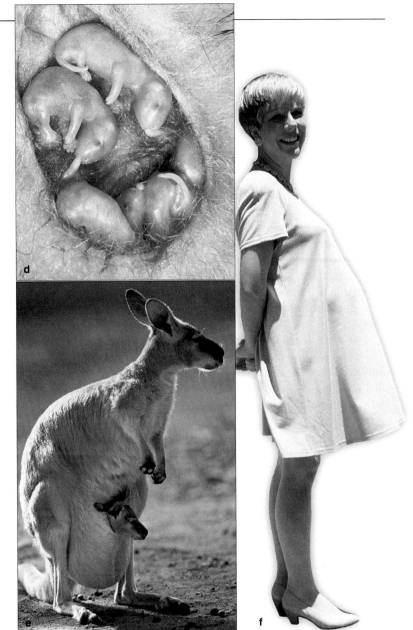

d

e   **f**

of sperm and egg *inside* a female's body. They invest energy to construct elaborate reproductive organs, such as a penis (by which a male deposits its sperm inside a female) and a uterus (a chamber inside the female where the embryo grows and develops).

Finally, animals set aside energy in forms that *nourish the developing individual* until it is developed enough to feed itself. For instance, nearly all animal eggs contain **yolk**, a protein-rich, lipid-rich substance that nourishes embryonic stages. The eggs of certain species have more yolk than others. Sea urchins make tiny eggs with very little yolk, release large numbers of them, and limit the biochemical investment in each one. Later, each fertilized egg develops into a self-feeding, free-moving larva in less than a day. Sea stars and other predators consume most of the eggs. So for sea urchins, bestowing as little as possible on as many gametes as possible does pay off, in terms of reproductive success.

By contrast, mother birds lay truly yolky eggs. Yolk nourishes the bird embryo through an extended period of development, inside an eggshell that forms after the egg is fertilized. Your mother put tremendous demands on her body to protect and nourish you through nine months of development inside her, starting from the

time you were an egg with almost no yolk. After you implanted yourself in her uterus, physical exchanges with her tissues supported you through the extended pregnancy (Figure 44.2*f*).

As these few examples suggest, animals show great diversity in reproduction and development. However, as you will see in the sections to follow, some patterns are widespread throughout the animal kingdom, and they can serve as a framework for our reading.

Separation into male and female sexes requires special reproductive cells and structures, neural and hormonal control mechanisms, and forms of behavior. A selective advantage—variation in traits among offspring—offsets the biological costs associated with the separation.

STAGES OF DEVELOPMENT—AN OVERVIEW

Embryos are a class of transitional forms on the road from a fertilized egg to an adult. Although they all start out as a single cell, embryos of different species often look different as they grow and develop. For example, you do not look like a frog now, and you did not look like one when you and the frog were early embryos, either. However, despite the differences in appearance, it is possible to identify certain patterns in the way that the embryos of nearly all animal species develop.

Figure 44.3 is an overview of the stages of animal development. During **gamete formation**, the first stage, eggs or sperm develop inside the reproductive organs of one parent's body. **Fertilization**, the second stage, starts when the plasma membrane of a sperm fuses with the plasma membrane of an egg. It is over when the egg nucleus and sperm nucleus fuse, thus forming a zygote.

The third stage, **cleavage**, is a program of mitotic cell divisions that *divide* the volume of egg cytoplasm into a number of smaller, nucleated cells called **blastomeres**. Cleavage only increases the number of cells; it does not change the original volume of egg cytoplasm.

As cleavage draws to a close, the pace of mitotic cell division slackens. The embryo enters **gastrulation**. This fourth stage of animal development is a time of major cellular reorganization. The newly formed cells become arranged into two or three primary tissues, often called germ layers. The cellular descendants of the primary tissues will form all tissues and organs of the adult:

1. **Ectoderm**. This is the *outermost* primary tissue layer, the one that forms first in the embryos of every animal. Ectoderm is the embryonic forerunner of the cell lineages that give rise to tissues of the nervous system and to the integument's outer layer.

2. **Endoderm**. Endoderm is the *innermost* primary tissue layer. It is the embryonic forerunner of the gut's inner lining and of organs derived from the gut.

3. **Mesoderm**. This *intermediate* primary tissue layer is the forerunner of muscle and most of the skeleton; of circulatory, reproductive, and excretory organs; and of connective tissue layers of the gut and integument. Mesoderm originated hundreds of millions of years ago, and it was a pivotal step in the evolution of nearly all large, complex animals.

After they have formed, the primary tissue layers give rise to subpopulations of cells. This marks the onset of **organ formation**. The subpopulations become unique in structure and function, and their descendants give rise to different kinds of tissues and organs.

During the final stage of animal development, **growth and tissue specialization**, organs increase in size and gradually assume their specialized functions. This stage continues into adulthood.

Figure 44.4 shows photographs and diagrams of several stages of embryonic development of a frog.

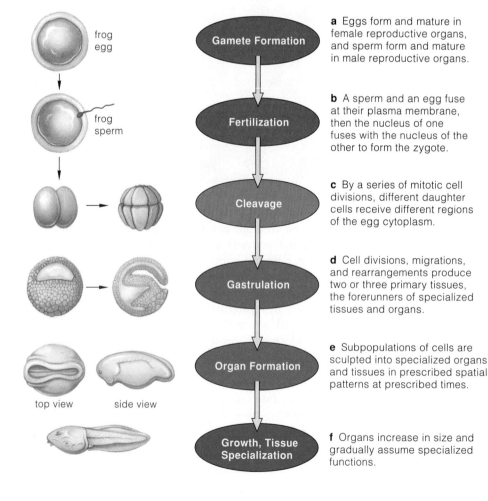

frog egg

frog sperm

top view   side view

a Eggs form and mature in female reproductive organs, and sperm form and mature in male reproductive organs.

Gamete Formation

b A sperm and an egg fuse at their plasma membrane, then the nucleus of one fuses with the nucleus of the other to form the zygote.

Fertilization

c By a series of mitotic cell divisions, different daughter cells receive different regions of the egg cytoplasm.

Cleavage

d Cell divisions, migrations, and rearrangements produce two or three primary tissues, the forerunners of specialized tissues and organs.

Gastrulation

e Subpopulations of cells are sculpted into specialized organs and tissues in prescribed spatial patterns at prescribed times.

Organ Formation

f Organs increase in size and gradually assume specialized functions.

Growth, Tissue Specialization

Figure 44.3 Overview of the stages of animal development. We use a few forms that appear in the frog life cycle as examples.

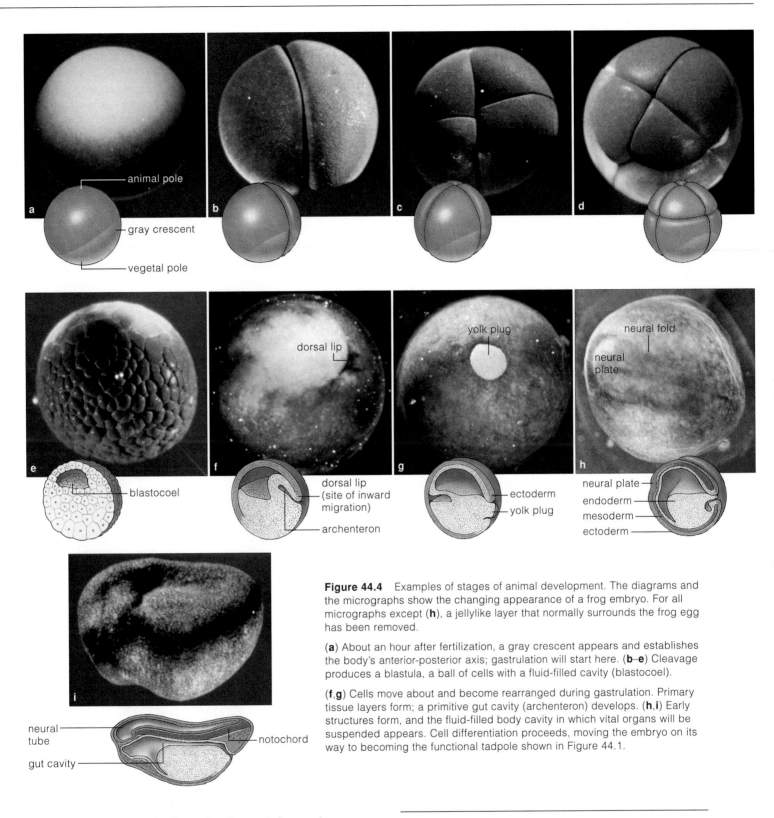

Take a moment to study them, for they reinforce a key concept. The structures that form during one stage of development serve as the foundation for the stage that comes after it. Successful development depends on the formation of all of those structures according to normal patterns, in prescribed sequence.

Figure 44.4 Examples of stages of animal development. The diagrams and the micrographs show the changing appearance of a frog embryo. For all micrographs except (**h**), a jellylike layer that normally surrounds the frog egg has been removed.

(**a**) About an hour after fertilization, a gray crescent appears and establishes the body's anterior-posterior axis; gastrulation will start here. (**b–e**) Cleavage produces a blastula, a ball of cells with a fluid-filled cavity (blastocoel).

(**f,g**) Cells move about and become rearranged during gastrulation. Primary tissue layers form; a primitive gut cavity (archenteron) develops. (**h,i**) Early structures form, and the fluid-filled body cavity in which vital organs will be suspended appears. Cell differentiation proceeds, moving the embryo on its way to becoming the functional tadpole shown in Figure 44.1.

Animal development proceeds from gamete formation and fertilization through cleavage, gastrulation, then organ formation, and finally growth and tissue specialization.

Development cannot proceed properly unless each stage is successfully completed before the next begins.

Information in the Egg Cytoplasm

You probably don't have an arm attached to your nose or toenails growing from your navel. The patterning of body parts for any complex animal, including yourself, is partly mapped out in the cytoplasm of an immature egg, or **oocyte**, even before a sperm enters the picture. A **sperm**, remember, consists of paternal DNA and a bit of equipment that helps the sperm reach and penetrate an egg. Compared to a sperm, an oocyte is much larger and more complex (Section 10.5).

As an oocyte matures, its volume increases. Enzymes, mRNA transcripts, and other factors become stockpiled in different parts of the cytoplasm and will be activated after fertilization. Typically they take part in early rounds of DNA replication and cell division. Also present are tubulin molecules and factors that will govern the angle and timing of their assembly into microtubules for a mitotic spindle. Such factors influence the pattern of cleavage. The cytoplasm also contains yolk, the amount and distribution of which will dictate where cleavage can proceed and how large the blastomeres will be.

We can find evidence of such regionally localized, "maternal messages" as early as fertilization. As a sperm fertilizes a frog egg, for example, it triggers structural reorganization of the egg's cortex, which includes the plasma membrane and the cytoplasm just under it. The cortex has pigment granules concentrated near one pole of the egg and yolk concentrated near the other. Upon fertilization, part of the cortex shifts away from the yolk. A crescent-shaped area of lighter colored cytoplasm is exposed. It is a **gray crescent**, a region of intermediate pigmentation, near the fertilized egg's equator. And it establishes the frog's anterior-posterior body axis.

In itself, the gray crescent is not evidence of regional differences in maternal messages. Such evidence comes from experiments of the sort shown in Figure 44.5b. It also comes from observing embryos in which localized cytoplasmic differences are pronounced enough to be tracked easily during development. For example, if you were to continue tracking the development of fertilized frog eggs, you would see that a gray crescent is always the site where gastrulation normally begins.

Cleavage—The Start of Multicellularity

Once fertilization is over, the zygote enters cleavage. Beneath the plasma membrane, its midsection bears a ring of microfilaments, made of the contractile protein actin. The ring tightens as microfilaments slide past one another and pinch in the cytoplasm. The cell surface above the tightening ring is pulled inward as a **cleavage furrow** (Section 9.5). It is the force of contraction that splits the cytoplasm into blastomeres. A new ring forms from actin subunits in each daughter cell.

Simply by virtue of where they form, blastomeres end up with different maternal messages. This outcome of cleavage, called **cytoplasmic localization**, helps seal

Figure 44.5 Examples of experiments that illustrate how the cytoplasm of a fertilized egg has localized differences that help determine the fate of cells in a developing embryo. The cortex of frog eggs contains granules of dark pigment, concentrated near one pole. At fertilization, a portion of the granule-containing cortex shifts toward the point of sperm entry. The shift exposes lighter colored, yolky cytoplasm in a crescent-shaped gray area:

pigmented cortex — gray crescent — yolk-rich cytoplasm — sperm penetrating frog egg

Normally, the first cleavage puts part of the gray crescent in both of the resulting blastomeres.

(**a**) For one experiment, the first two blastomeres that formed were separated from each other. Each still gave rise to a complete tadpole.

(**b**) For another experiment, a fertilized egg was manipulated so that the cut through the first cleavage plane missed the gray crescent entirely. Only one of the two blastomeres received the gray crescent. It alone developed into a normal tadpole. The blastomere deprived of maternal messages in the gray crescent gave rise to a ball of undifferentiated cells.

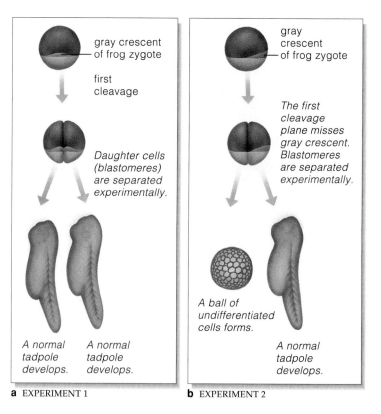

gray crescent of frog zygote

first cleavage

Daughter cells (blastomeres) are separated experimentally.

A normal tadpole develops. A normal tadpole develops.

a EXPERIMENT 1

gray crescent of frog zygote

The first cleavage plane misses gray crescent. Blastomeres are separated experimentally.

A ball of undifferentiated cells forms.

A normal tadpole develops.

b EXPERIMENT 2

Table 44.1 Examples of Cleavage Patterns

Yolk Distribution in the Egg	Cleavage Pattern	Representative Animals
COMPLETE CLEAVAGE		
Even distribution of sparse yolk	Radial	Lancelets, echinoderms
	Spiral	Flatworms, most mollusks, annelids
	Rotational	Mammals
Moderate amount of yolk at one end	Radial	Amphibians
INCOMPLETE CLEAVAGE		
Dense yolk concentrated at one end of egg	At small disk of yolk-free cytoplasm	Fishes, reptiles, birds
Dense yolk all through center	At periphery of yolk	Most insects, other arthropods

the developmental fate of each cell's descendants. Its cytoplasm alone may have molecules of a protein that can activate, say, the gene coding for a certain hormone. And its descendants alone will make the hormone.

Cleavage Patterns

In the simplest cleavage pattern, complete cuts occur after each nuclear division, and each cleavage plane is perpendicular to the mitotic spindle. In effect, the cuts parcel out a nucleus to each blastomere, which makes all of the blastomeres genetically equivalent.

Blastomeres do not grow in size before they divide again. Cleavage divides the volume of cytoplasm into increasingly smaller cells, often rapidly. A frog zygote becomes 37,000 cells in forty-three hours; a *Drosophila* zygote becomes 50,000 cells in only twelve. The decrease in the ratio of cytoplasmic to nuclear volume influences gene activation. For instance, in frogs, gene transcription starts after the twelfth division. If researchers double the amount of DNA in each blastomere, however, the transition will occur one division cycle ahead of time.

In most animals, the zygote's genes are silent during early cleavage. How fast the cuts proceed and how the blastomeres become arranged are under the control of the proteins and mRNAs stockpiled in the cytoplasm. In mammals, however, certain genes must be activated; cleavage cannot occur without the proteins they specify.

Table 44.1 correlates the amount and distribution of yolk with cleavage patterns of major animal groups. Yolk typically inhibits cleavage. If an egg has little yolk, it may be fully cleaved. If yolk is concentrated at one end, early cuts will be limited to part of the cytoplasm. Such eggs show polarity. The yolk-rich end is called the *vegetal* pole; the *animal* pole is the end closest to the nucleus. A frog egg is like this (Figure 44.4).

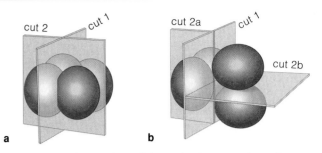

Figure 44.6 Comparison of the early cleavage planes for the egg of (**a**) sea urchins and (**b**) mammals. The cuts are radial in sea urchins and rotational in mammals.

Cleavage patterns differ among animals with little yolk in their eggs, so other, heritable factors also must influence the cuts. A sea urchin egg has little yolk and undergoes *radial* cleavage. That is, its cleavage furrows run horizontally and vertically with the animal-vegetal axis (Figure 44.6*a*). Successive cuts result in a **blastula**, a stage with blastomeres surrounding a fluid-filled cavity called a **blastocoel**. A sea urchin blastula has horizontal rows of blastomeres. Frog eggs also are cleaved radially, but yolk impedes cuts near the vegetal pole. Cleavage is faster and yields more, smaller size blastomeres near the animal pole. The blastocoel forms there, also (Figure 44.4*e*). Between the time that 16 to 64 blastomeres have formed, any amphibian blastula resembles a mulberry and is called a **morula**, which is Latin for mulberry.

By contrast, eggs of reptiles, birds, and most fishes undergo *incomplete* cleavage. The large volume of yolk restricts those early cuts to a small, caplike region near the animal pole. The result is two flattened layers of calls with a narrow cavity in between.

A mammalian egg undergoes *rotational* cleavage. The first cut passes through the egg's two poles. But then one of the resulting blastomeres is cut the same way and the other is cut horizontally (Figure 44.6*b*). The blastomeres divide slowly, at different times. And after the third cleavage, they abruptly huddle into a compact ball joined by tight junctions. The sixteen-cell stage is a morula. A blastocoel forms when descendants of the outer cells (trophoblasts) secrete fluid into the morula. As you will see in Chapter 45, trophoblasts will give rise to part of the placenta. The inner cells mass together, and they will give rise to the embryo proper.

The egg cytoplasm contains maternal instructions in the form of regionally distributed enzymes and other proteins, mRNAs, cytoskeletal elements, yolk, and other factors.

Cleavage divides a zygote into blastomeres, each with a localized part of maternal messages inherent in the egg cytoplasm. This outcome is called cytoplasmic localization.

Differences in the amount and distribution of yolk and other, heritable factors give rise to different patterns of cleavage among animal groups.

HOW SPECIALIZED TISSUES AND ORGANS FORM

Nearly all animals have a gut, with tissues and organs that function in digestion and absorption of nutrients. They have surface parts that protect internal parts and detect what is going on outside. In between, most have numerous organs, such as those dealing with structural support, movement, and blood circulation. Their three-layer body plan starts to emerge as cleavage ends and gastrulation begins. The embryo's size increases little, if any, but cells start migrating to new positions to form primary tissue layers—the ectoderm, endoderm, and mesoderm. Figure 44.7 shows how some outer cells of a sea urchin embryo migrate inward and form a lining for a cavity that will become the gut. Figure 44.8 shows the formation of a bird embryo's anterior-posterior axis, the outcome of cell migrations and other rearrangements. In every vertebrate, the anterior-posterior axis defines where a **neural tube**, the forerunner of the brain and spinal cord, will form. All such organs start forming by way of cell differentiation and morphogenesis.

Cell Differentiation

All cells of an embryo descend from the same zygote, so they have the same number and kinds of genes. They all activate the genes for histones, enzymes of glucose metabolism, and other proteins that are absolutely basic to cell survival. However, from gastrulation onward, certain groups of genes are activated in some cells but not others. When a cell selectively activates genes and synthesizes proteins not found in other cell types, we call this process **cell differentiation**. You read about the molecular basis of selective gene expression in Section 15.3. Basically, it results in proteins that are required for distinctive cell structures, products, and functions.

For example, when your eye lenses developed, some cells started synthesizing crystallin, a family of proteins

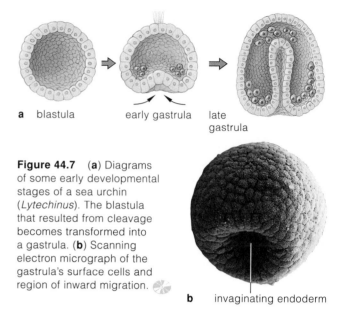

Figure 44.7 (**a**) Diagrams of some early developmental stages of a sea urchin (*Lytechinus*). The blastula that resulted from cleavage becomes transformed into a gastrula. (**b**) Scanning electron micrograph of the gastrula's surface cells and region of inward migration.

a blastula   early gastrula   late gastrula

b invaginating endoderm

that become incorporated in transparent fibers of the lens. Only those cells could activate the required genes. Long crystallin fibers formed in the cells and forced them to lengthen and flatten. Collectively, those differentiated cells impart unique optical properties to each eye's lens. And those crystallin-producing cells are only one of 150 or so differentiated cell types now present in your body.

As many experiments tell us, nearly all cells become fully differentiated without loss of genetic information. For example, John Gurdon removed the nucleus from unfertilized eggs of the African clawed frog (*Xenopus laevis*). Then he isolated intestinal cells from tadpoles of the same species. He ruptured their plasma membrane, left the nucleus and most of the cytoplasm intact, then inserted the nucleus into the enucleated egg. In some experiments, that nucleus directed the developmental steps leading to a complete frog! The intestinal cell had

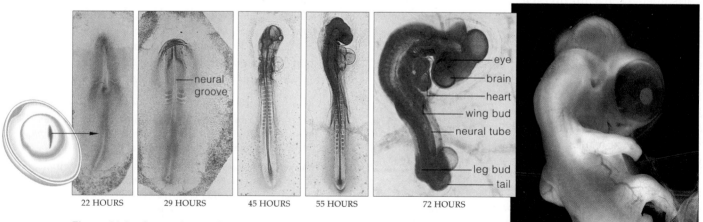

22 HOURS   29 HOURS   45 HOURS   55 HOURS   72 HOURS   168 HOURS (SEVEN DAYS OLD)

neural groove

eye
brain
heart
wing bud
neural tube
leg bud
tail

Figure 44.8 Onset of organ formation in a chick embryo during the first seven days of development. The heart begins to beat between thirty and thirty-six hours. You may have observed such embryos at the yolk surface of raw, fertilized eggs.

the same number and kinds of genes as the zygote, so it had all the genes necessary to form all of the required cell types. Its genes had not been somehow lost when it differentiated into an intestinal cell.

Human cells that normally contribute to only part of an embryo also retain the capacity to produce an entire individual. Consider the spontaneous separation of the two blastomeres of the two-cell human embryo. Their dissociation does not result in two half-embryos. The result is **identical twins**, two normal, fully developed individuals having the same genetic makeup.

Morphogenesis

Morphogenesis refers to a program of orderly changes in an embryo's size, shape, and proportions, the result being specialized tissues and early organs. As part of the program, cells divide, grow, migrate, and change in size. Tissues expand and fold, and cells in some of them die in controlled ways at prescribed locations.

Consider active cell migration. *Cells send out and use pseudopods that move them along prescribed routes.* When they reach their destination, they establish contact with cells already there. For example, forerunners of neurons interconnect this way as a nervous system is forming.

How do cells "know" where to move? They respond to adhesive cues, as when migrating Schwann cells stick to adhesion proteins on the surface of axons but not on blood vessels. And they respond to chemical gradients. Their migrations may be coordinated by the synthesis, release, deposition, and removal of specific chemicals in the extracellular matrix. Adhesive cues also tell the cells when to stop. Cells will migrate to regions of strongest adhesion, but once there, further migration is impeded. Section 23.1 describes such chemotactic behavior among the cells of a slime mold, *Dictyostelium discoideum*.

Besides this, as microtubules lengthen and as rings of microfilaments in cells constrict, *sheets of cells expand and fold inward and outward* (Figure 44.9). The assembly and disassembly of such components of the cytoskeleton are aspects of control mechanisms that bring about the changes. The size, shape, and proportion of body parts emerge through such controlled, localized events. We do not fully understand why some embryonic tissues expand more than others. Apparently, selective controls over the activity of regulatory genes play major roles.

Consider what happens after three primary tissues form in the embryos of amphibians, reptiles, birds, and mammals. At the embryo's midline, ectodermal cells elongate to form a neural plate, the first indication that a region of ectoderm is on its way to becoming nervous tissue. In some cells, microtubules lengthen. Other cells become wedge shaped as a microfilament ring contracts and constricts each cell at one end. Collectively, all the

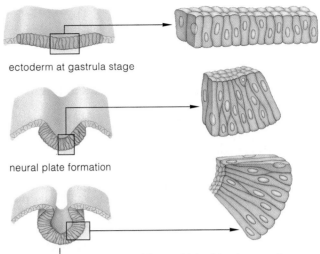

ectoderm at gastrula stage

neural plate formation

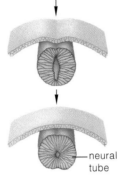

neural tube

Figure 44.9 Morphogenesis—an example. A neural tube forms by changes in the shapes of certain ectodermal cells. When gastrulation is over, ectoderm is a uniform sheet of cells. In some cells, microtubules lengthen, and the elongating cells form a neural plate. In other cells, microfilament rings at one end constrict and the cells become wedge shaped. *Their* part of the ectodermal sheet folds over the neural plate to form the tube.

changes in shape cause tissue flaps to fold and meet at the midline, thus forming a neural tube (Figure 44.9).

Finally, *programmed cell death helps sculpt body parts.* That is, cells that function for only limited periods in embryonic tissues can execute themselves. By this form of cell death, called **apoptosis**, molecular signals switch on weapons of self-destruction already stockpiled inside target cells. The Chapter 15 introduction describes the results. For now, think of a hand of a human embryo, which initially looks like a paddle (Section 9.5). Digits of cartilage are forming inside it. Cells in the tissue zones between digits have receptors for certain proteins that form by expression of certain genes. When the receptors bind the proteins, the cells commit mass suicide, and so the digits separate from one another. In some humans, a gene mutation blocks apoptosis and digits stay webbed.

In cell differentiation, a cell selectively uses certain genes and makes certain proteins, not found in other cell types, for distinctive cell structures, products, and functions.

Morphogenesis is a program of orderly changes in body size, shape, and proportions. It results in specialized tissues and organs at prescribed locations, at prescribed times.

Morphogenesis involves cell division, active cell migration, tissue growth and foldings, changes in cell size and shape, and programmed cell death by way of apoptosis.

TO KNOW A FLY

"Although small children have taboos against stepping on ants because such actions are said to bring on rain, there has never seemed to be a taboo against pulling off the legs or wings of flies. Most children eventually outgrow this behavior. Those who do not either come to a bad end or become biologists."

So wrote Vincent Dethier in *To Know a Fly*, a whimsical yet scientifically sound tribute to flies and their suitability for controlled laboratory experiments. Even before Dethier published his booklet in 1962, plenty of geneticists and developmental biologists had already come to the same conclusion. *Drosophila melanogaster* was and still is the fly of choice. It costs almost nothing to feed or house this tiny insect. It reproduces rapidly in bottles, it has a short life cycle, and disposing of spent bodies is a snap (Section 12.4).

For much of this century, the embryonic development of *Drosophila* has been studied in detail at the anatomical, cytological, biochemical, and genetic levels. The findings tell us much about the embryonic development, even the evolutionary history, of animals in general. They tell us something about our own development, even though fruit fly and mammalian embryos proceed through the stages of development according to different patterns.

SUPERFICIAL CLEAVAGE IN *DROSOPHILA* For example, mammalian eggs contain relatively little yolk, so cleavage furrows can cut through the entire egg. But the eggs of *Drosophila* (and those of most insects, fishes, reptiles, and birds) contain a large mass of yolk that restricts cleavage to a small portion of egg cytoplasm. The yolk mass of a *Drosophila* egg is centrally located and confines cleavage to the cytoplasm's periphery. This is one example of a *superficial* cleavage pattern. The egg nucleus repeatedly divides, but the cytoplasm does not; cleavage furrows do not form. The outcome is an early embryonic stage called a *syncytial* blastoderm (Figure 44.10).

Compared with a mammalian egg, which undergoes slow cuts every twelve to twenty-four hours, superficial cleavage of a fertilized *Drosophila* egg is rapid. Nuclear division proceeds every eight minutes until 256 nuclei are crowded together in the yolky cytoplasm. This stage of development is a type of *cellular* blastoderm. Each nucleus of the blastoderm has its own small array of microtubules and microfilaments, which will influence the forthcoming patterns of cell formation and elongation.

Most of the nuclei migrate toward the egg periphery. First, plasma membrane wraps around each nucleus that

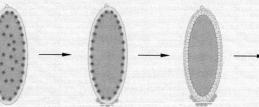
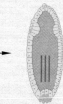

Figure 44.10
Superficial cleavage in the yolk-rich egg of *Drosophila*.

(a) A fertilized egg's *nucleus* repeatedly divides. *Cytoplasmic* division is delayed until after the nuclei move to the periphery of the cytoplasm.

(b) Pole cells form.
(c,d) Then other cells form; they are arrayed like a jacket around the central mass of yolk. This early stage of development is a type of blastoderm.

As the blastoderm forms, polar granules in the cytoplasm migrate and become isolated in pole cells that form at one end of the blastoderm. Pole cells are forerunners of germ cells that give rise to eggs or sperm in the adult fly.

| *Drosophila* zygote | repeated nuclear divisions without cytoplasmic division | migration of nuclei to periphery of the cytoplasm | formation of pole cells at posterior pole of egg | formation of a cell jacket around the central yolk mass | the cellular blastoderm |

a

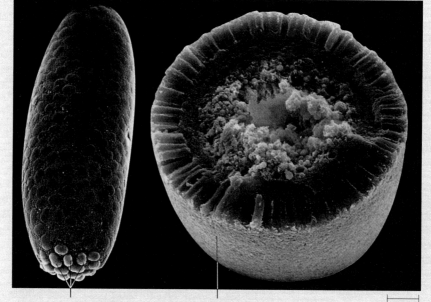

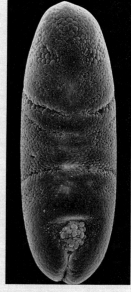

b Pole cells (forerunners of germ cells of the adult)

c Section through the blastoderm showing jacket of cells around yolk

200 μm

d Gastrulation, well under way

reaches the egg's posterior pole axis. The resulting pole cells are the forerunners of germ cells of the adult fly. This physical separation of the future reproductive cells from the rest of the blastoderm is one of the first events of insect development. After the separation occurs, the plasma membrane folds inward and forms a partition around each remaining nucleus and a bit of cytoplasm. Within four hours after fertilization, about 6,000 cells are arranged like a jacket around the yolky mass.

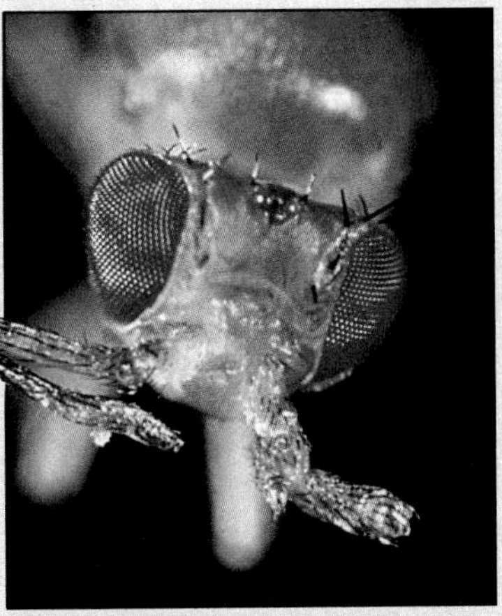

GENES AS MASTER ORGANIZERS   During the late 1940s, Edward Lewis of the California Institute of Technology discovered that the products of certain mutated genes introduce bizarre blips in pattern formation in *Drosophila*. To give one example, as a result of a single gene mutation, a leg grew out of the head region where an antenna should have grown, as shown by the photograph in Figure 44.11.

Look at the *Drosophila* zygote in Figure 44.12*a*. A **fate map** of its surface is a map that shows where each kind of differentiated cell in the forthcoming adult will originate. The embryo develops with a head and a tail end. In between, a series of body segments will form, each with its own gene-specified identity. In an adult fly, for example, the first segment (part of the thorax) has only legs, the next has legs and wings, the next has legs and balancing devices, and so on.

Researchers have put together a model to explain how polarity in the egg gives rise to segmented polarity of the adult body. First, *maternal effect* genes that specify regulatory proteins are transcribed, translated, or both (Figure 44.12*b*). The resulting mRNAs and proteins become localized in different parts of the egg cytoplasm. These gene products are activated in the zygote. They activate or suppress *gap* genes that map out broad regions of the body. As the product of one gap gene (hunchback) diffuses through the syncytial blastoderm, it creates a chemical gradient that influences the extent to which other gap genes are expressed. The different concentrations of gap gene products switch on *pair-rule* genes. The products of these genes accumulate in bands (stripes) corresponding to two body segments. And they activate *segment polarity* genes that divide the embryo into segment-size units.

Interactions among the products of the gap, pair-rule, and segment polarity genes control expression of another

class of genes—*homeotic* genes, which collectively govern the developmental fate of each segment.

When mutated, homeotic genes can transform one body segment into the likeness of another. Thus mutations in the *antennapedia* gene result in abnormal responses to regulatory proteins. In the example cited earlier, a mutated gene product activated the wrong set of homeotic genes in cells that are supposed to produce two antennae on the head. They gave rise to a pair of legs instead.

Researchers do not yet have a complete understanding of the multiple levels of gene interactions that specify body parts in *Drosophila*. But their studies of this tiny organism give us fascinating glimpses into the kinds of regulatory mechanisms that must come into play in other organisms as well. And think about *that* next time you swat at fruit flies, each not much bigger than a rounded pinhead, as they hover over a bowl of pungently ripe fruit.

Figure 44.11 Experimental evidence that certain genes are responsible for the specification of body parts. In a normal *Drosophila* larva, a cluster of cells gives rise to antennae on the head. If the larva carries a certain mutated form of the *antennapedia* gene, the adult fly will have legs instead of antennae on its head.

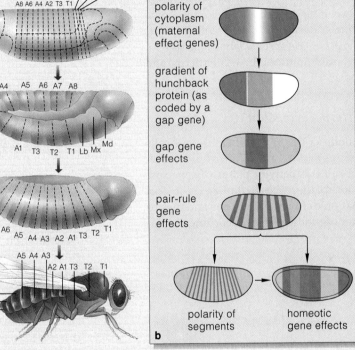

Figure 44.12 (**a**) Fate map of a *Drosophila* zygote. The dashed lines indicate regions that normally will develop into different body segments, with specialized structures and appendages. For this type of organism, each segment's fate is sealed at the time of cytoplasmic localization. (**b**) Generalized model of pattern formation in *Drosophila*, as described in the text.

polarity of cytoplasm (maternal effect genes)

gradient of hunchback protein (as coded by a gap gene)

gap gene effects

pair-rule gene effects

polarity of segments

homeotic gene effects

a

b

PATTERN FORMATION

The developmental fate of an embryonic cell depends partly on which bit of cytoplasm it acquired during cleavage. This is especially true of *Drosophila*, where gene products that were already localized in different parts of the egg cytoplasm give every cell its identity. Sidney Brenner calls this the "European style" of cell determination, in which lineage is everything.

By contrast, the "American style" of cell determination emphasizes the neighborhood. That is, the fate of each cell depends on whom it meets up with in the embryo. The cell retains the capacity to alter its fate, because it contains all of the instructions necessary to give rise to a whole animal. Remember, two blastomeres at the two-cell stage of a human embryo can each give rise to one of two identical twins. However, *when the cell is locked in at a particular place in the embryo, it becomes committed to starting the cell lineages that will form certain body parts in places where we expect those parts to form*. This style of cell determination predominates in echinoderm and vertebrate embryos.

A Theory of Pattern Formation

The sculpting of nondescript clumps of embryonic cells into specialized tissues and organs follows an ordered, spatial pattern. We call this activity **pattern formation**. As you have seen, local differences in the egg cytoplasm influence the formation of functional tissues and organs in all embryos. Later on, cells develop according to their

positions in the embryo. There, the cells interact by way of gene products that are sent, received, and used in prescribed sequences, at prescribed times. Increasingly diverse interactions complete the body plan.

Think about the preceding section, which described the sequence of gene interactions that map out the body plan of *Drosophila*. From studies of gene mutations in this organism and others, researchers have put together a theory of pattern formation. Here are its key points:

1. During development, classes of master genes are activated in orderly sequence, at prescribed times.

2. Regulatory proteins guide interactions among the master genes. The interactions result in the appearance of different gene products that are spatially organized relative to one another in the embryo.

3. Different genes are activated and suppressed in cells along the embryo's anterior-posterior axis and dorsal-ventral axis. Certain protein products of this selective gene expression create chemical gradients in the body of the embryo that help define each cell's identity.

4. **Homeotic genes** are a class of master genes that specify the development of specific body parts.

Embryonic Induction

A change in the developmental fate of an embryonic tissue, as brought about by exposure to a gene product released from an adjacent tissue, is called **embryonic induction**. Experimental modifications of a developing chick wing, as in Figure 44.13, yield examples. The chick wing forms, as a primordial bud, by way of mitotic cell divisions in mesoderm and ectoderm. The mesoderm induces the overlying ectodermal cells to elongate and form a ridge at its apex. This apical ectodermal ridge (AER) forms on the developing limbs of all birds and mammals. Surgically remove the AER before the bud grows fully, and further development ceases.

Rapid cell divisions in mesoderm beneath the AER generate the tissues that elongate the developing limb. As cells exit from this zone, they give rise to the tissues

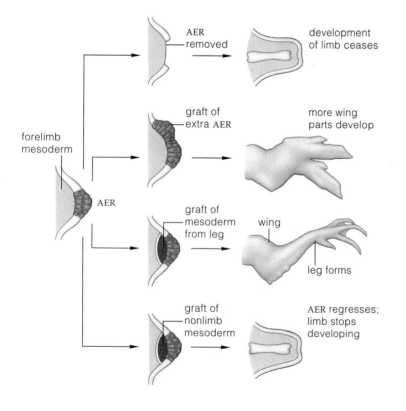

Figure 44.13 Some experiments that reveal an interaction between mesoderm and ectoderm during the formation of a chick wing. This mesoderm induces the overlying ectodermal cells to elongate and form a narrow ridge at its apex. An apical ectodermal ridge (AER) is a population of self-sustaining cells that do not mingle with surrounding cells. Remove AER when a wing bud is developing, and the bud will develop no more. Graft extra AER onto the bud, and the wing that forms has extra parts. Graft leg mesoderm just beneath the AER, and part of a leg (with toes) develops. Graft nonlimb mesoderm beneath the AER, and the AER regresses and wing development stops.

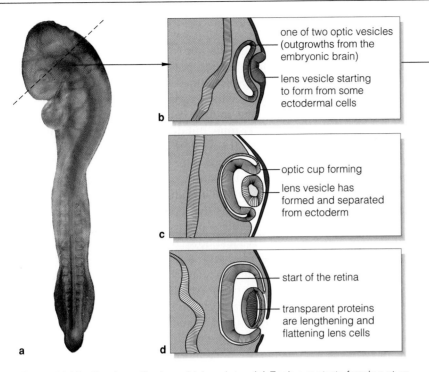

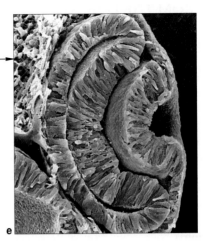

one of two optic vesicles (outgrowths from the embryonic brain)

lens vesicle starting to form from some ectodermal cells

b

optic cup forming

lens vesicle has formed and separated from ectoderm

c

start of the retina

transparent proteins are lengthening and flattening lens cells

a

d

e

Figure 44.14 Eye formation in a chick embryo. (**a**) Each eye starts forming at an optic vesicle, a fluid-filled pouch beneath ectoderm of the head. (**b**) When ectoderm cells come into contact with cells of the optic vesicle, they are induced to elongate, fold inward, and form a lens vesicle. (**c**) At the same time, the contact induces cells of the optic vesicle to sink inward, forming an optic cup from which most of the eyeball will form. (**d**) The cup's inner layer will produce the retina. (**e**) Scanning electron micrograph of a lens forming in the left optic cup of the embryo's head.

of the limb's outermost parts. Cells pushed out early on give rise to bones closest to the trunk of the body. Cells that depart much later are using their genes differently. They are biochemically different, and they respond to different signals. These cells give rise to the outermost bones and other structures of the limb.

A classic experiment by Hans Spemann, a pioneer in developmental biology, gave us the first stunning example of embryonic induction. In all vertebrates, the retina of the eye originates from the forebrain. The lens, which focuses light onto the retina, originates from the ectoderm (Figure 44.14). Spemann surgically removed retina-forming tissue from the forebrain region of a salamander embryo and inserted it into belly ectoderm. Afterward, a lens formed on the belly! As Spemann suspected, a molecular signal from forerunners of retina cells normally induces ectodermal cells to form a lens.

Today we know that slowly degradable proteins, now called **morphogens**, are among the gene products that form concentration gradients as they diffuse from an inducing tissue into adjoining tissues. The signals they represent are strongest at the start of the gradients, and they weaken with distance. Thus, cells at different positions along the concentration gradients are exposed to different chemical information, which guides selective expression of genes in different parts of the embryo.

Morphogens and other inducers switch on blocks of genes in sequence. Among the targets are homeotic genes, which occur in animals as evolutionarily distant as vertebrates and the roundworm *Caenorhabditis elegans*. While organs are first forming, the gene products interact with regulatory elements to activate or inhibit blocks of genes in similar ways among major animal groups. They direct the global inductive events that map out the body plan, including its major axes. Failure of induction when the organs are forming has global repercussions, as when a heart is constructed at the wrong location. After the basic plan is set, inductions are numerous—but localized. If a lens fails to form, for example, only the eye is affected.

It now appears that the first stages of development are open to evolutionary change, but that early formation of organs is not. This may be why we see no more than a few dozen body plans among all animals. We've known about the physical constraints (such as the surface-to-volume ratio) and architectural constraints (as imposed by the body's axes). Now we know of *phyletic* constraints on change, imposed on each lineage by master genes that operate at the time of early organ formation and govern induction of the basic body plan. Once inductive interactions have produced a structure, it is hard to start over again.

Pattern formation is the orderly, sequential sculpting of embryonic cells into specialized tissues and organs. Cells differentiate partly through cytoplasmic localization and, later, according to their position in the developing embryo.

Inductive interactions among classes of master genes map out the basic body plan and specify where and how specific body parts develop. Products of their selective expression create chemical gradients in the embryo that help seal each cell's developmental fate.

There are physical, architectural, and phyletic constraints on the evolution of animal body plans.

Post-Embryonic Development

What happens after an embryo is released or hatched from the parent? For many species, the new individual is a **juvenile**, a miniaturized version of the adult that simply changes in size and proportion until reaching sexual maturity. For others, **metamorphosis** occurs: the body form changes during the transition from embryo to adult. Metamorphosis requires growth in size, tissue reorganization, and remodeling of body parts, and it is under hormonal control. Section 37.8 gives examples of this transitional time for crustaceans and an insect.

For now, think back on Section 26.19, which showed examples of metamorphosis in insects. Larvae, nymphs, and pupae are examples of the transitional stages. Some insect nymphs are miniature adults. They go through episodes of **molting**, in which they shed a hardened cuticle and grow rapidly before the next cuticle hardens. Some insects show *incomplete* metamorphosis; they undergo gradual, partial change between the first and last molt. Bugs are like this. *Drosophila* is one of the insects that show *complete* metamorphosis. As described in Section 44.5, tissues of immature stages are destroyed and replaced before the adult emerges.

Similarly, frogs and other amphibians show striking and complete metamorphic changes that affect nearly all organs (Figure 44.1). In this case, thyroid hormones called thyroxine and triiodothyronine bring about the transformation of a tadpole into the adult. Increasing levels of these hormones cause some immature organs, such as those of the tail, to degenerate. Yet they bring about further development and differentiation of other organs, including the eye.

Aging and Death

Late in life, the body of all multicellular animals slowly deteriorates. Through processes collectively called **aging**, cells start to break down structurally and functionally, and this leads to structural changes and gradual loss of body functions. The introduction to Chapter 6 provides an example of aging in humans. All animals that show extensive cell differentiation undergo aging.

No one knows what causes aging, although research gives us interesting things to think about. For example, many years ago, Paul Moorhead and Leonard Hayflick cultured normal human embryonic cells. All of the cell lines proceeded to divide about fifty times, then the entire population died off. Hayflick took some cultured cells that were partway through the series of divisions and froze them for several years. Afterward, he thawed the cells and placed them in a culture medium. The cells proceeded to complete the cycle of fifty doublings—whereupon they all died on schedule.

Such experiments imply that normal types of cells have a limited division potential. That is, mitosis may be genetically programmed to decline at a certain time in the life cycle. This is certainly true of neurons. They stop undergoing mitotic cell division after birth, then deteriorate over time. But does the shutdown of mitosis *cause* aging or is it a *result* of aging?

DNA's capacity for self-repair may be gradually lost through an accumulation of gene mutations. As you read in the Chapter 6 introduction, free radicals—rogue molecular fragments that bombard DNA and the other biological molecules—are associated with aging. Also, researchers have just now correlated *Werner's syndrome*, an aging disorder, with a mutation in the gene coding for helicase. This enzyme is vital for DNA replication and repair operations. Whether by free radical attacks or by accumulated replication errors, *structural changes in DNA could endanger the production of functional enzymes and other proteins required for normal life processes.*

Think about how cells depend on smooth exchanges of materials between the cytoplasm and extracellular fluid. Also think about collagen, a structural component of bone and other connective tissues throughout the body. If something shuts down or mutates the collagen-encoding genes, the missing or altered gene product would affect the flow of oxygen, nutrients, hormones, and so forth to and from cells. The repercussions would ripple through the entire body. Finally, think about a deterioration in genes for membrane proteins that serve as self markers. If markers change, do T lymphocytes then perceive the body's own cells as foreign and attack them? If such autoimmune responses increase over time, they would invite the greater vulnerability to disease and stress associated with old age.

For humans, following a low-fat diet and aerobic, low-impact exercise program throughout adulthood is more likely to reduce some of the common effects of aging. Consistent exercise improves cardiovascular and respiratory functioning. It also minimizes bone loss in older adults by helping to maintain bone density.

After embryonic development, different animal groups show variation in the patterns by which they mature into the adult form.

Post-embryonic stages of some species resemble adults in miniature, in that they change little except in size and proportion before the adult form emerges.

Other species undergo metamorphosis. Post-embryonic stages undergo complete or partial changes in growth, tissue reorganization, and tissue remodeling before the adult emerges. Hormones control the changes.

All animals that show extensive cell differentiation also undergo a process of aging, which culminates in death.

Everything in the world dies, but we only know about it as a kind of abstraction. If you stand in a meadow, at the edge of a hillside, and look around carefully, almost everything you can catch sight of is in the process of dying, and most things will be dead long before you are. If it were not for the constant renewal and replacement going on before your eyes, the whole place would turn to stone and sand under your feet.

There are some creatures that do not seem to die at all; they simply vanish totally into their own progeny. Single cells do this. The cell becomes two, then four, and so on, and after a while the last trace is gone. It cannot be seen as death; barring mutation, the descendants are simply the first cell, living all over again. . . .

There are said to be a billion billion insects on the Earth at any moment, most of them with very short life expectancies by our standards. Someone estimated that there are 25 million assorted insects hanging in the air over every temperate square mile, in a column extending upward for thousands of feet, drifting through the layers of atmosphere like plankton. They are dying steadily, some by being eaten, some just dropping in their tracks, tons of them around the Earth, disintegrating as they die, invisibly.

Who ever sees dead birds, in anything like the huge numbers stipulated by the certainty of the death of all birds? A dead bird is an incongruity, more startling than an unexpected live bird, sure evidence to the human mind that something has gone wrong. Birds do their dying off somewhere, behind things, under things, never on the wing.

Animals seem to have an instinct for performing death alone, hidden. Even the largest, most conspicuous ones find ways to conceal themselves in time. If an elephant missteps and dies in an open place, the herd will not leave him there; the others will pick him up and carry the body from place to place, finally putting it down in some inexplicably suitable location. When elephants encounter the skeleton of an elephant in the open, they methodically take up each of the bones and distribute them, in a ponderous ceremony, over neighboring acres.

It is a natural marvel. All of the life on earth dies, all of the time, in the same volume as the new life that dazzles us each morning, each spring. All we see of this is the odd stump, the fly struggling on the porch floor of the summer house in October, the fragment on the highway. I have lived all my life with an embarrassment of squirrels in my backyard, they are all over the place, all year long, and I have never seen, anywhere, a dead squirrel.

I suppose that is just as well. If the Earth were otherwise, and all the dying were done in the open, with the dead there to be looked at, we would never have it out of our minds. We can forget about it much of the time, or think of it as an accident to be avoided, somehow. But it does make the process of dying seem more exceptional than it

really is, and harder to engage in at the times when we must ourselves engage.

In our way, we conform as best we can to the rest of nature. The obituary pages tell us of the news that we are dying away, while birth announcements in finer print, off at the side of the page, inform us of our replacements, but we get no grasp from this of the enormity of the scale. There are 4 billion of us on the Earth in this year, 1973, and all 4 billion must be dead, on a schedule, within this lifetime. The vast mortality, involving something over 50 million each year, takes place in relative secrecy. We can only really know of the deaths in our households, among our friends. These, detached in our minds from all the rest, we take to be unnatural events, anomalies, outrages. We speak of our own dead in low voices; struck down, we say, as though visible death can occur only for cause, by disease or violence, avoidably. We send off for flowers, grieve, make ceremonies, scatter bones, unaware of the rest of the 4 billion on the same schedule. All of that immense mass of flesh and bone and consciousness will disappear by absorption into the earth, without recognition by the transient survivors.

Less than half a century from now, our replacements will have more than doubled in numbers. It is hard to see how we can continue to keep the secret, with such multitudes doing the dying. We will have to give up the notion that death is a catastrophe, or detestable, or avoidable, or even strange. We will need to learn more about the cycling of life in the rest of the system, and about our connection in the process. Everything that comes alive seems to be in trade for everything that dies, cell for cell. There might be some comfort in the recognition of synchrony, in the information that we all go down together, in the best of company.

— LEWIS THOMAS, 1973

SUMMARY

1. For animals, sexual reproduction is the dominant reproductive mode. It requires specialized reproductive structures, control mechanisms, and forms of behavior that assist fertilization and support the offspring.

2. Among animals, embryonic development commonly proceeds through six stages:

a. Gamete formation, when oocytes (immature eggs) and sperm form and develop in reproductive organs. Molecular and structural components are stockpiled and become localized in different parts of the oocyte.

b. Fertilization, from the time a sperm penetrates an egg to the fusion of sperm and egg nuclei that results in a zygote (fertilized egg).

c. Cleavage, when mitotic cell divisions transform the zygote into smaller cells, or blastomeres. Cleavage does not increase the original volume of egg cytoplasm; it only increases the number of cells. It regionally divides the yolk, mRNAs, proteins, cytoskeletal elements, and other "maternal messages" among the new blastomeres, an outcome called cytoplasmic localization.

d. Gastrulation, when primary tissue layers (germ layers) form. Cell divisions, cell migrations, and other events lead to the formation of endoderm, ectoderm, and (in most species) mesoderm. All tissues of the adult body develop from primary tissue layers.

e. The onset of organ formation. The different organs start developing by a tightly orchestrated program of cell differentiation and morphogenesis.

f. Growth and tissue specialization, when organs enlarge overall and acquire their specialized chemical and physical properties. The maturation of tissues and organs continues into post-embryonic stages.

3. An embryo cannot develop properly unless each stage of development is successfully completed before the next stage begins.

4. In cell differentiation, a cell selectively uses certain genes and synthesizes proteins not found in other cell types. The outcome is subpopulations of specialized lineages of cells that differ from one another in their structure, biochemistry, and functioning.

5. Starting at the time of gastrulation, morphogenesis is a program by which the embryo changes in size, shape, and proportions. Cell divisions, active cell migrations, changes in cell size and shape, the growth and folding of tissues, and controlled cell death (by way of apoptosis) are involved.

6. Pattern formation is the emergence of the basic body plan and specific body parts in specific regions, in an orderly sequence. Cytoplasmic localization helps seal the fate of cells in this patterning. Among echinoderms and vertebrates, cell determination depends more on interactions among classes of master genes that produce certain products. The products are spatially organized and create chemical gradients that help seal the identity of each cell lineage in the embryo.

7. A change in the developmental fate of an embryonic cell lineage, as brought about by exposure to products released from an adjacent tissue, is called embryonic induction.

8. Following embryonic development, some animals grow directly into the adult form. Other animals show metamorphosis, in which reactivated growth and tissue reorganization produce larvae or some other immature forms before emergence of the sexually mature adult.

9. All animals that show extensive cell differentiation age; they undergo gradual changes in structure and a decline in efficiency.

Review Questions

1. What is the main benefit of sexual reproduction? What are some of its biological costs? *44.1*

2. Define the key events during gamete formation, fertilization, cleavage, gastrulation, and organ formation. At which stage is the frog embryo in the photograph at right? *44.2*

3. Does cleavage increase the volume of cytoplasm, the number of cells, or both, compared to the zygote? *44.2*

4. Define blastula. What is a blastocoel? *44.3*

5. Define cell differentiation and morphogenesis. *44.4*

6. Define cytoplasmic localization. At which stage does it occur? Does it play a larger role in cell determination in insects such as *Drosophila* than in vertebrates? *44.3, 44.5, 44.6*

7. Define pattern formation. Then explain the key points of the theory of pattern formation. *44.6*

Self-Quiz *(Answers in Appendix IV)*

1. Sexual reproduction among animals is _____ .
   a. biologically costly   c. evolutionarily advantageous
   b. diverse in its details   d. all of the above

2. Development cannot proceed properly unless each stage is successfully completed before the next begins, starting with
   _____ .
   a. gamete formation   d. gastrulation
   b. fertilization   e. organ formation
   c. cleavage   f. growth and tissue
                             specialization

3. During cleavage, the new blastomeres are allocated different regions of the fertilized egg's cytoplasm. This is called _____ .
   a. cytoplasmic localization   c. cell differentiation
   b. embryonic induction   d. morphogenesis

4. Primary tissue layers first appear _____ .
   a. in the egg cortex   c. in the gastrula
   b. during cleavage   d. in primary organs

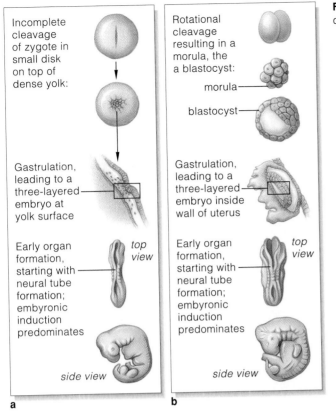

Incomplete cleavage of zygote in small disk on top of dense yolk:	Rotational cleavage resulting in a morula, the a blastocyst:
	morula
	blastocyst
Gastrulation, leading to a three-layered embryo at yolk surface	Gastrulation, leading to a three-layered embryo inside wall of uterus
Early organ formation, starting with neural tube formation; embyronic induction predominates	*top view*
	Early organ formation, starting with neural tube formation; embryonic induction predominates	*top view*
side view	*side view*

a   b

Figure 44.15 *Left*: Comparison of the pattern of embryonic development for (**a**) a chick and (**b**) a human.

Critical Thinking

1. Experimentally, it is possible to divide an amphibian egg so that the gray crescent is wholly within one of the two cells formed. If the two cells are separated from each other, only the cell with the gray crescent will form an embryo with a long axis, notochord, nerve cord, and back musculature. The other cell will form a shapeless mass of immature gut and blood cells. Reflect on Section 44.3. Does cytoplasmic localization or embryonic induction play a greater role in these outcomes?

2. Differences in the amount and distribution of yolk and other factors in the egg cytoplasm give rise to different cleavage patterns among animal groups. Think about the stages shown in Figure 44.15. Then speculate on how some of the differences might contribute to the pattern of incomplete cleavage of fertilized bird eggs (in this example, a *chick* egg) and to the pattern of complete, rotational cleavage of fertilized mammalian eggs (a human egg).

Also notice this: Despite obvious differences in the early stages of development, chick and human embryos resemble one another once early organ formation is under way. Using the theory of pattern formation, speculate on the role of embryonic induction in bringing about the similarities between these vertebrate lineages.

5. The astonishing internal complexity characteristic of most animals became possible following the evolution of _____ .
   a. ectoderm   c. myoderm
   b. mesoderm   d. endoderm

6. During development, the formation of subpopulations of different cell types is the outcome of _____ .
   a. selective gene expression   c. morphogenesis
   b. cell differentiation   d. a and b

7. As an embryo develops, tissues and organs change in size, shape, and proportions. This process is called _____ .
   a. gastrulation   c. morphogenesis
   b. pattern formation   d. metamorphosis

8. _____ is a program of orderly changes in an embryo's size, shape, and proportions that underlies organ formation.
   a. Gastrulation   c. Morphogenesis
   b. Pattern formation   d. Metamorphosis

9. In organisms ranging from fruit flies to humans, the body's organization relative to its long axis depends on _____ .
   a. morphogens   c. embryonic induction
   b. homeotic genes   d. all of the above

10. _____ refers to changes in the body form of transitional stages between the embryo and the adult.
    a. Gastrulation   c. Morphogenesis
    b. Pattern formation   d. Metamorphosis

11. Match the development stage with its description.
    ____ cleavage
    ____ gametogenesis
    ____ organ formation
    ____ growth, tissue specialization
    ____ gastrulation
    ____ fertilization

    a. egg and sperm mature in parents
    b. sperm nucleus, egg nucleus fuse
    c. formation of primary tissue layers
    d. cytoplasmic localization, not gene interactions, dominate in most animals (but not in mammals)
    e. organs, tissues increase in size, acquire specialized properties
    f. starts when primary tissue layers split into subpopulations of cells

Selected Key Terms

aging *44.7*
apoptosis *44.4*
asexual reproduction *44.1*
blastocoel *44.3*
blastomere *44.2*
blastula *44.3*
cell differentiation *44.4*
cleavage *44.2*
cleavage furrow *44.3*
cytoplasmic localization *44.3*
ectoderm *44.2*
embryo *44.2*
embryonic induction *44.6*
endoderm *44.2*
fate map *44.5*
fertilization *44.2*
gamete formation *44.2*
gastrulation *44.2*
gray crescent *44.3*

growth, tissue specialization *44.2*
homeotic gene *44.6*
identical twin *44.4*
internal fertilization *44.1*
juvenile *44.7*
mesoderm *44.2*
metamorphosis *44.7*
molting *44.7*
morphogen *44.6*
morphogenesis *44.4*
morula *44.3*
neural tube *44.4*
oocyte *44.3*
organ formation *44.2*
pattern formation *44.6*
sexual reproduction *44.1*
sperm *44.3*
yolk *44.1*
zygote *CI*

Readings

Caldwell, M. November 1992. "How Does a Single Cell Become a Whole Body?" *Discover* 13(11): 86–93.

Gilbert, S. 1994. *Developmental Biology*. Fourth edition. Sunderland, Massachusetts: Sinauer.

McGinnis, W., and M. Kuziora. February 1994. "The Molecular Architects of Body Design." *Scientific American* 270(2): 58–66.

Nusslein-Volhard, C. August 1996. "Gradients That Organize Embryonic Development. *Scientific American*, 54–61.

Raff, R., and T. Kaufman. 1983. *Embryos, Genes, and Evolution*. New York: Macmillan.

Web Site See *http://www.wadsworth.com/biology* for practice quiz questions, hypercontents, BioUpdates, and critical thinking. The Wadsworth Biology Resource Center provides a wealth of information fully organized and integrated by chapter.

45 HUMAN REPRODUCTION AND DEVELOPMENT

The Journey Begins

At first nothing appears to be happening to the egg, so recently fertilized in the billowing folds of an oviduct (Figure 45.1a). Before the clock ticks off twenty-four hours, however, a spectacular journey is under way.

Cleavage furrows herald the onset of a programmed series of cuts through the egg cytoplasm. Every twelve to twenty-four hours, a new cut carves up the cytoplasm until, by the fifth day, a fluid-filled ball of thirty-two tiny cells is tumbling along the moist, soft lining of the uterus. It is a type of blastula, an embryonic stage you read about in the preceding chapter. Its surface cells are organized as a thin layer that looks like the shell of a Ping-Pong ball. Inside, some cells huddle together against one part of the surface layer's wall. This is the inner cell mass, the forerunner of an adult body that will someday consist of trillions upon trillions of cells. Yet the whole blastula is smaller than the tip of a pin.

By week's end, molecular signals urge the blastula to anchor itself to the uterine lining and start burrowing into it. Soon afterward, some surface cells of the blastula send out fingerlike cytoplasmic projections. These grow into the domain of maternal blood vessels threading through the lining. The activity establishes functional connections with the mother's tissues that will metabolically support the developing individual through the months ahead.

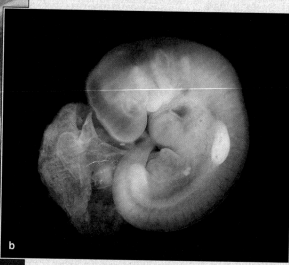

Figure 45.1 (**a**) Billowing folds on the inner, mucus-coated surface of an oviduct, part of a human female's reproductive system. An unfertilized egg, released earlier from a nearby ovary, was swept into the duct and was on its way to the uterus. A sperm entered the oviduct from the opposite direction. Somewhere in the mucosal folds, that sperm encountered and then penetrated the egg—and a remarkable developmental journey began. (**b**) A human embryo, as it appears just four weeks after the moment of fertilization.

Now, in an astonishing feat of self-organization, the inner cell mass transforms itself into a three-layered, oval disk. It does so as embryonic cells change in size and shape and migrate to new locations. Gradually, as an outcome of cell differentiation and morphogenesis, a pale, crescent-shaped embryo emerges. Day after day, the sculpting continues. Three weeks after fertilization the neural tube, forerunner of the central nervous system, starts forming. By the fourth week, tissue buds appear; they will develop into upper and lower limbs. A patch of cells in a tubular, primordial heart starts its rhythmic beating. From now on, through birth, adulthood, until the time of death, those cells will beat incessantly as the cardiac pacemaker.

By now, the developing embryo displays bilateral symmetry, cephalization, a tail, and other features that are the hallmarks of a vertebrate (Figure 45.1b). But is it a human? A minnow? A duckling? The answer does not become obvious until eight weeks into the journey, when the embryo is recognizably a human in the making. Later, in a body that is not much bigger than a peanut, complex cell interactions fill in the details of the body plan.

These events parallel your own beginning inside your mother. The story you are about to read is *your* story.

From the preceding chapter, you are acquainted with some principles that govern the development of animals in general. Now you will see how the principles apply to humans, starting with the formation of sperm and eggs in men and women. Gamete formation, remember, is the basis for sexual reproduction. And it is the first stage of animal development.

For both men and women, the reproductive system consists of a pair of gonads (the primary reproductive organs), accessory glands, and ducts. Male gonads are **testes** (singular, testis), and female gonads are **ovaries**. Testes produce sperm, and ovaries produce eggs. Both also secrete sex hormones that influence reproductive functions as well as the development of **secondary sexual traits**. Such traits are features that we associate with maleness and femaleness, although they do not play a direct role in reproduction.

Recall, from Section 12.3, that gonads have the same appearance in all early human embryos. After seven weeks of development, activation of genes on the sex chromosomes as well as hormone secretions trigger their development into testes or ovaries. As you will see, gonads and accessory organs are already formed at birth, but they do not reach their full size and become functional until the individual is a teenager.

KEY CONCEPTS

1. The human reproductive system consists of a pair of primary reproductive organs, or gonads, as well as a number of accessory glands and ducts. Testes are sperm-producing male gonads, and ovaries are egg-producing female gonads.

2. In response to signals from the hypothalamus and the pituitary gland, gonads also release sex hormones that orchestrate reproductive functions and the development of secondary sexual traits.

3. Human males continually produce sperm from puberty onward. The hormones testosterone, LH, and FSH control male reproductive functions.

4. Human females are fertile on a cyclic basis. Each month during their reproductive years, an egg is released from one of a pair of ovaries, and the lining of the uterus is primed for pregnancy. Estrogen, progesterone, FSH, and LH are the hormones that dominate this cyclic activity.

5. Human embryonic development starts with gamete formation and proceeds through fertilization, cleavage, gastrulation, organ formation, and growth and tissue specialization.

Figure 45.2 shows the organs of the male reproductive system and Table 45.1 lists their functions. Again, an adult male has a pair of gonads, of a type called testes. These are *primary* reproductive organs, the equivalent of ovaries in females. They produce sperm and the sex hormones that govern male reproductive functions as well as the development of *secondary* sexual traits. Such traits do not play a direct role in reproduction, but they are still associated in distinct ways with maleness (or femaleness). Examples are the amount and distribution of body fat, hair, and skeletal muscle.

Where Sperm Form

In an embryo that is destined to become male, a pair of testes form on the abdominal cavity wall. Before birth, his testes descend from the abdominal cavity into the scrotum, an outpouching of skin that hangs below the pelvic region. By the time of birth, testes are fully formed miniatures of the adult organs. They start producing sperm at puberty, the time of sexual maturation. Boys typically enter puberty between ages twelve and sixteen.

Figure 45.2a shows the position of the scrotum in an adult male. If sperm cells are to develop as they should, the temperature inside the scrotum must remain a few degrees cooler than the rest of the body's normal core temperature. A control mechanism, which operates by stimulating or inhibiting the contraction of smooth muscles in the scrotum's wall, helps assure that the internal temperature does not stray far from 95°F. When the air just outside the body becomes cold, contractions draw the pouch closer to the warmer body mass. When it is warm outside, muscles relax and thereby lower the pouch.

Packed inside each testis are a great number of small, highly coiled tubes, the **seminiferous tubules**. Sperm formation begins in the tubules, in the manner described in Section 45.2.

Where Semen Forms

Mammalian sperm travel from the testes through a series of ducts leading to the urethra. When they enter one of two long, coiled ducts, the sperm are not quite mature. These first ducts are the epididymides (singular, epididymis). Secretions from glandular cells of the duct wall trigger activities that put the finishing touches on sperm maturation. Until the time that fully mature sperm depart from the body, they are stored in the last stretch of each epididymis.

When the male is sexually aroused, contractions of muscles in the walls of the reproductive organs propel mature sperm into and through a pair of thick-walled tubes called the vas deferentia (singular, vas deferens).

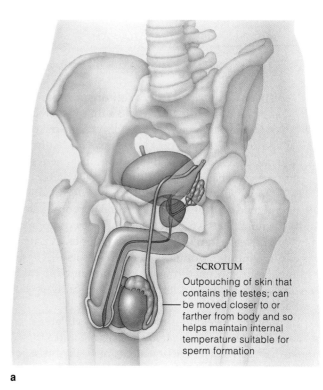

SCROTUM
Outpouching of skin that contains the testes; can be moved closer to or farther from body and so helps maintain internal temperature suitable for sperm formation

a

Figure 45.2 *Above and facing page*: Components of the reproductive system of the human male and their functions.

Table 45.1	Organs and Accessory Glands of the Male Reproductive Tract
REPRODUCTIVE ORGANS:	
Testis (2)	Production of sperm, sex hormones
Epididymis (2)	Sperm maturation site and sperm storage
Vas deferens (2)	Rapid transport of sperm
Ejaculatory duct (2)	Conduction of sperm to penis
Penis	Organ of sexual intercourse
ACCESSORY GLANDS:	
Seminal vesicle (2)	Secretion of large part of semen
Prostate gland	Secretion of part of semen
Bulbourethral gland (2)	Production of lubricating mucus

From there, contractions propel sperm through a pair of ejaculatory ducts, then on through the urethra. This last tube extends through the penis, the male sex organ, and opens at its tip. The urethra, recall, is a duct that also functions in urine excretion.

Glandular secretions become mixed with sperm as they travel to the urethra. The result is **semen**, a thick fluid that eventually is expelled from the penis during sexual activity. Early in the formation of semen, a pair

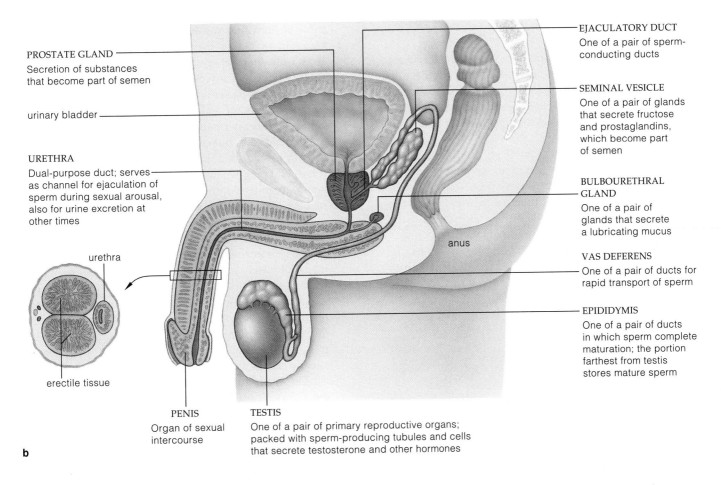

PROSTATE GLAND
Secretion of substances that become part of semen

urinary bladder

URETHRA
Dual-purpose duct; serves as channel for ejaculation of sperm during sexual arousal, also for urine excretion at other times

urethra

erectile tissue

PENIS
Organ of sexual intercourse

TESTIS
One of a pair of primary reproductive organs; packed with sperm-producing tubules and cells that secrete testosterone and other hormones

EJACULATORY DUCT
One of a pair of sperm-conducting ducts

SEMINAL VESICLE
One of a pair of glands that secrete fructose and prostaglandins, which become part of semen

BULBOURETHRAL GLAND
One of a pair of glands that secrete a lubricating mucus

anus

VAS DEFERENS
One of a pair of ducts for rapid transport of sperm

EPIDIDYMIS
One of a pair of ducts in which sperm complete maturation; the portion farthest from testis stores mature sperm

b

of seminal vesicles secrete fructose. The sperm use this sugar as an energy source. Seminal vesicles also secrete certain kinds of prostaglandins that can induce muscle contractions. Possibly these signaling molecules take effect during sexual activity. At that time they might induce contractions in the female's reproductive tract and thereby assist sperm movement through it.

The secretions from a prostate gland probably help buffer the acidic conditions that sperm must encounter within the female tract. The vaginal pH is about 3.5–4.0, but the motile capacity of sperm improves at pH 6. Two bulbourethral glands secrete some mucus-rich fluid into the urethra when the male is sexually aroused.

Cancers of the Prostate and Testis

Until recently, cancers of the male reproductive tract did not get much media coverage in the United States, even though *prostate cancer* alone kills 40,000 older men annually. This is surprising, given that the mortality rate for breast cancer, which is highly publicized, is not much higher. You also may be surprised to learn that *testicular cancer* is a frequent cause of death among young men. About 5,000 cases are diagnosed each year in the United States alone.

Both cancers are painless during their early stages. If they are not detected in time, they can spread silently into the lymph nodes of the abdomen, chest, neck, and eventually the lungs. Once the cancer has metastasized, prospects are not good. Testicular cancer, for example, kills as many as 50 percent of those stricken.

Doctors are able to detect prostate cancer in older men through physical examination and blood tests for a rise in prostate-specific antigen. In addition, once a month from high school onward, men should examine each testis after a warm bath or shower, when scrotal muscles are relaxed. The testis should be rolled gently between the thumb and the forefinger to check for any enlargement, hardening, or lump. Such changes might or might not cause noticeable discomfort. But they must be reported so that a physician can order a complete examination. Treatment of testicular cancer has one of the highest success rates, provided the cancer is caught before it starts to spread.

Human males have a pair of testes—primary reproductive organs that produce sperm and sex hormones—as well as accessory glands and ducts. The hormones influence sperm formation and the development of secondary sexual traits.

Sperm Formation

Each testis is about 5 centimeters (less than 2 inches) long. Even so, 125 meters of seminiferous tubules are packed inside it! As many as 300 wedge-shaped lobes, of the sort shown in Figures 45.3 and 45.4, partition its interior. Two or three coiled tubules press against each other inside each lobe.

Just inside each tubule's wall are undifferentiated cells called spermatogonia (singular, spermatogonium). Ongoing cell divisions force them away from the wall, toward the interior. During their forced departure, the cells undergo mitotic cell division. Their daughter cells, primary spermatocytes, are the ones that enter meiosis.

As Figure 45.4a indicates, the nuclear divisions are accompanied by *incomplete* cytoplasmic divisions. Thus the developing cells are interconnected by cytoplasmic

bridges. Ions and molecules required for development move freely across the bridges, so successive divisions from each spermatogonium produce clones of cells that mature in synchrony. **Sertoli cells**, the only other type of cell located in the seminiferous tubules, provide the cells with nourishment and molecular signals.

Secondary spermatocytes form during meiosis I. They are haploid cells, but each one of their chromosomes is in the duplicated state; it consists of two sister chromatids. (Here you may wish to review the general discussion of spermatogenesis in Section 10.5.) The sister chromatids of each chromosome separate from each other during meiosis II. The daughter cells are haploid spermatids. These gradually develop into sperm, the male gametes.

Each mature sperm is a flagellated cell with a core of microtubules in its long tail. An enzyme-containing cap, the acrosome, covers most of the head region. The head is mainly a DNA-packed nucleus (Figure 45.4b). During fertilization, enzymes of the cap itself help the sperm penetrate the extracellular material around an egg. In a midpiece just behind the head, arrays of mitochondria supply energy for the tail's whiplike movements.

The entire process of sperm formation takes nine to ten weeks. From puberty onward, human males produce sperm continuously, so that many millions of cells are in different stages of development on any given day.

Hormonal Controls

Coordinated secretions of LH, FSH, and testosterone govern reproductive function in males. Take a look at Figure 45.3b, which shows the location of **Leydig cells** in tissue between the lobes in testes. Leydig cells secrete **testosterone**, a steroid hormone. Testosterone is central to the growth, form, and functions of the reproductive tract in males. Besides having the main role in sperm formation, it stimulates sexual and aggressive behavior. It promotes development of secondary sexual traits in

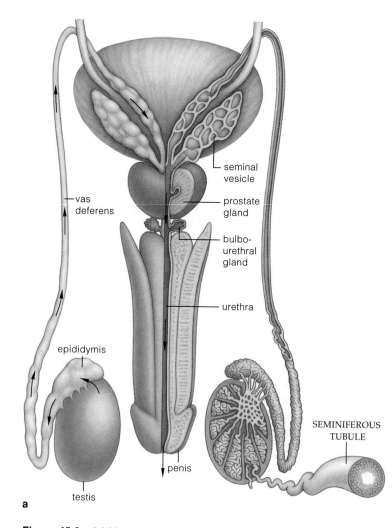

seminal vesicle

vas deferens

prostate gland

bulbo-urethral gland

urethra

epididymis

SEMINIFEROUS TUBULE

penis

testis

a

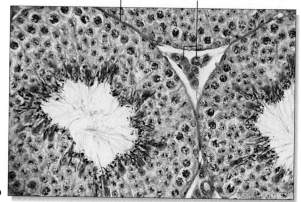

wall of seminiferous tubule   Leydig cells between tubules

b

Figure 45.3 (**a**) Male reproductive tract, posterior view. The arrows show the route that sperm take before ejaculation from a sexually aroused male. (**b**) Light micrograph of cells inside three adjacent seminiferous tubules, cross-section. Leydig cells occupy tissue spaces between the tubules.

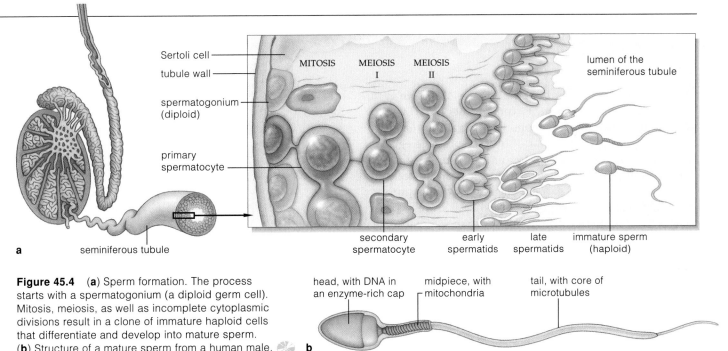

Sertoli cell
tubule wall
spermatogonium (diploid)
primary spermatocyte

MITOSIS   MEIOSIS I   MEIOSIS II

lumen of the seminiferous tubule

secondary spermatocyte   early spermatids   late spermatids   immature sperm (haploid)

a   seminiferous tubule

Figure 45.4 (**a**) Sperm formation. The process starts with a spermatogonium (a diploid germ cell). Mitosis, meiosis, as well as incomplete cytoplasmic divisions result in a clone of immature haploid cells that differentiate and develop into mature sperm. (**b**) Structure of a mature sperm from a human male.

head, with DNA in an enzyme-rich cap   midpiece, with mitochondria   tail, with core of microtubules

b

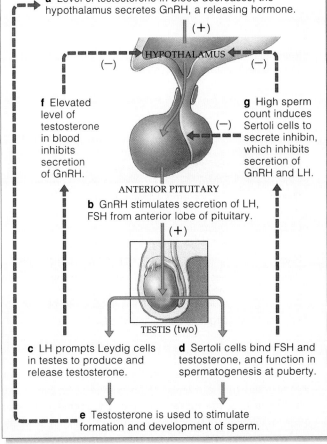

a Level of testosterone in blood decreases; the hypothalamus secretes GnRH, a releasing hormone.

(+)

HYPOTHALAMUS

(−)   (−)

f Elevated level of testosterone in blood inhibits secretion of GnRH.

(−)

g High sperm count induces Sertoli cells to secrete inhibin, which inhibits secretion of GnRH and LH.

ANTERIOR PITUITARY

b GnRH stimulates secretion of LH, FSH from anterior lobe of pituitary.

(+)

TESTIS (two)

c LH prompts Leydig cells in testes to produce and release testosterone.

d Sertoli cells bind FSH and testosterone, and function in spermatogenesis at puberty.

e Testosterone is used to stimulate formation and development of sperm.

Figure 45.5 Negative feedback loops to the hypothalamus and to the anterior lobe of the pituitary gland from the testes. Through these loops, excess testosterone production shuts off the mechanisms leading to its production. This helps maintain the testosterone level in amounts required for sperm formation.

males, including an obvious increase in the growth of facial hair and a deepening of the voice at puberty.

LH and **FSH**, remember, are secreted by the anterior lobe of the pituitary gland (Section 37.3). Initially, both hormones were named for their effects in females. (One abbreviation stands for *Luteinizing Hormone*, the other for *Follicle-Stimulating Hormone*.) Later on, researchers discovered that the molecular structure of LH and FSH is identical in both males and females.

The hypothalamus, a part of the forebrain, controls sperm formation by controlling secretion of LH, FSH, and testosterone. As Figure 45.5 shows, in response to low blood levels of testosterone and other factors, the hypothalamus secretes **GnRH**. This releasing hormone stimulates the anterior lobe of the pituitary gland to release LH and FSH, which have targets in the testes. LH prods Leydig cells in testes to secrete testosterone, which helps stimulate the formation and development of sperm. Sertoli cells have receptors for FSH, which is crucial to establishing spermatogenesis at puberty, but researchers don't know whether this hormone has roles in the normal functioning of mature human testes.

A high testosterone level in blood has an inhibitory effect on GnRH release. Also, when the sperm count is high, Sertoli cells release inhibin. This protein hormone acts on the hypothalamus and pituitary to inhibit the release of GnRH and FSH. In this way, feedback loops to the hypothalamus kick in, with a resulting decrease in testosterone secretion and sperm formation.

Sperm formation depends on the hormones LH, FSH, and testosterone. Negative feedback loops from the testes to the hypothalamus and pituitary gland control their secretion.

The Reproductive Organs

We turn now to the reproductive system of a human female. Figure 45.6 shows its components and Table 45.2 summarizes their functions. The female's *primary* reproductive organs, her pair of ovaries, produce eggs and secrete sex hormones. After an immature egg, or **oocyte**, is released from an ovary, it enters the adjacent entrance of an oviduct, part of which is shown in Figure 45.1*a*. A female has a pair of oviducts. Each serves as a channel to the **uterus**, a hollow, pear-shaped organ in which embryos can grow and develop.

The uterine wall is primarily a thick layer of smooth muscle called myometrium. As you will see, the wall's inner lining, the **endometrium**, is central to embryonic development. It consists of connective tissues, glands, and blood vessels. The narrowed portion of the uterus is the cervix. A muscular tube, the vagina, extends from the cervix to the surface of the body. The vagina receives sperm and functions as part of the birth canal.

At the body surface are external genitals (vulva) that include organs for sexual stimulation. Outermost are a pair of fat-padded skin folds, the labia majora. They enclose a smaller pair of skin folds (labia minora) that are highly vascularized but have no fatty tissue. The smaller folds partly enclose the clitoris, a sex organ that is sensitive to stimulation. At the body's surface, the urethra's opening is positioned about midway between the clitoris and the vaginal opening.

Overview of the Menstrual Cycle

Most mammalian females follow an *estrous* cycle. That is, they are fertile and in heat (sexually receptive to males) only at certain times of year. By contrast, female primates, including humans, follow a **menstrual cycle**. Such females are fertile intermittently, on a cyclic basis, and heat is not synchronized with their fertile periods. Said another way, female primates of reproductive age can become pregnant only at certain times of year, but they may be receptive to sex at any time.

During each menstrual cycle, an oocyte matures and escapes from an ovary. Also, the endometrium becomes primed to receive and nourish a forthcoming embryo *if* fertilization takes place. However, if the oocyte does not become fertilized, a total of four to six tablespoons of blood-rich fluid starts flowing out through the vaginal canal. The recurring blood flow is called menstruation. It means "there is no embryo at this time," and it marks the first day of a new cycle. The uterine nest is being sloughed off and is about to be constructed once again.

The events just sketched out proceed through three phases. The cycle starts with a **follicular phase**, a time of menstruation, endometrial breakdown and rebuilding,

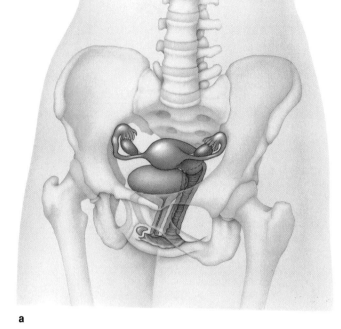

a

Figure 45.6   *Above and facing page*: Components of the human female reproductive system and their functions.

Table 45.2	Female Reproductive Organs
Ovaries	Oocyte production and maturation, sex hormone production
Oviducts	Ducts for conducting oocyte from ovary to uterus; fertilization normally occurs here
Uterus	Chamber in which new individual develops
Cervix	Secretion of mucus that enhances sperm movement into uterus and (after fertilization) reduces embryo's risk of bacterial infection
Vagina	Organ of sexual intercourse; birth canal

and maturation of an oocyte. The next phase, **ovulation**, is restricted to the release of an oocyte from the ovary. Finally, during the **luteal phase** of the menstrual cycle, an endocrine structure (corpus luteum) forms, and the endometrium is primed for pregnancy (Table 45.3).

All three phases are governed by feedback loops to the hypothalamus and pituitary gland from the ovaries. FSH and LH promote cyclic changes in the ovaries. As you will see, the FSH and LH also stimulate the ovaries to secrete sex hormones—**estrogens** and **progesterone**.

OVARY
One of a pair of primary reproductive organs in which oocytes (immature eggs) form and mature; produces hormones (estrogens and progesterone), which stimulate maturation of oocytes, formation of corpus luteum (a glandular structure), and preparation of the uterine lining for pregnancy

OVIDUCT
One of a pair of ciliated channels through which oocytes are conducted from an ovary to the uterus; usual site of fertilization

UTERUS
Chamber in which embryo develops; its narrowed-down portion (the cervix) secretes mucus that helps sperm move into uterus and that bars many bacteria

MYOMETRIUM
Thick muscle layers of uterus that stretch enormously during pregnancy

ENDOMETRIUM
Inner lining of uterus; site of implantation of blastocyst (early embryonic stage); becomes thickened, nutrient-packed, highly vascularized tissue during a pregnancy; gives rise to maternal portion of placenta, an organ that metabolically supports embryonic and fetal development

urinary bladder

urethra

opening of cervix

CLITORIS
Small organ responsive to sexual stimulation

LABIUM MINOR
One of a pair of inner skin folds of external genitals

LABIUM MAJOR
One of a pair of outermost, fat-padded skin folds of external genitals

anus

VAGINA
Organ of sexual intercourse; also serves as birth canal

b

Table 45.3	Events of the Menstrual Cycle	
Phase	Events	Days of Cycle*
Follicular phase	Menstruation; endometrium breaks down	1–5
	Follicle matures in ovary; endometrium rebuilds	6–13
Ovulation	Oocyte released from ovary	14
Luteal phase	Corpus luteum forms, secretes progesterone; the endometrium thickens and develops	15–28

* Assuming a 28-day cycle.

The secretions of FSH and LH also promote the cyclic changes in the structure of the endometrium.

A human female's menstrual cycles start between ages ten and sixteen. Each cycle lasts for about twenty-eight days, but this is merely the average. It runs longer for some women and shorter for others. The menstrual cycles continue until a woman is in her late forties or early fifties, when her supply of eggs is dwindling and hormonal secretions slow down. This is the onset of *menopause*, the twilight of reproductive capacity.

You may have heard about *endometriosis*, a condition that arises when endometrial tissue abnormally spreads and grows outside the uterus. Endometrial scar tissue may form on ovaries or oviducts and lead to infertility. In the United States, 10 million women may be affected annually. Possibly the condition arises when menstrual flow backs up through the oviducts and spills into the pelvic cavity. Or perhaps some embryonic cells became positioned in the wrong place before birth and were stimulated to grow during puberty, when sex hormones became active. Whatever the case, estrogen still acts on cells in the mislocated tissue. The resulting symptoms include pain during menstruation, sex, or urination.

Ovaries, the female primary reproductive organs, produce oocytes (immature eggs) and sex hormones. Endometrium lines the uterus, a chamber in which embryos develop.

Secretion of the sex hormones estrogen and progesterone is coordinated on a cyclic basis through the reproductive years.

A cycle starts with menstruation, breakdown and rebuilding of the endometrium, and maturation of an oocyte. After release of a mature egg (ovulation), it ends when a corpus luteum forms and the endometrium is primed for pregnancy.

Cyclic Changes in the Ovary

Take a moment to review Figure 8.9, the generalized picture of meiosis in an oocyte. A normal baby girl has about 2 million *primary* oocytes in her ovaries. By the time she is seven years old, about 300,000 remain; her body resorbed the rest. Primary oocytes already have entered meiosis I, but the nuclear division process was arrested in a genetically programmed way. Meiosis will resume in one oocyte at a time, starting with the first menstrual cycle. Only about 400 to 500 oocytes will be released during her reproductive years.

Figure 45.7 shows a primary oocyte located near the surface of an ovary. A layer of cells, the granulosa cells, surrounds and nourishes it. Together, a primary oocyte and the cell layer around it are a **follicle**. At the start of a menstrual cycle, the hypothalamus is secreting GnRH, which stimulates the anterior pituitary to release FSH and LH (Figure 45.8). In females, these hormones stimulate the follicle to grow. (FSH, remember, is the abbreviation for *F*ollicle-*S*timulating *H*ormone.)

The oocyte starts to increase in size, and more layers of cells form around it. Glycoprotein deposits build up between the oocyte and these cell layers. As they do, they widen the space between them. In time the deposits form a noncellular coating, the **zona pellucida**, around the oocyte.

FSH and LH cause cells outside the zona pellucida to secrete estrogens. An estrogen-containing fluid accumulates in the follicle, and estrogen levels in the blood start to rise. About eight to ten hours before its release from the ovary, the oocyte completes meiosis I. Then its cytoplasm divides, forming two cells. One cell, the **secondary oocyte**, ends up with nearly all the cytoplasm. The other cell is the first of three **polar bodies**. The meiotic parceling of chromosomes between the cells gives the secondary oocyte a haploid chromosome number, which is the exact number required for gametes and for sexual reproduction.

About halfway through the cycle, the pituitary gland detects the rising blood level of estrogens. It responds with a brief outpouring of LH. The LH triggers cellular contractions that cause the fluid-engorged follicle to balloon outward, then rupture. The fluid escapes and carries the secondary oocyte with it (Figure 45.7). *Thus the midcycle surge of LH triggers ovulation—the release of a secondary oocyte from the ovary.*

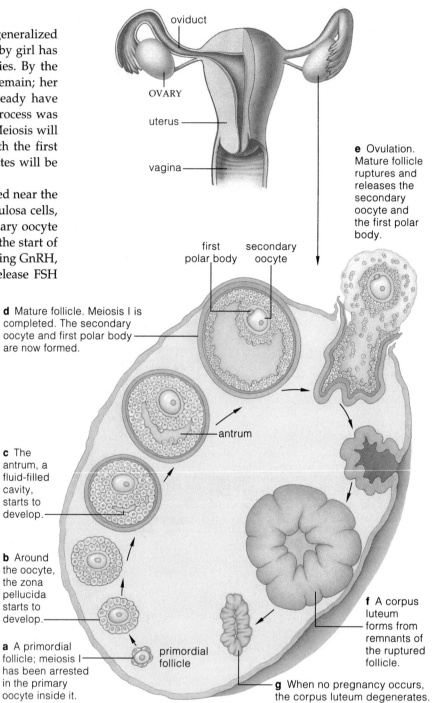

e Ovulation. Mature follicle ruptures and releases the secondary oocyte and the first polar body.

d Mature follicle. Meiosis I is completed. The secondary oocyte and first polar body are now formed.

first polar body   secondary oocyte

c The antrum, a fluid-filled cavity, starts to develop.

antrum

b Around the oocyte, the zona pellucida starts to develop.

a A primordial follicle; meiosis I has been arrested in the primary oocyte inside it.

primordial follicle

f A corpus luteum forms from remnants of the ruptured follicle.

g When no pregnancy occurs, the corpus luteum degenerates.

oviduct

OVARY

uterus

vagina

Figure 45.7 Cyclic events in a human ovary, cross-section.

A follicle at a given location in the ovary remains there all through the menstrual cycle. It does not move around as in this sketch, which only shows the *order* in which events occur. During the cycle's first phase, the follicle grows and matures. At ovulation, the second phase, the mature follicle ruptures and releases a secondary oocyte. During the third phase, a corpus luteum forms from the follicle's remnants. If pregnancy does not occur, the corpus luteum self-destructs.

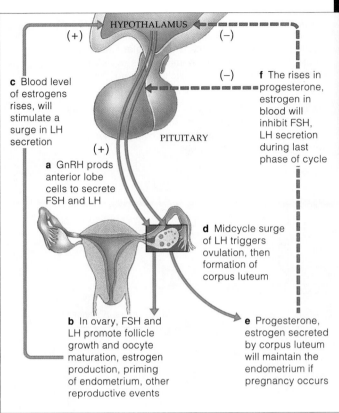

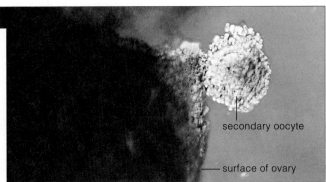

secondary oocyte

surface of ovary

c Blood level of estrogens rises, will stimulate a surge in LH secretion

f The rises in progesterone, estrogen in blood will inhibit FSH, LH secretion during last phase of cycle

a GnRH prods anterior lobe cells to secrete FSH and LH

d Midcycle surge of LH triggers ovulation, then formation of corpus luteum

b In ovary, FSH and LH promote follicle growth and oocyte maturation, estrogen production, priming of endometrium, other reproductive events

e Progesterone, estrogen secreted by corpus luteum will maintain the endometrium if pregnancy occurs

HYPOTHALAMUS

PITUITARY

Figure 45.8 Feedback control of hormonal secretion during a menstrual cycle. A positive feedback loop from an ovary to the hypothalamus causes a surge in LH secretion. The surge triggers ovulation. The light micrograph shows a secondary oocyte being released from an ovary at this time. Afterward, negative feedback loops to the hypothalamus and pituitary inhibit FSH secretion. They prevent another follicle from maturing until the cycle is completed.

Cyclic Changes in the Uterus

The estrogens released during the early phase of the menstrual cycle also help pave the way for pregnancy. They stimulate the growth of the endometrium and its glands. Just before the midcycle surge of LH, cells of the follicle start to secrete some progesterone as well as estrogens. Blood vessels grow rapidly in the thickened endometrium. Then, at the time of ovulation, estrogens act on tissue around the cervical canal, the narrowed-down portion of the uterus that leads to the vagina. The cervix now starts to secrete large quantities of a thin, clear mucus—an ideal medium for sperm travel.

Following ovulation, another structure dominates the cycle. The granulosa cells left behind in the follicle differentiate and form a yellowish glandular structure, the **corpus luteum**. (The name means "yellow body.") The corpus luteum forms as an outcome of the midcycle surge of LH. Hence the name *Luteinizing Hormone*.

The corpus luteum secretes progesterone and some estrogen. Progesterone prepares the reproductive tract for the arrival of a blastocyst, which is the mammalian blastula that forms from a fertilized egg (Section 44.3). For example, this hormone causes cervical mucus to get thick and sticky. The mucus may keep normal bacterial inhabitants of the vagina out of the uterus. Progesterone also maintains the endometrium during a pregnancy.

A corpus luteum persists for about twelve days. All the while, the hypothalamus is calling for minimal FSH secretion and so stops other follicles from developing. If a blastocyst does not burrow into the endometrium, the corpus luteum self-destructs during the last days of the cycle. It does so by secreting certain prostaglandins that apparently disrupt its own functioning.

After this, progesterone and estrogen levels in blood decline rapidly, and so the endometrium starts to break down. Deprived of oxygen and nutrients, the lining's blood vessels constrict and tissues die. Blood escapes as the walls of weakened capillaries start rupturing. The blood and the sloughed endometrial tissues make up a menstrual flow, which continues for three to six days. Then the cycle begins anew, with rising estrogen levels stimulating the repair and growth of the endometrium.

By menopause, the supply of oocytes is dwindling, hormone secretions slow down, and in time menstrual cycles (and fertility) will be over. The eggs that are still to be released late in a woman's life are at some risk of alterations in chromosome number or structure when meiosis resumes. A newborn with Down syndrome, as described in Section 12.9, is one possible outcome.

During a menstrual cycle, FSH and LH stimulate growth of an ovarian follicle (a primary oocyte and the surrounding layer of cells). The first meiotic cell division results in a secondary oocyte and the first polar body.

A midcycle surge of LH triggers ovulation, the release of the secondary oocyte and the polar body from the ovary.

Early on, estrogens call for endometrial repair and growth. Then estrogens and progesterone prepare the endometrium and other parts of the reproductive tract for pregnancy.

VISUAL SUMMARY OF THE MENSTRUAL CYCLE

By now, you probably have come to the conclusion that the menstrual cycle is not a simple tune on a biological banjo. It is a full-blown hormonal symphony!

Before continuing with your reading, take a moment to review Figure 45.9. It may leave you with a better understanding of the coordinated changes in hormone levels during each menstrual cycle, as correlated with the cyclic changes that they bring about in the ovary and uterus.

Figure 45.9 Changes that occur in the ovary and in the uterus, correlated with changing hormone levels during each turn of the menstrual cycle. *Green* arrows indicate which hormones dominate the cycle's first phase (the time when the follicle matures), then the second phase (when the corpus luteum forms).

(**a,b**) FSH and LH secretions bring about the changes in ovarian structure and function. (**c,d**) Estrogen and progesterone secretions from the ovary stimulate the changes in the endometrium.

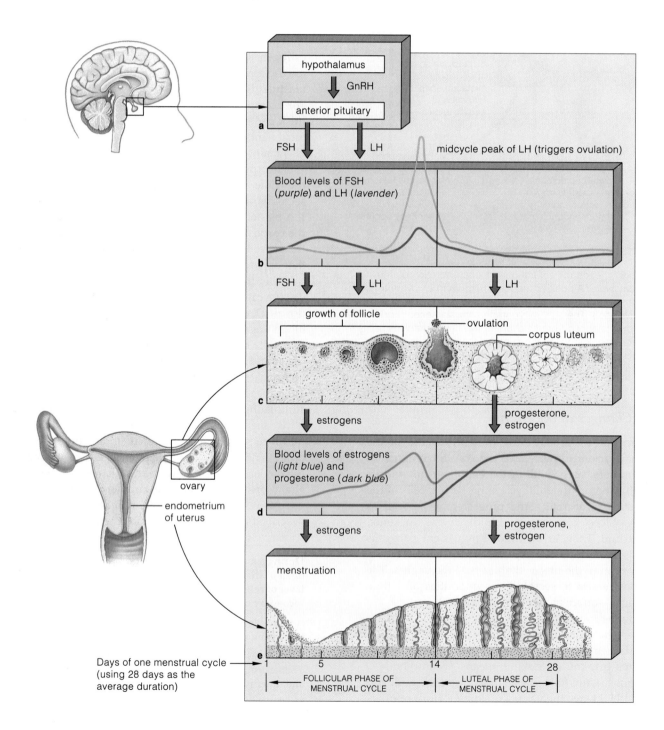

Sexual Intercourse

Suppose a secondary oocyte is on its way down an oviduct when a female and male are engaged in sexual intercourse, or **coitus**. The male sex act requires *erection*, whereby the normally limp penis stiffens and lengthens, and *ejaculation*, a forceful expulsion of semen into the urethra and out from the penis. As Figure 45.2 shows, the penis incorporates cylinders of spongy tissue. Many friction-activated sensory receptors occur on the glans penis, the mushroom-shaped tip of one cylinder. In sexually unaroused males, the large blood vessels that lead into the cylinders are vasoconstricted. In aroused males, these vasodilate, so blood flows into the cylinders faster than it flows out. Blood collects in the spongy tissue, and the engorgement stiffens and lengthens the organ, this being helpful for penetration into the vaginal canal.

During coitus, pelvic thrusts stimulate the penis as well as the female's clitoris and vaginal wall. The mechanical stimulation triggers involuntary contractions inside the male reproductive tract. They rapidly force sperm out of each epididymis. They force the contents of seminal vesicles and the prostate gland into the urethra. The resulting mixture, semen, is ejaculated into the vagina. During ejaculation, a sphincter closes off the neck of the bladder and thereby prevents urination.

Emotional intensity, hard breathing, heart pounding, as well as generalized skeletal muscle contractions accompany the rhythmic throbbing of the pelvic muscles. At **orgasm**, the end of the sex act, strong sensations of release, warmth, and relaxation dominate. Similar sensations typify female orgasm. It is a common misconception that a female cannot become pregnant if she doesn't reach orgasm. Don't believe it.

Fertilization

Now sperm are in the vagina. A single ejaculation can put 150 million to 350 million there. If they arrive a few days before or after ovulation or anytime in between, fertilization may be the outcome. Within thirty minutes after ejaculation, muscle contractions move the sperm deeper into the female reproductive tract. Only a few hundred sperm actually reach the upper portion of the oviduct, where fertilization usually takes place.

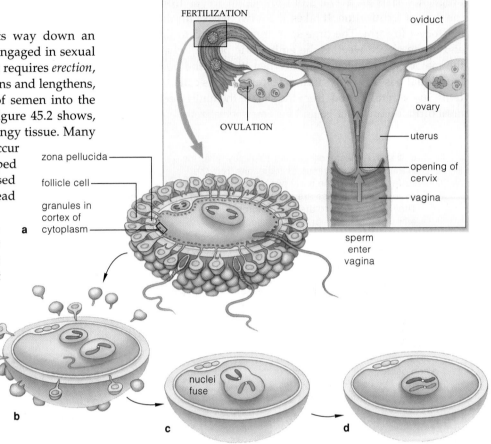

Figure 45.10 Fertilization. (**a**) Many sperm surround a secondary oocyte. Acrosomal enzymes clear a path through the zona pellucida. (**b**) When a sperm does penetrate the secondary oocyte, granules in the egg cortex release substances that make the zona pellucida impenetrable to other sperm. Penetration also stimulates meiosis II of the oocyte's nucleus. (**c**) The sperm's tail degenerates and its nucleus enlarges and fuses with the oocyte nucleus. (**d**) With fusion, fertilization is over. The zygote has formed.

The stunning micrograph that opens Unit II shows living sperm around a secondary oocyte. When sperm contact an oocyte, their cap releases acrosomal enzymes that clear a path through the zona pellucida (Figure 45.10). Although many sperm might get this far, usually only one fuses with the oocyte. Only its nucleus and centrioles do not degenerate in the oocyte's cytoplasm. Penetration induces the secondary oocyte and the first polar body to complete meiosis II. There are now three polar bodies and a mature egg, or **ovum** (plural, ova). As the sperm and egg nuclei fuse, their chromosomes restore the diploid number for a brand new zygote.

The intense physiological events that accompany coitus have one function: to put sperm on a collision course with an egg. Fertilization is over with the fusion of a sperm nucleus and egg nucleus, which results in a diploid zygote.

Pregnancy lasts an average of thirty-eight weeks from the time of fertilization. It takes about two weeks for the blastocyst to form. The time span from the third to the end of the eighth week is the **embryonic** period, when the major organ systems form. When it ends, the new individual has distinctly human features and is called a **fetus**. In the **fetal period**, from the start of the ninth week until birth, organs enlarge and become specialized.

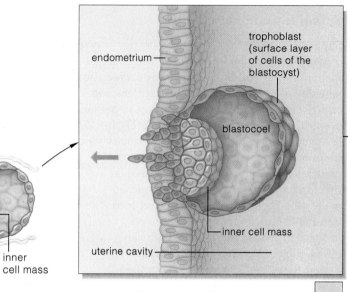

d DAY 5. A blastocoel (a fluid-filled cavity) forms in the morula as a result of secretions from the surface cells. By the thirty-two-cell stage, cells of an inner cell mass are already differentiating. They will give rise to the embryo proper. This embryonic stage is called a blastocyst.

c DAY 4. By 96 hours, there is a ball of sixteen to thirty-two cells that is shaped like a mulberry. It is a morula (after *morum*, Latin for mulberry). Cells of the surface layer will function in implantation and will give rise to a membrane, the chorion.

b DAY 3. After the third cleavage, the cells suddenly huddle together into a compacted ball, which becomes stabilized by numerous tight junctions among the outer cells. Gap junctions form among the interior cells and enhance intercellular communication.

a DAYS 1–2. Cleavage begins within 24 hours after fertilization. The first cleavage furrow extends between the two polar bodies. Subsequent cuts are rotational, so the resulting cells are not symmetrically arranged (compare Section 44.3). Until the eight-cell stage forms, the cells are loosely arranged, with considerable space between them.

e DAYS 6–7. Surface cells of the blastocyst attach to the endometrium and start to burrow into it. Implantation is under way.

actual size

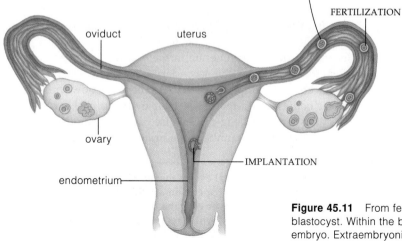

We typically call the first three months of pregnancy the *first* trimester. The *second* trimester extends from the start of the fourth month to the end of the sixth. The *third* trimester extends from the seventh month until birth. Beginning with Figure 45.11, the next series of illustrations show the characteristic features of the new individual at progressive stages of development.

Cleavage and Implantation

Three to four days after fertilization, the zygote is already in the oviduct, and cleavage is under way. Genes are already being expressed; the early divisions depend on their products. At the eight-cell stage, the cells huddle into a compacted ball. By the fifth day, there is a surface layer of cells (the trophoblast), a cavity filled with their secretions (blastocoel), and a cluster of interior cells (the inner cell mass). These are the defining features of the embryonic stage called the **blastocyst** (Figure 45.11*d*).

Six or seven days after fertilization, **implantation** is under way. By this process, the blastocyst adheres to the uterine lining, some of its cells send out projections that invade the mother's tissues, and connections start forming that will metabolically support the developing embryo through the months ahead. While the invasion is proceeding, the inner cell mass develops into two cell

Figure 45.11 From fertilization through implantation. Early on, cleavage produces the blastocyst. Within the blastocyst, the inner cell mass gives rise to the disk-shaped early embryo. Extraembryonic membranes (the amnion, chorion, and yolk sac) start forming.

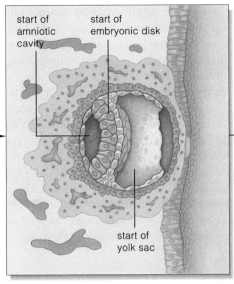

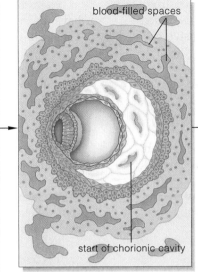

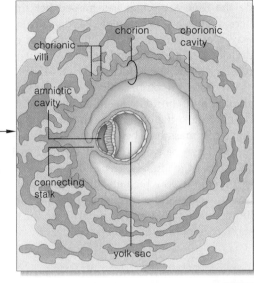

f DAYS 10–11. The yolk sac, embryonic disk, and amniotic cavity have started to form from parts of the blastocyst.

actual size

g DAY 12. Blood-filled spaces form in maternal tissue. The chorionic cavity starts to form.

actual size

h DAY 14. A connecting stalk has formed between the embryonic disk and chorion. Chorionic villi, which will be features of a placenta, start to form.

actual size

layers of a flattened, somewhat circular shape. Together, they represent the embryonic disk—and in short order they will give rise to the embryo proper.

Extraembryonic Membranes

As implantation progresses, membranes start to form outside the embryo. First a fluid-filled, *amniotic* cavity opens up between the embryonic disk and part of the blastocyst's surface (Figure 45.11*f*). Then cells migrate around the wall of the cavity and form the **amnion**, a membrane that will enclose the embryo. Fluid inside the cavity will function as a buoyant cradle where the embryo can grow, move freely, and be protected from abrupt temperature changes and mechanical impacts.

While the amnion forms, other cells migrate around the inner wall of the blastocyst's first cavity. They form a lining that becomes the **yolk sac**. This extraembryonic membrane speaks of the evolutionary heritage of land vertebrates (Section 27.7). For most animal species that produce shelled eggs, the sac holds nutritive yolk. In humans, some of the yolk sac becomes a site of blood cell formation, and some will give rise to germ cells, the forerunners of gametes.

Before the blastocyst is fully implanted, spaces open in maternal tissues and fill with the blood seeping in from ruptured capillaries. Inside the blastocyst, another cavity opens around the amnion and yolk sac. Fingerlike projections start to form on the cavity's lining, which is the **chorion**. This new membrane will become part of a spongy, blood-engorged tissue called the placenta.

After the blastocyst is completely implanted, another extraembryonic membrane will form as an outpouching of the yolk sac. This third membrane will become the **allantois**. An allantois has different functions in different animal groups. Among reptiles, birds, and some of the mammals, an allantois serves in respiration and storage of metabolic wastes. In humans, the urinary bladder as well as blood vessels for the placenta form from it.

One more point should be made here. Cells of the blastocyst secrete the hormone **HCG** (*Human Chorionic Gonadotropin*), which stimulates the corpus luteum to keep on secreting progesterone and estrogen. Thus the blastocyst itself prevents menstrual flow and works to avoid being sloughed off until the placenta takes over the task, some eleven weeks later. By the start of the third week, HCG can be detected in the mother's blood or urine. At-home *pregnancy tests* use a treated "dip-stick" that changes color when HCG is present in urine.

A human blastocyst consists of a surface layer of cells around a fluid-filled cavity (blastocoel) and an inner cell mass, which will give rise to the embryo proper.

Six or seven days after fertilization, the blastocyst implants itself in the endometrium. Projections from its surface invade maternal tissues, and connections start to form that in time will metabolically support the developing embryo.

Some parts of the blastocyst give rise to an amnion, yolk sac, chorion, and allantois. These extraembryonic membranes have different roles. But together they are vital for the structural and functional development of the embryo.

EMERGENCE OF THE VERTEBRATE BODY PLAN

By the time a woman has missed her first menstrual period, cleavage is completed and gastrulation is under way. **Gastrulation**, recall, is the stage of extensive cell divisions, migrations, and rearrangements by which the primary tissue layers form (Section 44.4).

By now, the embryonic disk is surrounded by the amnion and chorion, except at the point where a stalk connects it to the chorion's inner wall. Before then, the inner cell mass had behaved *as if* it were still perched on the large ball of yolk that evolved in the reptilian ancestors of mammals. Hence the flattened, two-layered embryonic disk, which also forms in reptiles and birds. Cell divisions and migrations of one layer produced the lining of the yolk sac. The other layer, the epiblast, now starts to generate the embryo proper.

For example, by the eighteenth day, two neural folds have appeared on the embryonic disk (Figure 45.12*b*). They will merge to form the neural tube. Early in the third week, an allantois, a sausage-shaped outpouching, appears on the yolk sac (Greek *allas*, meaning sausage). In humans, remember, the allantois serves only in the formation of the urinary bladder and of blood vessels for the placenta. Some mesoderm folds into a tube that becomes the notochord. A human notochord is only a structural framework; bony segments of the vertebral column soon form around it. Toward the end of the third week, some mesoderm gives rise to the **somites**. These paired segments are the embryonic source of most bones and skeletal muscles of the head and trunk, as well as the dermis overlying all of those regions. Pharyngeal

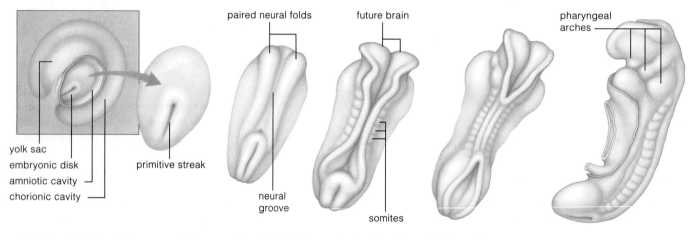

a DAY 15. A faint band appears around a depression along the axis of the embryonic disk. This is the primitive streak, and it marks the onset of gastrulation in vertebrate embryos.

b DAYS 18–23. Cell divisions and migrations, tissue folding, and other events of morphogenesis lead to early organ formation. Neural folds will merge to form the neural tube. Somites (bumps of mesoderm) appear near the embryo's dorsal surface. They will give rise to most of the skeleton's axial portion, skeletal muscles, and much of the dermis.

c DAYS 24–25. By now, some embryonic cells have given rise to pharyngeal arches. These will contribute to the formation of the face, neck, mouth, nasal cavities, larynx, and pharynx.

Figure 45.12 Hallmarks of the embryonic period of humans and other vertebrates. A primitive streak appears, then a notochord. Neural folds appear, then somites and pharyngeal arches. These are dorsal views (of the embryo's back). Compare the diagrams with the photographs in Figure 45.14.

Through cell divisions and migrations, the midline of that layer thickens faintly around a depression at its surface. This "primitive streak" lengthens and thickens further the next day. It marks the onset of gastrulation (Figure 45.12*a*). It also defines the anterior-posterior axis of the embryo and, in time, its bilateral symmetry. Cells migrating inward at this site give rise to endoderm and mesoderm. Now pattern formation begins, as described in Section 44.6. Interactions among the classes of master genes map out the basic body plan characteristic of all vertebrates. Through embryonic inductions, specialized tissues and organs start forming in orderly sequence in prescribed parts of the embryo.

arches start to form. They will help form the face, neck, mouth, nose, larynx, and pharynx. Small spaces open up in parts of the mesoderm. In time, all of these spaces will interconnect and form the coelomic cavity.

During the third week after fertilization, a time when the woman has missed her first menstrual period, the basic vertebrate body plan emerges in the new individual.

A primitive streak, neural tube, somites, and pharyngeal arches form during the embryonic period of all vertebrates. Formation of the primitive streak establishes the body's anterior-posterior axis and its bilateral symmetry.

ON THE IMPORTANCE OF THE PLACENTA

Even before the onset of the embryonic period, the extraembryonic membranes have been collaborating with the uterus to sustain the embryo's rapid growth. By the third week, tiny fingerlike projections from the chorion have grown profusely into the maternal blood that has pooled in endometrial spaces. The projections, the chorionic villi, enhance the exchange of substances between the mother and the new individual. They are functional components of the **placenta**. This blood-engorged organ consists of both endometrial tissue and extraembryonic membranes. At full term, the placenta will make up a fourth of the inner surface of the uterus (Figure 45.13).

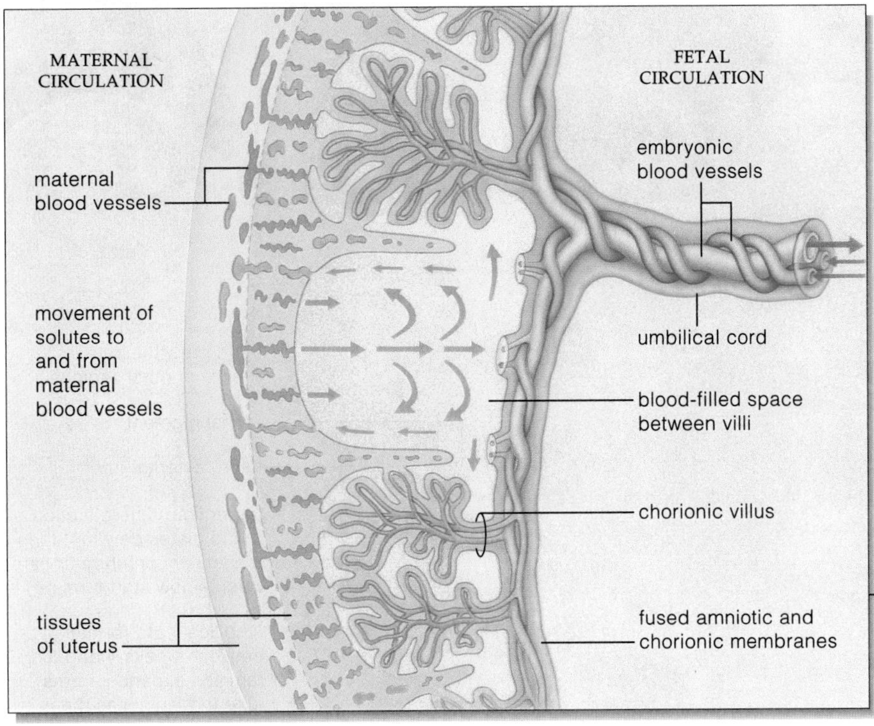

4 weeks

8 weeks

12 weeks

appearance of the placenta at full term

MATERNAL CIRCULATION

FETAL CIRCULATION

maternal blood vessels

embryonic blood vessels

movement of solutes to and from maternal blood vessels

umbilical cord

blood-filled space between villi

chorionic villus

tissues of uterus

fused amniotic and chorionic membranes

a

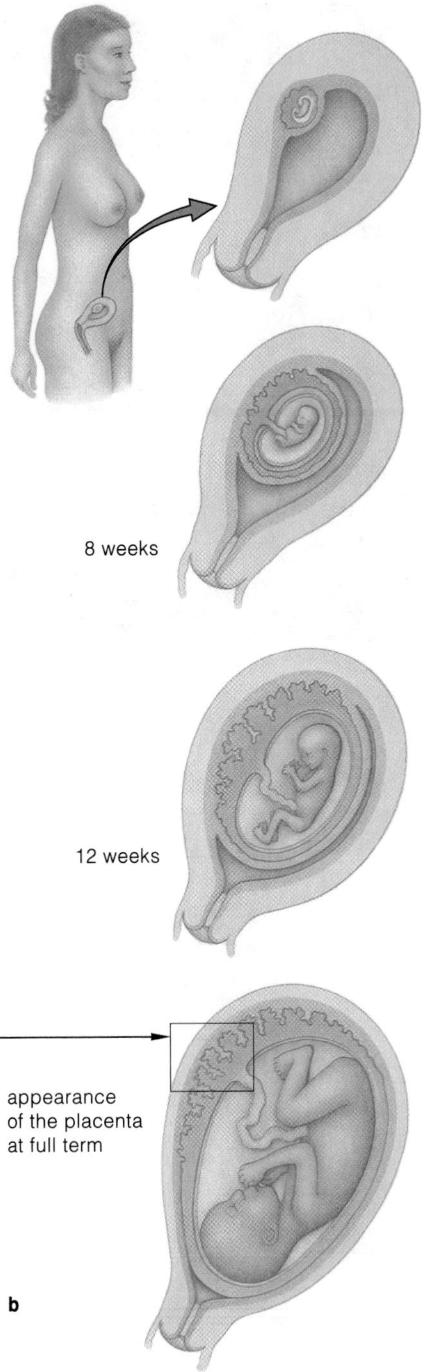

b

Figure 45.13 (**a**) Relationship between fetal and maternal blood circulation in a full-term placenta (**b**). Blood vessels extend from the fetus, through the umbilical cord, and into chorionic villi. Maternal blood spurts into spaces between the villi. Oxygen, carbon dioxide, and other small solutes diffuse across the placental membrane surface. There is no gross intermingling of the two bloodstreams.

The placenta is the body's way of sustaining the new individual while allowing its blood vessels to develop apart from the mother's. Oxygen and nutrients diffuse out of the maternal blood vessels, across the placenta's blood-filled spaces, then into embryonic blood vessels. (These vessels converge in the umbilical cord, the lifeline between the placenta and the new individual.) Carbon dioxide and other wastes diffuse in the other direction. The mother's lungs and kidneys quickly dispose of them.

After the third month, the placenta secretes progesterone and estrogens—and so maintains the uterine lining.

The placenta is a blood-engorged organ of endometrial and extraembryonic membranes. It permits exchanges between the mother and the new individual without intermingling their bloodstreams. Thus it sustains the individual and allows its blood vessels to develop apart from the mother's.

EMERGENCE OF DISTINCTLY HUMAN FEATURES

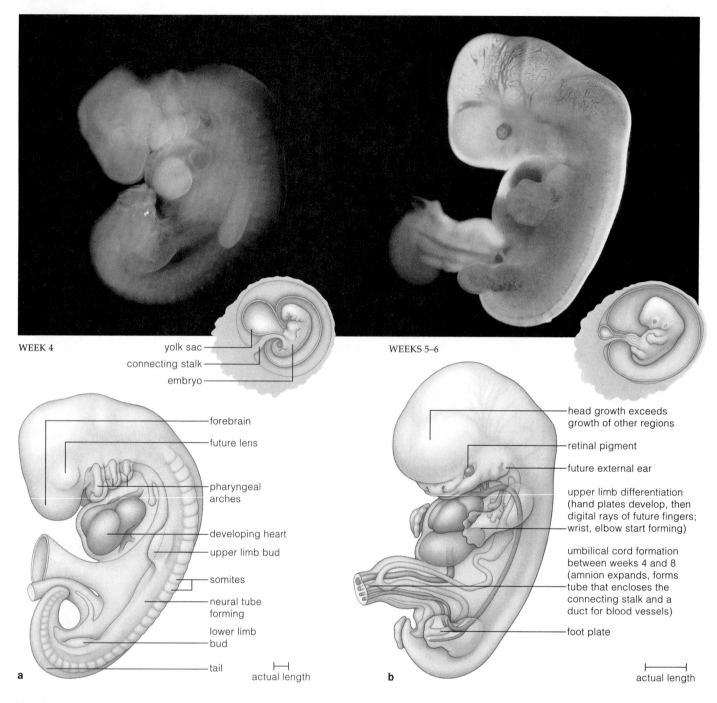

WEEK 4

yolk sac
connecting stalk
embryo

WEEKS 5–6

forebrain

future lens

pharyngeal arches

developing heart

upper limb bud

somites

neural tube forming

lower limb bud

tail

a

actual length

head growth exceeds growth of other regions

retinal pigment

future external ear

upper limb differentiation (hand plates develop, then digital rays of future fingers; wrist, elbow start forming)

umbilical cord formation between weeks 4 and 8 (amnion expands, forms tube that encloses the connecting stalk and a duct for blood vessels)

foot plate

b

actual length

By the end of the fourth week of the embryonic period, the embryo has grown to 500 times its original size. The placenta has been sustaining the growth spurt, but now the pace slows as details of organs fill in. Limbs form, fingers and toes emerge from embryonic paddles, and the umbilical cord forms. The circulatory system becomes intricate. Growth of the all-important head now surpasses that of any other body region (Figure 45.14). The embryonic period ends as the eighth week closes. No longer is

Figure 45.14 (**a**) Human embryo four weeks after fertilization. Like all vertebrates, it has a tail and pharyngeal arches. (**b**) Embryo at five to six weeks. (**c**) Embryo at the boundary between the embryonic and fetal periods. It now has human features. It floats in amniotic fluid. The chorion covers the amniotic sac but has been pulled aside here. (**d**) Fetus at sixteen weeks. Movements begin as nerves make functional connections with forming muscles. Legs kick, arms wave, fingers grasp, the mouth puckers. These reflex actions will be vital skills in the world outside the uterus.

the embryo merely "a vertebrate." By now its features clearly define it as a human fetus.

During the second trimester, the fetus is moving facial muscles. It frowns; it squints. It busily practices

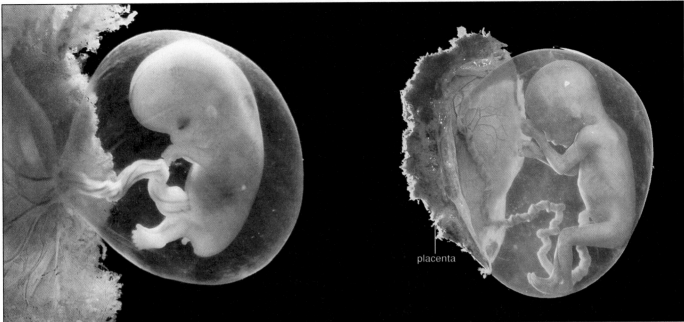
placenta

WEEK 8

final week of embryonic
period; embryo looks
distinctly human
compared to other
vertebrate embryos

upper and lower limbs well
formed; fingers and then
toes have separated

primordial tissues of
all internal, external
structures now developed

tail has become stubby

c |—— actual length ——|

WEEK 16 ——
Length: 16 centimeters
           (6.4 inches)
Weight: 200 grams
           (7 ounces)

WEEK 29
Length: 27.5 centimeters
           (11 inches)
Weight: 1,300 grams
           (46 ounces)

WEEK 38 (full term) ——→
Length: 50 centimeters
           (20 inches)
Weight: 3,400 grams
           (7.5 pounds)

During fetal period, length
measurement extends
from crown to heel (for
embryos, it is the longest
measurable dimension, as
from crown to rump).

d

the sucking reflex, as shown in the next section. Now the mother can easily sense movements of the fetal arms and legs. When the fetus is five months old, she can hear its heart through a stethoscope positioned on her abdomen. Soft, fuzzy hair (the lanugo) covers the fetal body. The skin is wrinkled, reddish, and protected from abrasion by a thick, cheesy coating. In the sixth month, delicate eyelids and eyelashes form. During the seventh month, the eyes open.

A fetus born prematurely before twenty-two weeks have passed cannot survive. The situation also is grave for births before twenty-eight weeks, mainly because lungs have not developed sufficiently. Even with the best medical care, premature infants have difficulty breathing and maintaining a normal core temperature. By the ninth month, however, the rate of survival has increased to 95 percent.

During the fetal period, primordial tissues that formed in the early embryo become sculpted into distinctly human features.

MOTHER AS PROVIDER, PROTECTOR, POTENTIAL THREAT

A woman who decides to become pregnant is committing a large part of her body's resources and functions to the development of a new individual. From fertilization until birth, her future child is absolutely at the mercy of her diet, health habits, and life-style (Figure 45.15).

SOME NUTRITIONAL CONSIDERATIONS How does a pregnant woman best provide nutrients for the embryo, then the fetus? The same balanced diet that is good for her should provide her future child with all of the required carbohydrates, lipids, and proteins. (Chapter 42 is one starting point for an understanding of human nutritional requirements.) The woman's own demands for vitamins and minerals increases during pregnancy. The placenta absorbs enough for her embryo from the bloodstream, except for folate (folic acid). She can reduce the risk that her embryo will develop severe neural tube defects by taking more B-complex vitamins (under supervision by her doctor) before conception and during early pregnancy. Besides this, one study showed that women who smoked while pregnant had depressed blood levels of vitamin C even when their vitamin C intake was identical to that of a

control group. *And so did their fetuses.* Smoking may affect utilization of other nutrients as well.

A pregnant woman also must eat enough so that her body weight increases by between 20 and 25 pounds, on the average. If her weight gain is a great deal lower than that, she is stacking the deck against the fetus. Compared to newborns of normal weight, significantly underweight newborns have more postdelivery complications. They are also at greater risk of having impaired mental functions later in life.

As birth approaches, a fetus makes greater nutritional demands of the mother. Clearly her diet will profoundly influence the remaining developmental events. The brain, like most of the other fetal organs, is especially vulnerable in the weeks just before and after birth, when it undergoes its greatest expansion. By now, all of its neurons have formed. Poor nutrition now will have repercussions on intelligence and other neural functions later in life.

RISK OF INFECTIONS The antibodies circulating in a pregnant woman's blood continually move across the placenta. They protect the new individual from all but

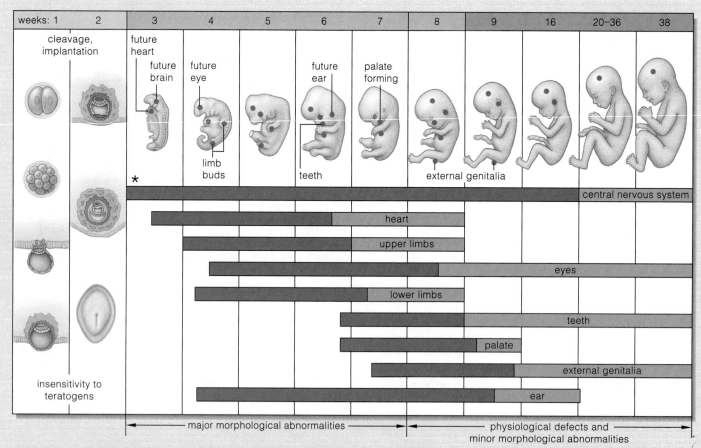

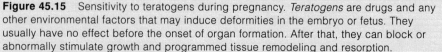

Figure 45.15 Sensitivity to teratogens during pregnancy. *Teratogens* are drugs and any other environmental factors that may induce deformities in the embryo or fetus. They usually have no effect before the onset of organ formation. After that, they can block or abnormally stimulate growth and programmed tissue remodeling and resorption.

★ *dark blue* denotes highly sensitive period; *light blue* denotes less severe sensitivity to teratogens

the most serious bacterial infections. Some virus-induced diseases can be dangerous during the first six weeks after fertilization, which is a critical time of organ formation. Suppose a pregnant woman contracts rubella (German measles) during this critical period. There is a 50 percent chance that some organs of her embryo will not form properly. For example, if she becomes infected when embryonic ears are forming, her newborn may be deaf. If she becomes infected during or after the fourth month of pregnancy, the disease will have no notable effect. However, a woman can avoid this risk entirely, because vaccination before pregnancy can prevent rubella.

EFFECTS OF PRESCRIPTION DRUGS A pregnant woman absolutely should not take any drugs unless she is under close medical supervision. Think about what happened when the tranquilizer *thalidomide* was being routinely prescribed in Europe. Women who had used thalidomide during the first trimester gave birth to infants who were either missing arms and legs or had grossly deformed ones. As soon as its connection with the deformities was apparent, thalidomide was withdrawn from the market.

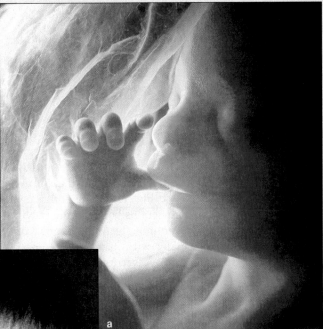

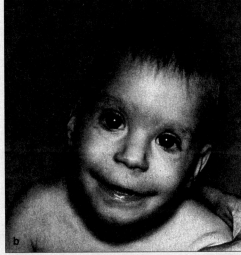

However, other tranquilizers, sedatives, and barbiturates are still being prescribed. There is a risk that they may cause similar, although less severe, damage. Even certain *anti-acne drugs* increase the risk of facial and cranial deformities. Or consider two overprescribed antibiotics. One of these, tetracycline, yellows teeth. Streptomycin causes hearing problems and may adversely affect the nervous system.

Figure 45.16 (**a**) The fetus at eighteen weeks. (**b**) An infant with fetal alcohol syndrome, or FAS. Obvious symptoms are low and prominent ears, improperly developed cheekbones, as well as an abnormally wide, smooth upper lip. The child can expect growth problems and abnormalities of the nervous system. This disorder currently affects about 1 in 750 newborns in the United States.

EFFECTS OF ALCOHOL As a fetus grows, its physiology becomes increasingly like the mother's. Alcohol passes freely across the placenta. Even one drink has the same effects on the fetus that it has on her. Excessive alcohol intake during pregnancy invites a set of deformities called *fetal alcohol syndrome*, or FAS. Symptoms include reduced brain and head size, mental retardation, facial deformities, poor growth and coordination, and often heart defects (Figure 45.16*b*). In 1992, nearly 4 of every 10,000 newborns were diagnosed with FAS, and the rate is rising. About 60 to 70 percent of newborns of alcoholic women have FAS.

Some researchers suspect that any alcohol at all may be harmful to the fetus. Increasingly, doctors are urging near- or total abstinence during pregnancy.

EFFECTS OF COCAINE A pregnant woman who uses cocaine, crack especially, disrupts the nervous system of her future child as well as her own. Here you may wish to

read again the start of Chapter 35, which describes the kind of future that offspring of a crack addict face.

EFFECTS OF CIGARETTE SMOKE Cigarette smoke impairs fetal growth and development. Also, a long-term study at Toronto's Hospital for Sick Children showed that toxic elements in tobacco accumulate even in the fetuses of pregnant nonsmokers who are exposed to *secondhand smoke* at home or work.

Daily smoking during pregnancy leads to underweight newborns even if the woman's weight, nutritional status, and all other relevant variables match those of pregnant nonsmokers. Smoking has other effects. In Great Britain, all infants born the same week were tracked for seven years. Those of smokers were smaller, died of more post-delivery complications, and had twice as many heart abnormalities. At age seven, their average "reading age" was nearly half a year behind children of nonsmokers.

The mechanisms by which smoking affects a fetus are not known. Its demonstrated effects are evidence that the placenta, marvelous structure that it is, cannot prevent all assaults on the fetus that the human mind can dream up.

Giving Birth

On average, pregnancy ends thirty-eight weeks after fertilization, give or take a few weeks. The birth process is called **labor** (or parturition, or delivery). It requires dilation of the cervical canal, so that the fetus can move from the uterus, through the vagina, and out into the world. It also requires very strong uterine contractions as the driving force behind the expulsion.

In the last trimester, mild uterine contractions begin and the cervix softens as its connective tissues loosen. Relaxin, a peptide hormone secreted from the corpus luteum and placenta, may bring about the softening. Relaxin also makes the connections between the pelvic bones loosen up. In the meantime, the fetus "drops," or is shifted downward, usually with its head in contact with the cervix (Figure 45.17a). A *breech birth* is the likely outcome when any part of the fetus other than the head is positioned first near the birth canal.

Rhythmic, usually painless contractions herald the onset of labor. They increase in frequency and intensity over the next two to eighteen hours. What triggers the change in contractility? According to one hypothesis, strong contractions are induced when the concentration of cell receptors for **oxytocin**, another peptide hormone, increases enormously in muscle cells of the uterine wall. Oxytocin, remember, is a hypothalamic hormone that is stored in the posterior pituitary. Some studies indicate that oxytocin may also be produced locally, in the uterus itself, before labor. However, the signal that starts all this activity has not been identified.

Just before birth, the amnion typically ruptures, and "water" (amniotic fluid) gushes from the vagina. Usually contractions expel the fetus in less than an hour after the cervix has fully dilated. Fifteen to thirty minutes later, the contractions cause the placenta to detach from the myometrium; it is expelled as the "afterbirth." The contractions also constrict blood vessels at the placental attachment site and so prevent hemorrhaging. Someone ties and severs the umbilical cord; a few days after it shrivels, the stump will form the newborn's navel. Once the lifeline to the mother has been severed, the newborn embarks on a course of post-embryonic development, starting with the extended period of nurtured existence and learning that is typical of primates.

Many new mothers commonly sink into postpartum depression, or *afterbaby blues*. Apparently corticotropin releasing hormone (CRH) that the placenta produces during pregnancy has a role in this. Its level in blood rises by as much as three times during pregnancy and may influence the timing of labor. Possibly cortisol, a stress hormone, is secreted in response to the elevated CRH levels to help the mother cope with extraordinary stresses that pregnancy and labor place on her body. Once the placenta is expelled, her CRH levels crash to levels that are typical of some clinical depressions. The placental CRH may suppress a feedback loop concerned with the release of CRH from the hypothalamus (Section 37.6). Once the baby has been born, it takes time before the hypothalamus can rebound. The mother's afterbaby blues continue until the hypothalamic secretion of CRH returns to normal.

Nourishing the Newborn

Survival of the newborn depends on an ongoing supply of milk or its nutritional equivalent. Milk production, or **lactation**, occurs in mammary glands in the mother's breasts. Before pregnancy, the breasts consist mainly of adipose tissue and an undeveloped duct system (Figure 45.18a). Their size depends on how much fat they hold, not on milk-producing ability. During pregnancy, they respond to estrogen and progesterone, and a glandular system of milk production develops (Figure 45.18b).

For the first few days after birth, the glands produce a fluid rich in proteins and lactose. Then the anterior

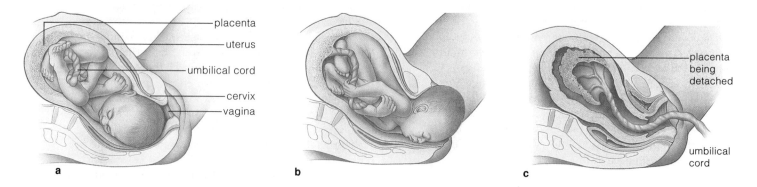

a   placenta   uterus   umbilical cord   cervix   vagina   b   c   placenta being detached   umbilical cord

Figure 45.17 Expulsion of a human fetus during the process of birth. Contractions result in the expulsion of the afterbirth (the placenta, tissue fluid, and blood).

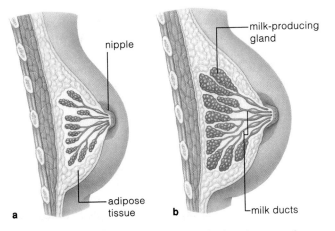

nipple
milk-producing gland
adipose tissue
milk ducts
a
b

Figure 45.18 (**a**) Breast of a woman who is not pregnant. (**b**) Breast of a lactating woman.

pituitary secretes **prolactin**, a hormone that induces the synthesis of the enzymes for milk production (Section 37.3). As a newborn suckles, the pituitary also releases oxytocin. This hormone triggers contractions that force milk into breast tissue ducts and that shrink the uterus back to normal size.

Regarding Breast Cancer

Each year in the United States alone, well over 100,000 women develop *breast cancer*. Obesity, high cholesterol, and high estrogen levels contribute to the cancerous transformation. With early detection and treatment, the chances for cure are excellent. Once a month, about a week after menstruating, a woman should examine her breasts. She can contact her physician or the American Cancer Society (listed in the telephone directory) for a pamphlet on recommended examination procedures.

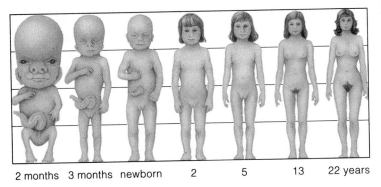

2 months   3 months   newborn   2   5   13   22 years

Figure 45.19 Observable, proportional changes in the human body during prenatal and postnatal growth. Changes in overall physical appearance are gradual but quite noticeable until the teenage years. For example, the head becomes proportionally smaller, compared to what it was during the embryonic period. The legs become longer and the trunk becomes shorter.

Table 45.4 Stages of Human Development	
PRENATAL PERIOD	
Zygote	Single cell resulting from fusion of sperm nucleus and egg nucleus at fertilization
Morula	Solid ball of cells produced by cleavages
Blastocyst	Ball of cells with surface layer, fluid-filled cavity, and inner cell mass (the mammalian blastula)
Embryo	All developmental stages from two weeks after fertilization until end of eighth week
Fetus	All developmental stages from the ninth week to birth (about thirty-eight weeks after fertilization)
POSTNATAL PERIOD	
Newborn	Individual during the first two weeks after birth
Infant	Individual from two weeks to about fifteen months after birth
Child	Individual from infancy to about ten or twelve years
Pubescent	Individual at puberty, when secondary sexual traits develop; girls between ten and fifteen years, boys between twelve and sixteen years
Adolescent	Individual from puberty until about three or four years later; physical, mental, emotional maturation
Adult	Early adulthood (between eighteen and twenty-five years); bone formation and growth finished. Changes proceed very slowly afterward
Old age	Aging follows late in life

Postnatal Development, Aging, and Death

After birth, a new individual follows a course of further growth and development that leads to the adult, the mature form of the species. Table 45.4 summarizes the *prenatal* (before birth) stages and *postnatal* (after birth) stages. Figure 45.19 is an example of the proportional changes that occur in the body as the life cycle unfolds. Postnatal growth is most rapid between ages thirteen and nineteen. Then, secretions of sex hormones step up and bring about the emergence of secondary sexual traits as well as sexual maturity. Not until adulthood are the bones fully mature. Body tissues normally are maintained in top condition during early adulthood. As the years pass, it becomes more and more difficult to maintain and repair existing tissues, and so the body gradually deteriorates. By processes collectively known as *aging*, the body's cells gradually start to break down. As Section 44.7 indicates, the processes are not fully understood, but they lead to structural changes and gradual loss of bodily functions. All animals that show extensive cell differentiation undergo aging.

The human life cycle flows naturally from the time of birth, growth, and development, to production of the individual's own offspring, and on through aging to the time of death.

Some Ethical Considerations

The transformation of a zygote into an adult of intricate detail raises profound questions. *When does development begin?* As you have seen, major developmental events unfold even before fertilization. *When does life begin?* During her lifetime, a woman can produce as many as 500 eggs, all of which are alive. During one ejaculation, a man can release a quarter of a billion sperm, which are alive. Even before sperm and egg merge by chance and establish the genetic makeup of a new individual, they are as much alive as any other form of life. It is scarcely tenable, then, to say life begins at fertilization. *Life began billions of years ago; and every gamete, every zygote, every sexually mature individual is but a fleeting stage in the continuation of that beginning.*

This greater perspective on life cannot diminish the meaning of conception. It is no small thing to entrust a new individual with the gift of life, wrapped in the unique evolutionary threads of our species and handed down through an immense sweep of time.

Yet how can we reconcile the marvel of individual birth with growing awareness of the astounding birth rate for our whole species? While this book is being written, an average of 10,700 newborns enter the world every single hour. By the time you go to bed tonight, there will be 257,000 more people on Earth than there were last night at that hour. Within a week, the number will reach 1,800,000—about as many people as there are now in the entire state of Massachusetts. *Within one week.* Worldwide, human population growth is rapidly outstripping resources, and each year many millions face the horrors of starvation. Living as we do on one of the most productive continents on Earth, few of us know what it means to give birth to a child, to give it the gift of life, and have no food to keep it alive.

And how can we reconcile the marvel of birth with the confusion that surrounds unwanted pregnancies? Even highly developed countries do not have adequate educational programs concerning fertility. And a great number of people are not inclined to exercise control. Each year in the United States alone, there are 1 million teenage pregnancies. (Many parents actually encourage early boy-girl relationships without thinking through the great risks of premarital intercourse and unplanned pregnancy. Advice is often condensed to a terse "Don't do it. But if you do it, be careful!") And each year, there are 1.6 million abortions among all age groups.

The motivation to engage in sex has been evolving for more than 500 million years. A few centuries of moral and ecological arguments for its suppression have not stopped all that many unwanted pregnancies. Besides this, complex social factors have contributed to a population growth rate that is out of control.

How will we reconcile our biological past and the need for a stabilized cultural present? Whether and how human fertility is to be controlled is one of the most volatile issues of our time. We will return to this issue in the next chapter, in the context of principles that govern the growth and stability of populations. Here, we briefly consider some control options.

Birth Control Options

The most effective method of birth control is complete *abstinence*, no sexual intercourse whatsoever. Data show that it is unrealistic to expect many people to practice it.

A less reliable variation of abstinence is the *rhythm method*. The idea is to avoid intercourse in the woman's fertile period, starting a few days before ovulation and ending a few days after. The woman identifies and tracks her fertile period. She also keeps records of the length of her menstrual cycles, takes her temperature each morning when she wakes up, or both. (The body's core temperature rises by one-half to one degree just before a fertile period.) But ovulation may not be regular, and miscalculations are frequent. Also, sperm deposited in the vagina a few days before ovulation may survive until ovulation. The rhythm method *is* inexpensive; it costs nothing after you buy a thermometer. It does not require fittings and periodic medical checkups. But its practitioners run a risk of pregnancy (Figure 45.20).

Withdrawal, or removing the penis from the vagina before ejaculation, dates back at least to biblical times. But withdrawal requires very strong willpower, and the practice may fail anyway. Fluid released from the penis just before ejaculation may contain some sperm.

Douching (rinsing the vagina with a chemical right after intercourse) is next to useless. Sperm move past the cervix and out of reach of the douche within ninety seconds after ejaculation.

Controlling fertility by surgical intervention is less chancy. In *vasectomy*, a physician makes a tiny incision in a man's scrotum, then severs and ties off each vas deferens. The operation takes only twenty minutes and requires only a local anesthetic. After the operation, sperm cannot leave the testes and cannot be present in semen. So far, there is no firm evidence that vasectomy disrupts hormonal functions or adversely affects sexual activity. Vasectomies can be reversed. But half of those who submit to surgery later develop antibodies against sperm and may not be able to regain fertility.

Females may have a *tubal ligation*. By this surgical intervention, oviducts are cauterized or cut and tied off, usually in a hospital. When the operation is performed correctly, tubal ligation is the most effective means of birth control. A few women have recurring pain in the pelvic area. The operation sometimes can be reversed.

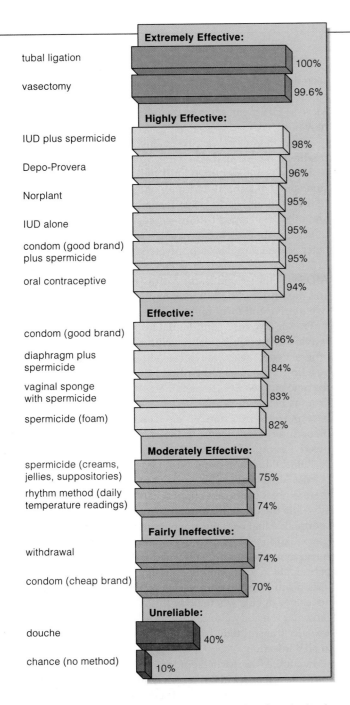

Extremely Effective:

tubal ligation — 100%
vasectomy — 99.6%

Highly Effective:

IUD plus spermicide — 98%
Depo-Provera — 96%
Norplant — 95%
IUD alone — 95%
condom (good brand) plus spermicide — 95%
oral contraceptive — 94%

Effective:

condom (good brand) — 86%
diaphragm plus spermicide — 84%
vaginal sponge with spermicide — 83%
spermicide (foam) — 82%

Moderately Effective:

spermicide (creams, jellies, suppositories) — 75%
rhythm method (daily temperature readings) — 74%

Fairly Ineffective:

withdrawal — 74%
condom (cheap brand) — 70%

Unreliable:

douche — 40%
chance (no method) — 10%

Figure 45.20 Comparison of the effectiveness of methods of contraception in the United States. Percentages reflect the number of unplanned pregnancies per 100 couples who used only that method of birth control for a year. For example, "94% effectiveness" for oral contraceptives (the Pill) means that 6 of every 100 women will still become pregnant, on the average.

Other, less drastic methods of controlling fertility involve physical or chemical barriers to prevent sperm from entering the uterus and moving to the ovarian ducts. *Spermicidal foam* and *spermicidal jelly* are toxic to sperm. They are transferred from an applicator into the vagina just before intercourse. These products are not always reliable unless used with another device, such as a diaphragm or condom.

A *diaphragm* is a flexible, dome-shaped device. It is inserted into the vagina and positioned over the cervix before intercourse. It is relatively effective when fitted initially by a doctor, used with foam or jelly before each sexual contact, inserted correctly each time, and left in place for a prescribed length of time.

Good brands of *condoms*—thin, tight-fitting sheaths worn over the penis during intercourse—are up to 95% effective if used with a spermicide. Only those made of latex provide protection against sexually transmitted diseases (Section 34.19). However, condoms often tear and leak, at which time they become absolutely useless.

The *birth control pill* is an oral contraceptive made of synthetic estrogens and progestins (progesterone-like hormones). Taken daily except for the last five days of a menstrual cycle, it suppresses oocyte maturation and ovulation. Often it corrects erratic menstrual cycles and reduces cramps, but in some women it causes nausea, weight gain, tissue swelling, and headaches. With more than 50 million women taking it, "the Pill" is the most-used fertility control method. It is 94 percent effective. Earlier formulations were linked to blood clots, high blood pressure, and maybe breast cancer. Lower doses of newer formulations might lessen the risk of breast and endometrial cancer. The possibility of a correlation between the Pill and breast cancer is still under study.

Progestin injections or implants inhibit ovulation. A *Depo-Provera* injection works for three months and is 96 percent effective. *Norplant* (six rods implanted under the skin) works for five years and is 95 percent effective. Both may cause sporadic, heavy bleeding, and doctors have some trouble surgically removing Norplant rods.

A pregnancy test doesn't register positive until after implantation. According to one view, a woman is not pregnant until that time. From this viewpoint, RU-486, the *morning-after pill*, intercepts pregnancy. It interferes with hormonal signals that control the events between ovulation and implantation. Three pills, taken within seventy-two hours after intercourse, block fertilization or prevent a blastocyst from burrowing into the uterine lining. RU-486 has side effects (nausea, vomiting, and breast tenderness, for the most part), but not all women experience them. However, RU-486 disrupts complex hormonal interactions and should only be taken under medical supervision. As is true of oral contraceptives, it may trigger elevated blood pressure, blood clots, or breast cancer in some women. At this writing, RU-486 is available in Europe. In the United States, its use is still a subject of controversy.

Whether and how human fertility is to be controlled is a volatile issue. It centers on reconciling our biological past with seriously divergent views about our cultural present.

45.14 SEXUALLY TRANSMITTED DISEASES

At some point in their life, at least one of every four people in the United States who engage in sexual intercourse will probably be infected by the pathogens that cause **sexually transmitted diseases** (STDs). STDs have reached epidemic proportions; more than 56 million are already infected. Two-thirds of those are under age twenty-five; one-fourth are teenagers. Women and children are hardest hit. Worse yet, antibiotic-resistant strains of bacteria are on the rise, and some viral diseases simply cannot be cured.

Urban poverty, prostitution, intravenous drug abuse, and sex-for-drugs are fanning the epidemic. Yet STDs also are rampant in high schools and colleges, where students too often think, "It can't happen to me." They reject the idea that no sex—abstinence—is the only safe sex. In one poll of high school students, two-thirds of the respondents said they don't use condoms. More than 40 percent of them have two or more sex partners.

The economics of this health problem are staggering. In 1993, the Centers for Disease Control related to us the annual cost of treating the most prevalent STDs: *Herpes*, $759 million; gonorrhea, $1 billion; chlamydial infection, $2.4 billion; and pelvic inflammatory disease, $4.2 billion. This does not include the accelerating cost of treating patients with AIDS. In many developing countries, AIDS alone threatens to overwhelm health-care delivery systems and to unravel decades of economic progress.

The social consequences are sobering. Mothers bestow a chlamydial infection on one of every twenty newborns in the United States. They bestow type II *Herpes* virus on 1 in 10,000 newborns; one-half of the infected babies die, and one-fourth have severe neural defects. Every year, 1 million women develop pelvic inflammatory disease, and 100,000 to 150,000 become infertile.

AIDS Someone can become infected by HIV, the human immunodeficiency virus, and not even know it. However, the infection marks the start of a titanic battle that the immune system almost certainly will lose (Section 40.11). At first there may be no outward symptoms. Five to ten years later, a set of chronic disorders develops. They are evidence of *AIDS* (acquired immune deficiency syndrome). When the immune system finally does give up, the stage is set for opportunistic infections. Normally harmless, resident bacteria are the first to take advantage of the lowered resistance. Dangerous pathogens also take their toll. In time, the immune-compromised person simply is overwhelmed.

Most commonly, HIV spreads by vaginal, anal, and oral intercourse and by IV drug users. Most of the infections occur by the transfer of blood, semen, urine, or vaginal secretions between people. Cuts or abrasions on the penis, vagina, rectum, and maybe in membranes in the mouth serve as HIV's entrances to the internal environment of a new host.

Today there is no vaccine against HIV and no effective treatment for AIDS. If you get it, you die. *There is no cure.*

HIV was not identified until 1981, but it was present in some parts of Central Africa for at least several decades. In the 1970s and early 1980s, it spread to the United States and other developed countries. Most of those initially infected were male homosexuals. Now a large part of the heterosexual population is infected or at risk. Worldwide, 22.2 million are known to be HIV infected. By early 1992, AIDS was the second leading cause of death among men and the fifth leading cause among women between ages twenty-five and forty-four.

Free or low-cost, confidential testing for HIV exposure is available at public health facilities and at many doctor's offices. People who are worried should know there may be a time lag between their first exposure and the first test to come out positive. It takes a few weeks to six months or more before detectable amounts of antibodies will form in response to infection. (The presence of antibodies in the blood indicates exposure to the virus.) Anyone who tests positive is capable of spreading the virus.

Public education programs attempt to stop the spread of HIV. Most health-care workers advocate safe sex, yet there is great confusion about what "safe" means. Many advocate the use of high-quality latex condoms *and* a spermicide that contains nonoxynol-9 to help block the transmission of the virus. Even then, there is a slight risk. As an unfortunate couple learned in 1997, no one should participate in open-mouthed kissing with someone who tests positive for HIV. Caressing is not risky *if* there are no lesions or cuts where body fluids that harbor the virus can enter the body. Pronounced lesions caused by other sexually transmitted diseases may increase susceptibility to HIV infection.

In sum, AIDS reached epidemic proportions mainly for three reasons. First, it took a while to discover that the virus can travel in semen, blood, and vaginal fluid and that *behavioral* controls can limit its spread. Second, it took time to develop ways to test symptom-free carriers, who infect others. Third, many still don't realize that the medical, economic, and social consequences affect everyone.

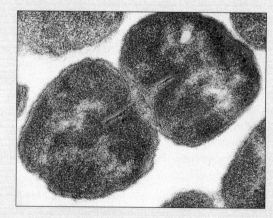

Figure 45.21 Bacterial agents of gonorrhea.

GONORRHEA During intercourse, *Neisseria gonorrhoeae* (Figure 45.21) can enter the body at mucous membranes of the urethra, cervix, and anal canal. This bacterium causes *gonorrhea*. An infected female might only notice a slight vaginal discharge or burning

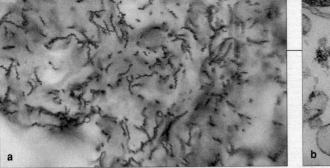
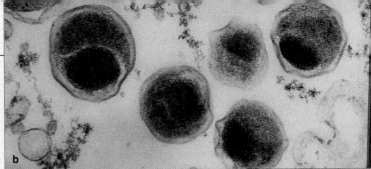

Figure 45.22
The bacterial agents of (**a**) syphilis and (**b**) chlamydia.

sensation while urinating. If the infection spreads to her oviducts, it may induce severe cramps, fever, vomiting, and scar tissue formation, which may cause sterility. Males have more obvious symptoms. Within a week after infection, the penis discharges yellow pus. Urinating is more frequent and may be painful.

This STD is rampant even though prompt treatment quickly cures it. Why? Women experience no troubling symptoms during early stages. Also, infection does not confer immunity to the bacterium, perhaps because there are sixteen or more different strains of it. Contrary to common belief, someone can get gonorrhea over and over again. Also, oral contraceptives, used so widely, promote infection by altering vaginal pH. Populations of resident bacteria decline—and *N. gonorrhoeae* is free to move in.

SYPHILIS   *Syphilis*, a dangerous STD, is caused by a spirochete, *Treponema pallidum* (Figure 45.22*a*). Sex with an infected partner puts the motile, spiral bacterium on the surface of genitals or into the cervix, vagina, or oral cavity. It can enter the body through tiny epidermal cuts. One to eight weeks later, new treponemes are twisting about in a flattened, painless chancre (local ulcer). This first chancre is a symptom of the primary stage of syphilis. By then, treponemes are in the blood. The treponemes can cross the placenta of an infected, pregnant woman. The result will be a miscarriage, stillbirth, or syphilitic newborn.

Usually a chancre heals, but in mucous membranes, joints, bones, eyes, spinal cord, and brain, treponemes are multiplying. More chancres and a skin rash develop in this infectious, secondary stage of syphilis. Symptoms subside; immune responses counter the disease in about 25 percent of the cases. Another 25 percent remain infected but symptom-free. In the rest, minor to significant lesions and scars appear in the skin and liver, bones, and other internal organs. Few treponemes occur in this tertiary stage. But the host's immune system is hypersensitive to them. Chronic immune reactions may severely damage the brain and spinal cord and lead to general paralysis.

Probably because the symptoms are so alarming, more people seek early treatment for syphilis than for gonorrhea. Later stages require prolonged treatment.

PELVIC INFLAMMATORY DISEASE   *Pelvic inflammatory disease* (PID) is one of the most serious complications of gonorrhea, chlamydial infections, and other STDs. It also can arise when normal bacterial residents of the vagina ascend to the pelvic region. Most commonly, the uterus, oviducts, and ovaries are affected. There are bleeding and vaginal discharge. Pain in the lower abdomen may be as

severe as an acute appendicitis attack. The oviducts may become scarred and invite abnormal pregnancies as well as sterility.

GENITAL HERPES   About 25 million Americans have *genital herpes*, caused by the type II *Herpes simplex* virus. Infection requires direct contact with active *Herpes* viruses or sores that contain them. Mucous membranes of the mouth or the genitals are highly susceptible to invasion. Symptoms often are mild or absent. Among infected women, small, painful blisters may appear on the vulva, cervix, urethra, or anal tissues. Among men, blisters form on the penis and anal tissues. Within three weeks, the virus enters latency, and the sores crust over and heal.

This virus is reactivated sporadically. Each time, it causes new, painful sores at or near the original site of infection. Sexual intercourse, menstruation, emotional stress, or other infections can trigger *Herpes* infections. Acyclovir, an antiviral drug, decreases the healing time and often decreases the pain and viral shedding.

GENITAL WARTS   More than sixty types of the human papillomaviruses (HPV) are known. A few cause *genital warts*, or benign, bumplike growths. HPV infection of the genitals and anus has become the most prevalent STD in the United States. Type 16 HPV does not usually cause obvious warts but may be linked to precancerous sores and cancers of the cervix, vagina, vulva, penis, and anus. In one Seattle study, 22 percent of female college students who were examined tested positive for the virus.

CHLAMYDIAL INFECTION   *Chlamydia trachomatis* (Figure 45.22*b*), a parasitic bacterium, spends part of its life cycle in cells of the genital and urinary tracts. It causes several diseases, including *NGU* (chlamydial nongonococcal urethritis). NGU is far more common than syphilis or gonorrhea. *N. gonorrhoeae* and *C. trachomatis* often are transmitted at the same time. Prompt doses of penicillin cure the gonorrhea—but not NGU, which requires both tetracycline and sulfonamide.

NGU leads to inflammation of the cervix and, in both sexes, the urethra. Infected people may notice a burning sensation while urinating. But when they are symptom-free, they don't seek treatment and complications develop. In males, the prostate becomes swollen and inflamed. In females, the infection may spread into the uterus and the oviducts to cause pelvic inflammatory disease.

Symptom-free infected individuals are unwittingly spreading destructive chlamydial infections through all ethnic groups, among poor and affluent alike.

Because of sterility or infertility, about 15 percent of all couples in the United States cannot conceive. For example, hormonal imbalances in the female body may prevent ovulation, or the male's sperm count may be so low that fertilization would be next to impossible without medical intervention.

In vitro fertilization is conception outside the body (literally "in glass" petri dishes or test tubes). It may occur if the couple can produce normal sperm and oocytes. First the woman receives injections of a hormone that prepares her ovaries for ovulation. Afterward, the preovulatory oocyte is removed from her with a suction device. In the meantime, sperm from the male are placed in a solution that simulates fluid in the oviducts. A few hours after the sperm and suctioned oocyte make contact in a petri dish, fertilization may result. Twelve hours later, the zygote is transferred to a solution that can sustain it through the initial cleavages. Two to four days later, the resulting ball of cells is transferred to the female's uterus.

Each attempt at in vitro fertilization costs about 8,000 dollars, and most attempts aren't successful. In 1994, each "test-tube" baby cost the nation's health-care system about 60,000 to 100,000 dollars, on average. The childless couple may believe no cost is too great. But in an era of increased population growth and shrinking medical coverage, is the cost too great for society to bear? Court battles are being waged over this issue.

At the other extreme is **abortion**, the dislodging and removal of the blastocyst, embryo, or fetus from the uterus. At one time in the United States, abortions were forbidden by law unless the pregnancy endangered the mother's life. Later the Supreme Court ruled that the government does not have the right to forbid abortions during the early stages of pregnancy (typically up to five months). Before this ruling, there were dangerous, traumatic, and often fatal attempts to abort embryos, either by pregnant women themselves or by quacks.

From a clinical standpoint, vacuum suctioning and other methods make abortion painless for the woman, rapid, and free of complications when performed during the first trimester. Abortions performed in the second and third trimesters are extremely controversial unless the mother's life is threatened. Even so, for both medical and humanitarian reasons, the majority of people in this country generally agree that the preferred route to birth control is not through abortion. Rather, it is through sexually responsible behavior that prevents unwanted pregnancy from happening in the first place.

This biology textbook cannot offer you the "right" answer to a moral question because of the reasons given in Section 1.7. It *can* provide you with a serious, detailed description of how a new human individual develops. Your choice of the "right" answer to the question of the morality of abortion will be just that—your choice—and one that can be based on objective insights into the nature of life.

1. Humans have a pair of primary reproductive organs (sperm-producing testes in males and egg-producing ovaries in females), accessory ducts, and glands. Testes and ovaries also produce sex hormones that influence reproductive functions and secondary traits.

a. The hormones LH, FSH, and testosterone control sperm formation. They are part of feedback loops from the testes to the hypothalamus and anterior lobe of the pituitary gland.

b. The hormones estrogen, progesterone, FSH, and LH control the maturation and release of oocytes from the ovary, as well as cyclic changes in the endometrium (inner lining of the uterus). Hormonal secretions are part of feedback loops from the ovaries to the hypothalamus and anterior lobe of the pituitary gland.

2. A menstrual cycle is a recurring cycle of intermittent fertility in the reproductive years of female humans and other primates. These events occur during each cycle:

a. Follicular phase: One of many follicles matures inside an ovary. Each follicle is an oocyte and the cell layer surrounding it. Meanwhile, the endometrium is prepared for a possible pregnancy. It breaks down each cycle if pregnancy does not occur.

b. Ovulation: A midcycle surge of the LH level in blood triggers ovulation, or the release of a secondary oocyte from the ovary.

c. Luteal phase: After ovulation a glandular structure, the corpus luteum, develops from the remnants of the follicle. It secretes progesterone and some estrogen that prime the endometrium for fertilization. If fertilization occurs, the corpus luteum is maintained.

3. After fertilization, cleavage produces a blastocyst that becomes implanted in the endometrium. Three primary tissue layers form that will give rise to all organs. They are ectoderm, endoderm, and mesoderm.

4. Four extraembryonic membranes form as embryos of humans (and other vertebrates) develop:

a. The amnion becomes a fluid-filled sac around the embryo, which it protects from mechanical shocks, abrupt temperature changes, and drying out.

b. The yolk sac stores nutritive yolk in most shelled eggs, but in humans, part becomes a major site of blood formation and some of its cells give rise to germ cells. (Eventually germ cells give rise to sperm and eggs.)

c. The chorion is protective membrane around the embryo and the other membranes. It also becomes a major component of the placenta.

d. In humans, the blood vessels for the placenta arise from the allantois, as does the urinary bladder.

5. The placenta, a blood-engorged organ, is composed of endometrium and extraembryonic membranes. The placenta allows the embryonic blood vessels to develop

independently of the mother's, even while allowing oxygen, nutrients, and wastes to diffuse between them.

6. To some extent, the placenta serves as a protective barrier for the fetus. But it cannot protect the fetus from harmful effects of the mother's nutritional deficiencies, infections, intake of prescription drugs, illegal drugs, alcohol, and cigarette smoking.

7. During pregnancy, both estrogen and progesterone stimulate the growth and development of mammary glands. During labor, strong uterine contractions expel the fetus and the afterbirth. After delivery, nursing stimulates the release of prolactin and oxytocin, which in turn stimulate milk production and release.

8. Whether and how to control human sexuality and fertility are major ethical issues of our time. There is widespread agreement that sexually responsible behavior rather than abortion is the preferred route to birth control.

Review Questions

1. Name the two primary reproductive organs of the human male and where sperm formation starts inside them. Does semen consist of sperm, glandular secretions, or both? *45.1*

2. Does sperm formation require mitosis, meiosis, or both? *45.2*

3. Study Figure 45.5. Then, on your own, sketch the feedback loops to the hypothalamus and anterior pituitary from the testes that govern sperm formation. Include the names of the releasing hormone, hormones, and cells in the testes involved in these loops. *45.2*

4. Name the two primary reproductive organs of the human female. What is the endometrium? *45.3*

5. Distinguish between:
   a. oocyte and ovum *45.3, 45.6*
   b. ollicular phase and luteal phase of menstrual cycle *45.3*
   c. follicle and corpus luteum *45.4*
   d. ovulation and implantation *45.3, 45.7*

6. What is the menstrual cycle? Name four of the hormones that influence this cycle. Which one triggers ovulation? *45.3, 45.4*

7. Study Figure 45.8. Then, on your own, sketch the feedback loops to the hypothalamus and anterior pituitary from the ovaries that govern the menstrual cycle. Include the names of the releasing hormone, hormones, and ovarian structures involved in these loops. *45.4*

8. Describe the cyclic changes that occur in the endometrium during the menstrual cycle. What role does the corpus luteum play in the changes? *45.4*

9. Distinguish between the embryonic period and fetal period of human development. Then describe the general organization of a human blastocyst. *45.7*

10. State the embryonic source of the amnion, yolk sac, chorion, and allantois. Then state the role that each extraembryonic membrane plays in the structure or functioning of the developing individual. *45.7*

11. Name some of the early organs that are the hallmark of the embryonic period of humans and other vertebrates. *45.8*

12. Describe the placenta's structure and function. Do maternal and fetal bloodstreams grossly intermingle in this organ? *45.9*

13. Name the releasing hormone with a key role in labor. Then state the role of estrogen, progesterone, prolactin, and oxytocin in assuring that the newborn will have an ongoing supply of milk. *45.12*

14. Label the components of the human male reproductive system and state their functions: *45.1, 45.2*

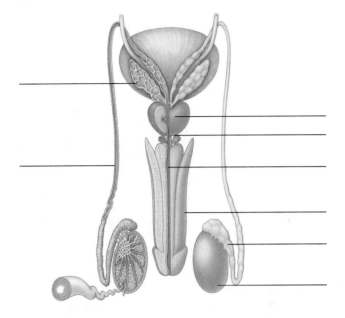

15. Label the components of the human female reproductive system and state their functions: *45.3, 45.4*

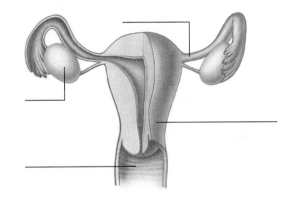

Self-Quiz (Answers in Appendix IV)

1. The _____ and the pituitary gland control secretion of sex hormones from the testes and the ovaries.

2. _____ secretions govern reproductive function in males.
   a. FSH   d. both a and b
   b. LH   e. both b and c
   c. Testosterone f. all of the above

3. During a menstrual cycle, a midcycle surge of _____ triggers ovulation.
   a. estrogen   b. progesterone   c. LH   d. FSH

4. During a menstrual cycle, _____ and _____ secreted by the corpus luteum prime the uterus for pregnancy.
   a. FSH; LH   c. estrogen; progesterone
   b. FSH; testosterone d. estrogens only

5. In implantation, a _____ burrows into the endometrium.
   a. zygote   c. blastocyst
   b. gastrula   d. morula

6. The _____, a fluid-filled sac, surrounds and protects the embryo from mechanical shocks and keeps it from drying out.
    a. yolk sac   b. allantois   c. amnion   d. chorion

7. At full term, a placenta _____ .
    a. is composed of extraembryonic membranes
    b. directly connects maternal and fetal blood vessels
    c. keeps maternal and fetal blood vessels separated

8. (A) _____ form(s) in all vertebrate embryos.
    a. neural plate   c. pharyngeal arches   e. a through c
    b. somites   d. primitive streak   f. a through d

9. Distinctly human features emerge in the embryo by the end of the _____ week after fertilization.
    a. second   c. fourth   e. eighth
    b. third   d. fifth   f. sixteenth

10. Match the term with the most suitable description.
    _____ seminiferous   a. glandular structure formed
          tubule   from follicle remnants
    _____ allantois   b. most bones, skeletal muscles of
    _____ corpus   head and trunk form from them
          luteum   c. helps form urinary bladder, blood
    _____ somites   vessels for placenta in humans
    _____ yolk sac   d. blood cell formation site, source
                                 of germ cells
                              e. sperm form here

Critical Thinking

1. Suppose that, at the time a human zygote is undergoing cleavage, the first two blastomeres that form completely separate from each other. Both blastomeres and their cellular descendants continue to divide on schedule, on the prescribed developmental program. The result is *identical twins*, or two normal, genetically identical individuals. *Nonidentical twins* form when two different secondary oocytes are fertilized at the same time by two different sperm. On the basis of this information, explain why nonidentical twins show considerable genetic variability and identical twins do not.

2. Infection by the rubella virus apparently has an inhibitory effect on mitosis. Serious birth defects result when a woman is infected during the first trimester of pregnancy, but not later. Review the developmental events that unfold during pregnancy and explain why this might be so.

3. Imagine you are an obstetrician advising a woman who has just learned she is pregnant. What instructions would you provide concerning her diet and behavior during pregnancy?

4. In the United States, teenage pregnancies and STD infections are rampant. Suppose the office of the Surgeons General asks you to take part in a task force that will recommend practices that might reduce the incidence of both. What practices might meet with the greatest success? What kind of enthusiasm or resistance might they provoke among the teenagers in your community? Among adults? Explain why.

Selected Key Terms

abortion *45.15*
allantois *45.7*
amnion *45.7*
blastocyst, human *45.7*
chorion *45.7*
coitus *45.6*
corpus luteum *45.4*
endometrium *45.3*
estrogen *45.3*
fetus *45.7*
follicle *45.4*
follicular phase
    (menstrual cycle) *45.3*
FSH *45.2*
gastrulation *45.8*
GnRH *45.2*
HCG *45.7*
implantation *45.7*
in vitro fertilization *45.15*
labor (birth process) *45.12*
lactation *45.12*
Leydig cell *45.2*
LH *45.2*
luteal phase (menstrual cycle) *45.3*

menstrual cycle *45.3*
oocyte *45.3*
orgasm *45.6*
ovary *CI*
ovulation *45.3*
ovum *45.6*
oxytocin *45.12*
placenta *45.9*
polar body *45.4*
progesterone *45.3*
prolactin *45.12*
secondary oocyte *45.4*
secondary sexual trait *CI*
semen *45.1*
seminiferous tubule *45.1*
Sertoli cell *45.2*
sexually transmitted
    disease (STD) *45.14*
somite *45.8*
testis (testes) *CI*
testosterone *45.2*
uterus *45.3*
yolk sac *45.7*
zona pellucida *45.4*

Readings

Aral, S., and K. Holmes. February 1991. "Sexually Transmitted Diseases in the AIDS Era." *Scientific American* 2(264): 62–69.

Caldwell, M. November 1992. "How Does a Single Cell Become a Whole Body?" *Discover* 13(11): 86–93.

Cohen, J., and I. Stewart. April 1994. "Our Genes Aren't Us." *Discover* 15(4): 78–84. Argument that DNA means nothing in the absence of a developmental context.

Gilbert, S. 1994. *Developmental Biology*. Fourth edition. Sunderland, Massachusetts: Sinauer.

Larsen, W. 1993. *Human Embryology*. New York: Churchill Livingston. Paperback. Recommended for the serious student, but maybe not for the fainthearted.

McGinnis, W., and M. Kuziora. February 1994. "The Molecular Architects of Body Design." *Scientific American* 270(2): 58–66.

Nilsson, L., et al. 1986. *A Child Is Born*. New York: Delacorte Press/Seymour Lawrence.

Sherwood, L. 1997. *Human Physiology*. Fourth edition. Belmont, California: Wadsworth.

Zack, B. July 1981. "Abortion and the Limitations of Science." *Science* 213: 291.

Web Site See *http://www.wadsworth.com/biology* for practice quiz questions, hypercontents, BioUpdates, and critical thinking. The Wadsworth Biology Resource Center provides a wealth of information fully organized and integrated by chapter.

FACING PAGE: *Two organisms—a fox in the shadows cast by a snow-dusted spruce tree. What are the nature and consequences of their interactions with each other, with other organisms, and with their environment? By the end of this last unit, you possibly will see worlds within worlds in such photographs.*

APPENDIX I.   BRIEF CLASSIFICATION SCHEME

The classification scheme that follows is a composite of several that microbiologists, botanists, and zoologists use. The major groupings are agreed upon, more or less. There is not always agreement, however, on what to call a given grouping or where it might fit within the overall hierarchy. There are several reasons for the lack of total consensus.

First, the fossil record varies in its quality and in its completeness. Therefore, the phylogenetic relationship of one group to others is sometimes open to interpretation. Comparative studies at the molecular level are firming up the picture, but this work is still under way.

Second, ever since the time of Linnaeus, classification schemes have been based on the perceived morphological similarities and differences among organisms. Although some original interpretations are now open to question, we are so used to thinking about organisms in certain ways that reclassification often proceeds slowly. Traditionally, for example, birds and reptiles have been considered to be separate classes (Reptilia and Aves). And yet there are now compelling arguments for grouping the lizards and snakes together in one class, and the crocodilians, dinosaurs, and birds in a different class.

Third, researchers in microbiology, mycology, botany, zoology, and the other fields of biological inquiry have inherited a wealth of literature, based on classification schemes that were developed over time in each of those fields. Many see no good reason to give up the established terminology and thereby disrupt access to the past. Until recently, for example, microbiologists and botanists have been using *division*, and zoologists *phylum*, for taxa that are equivalent in the hierarchy of classification. Many still do. Opinions are still polarized with respect to the kingdom Protista, certain members of which could just as easily be grouped in the kingdoms of plants, fungi, or animals. Indeed, the term protozoan is a holdover from an earlier scheme in which amoebas and certain other single-celled organisms were ranked as simple animals.

Given the problems, why do we even bother imposing artificial frameworks on the history of life? We do this for the same reason that a writer might decide to break up the history of civilization into several volumes, a number of chapters, and many paragraphs. Both efforts are attempts to impart obvious structure to what might otherwise be an overwhelming body of knowledge and to enhance the retrieval of information from it.

Finally, bear in mind that we include this classification scheme primarily for your reference purposes. Besides being open to revision, it also is by no means complete. Numerous existing and extinct organisms of the so-called lesser phyla are not represented here. Our strategy is to focus mainly on organisms mentioned in the text. A few examples of organisms also are listed under the entries.

SUPERKINGDOM PROKARYOTA.   Prokaryotes. Almost all microscopic species with DNA concentrated in a region of cytoplasm, not inside a membrane-bounded nucleus. All are bacteria, either single cells or simple associations of cells. Autotrophs and heterotrophs (Table 22.2). Reproduce by prokaryotic fission, sometimes by budding and by bacterial conjugation. *Bergey's Manual of Systematic Bacteriology*, the authoritative reference in the field, calls this "a time of taxonomic transition." It groups bacteria mostly by numerical taxonomy (Section 22.6), not on phylogeny. The scheme presented here reflects strong evidence of evolutionary relationships for at least some bacterial groupings.

KINGDOM EUBACTERIA.   Gram-negative and gram-positive forms. Peptidoglycan in cell wall. Photosynthetic autotrophs, chemosynthetic autotrophs, and heterotrophs.

PHYLUM GRACILICUTES.   Typical Gram-negative, thin wall. Autotrophs (photosynthetic and chemosynthetic) and heterotrophs. *Anabaena* and other cyanobacteria. *Escherichia, Pseudomonas, Neisseria, Myxococcus.*

PHYLUM FIRMICUTES.   Typical Gram-positive, thick wall. Heterotrophs. *Bacillus, Staphylococcus, Streptococcus, Clostridium, Actinomycetes.*

PHYLUM TENERICUTES.   Gram-negative, wall absent. Heterotrophs (saprobes, pathogens). *Mycoplasma.*

KINGDOM ARCHAEBACTERIA.   Methanogens, halophiles, thermophiles. Strict anaerobes, distinct from other bacteria in cell wall, membrane lipids, ribosomes, and RNA sequences. *Methanobacterium, Halobacterium, Sulfolobus.*

SUPERKINGDOM EUKARYOTA.   Eukaryotes. Single-celled and multicelled species. Cells start out life with a nucleus (encloses the DNA) and usually other membrane-bound organelles. Chromosomes with numerous proteins attached.

KINGDOM PROTISTA.   Diverse single-celled, colonial, and multicelled eukaryotic species, currently most easily classified by what they are *not* (not bacteria, fungi, plants, or animals). Autotrophs, heterotrophs, or both (Table 23.3). Reproduce sexually and asexually (by meiosis, mitosis, or both). Many related evolutionarily to plants, fungi, and possibly animals.

PHYLUM CHYTRIDIOMYCOTA.   Chytrids. Heterotrophs; saprobic decomposers or parasites. *Chytridium.*

PHYLUM OOMYCOTA.   Water molds. Heterotrophs. Decomposers, some parasites. *Saprolegnia, Phytophthora, Plasmopara.*

PHYLUM ACRASIOMYCOTA.   Cellular slime molds. Heterotrophs with free-living, phagocytic amoeboid cells and spore-bearing stages. *Dictyostelium.*

PHYLUM MYXOMYCOTA.   Plasmodial slime molds. Heterotrophs with free-living, phagocytic amoeboid cells and spore-bearing stages. Aggregate into streaming mass of cells that discard plasma membranes. *Physarum.*

PHYLUM SARCODINA.   Amoeboid protozoans. Heterotrophs, free-living or endosymbiotic, some pathogens. Soft-or shelled bodies, locomotion by pseudopods. The rhizopods (naked amoebas, foraminiferans), *Amoeba proteus, Entomoeba.* Also the actinopods (radiolarians, heliozoans).

PHYLUM MASTIGOPHORA.   Animal-like flagellated protozoans. Heterotrophs, free-living, many internal parasites. All with one to several flagella. *Trypanosoma, Trichomonas, Giardia.*

APICOMPLEXA.   Heterotrophs, many parasitic. Complex of rings, tubules, other structures at head end. Most familiar members called sporozoans. *Plasmodium, Toxoplasma.*

PHYLUM CILIOPHORA.   Ciliated protozoans. Heterotrophs, predators or symbionts, some parasitic. All have cilia. Free-living, sessile, or motile. *Paramecium*, hypotrichs.

PHYLUM EUGLENOPHYTA.   Euglenoids. Mostly heterotrophs, some autotrophs (photosynthetic). Flagellated. *Euglena*.

PHYLUM PYRRHOPHYTA.   Dinoflagellates. Photosynthetic, mostly, but some heterotrophs. *Gymnodinium breve*.

PHYLUM CHRYSOPHYTA.   Golden algae, yellow-green algae, diatoms. Photosynthetic. Some flagellated, others not. *Mischococcus, Synura, Vaucheria*.

PHYLUM RHODOPHYTA.   Red algae. Photosynthetic, some parasitic. Nearly all marine, some freshwater. *Porphyra. Bonnemaisonia, Euchema*.

PHYLUM PHAEOPHYTA.   Brown algae. Photosynthetic, nearly all in temperate or marine waters. *Macrocystis, Fucus, Sargassum, Ectocarpus, Postelsia*.

PHYLUM CHLOROPHYTA.   Green algae. Mostly photosynthetic, some parasitic. Most freshwater, some marine or terrestrial. *Chlamydomonas, Spirogyra, Ulva, Volvox, Codium, Halimeda*.

KINGDOM FUNGI.   Nearly all multicelled eukaryotic species. Heterotrophs; mostly saprobic decomposers, some parasites. Nutrition based on extracellular digestion of organic matter and absorption of nutrients by individual cells. Multicelled species form absorptive mycelium within substrates and structures that produce asexual spores (and sometimes sexual spores).

PHYLUM ZYGOMYCOTA.   Zygomycetes. Zygosporangia (zygote inside thick wall) formed by sexual reproduction. Bread molds, related forms. *Rhizopus, Philobolus*.

PHYLUM ASCOMYCOTA.   Ascomycetes. Sac fungi. Sac-shaped cells form sexual spores (ascospores). Most yeasts and molds, morels, truffles. *Saccharomycetes, Morchella, Neurospora, Sarcoscypha. Claviceps, Ophiostoma*.

PHYLUM BASIDIOMYCOTA.   Basidiomycetes. Club fungi. Most diverse group. Produce basidiospores inside club-shaped structures. Mushrooms, shelf fungi, stinkhorns. *Agaricus, Amanita, Puccinia, Ustilago*.

IMPERFECT FUNGI.   Sexual spores absent or undetected. The group has no formal taxonomic status. If better understood, a given species might be grouped with sac fungi or club fungi. *Verticillium, Candida, Microsporum, Histoplasma*.

LICHENS.   Mutualistic interactions between fungal species and a cyanobacterium, green alga, or both. *Usnea, Cladonia*.

KINGDOM PLANTAE.   Multicelled eukaryotes. Nearly all photosynthetic autotrophs with chlorophylls *a* and *b*. Some parasitic. Nonvascular and vascular species, generally with well-developed root and shoot systems. Nearly all adapted in form and function to survive dry conditions in land habitats; a few in aquatic habitats. Sexual reproduction predominant; also asexual reproduction by vegetative propagation.

PHYLUM RHYNIOPHYTA.   Earliest known vascular plants; muddy habitats. Extinct. *Cooksonia, Rhynia*.

PHYLUM PROGYMNOSPERMOPHYTA.   Progymnosperms. Ancestral to early seed-bearing plants; extinct. *Archaeopteris*.

PHYLUM PTERIDOSPERMOPHYTA.   Seed ferns. Fernlike gymnosperms; extinct. *Medullosa*

PHYLUM CHAROPHYTA.   Stoneworts.

PHYLUM BRYOPHYTA.   Bryophytes: mosses, liverworts, hornworts. Seedless, nonvascular, haploid dominance. *Marchantia, Polytrichum, Sphagnum*.

PHYLUM PSILOPHYTA.   Whisk ferns. Seedless, vascular. No obvious roots, leaves on sporophyte. *Psilotum*.

PHYLUM LYCOPHYTA.   Lycophytes, club mosses. Seedless, vascular. Leaves, branching rhizomes, vascularized roots and stems. *Lycopodium, Selaginella*.

PHYLUM SPHENOPHYTA.   Horsetails. Seedless, vascular. Some sporophyte stems photosynthetic, others nonphotosynthetic, spore-producing. *Equisetum*.

PHYLUM PTEROPHYTA.   Ferns. Largest group of seedless vascular plants (12,000 species), mainly tropical, temperate habitats.

PHYLUM CYCADOPHYTA.   Cycads. Type of gymnosperm (vascular, bears "naked" seeds). Tropical, subtropical. Palm-shaped leaves, simple cones on male and female plants. *Zamia*.

PHYLUM GINKGOPHYTA.   Ginkgo (maidenhair tree). Type of gymnosperm. Seeds with fleshy outer layer. *Ginkgo*.

PHYLUM GNETOPHYTA.   Gnetophytes. Only gymnosperms with vessels in xylem and double fertilization (but endosperm does not form). *Ephedra, Welwitchia*.

PHYLUM CONIFEROPHYTA.   Conifers. Most common and familiar gymnosperms. Generally cone-bearing with needle-like or scale-like leaves.

Family Pinaceae. Pines, firs, spruces, hemlock, larches, Douglas firs, true cedars. *Pinus*.

Family Cupressaceae. Junipers, cypresses. *Juniperus*.

Family Taxodiaceae. Bald cypress, redwoods, Sierra bigtree, dawn redwood. *Sequoia*.

Family Taxaceae. Yews.

PHYLUM ANTHOPHYTA.   Angiosperms (flowering plants). Largest group of vascular seed-bearing plants. Only organisms that produce flowers, fruits.

Class Dicotyledonae. Dicotyledons (dicots). Some families of several different orders are listed:

Family Nymphaeaceae. Water lilies.
Family Papaveraceae. Poppies.
Family Brassicaceae. Mustards, cabbages, radishes.
Family Malvaceae. Mallows, cotton, okra, hibiscus.
Family Solanaceae. Potatoes, eggplant, petunias.
Family Salicaceae. Willows, poplars.
Family Rosaceae. Roses, apples, almonds, strawberries.
Family Fabaceae. Peas, beans, lupines, mesquite.
Family Cactaceae. Cacti.
Family Euphorbiaceae. Spurges, poinsettia.
Family Cucurbitaceae. Gourds, melons, cucumbers, squashes.
Family Apiaceae. Parsleys, carrots, poison hemlock.
Family Aceraceae. Maples.
Family Asteraceae. Composites. Chrysanthemums, sunflowers, lettuces, dandelions.

Class Monocotyledonae. Monocotyledons (monocots). Some families of several different orders are listed:

Family Liliaceae. Lilies, hyacinths, tulips, onions, garlic.
Family Iridaceae. Irises, gladioli, crocuses.
Family Orchidaceae. Orchids.
Family Arecaceae. Date palms, coconut palms.
Family Cyperaceae. Sedges.
Family Poaceae. Grasses, bamboos, corn, wheat, sugarcane.
Family Bromeliaceae. Bromeliads, pineapples, Spanish moss.

KINGDOM ANIMALIA.   Multicelled eukaryotes, nearly all with tissues, organs, and organ systems and with motility during at least part of the life cycle. Heterotrophs; predators (herbivores, carnivores, omnivores), parasites, detritivores. Reproduce sexually (and asexually in many species). Continuous stages of embryonic development.

PHYLUM PLACOZOA.   Marine. Simplest known animal. Two cell layers, no mouth, no organs. *Trichoplax*.

PHYLUM MESOZOA.   Ciliated, wormlike parasites, about the same level of complexity as *Trichoplax*.

PHYLUM PORIFERA.   Sponges. No symmetry, tissues, or organs.

PHYLUM CNIDARIA.   Radial symmetry, tissues, nematocysts.

Class Hydrozoa. Hydrozoans. *Hydra, Obelia, Physalia*.
Class Scyphozoa. Jellyfishes. *Aurelia*.
Class Anthozoa. Sea anemones, corals. *Telesto*.

PHYLUM CTENOPHORA.   Comb jellies. Modified radial symmetry.

PHYLUM PLATYHELMINTHES.   Flatworms. Bilateral, cephalized; simplest animals with organ systems. Saclike gut.

Class Turbellaria. Triclads (planarians), polyclads. *Dugesia*.
Class Trematoda. Flukes. *Schistosoma*.
Class Cestoda. Tapeworms. *Taenia*.

PHYLUM NEMERTEA. Ribbon worms.

PHYLUM NEMATODA. Roundworms. *Ascaris, Trichinella.*

PHYLUM ROTIFERA. Rotifers.

PHYLUM MOLLUSCA. Mollusks.
  Class Polyplacophora. Chitons.
  Class Gastropoda. Snails (periwinkles, whelks, limpets, abalones, cowries, conches, nudibranchs, tree snails, garden snails), sea slugs, land slugs.
  Class Bivalvia. Clams, mussels, scallops, cockles, oysters, shipworms.
  Class Cephalopoda. Squids, octopuses, cuttlefish, nautiluses. *Loligo.*

PHYLUM BRYOZOA. Bryozoans (moss animals).

PHYLUM BRACHIOPODA. Lampshells.

PHYLUM ANNELIDA. Segmented worms.
  Class Polychaeta. Mostly marine worms.
  Class Oligochaeta. Mostly freshwater and terrestrial worms, but many marine. *Lumbricus* (earthworms).
  Class Hirudinea. Leeches.

PHYLUM TARDIGRADA. Water bears.

PHYLUM ONYCHOPHORA. Onychophorans. *Peripatus.*

PHYLUM ARTHROPODA.
  Subphylum Trilobita. Trilobites; extinct.
  Subphylum Chelicerata. Chelicerates. Horseshoe crabs, spiders, scorpions, ticks, mites.
  Subphylum Crustacea. Shrimps, crayfishes, lobsters, crabs, barnacles, copepods, isopods (sowbugs).
  Subphylum Uniramia.
    Superclass Myriapoda. Centipedes, millipedes.
    Superclass Insecta.
      Order Ephemeroptera. Mayflies.
      Order Odonata. Dragonflies, damselflies.
      Order Orthoptera. Grasshoppers, crickets, katydids.
      Order Dermaptera. Earwigs.
      Order Blattodea. Cockroaches.
      Order Mantodea. Mantids.
      Order Isoptera. Termites.
      Order Mallophaga. Biting lice.
      Order Anoplura. Sucking lice.
      Order Homoptera. Cicadas, aphids, leafhoppers, spittlebugs.
      Order Hemiptera. Bugs.
      Order Coleoptera. Beetles.
      Order Diptera. Flies.
      Order Mecoptera. Scorpion flies. *Harpobittacus.*
      Order Siphonaptera. Fleas.
      Order Lepidoptera. Butterflies, moths.
      Order Hymenoptera. Wasps, bees, ants.

PHYLUM ECHINODERMATA. Echinoderms.
  Class Asteroidea. Sea stars.
  Class Ophiuroidea. Brittle stars.
  Class Echinoidea. Sea urchins, heart urchins, sand dollars.
  Class Holothuroidea. Sea cucumbers.
  Class Crinoidea. Feather stars, sea lilies.
  Class Concentricycloidea. Sea daisies.

PHYLUM HEMICHORDATA. Acorn worms.

PHYLUM CHORDATA. Chordates.
  Subphylum Urochordata. Tunicates, related forms.
  Subphylum Cephalochordata. Lancelets.
  Subphylum Vertebrata. Vertebrates.
    Class Agnatha. Jawless vertebrates (lampreys, hagfishes).
    Class Placodermi. Jawed, heavily armored fishes; extinct.
    Class Chondrichthyes. Cartilaginous fishes (sharks, rays, skates, chimaeras).
    Class Osteichthyes. Bony fishes.
      Subclass Dipnoi. Lungfishes.
      Subclass Crossopterygii. Coelacanths, related forms.
      Subclass Actinopterygii. Ray-finned fishes.
        Order Acipenseriformes. Sturgeons, paddlefishes.
        Order Salmoniformes. Salmon, trout.
        Order Atheriniformes. Killifishes, guppies.
        Order Gasterosteiformes. Seahorses.
        Order Perciformes. Perches, wrasses, barracudas, tunas, freshwater bass, mackerels.
        Order Lophiiformes. Angler fishes.
    Class Amphibia. Mostly tetrapods; embryo enclosed in amnion.
      Order Caudata. Salamanders.
      Order Anura. Frogs, toads.
      Order Apoda. Apodans (caecilians).
    Class Reptilia. Skin with scales, embryo enclosed in amnion.
      Subclass Anapsida. Turtles, tortoises.
      Subclass Lepidosaura. *Sphenodon,* lizards, snakes.
      Subclass Archosaura. Dinosaurs (extinct), crocodiles, alligators.
    Class Aves. Birds. (In more recent schemes, dinosaurs, crocodilians, and birds are often grouped in the same category.)
      Order Struthioniformes. Ostriches.
      Order Sphenisciformes. Penguins.
      Order Procellariiformes. Albatrosses, petrels.
      Order Ciconiiformes. Herons, bitterns, storks, flamingoes.
      Order Anseriformes. Swans, geese, ducks.
      Order Falconiformes. Eagles, hawks, vultures, falcons.
      Order Galliformes. Ptarmigan, turkeys, domestic fowl.
      Order Columbiformes. Pigeons, doves.
      Order Strigiformes. Owls.
      Order Apodiformes. Swifts, hummingbirds.
      Order Passeriformes. Sparrows, jays, finches, crows, robins, starlings, wrens.
    Class Mammalia. Skin with hair; young nourished by milk-secreting glands of adult.
      Subclass Prototheria. Egg-laying mammals (duckbilled platypus, spiny anteaters).
      Subclass Metatheria. Pouched mammals or marsupials (opossums, kangaroos, wombats).
      Subclass Eutheria. Placental mammals.
        Order Insectivora. Tree shrews, moles, hedgehogs.
        Order Scandentia. Insectivorous tree shrews.
        Order Chiroptera. Bats.
        Order Primates.
          Suborder Strepsirhini (prosimians). Lemurs, lorises.
          Suborder Haplorhini (tarsioids and anthropoids).
            Infraorder Tarsiiformes. Tarsiers.
            Infraorder Platyrrhini (New World monkeys).
              Family Cebidae. Spider monkeys, howler monkeys, capuchin.
            Infraorder Catarrhini (Old World monkeys and hominoids).
              Superfamily Cercopithecoidea. Baboons, macaques, langurs.
              Superfamily Hominoidea. Apes and humans.
                Family Hylobatidae. Gibbon.
                Family Pongidae. Chimpanzees, gorillas, orangutans.
                Family Hominidae. Existing and extinct human species (*Homo*) and australopiths.
        Order Carnivora. Carnivores.
          Suborder Feloidea. Cats, civets, mongooses, hyenas.
          Suborder Canoidea. Dogs, weasels, skunks, otters, raccoons, pandas, bears.
        Order Proboscidea. Elephants; mammoths (extinct).
        Order Sirenia. Sea cows (manatees, dugongs).
        Order Perissodactyla. Odd-toed ungulates (horses, tapirs, rhinos).
        Order Artiodactyla. Even-toed ungulates (camels, deer, bison, sheep, goats, antelopes, giraffes).
        Order Edentata. Anteaters, tree sloths, armadillos.
        Order Tubulidentata. African aardvarks.
        Order Cetacea. Whales, porpoises.
        Order Rodentia. Most gnawing animals (squirrels, rats, mice, guinea pigs, porcupines).

Metric-English Conversions

Length

English		Metric
inch	=	2.54 centimeters
foot	=	0.30 meter
yard	=	0.91 meter
mile (5,280 feet)	=	1.61 kilometer

To convert	multiply by	to obtain
inches	2.54	centimeters
feet	30.00	centimeters
centimeters	0.39	inches
millimeters	0.039	inches

Weight

English		Metric
grain	=	64.80 milligrams
ounce	=	28.35 grams
pound	=	453.60 grams
ton (short) (2,000 pounds)	=	0.91 metric ton

To convert	multiply by	to obtain
ounces	28.3	grams
pounds	453.6	grams
pounds	0.45	kilograms
grams	0.035	ounces
kilograms	2.2	pounds

Volume

English		Metric
cubic inch	=	16.39 cubic centimeters
cubic foot	=	0.03 cubic meter
cubic yard	=	0.765 cubic meters
ounce	=	0.03 liter
pint	=	0.47 liter
quart	=	0.95 liter
gallon	=	3.79 liters

To convert	multiply by	to obtain
fluid ounces	30.00	milliliters
quart	0.95	liters
milliliters	0.03	fluid ounces
liters	1.06	quarts

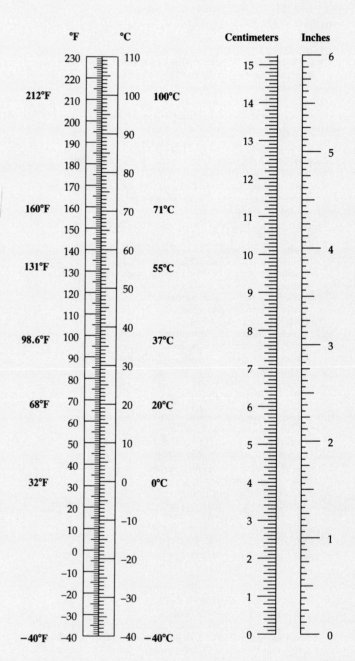

APPENDIX III. ANSWERS TO SELF-QUIZZES

CHAPTER 33
1. a
2. b
3. b
4. b
5. c
6. d
7. c
8. d
9. c
10. a
11. receptors, integrators, effectors
12. b, a, c, f, g, e, d

CHAPTER 34
1. d
2. b
3. d
4. a
5. d
6. d
7. c
8. b, d, c, a

CHAPTER 35
1. nerve net
2. a
3. false
4. true
5. c
6. b
7. b
8. d
9. a
10. b
11. d
12. e, d, b, c, a

CHAPTER 36
1. stimulus
2. sensation
3. perception
4. d
5. b
6. b
7. e
8. d
9. c
10. d
11. b
12. c
13. e, c, b, d, a

CHAPTER 37
1. f
2. d
3. b
4. d
5. e
6. b
7. b
8. c
9. d, f, c, e, a b

CHAPTER 38
1. integumentary
2. skeletal; muscle
3. smooth; cardiac; skeletal
4. c
5. d
6. d
7. b
8. d
9. f
10. h, f, g, e, a, c, b, i, d

CHAPTER 39
1. c
2. d
3. d
4. c
5. c
6. d
7. b
8. c
9. c
10. e, b, a, d, c
11. d, e, a, f, c, b

CHAPTER 40
1. e
2. c
3. c
4. d
5. a
6. d
7. d
8. b
9. a
10. c, b, a, e, d, f

CHAPTER 41
1. d
2. d
3. d
4. b
5. c
6. e
7. d
8. c
9. d
10. d, b, f, e, a, c

CHAPTER 42
1. e
2. d
3. caloric; energy
4. a
5. d
6. b
7. c
8. a
9. b
10. d, c, e, g, f, b, a

CHAPTER 43
1. d
2. f
3. a
4. b
5. c
6. d
7. d
8. d
9. d, e, a, c, b
10 f, e, g, b, c, d, a

CHAPTER 44
1. d
2. a
3. a
4. c
5. b
6. d
7. c
8. b
9. d
10. d
11. d, a, f, e, c, b

CHAPTER 45
1. hypothalamus
2. f
3. LH
4. c
5. c
6. c
7. c
8. f
9. e
10. e, c, a, b, d

A GLOSSARY OF BIOLOGICAL TERMS

ABO blood typing Method of using two surface proteins (A, B, or both) of red blood cells to characterize an individual's blood. O signifies the absence of both proteins.

abortion The spontaneous expulsion or the dislodging of an embryo or a fetus from the uterus.

abscisic acid (ab-SISS-ik) Plant hormone that promotes stomatal closure, bud dormancy, and seed dormancy.

abscission (ab-SIH-zhun) [L. *abscindere*, to cut off] The dropping of leaves, flowers, fruits, or other plant parts due to hormonal action.

absorption Of most animals, movement of nutrients, fluid, and ions across the gut lining and into the internal environment.

accessory pigment Light-trapping pigment molecule; it contributes to photosynthesis by extending the range of usable wavelengths beyond those absorbed by the chlorophylls.

acid [L. *acidus*, sour] A substance that releases hydrogen ions when dissolved in water.

acid rain The falling to Earth of rain (or snow) that contains sulfur and nitrogen oxides. Also called wet acid deposition (as opposed to dry acid deposition of airborne particles of sulfur and nitrogen oxides).

acoelomate (ay-SEE-luh-mate) Of some of the invertebrates, having no fluid-filled cavity between the gut and body wall.

acoustical signal Sounds that are normally used as a form of communication between animals of the same species.

actin (AK-tin) One of the motor proteins with roles in contraction. Interacts with myosin in muscle cells.

action potential Abrupt, brief reversal in the steady voltage difference across the plasma membrane (the resting membrane potential) of a neuron and other excitable cells.

activation energy The minimum amount of collision energy necessary to drive reactant molecules to an activated state (the transition state) at which a given chemical reaction will proceed spontaneously.

activator A regulatory protein having a role in a positive control system that promotes gene transcription.

active site A crevice in the surface of an enzyme molecule where a specific reaction is catalyzed, or made to proceed far faster than it would spontaneously.

active transport A pumping of one or more specific solutes through the interior of a transport protein that spans the lipid bilayer of a cell membrane. The solute is transported against its concentration gradient. An energy boost, as from ATP, activates the protein.

adaptation [L. *adaptare*, to fit] Of evolution, being adapted (or becoming more adapted) to a given set of environmental conditions. Of a sensory neuron, decreasing frequency of action potentials (or their cessation) even if a stimulus is maintained at constant strength.

adaptive behavior A behavior that promotes propagation of an individual's genes and that tends to increase in frequency in the population over time.

adaptive radiation A burst of divergences from a single lineage that gives rise to many new species, each adapted to an unoccupied or new habitat or to using a novel resource.

adaptive trait Any aspect of form, function, or behavior that helps an individual survive and reproduce under prevailing conditions.

adaptive zone A way of life available for organisms that are physically, ecologically, and evolutionarily equipped to live it, such as "catching insects in the air at night."

adenine (AH-de-neen) A purine; a nitrogen-containing base in certain nucleotides.

adenosine diphosphate (ah-DEN-uh-seen die-FOSS-fate) ADP, an organic compound that can make energy transfers in cells; typically formed by hydrolysis of ATP.

adenosine phosphate Any of a number of relatively small organic compounds, some of which function as chemical messengers within and between cells, and others (such as ATP) that function as energy carriers.

ADH Antidiuretic hormone. Hypothalamic hormone that induces water conservation as required during control of extracellular fluid volume and solute concentrations.

adhesion protein A protein that helps cells of the same type locate one another during development, adhere, and remain in the proper position in the proper tissue.

adipose tissue A type of connective tissue having an abundance of fat-storing cells and blood vessels for transporting fats.

ADP Adenosine diphosphate. A nucleotide coenzyme that accepts unbound phosphate or a phosphate group to become ATP.

ADP/ATP cycle In cells, a mechanism of ATP renewal. During a phosphate-group transfer, ATP reverts to ADP, then it forms again by phosphorylation of ADP.

aerobic respiration (air-OH-bik) [Gk. *aer*, air, + *bios*, life] The main pathway of ATP formation, for which oxygen is the final acceptor of electrons stripped from glucose or another organic compound. It proceeds from glycolysis through the Krebs cycle and electron transport phosphorylation. For each glucose molecule, a typical net yield is 36 ATP.

age structure The number of individuals in each of several or many age categories for a population.

agglutination (ah-glue-tin-AY-shun) In this defensive response, antibodies circulating in blood act against foreign cells (such as transfused red blood cells of the wrong type) and cause them to clump together.

aging A range of processes, including the breakdown of cell structure and function, by which the body gradually deteriorates. All multicelled species that show extensive cell differentiation undergo aging.

AIDS Short for acquired immunodeficiency syndrome. A set of chronic disorders that arises following infection by the human immunodeficiency virus (HIV), which destroys key cells of the immune system.

alcohol An organic compound that has one or more hydroxyl groups (—OH) and readily dissolves in water. Sugars are examples.

alcoholic fermentation Anaerobic pathway of ATP formation. Pyruvate from glycolysis is degraded to acetaldehyde, which accepts electrons from NADH to form ethanol with a net yield of two ATP. NAD$^+$ is regenerated.

aldosterone (al-DOSS-tuh-rohn) Hormone secreted by the adrenal cortex that helps regulate sodium reabsorption.

allantois (ah-LAN-twahz) [Gk. *allas*, sausage] One of four extraembryonic membranes that functions in respiration and in storing the metabolic wastes of embryos of reptiles, birds, and some mammals. In humans, it gives rise to blood vessels for the placenta and to the urinary bladder.

allele (uh-LEEL) At a given gene locus on a chromosome, one of two or more slightly different molecular forms of a gene that arise through mutation and that code for different versions of the same trait.

allele frequency The abundance of each kind of allele in the population as a whole.

allergen Normally harmless substance that provokes inflammation, excessive mucus secretion, and other defense responses.

allergy A response to an allergen.

allopatric speciation [Gk. *allos*, different, L *patria*, native land] Speciation that follows the end of gene flow between populations or subpopulations of a species as a result of their geographic isolation from each other.

allosteric control (AL-oh-STARE-ik) Form of control over a metabolic reaction or pathway that operates by binding a specific substance at a control site on a specific enzyme.

altruism (al-true-ISS-tik) Self-sacrificing behavior; an individual behaves in a way that helps others but decreases its own chances of reproductive success.

alveolus (ahl-VEE-uh-lus), plural **alveoli** [L. *alveus*, small cavity] One of the cupped, thin-walled outpouchings of respiratory bronchioles. A site where oxygen diffuses from air in the lungs to blood, and carbon dioxide diffuses from blood to the lungs.

amino acid (uh-MEE-no) A small organic molecule with a hydrogen atom, an amino group, an acid group, and an R group bonded covalently to a central carbon atom; the subunit of polypeptide chains.

ammonification (uh-moan-ih-fih-KAY-shun) A process by which certain soil bacteria and fungi break down nitrogenous wastes and remains of organisms; part of the nitrogen cycle.

amnion (AM-nee-on) Of land vertebrates, one of four extraembryonic membranes; the boundary layer of a fluid-filled sac that allows the embryo to grow in size, move freely, and be protected from sudden impacts and temperature shifts.

amniote egg An egg, often with a leathery or calcified shell, that has extraembryonic membranes, including the amnion.

amphibian A type of vertebrate somewhere between fishes and reptiles in body plan and reproductive mode; salamanders, frogs and toads, and caecilians are existing groups.

anaerobic pathway (an-uh-ROW-bik) [Gk. *an*, without, + *aer*, air] Metabolic pathway in which a substance other than oxygen serves as the final acceptor of electrons that have been stripped from substrates.

analogous structures (ann-AL-uh-gus) [Gk. *analogos*, similar to one another] Body parts that once differed in evolutionarily distant lineages, then converged in structure and function as the lineages responded to similar environmental pressures.

anaphase (AN-uh-faze) Of anaphase I of meiosis, the stage when each homologous chromosome separates from its partner and both move to opposite spindle poles. Of mitosis and of anaphase II of meiosis, sister chromatids of each chromosome separate and move to opposite poles.

aneuploidy (AN-yoo-ploy-dee) Having one more chromosome or one less relative to the parental chromosome number.

angiosperm (AN-gee-oh-spurm) [Gk. *angeion*, vessel, and *spermia*, seed] A flowering plant.

animal A multicelled heterotroph that feeds on other organisms, that is motile for at least part of the life cycle, that develops by a series of embryonic stages, and that usually has tissues, organs, and organ systems.

annelid A type of invertebrate classified as a segmented worm; an oligochaete (such as an earthworm), leech, or polychaete.

annual A flowering plant that completes its life cycle in one growing season.

anther [Gk. *anthos*, flower] A pollen-bearing part of a stamen.

antibiotic One of many metabolic products of certain microorganisms that can kill their bacterial competitors for nutrients in soil.

antibody [Gk. *anti*, against] One of a diverse array of antigen-binding receptors. Only B cells make antibody molecules and position them at their surface or secrete them.

anticodon A sequence of three nucleotide bases in a tRNA molecule that can base-pair with a codon in an mRNA molecule.

antigen (AN-tih-jen) [Gk. *anti*, against, + *genos*, race, kind] A molecular configuration that white blood cells recognize as foreign and that triggers an immune response. Most antigens are proteins at the surface of pathogens or tumor cells.

antigen-MHC complex Processed antigen fragments bound with a suitable MHC molecule and displayed at the cell surface; basis of antigen recognition that promotes lymphocyte cell divisions.

antigen-presenting cell Any cell displaying antigen-MHC complexes at its surface.

aorta (ay-OR-tah) [Gk. *airein*, to lift, heave] Main artery of systemic circulation; carries oxygenated blood away from the heart to all body regions except the lungs.

apical dominance Inhibitory influence of a terminal bud on growth of lateral buds.

apical meristem (AY-pih-kul MARE-ih-stem) [L. *apex*, top, + Gk. *meristos*, divisible] A mass of self-perpetuating cells responsible for primary growth at root and shoot tips.

apoptosis (APP-oh-TOE-sis) Of multicelled organisms, a form of cell death; molecular signals activate weapons of self-destruction already stockpiled in target cells. It occurs when a cell has completed its prescribed function or becomes altered, as by infection or cancerous transformation.

appendicular skeleton (ap-en-DIK-yoo-lahr) Bones of the limbs, hips, and shoulders.

Archaebacteria A kingdom of prokaryotes; encompasses methanogens, halophiles, and thermophiles, all of which differ from the eubacteria in chemical composition and in cell wall and membrane characteristics.

archipelago An island chain some distance away from a continent.

area effect Larger islands tend to support more species than smaller ones at equivalent distances from sources of colonizing species.

arteriole (ar-TEER-ee-ole) A blood vessel between an artery and capillary; a control point where the blood volume delivered to a given body region can be adjusted.

artery A large-diameter, rapid-transport blood vessel with a thick, muscular wall that smooths out the pulsations in blood pressure caused by heart contractions.

arthropod An invertebrate with a hardened exoskeleton, specialized body segments, and jointed appendages. Spiders, crabs, and insects are examples.

artificial selection Selection of traits among individuals of a population that occurs in an artificial environment, under contrived, manipulated conditions.

asexual reproduction Any of a number of modes of reproduction by which offspring arise from a single parent and inherit the genes of that parent only.

atmosphere A volume of gases, airborne particles, and water vapor that envelops the Earth; 80 percent of its mass is distributed within seventeen miles of the Earth's surface.

atmospheric cycle A biogeochemical cycle in which the atmosphere is the largest reservoir of an element. The carbon cycle and nitrogen cycle are examples.

atom The smallest particle unique to a given element; it has one or more positively charged protons, electrons, and (except for hydrogen), neutrons.

atomic number The number of protons in the nucleus of each atom of an element; the number differs for each element.

ATP Adenosine triphosphate (ah-DEN-uh-seen try-FOSS-fate). A nucleotide of adenine, ribose, and three phosphate groups that acts as an energy carrier. Its phosphate-group transfers drive nearly all energy-requiring metabolic reactions.

australopith (OHSS-trah-low-pith) [L. *australis*, southern, + Gk. *pithekos*, ape] Any of the earliest known hominids; a primate species on or near the evolutionary road that led to modern humans.

autoimmune response Misdirected immune response in which lymphocytes mount an attack against normal body cells.

autonomic nervous system (auto-NOM-ik) All nerves from the central nervous system to the smooth muscle, cardiac muscle, and glands of the viscera (internal organs and structures) of the vertebrate body.

autosome Any of the pairs of chromosomes that are the same in both males and females of the species.

autotroph (AH-toe-trofe) [Gk. *autos*, self, + *trophos*, feeder] Organism that synthesizes its own organic compounds using carbon dioxide (as the carbon source) and energy from the physical environment (such as sunlight energy). *Compare* heterotroph.

auxin (AWK-sin) A plant hormone that influences growth, such as stem elongation.

axial skeleton (AX-ee-uhl) Of a vertebrate skeleton, the skull, backbone, ribs, and breastbone (sternum).

axon A cylindrical extension from the cell body of a neuron, often with finely branched endings, that is specialized for the rapid propagation of action potentials.

B lymphocyte (B cell) The only white blood cell that produces antibodies, then positions them at the cell surface or secretes them as weapons in immune responses.

bacterial conjugation A transfer of plasmid DNA from one bacterial cell to another.

bacterial flagellum Of many bacterial cells, a whiplike motile structure that does not contain a core of microtubules.

bacteriophage (bak-TEER-ee-oh-fahj) [Gk. *baktērion*, small staff, rod, + *phagein*, to eat] Category of viruses that infect bacterial cells.

balancing selection All forms of selection that are maintaining two or more alleles for a trait in a population. When the resulting genetic variation persists, this is a case of balanced polymorphism.

Barr body In cells of female mammals, one of two X chromosomes that was randomly condensed so that its genes are inactivated.

basal body A centriole which, after giving rise to microtubules of a flagellum or cilium, remains attached to its base in the cytoplasm.

base Any substance that accepts hydrogen ions when dissolved in water.

base pair Two nucleotide bases that are located in two adjacent strands of DNA or RNA and that are hydrogen-bonded to each other.

base sequence The particular order in which one nucleotide base follows the next in a strand of DNA or RNA. The order is unique in at least some regions for each species.

basophil Fast-acting white blood cell that secretes histamine and other substances to maintain an inflammatory response.

behavior, animal A response to external and internal stimuli that requires sensory, neural, endocrine, and effector components. Behavior has a genetic basis and can evolve; it also can be modified through learning.

benthic province All sediments and rocky formations of the ocean bottom.

biennial (bi-EN-yul) A flowering plant that lives through two growing seasons.

bilateral symmetry Body plan in which the left and right halves of an animal are mirror-images of each other.

binary fission Of flatworms and some other animals, a mode of asexual reproduction; growth by way of mitotic cell divisions is followed by division of the whole body into two parts of the same or different sizes. Not the same as prokaryotic fission, by which bacteria reproduce.

biogeochemical cycle The movement of an element from the environment to organisms, then back to the environment.

biogeographic realm [Gk. *bios*, life, + *geographein*, to describe the Earth's surface] One of six major land divisions, each with distinguishing plants and animals; it retains its general identity because of climate and geographic barriers to gene flow.

biological clock Internal time-measuring mechanism that helps adjust an organism's daily activities, seasonal activities, or both in response to environmental cues.

biological magnification The increasing concentration of a nondegradable or slowly degradable substance in body tissues as it is passed along food chains.

biological species concept A species is one or more populations of individuals that are interbreeding under natural conditions and producing fertile offspring, and that are reproductively isolated from other such populations. The concept applies only to sexually reproducing species.

bioluminescence A flashing of light that emanates from an organism when excited electrons of luciferins (highly fluorescent substances) return to a lower energy level.

biomass Combined weight of all organisms at a given trophic level in an ecosystem.

biome A broad, vegetational subdivision of a biogeographic realm; shaped by climate, topography, and composition of regional soils.

biosphere [Gk. *bios*, life, + *sphaira*, globe] All regions of the Earth's waters, crust, and atmosphere in which organisms live.

biosynthetic pathway A metabolic pathway by which organic compounds necessary for life are synthesized.

biotic potential Of population growth for a given species, the maximum rate of increase per individual under ideal conditions.

bipedalism Habitually walking on two feet, as by ostriches and humans.

bird The only vertebrate that produces feathers and that has strong resemblances and evolutionary connections to reptiles.

blastocyst (BLASS-tuh-sist) [Gk. *blastos*, sprout, + *kystis*, pouch] Outcome of cleavage of a fertilized mammalian egg; a surface layer of blastomeres around a cavity filled with their secretions (the blastocoel), and an inner cell mass (several blastomeres huddled together against the cavity's inner surface).

blastomere One of the small, nucleated cells produced during the cleavage stage of animal development.

blastula (BLASS-chew-lah) Outcome of one type of cleavage pattern; blastomeres formed by the successive cuts surround a fluid-filled cavity (blastocoel).

blood A fluid connective tissue composed of water, solutes, and formed elements (blood cells and platelets); it carries substances to and from cells and helps maintain an internal environment favorable for cell activities.

blood pressure Fluid pressure, generated by heart contractions, that circulates blood.

blood-brain barrier Mechanism that exerts some control over which solutes enter the cerebrospinal fluid and thus helps protect the brain and spinal cord.

bone Mineral-hardened connective tissue of bones; one of the organs of the vertebrate skeleton that help move the body and its parts, protect of other organs, store minerals, and (in some) produce blood cells.

bottleneck A severe reduction in population size, as brought about by intense selection pressure or a natural calamity.

Bowman's capsule The cup-shaped portion of a nephron that receives water and solutes being filtered from the blood in the kidneys.

brain The most complex integrating center of most nervous systems; it receives, processes, and integrates sensory input and issues coordinated commands for response by muscles and glands.

brain stem The vertebrate nervous tissue that evolved first and that still persists in the hindbrain, midbrain and forebrain.

bronchiole A component of the finely branched bronchial tree inside each lung.

bronchus, plural **bronchi** (BRONG-cuss, BRONG-kee) [Gk. *bronchos*, windpipe] Tube-like branchings of the trachea that lead into the lungs of most vertebrates.

brown alga A photoautotrophic protistan with a notable abundance of xanthophyll pigments; such algae occupy marine habitats.

bryophyte A nonvascular land plant, such as a moss.

bud An undeveloped shoot of meristematic tissue, primarily; often covered and protected by scales (modified leaves).

buffer system A partnership between a weak acid and the base that forms when it dissolves in water. The two work as a pair to counter slight shifts in pH.

bulk A volume of fiber and other undigested material that absorption processes in the small intestine cannot decrease.

bulk flow In response to a pressure gradient, the movement of more than one kind of molecule in the same direction in the same medium (as in blood, sap, or air).

C4 pathway A pathway of photosynthesis in which carbon dioxide is fixed twice, in two different cell types. Carbon dioxide accumulates in the leaf and helps counter photorespiration. The first compound formed is the 4-carbon oxaloacetate.

Calvin-Benson cycle Cyclic reactions that are the *synthesis* part of the light-independent reactions of photosynthesis. In plants, RuBP or some other compound to which carbon has been affixed undergoes rearrangements; a sugar phosphate forms and the RuBP is regenerated. The cycle runs on ATP and NADPH from the light-dependent reactions.

CAM plant A plant that conserves water by opening stomata only at night, when it fixes carbon dioxide by way of a C4 pathway.

cambium (KAM-bee-um), plural **cambia** One of two types of meristems responsible for secondary growth (increases in stem and root diameter). Vascular cambium gives rise to secondary xylem and secondary phloem; cork cambium gives rise to periderm.

camouflage Adaptations in body color, form, or patterning, and in behavior that function in predator avoidance; they help an organism hide in the open (blend with its surroundings) and escape detection.

cancer A malignant tumor; its cells show gross abnormalities in the plasma membrane and cytoplasm, skewed growth and division, and weakened capacity for adhesion within the parent tissue (leading to metastasis). Unless eradicated, cancer is lethal.

capillary, blood [L. *capillus*, hair] A thin-walled vessel that functions in the exchange of carbon dioxide, oxygen, and some other substances between blood and interstitial fluid, which bathes living cells.

capillary bed A diffusion zone, consisting of a great number of capillaries, where substances are exchanged between blood and interstitial fluid.

carbohydrate [L. *carbo*, charcoal, + *hydro*, water] A molecule that consists of carbon, hydrogen, and oxygen in a 1:2:1 ratio (there are exceptions). All cells use carbohydrates as structural materials, energy reservoirs, and transportable forms of energy. The monosaccharides, oligosaccharides, and polysaccharides are three classes.

carbon cycle A biogeochemical cycle in which carbon moves from the atmosphere (its largest reservoir), through the ocean and organisms, then to the atmosphere.

carbon dioxide fixation Of photosynthesis, the first enzyme-mediated step of the light-independent reactions. Carbon (from CO_2) is affixed to RuBP or another compound for entry into the Calvin-Benson cycle.

carcinogen (kar-SIN-uh-jen) A substance or an agent, such as ultraviolet radiation, that can trigger cancer.

cardiac cycle (KAR-dee-ak) [Gk. *kardia*, heart, + *kyklos*, circle] The sequence of muscle contraction and relaxation for one heartbeat.

cardiac pacemaker Sinoatrial (SA) node; the basis of the normal rate of heartbeat. The self-excitatory cardiac muscle cells that spontaneously generate rhythmic waves of excitation over the heart chambers.

cardiovascular system Of most animals, an organ system of blood, one or more hearts, and blood vessels that functions in the rapid transport of substances to and from cells.

carnivore [L. *caro, carnis*, flesh, + *vorare*, to devour] An animal that eats other animals; a type of heterotroph.

carotenoid (kare-OTT-en-oyds) A light-sensitive, accessory pigment that transfers absorbed energy to chlorophylls. Different types absorb violet and blue wavelengths and transmit red, orange, and yellow.

carpel (KAR-pul) The female reproductive part of a flower; sometimes called a pistil. The lower portion of a single carpel (or of a structure composed of two or more) is an ovary. The upper portion has a stigma (a pollen-capturing surface tissue) and often a style (slender extension of the ovary wall).

carrying capacity The maximum number of individuals in a population (or species) that can be sustained indefinitely by a given environment.

cartilage A type of connective tissue with solid yet pliable intercellular material that resists compression.

Casparian strip A waxy band that is an impermeable barrier between the walls of abutting cells making up the endodermis (and exodermis, if present) inside roots.

cDNA Any DNA molecule copied from a mature mRNA transcript by way of reverse transcription.

cell [L. *cella*, small room] The smallest living unit; an organized unit that can survive and reproduce on its own, given suitable DNA instructions and environmental resources—notably energy and raw materials.

cell count The number of cells of a given type in a microliter of blood.

cell cycle Events by which a cell increases in mass, roughly doubles its number of cytoplasmic components, duplicates its DNA, then undergoes nuclear and cytoplasmic division. It extends from the time a new cell is produced until it completes division.

cell differentiation Developmental process in which different cell populations activate and suppress a fraction of their genes in different ways and so become specialized in composition, structure, and function.

cell junction Of multicelled organisms, a point of contact that links two adjoining cells physically, functionally, or both.

cell plate A disklike structure that forms from remnants of a microtubular spindle when a plant cell divides; it develops into a crosswall that partitions the cytoplasm.

cell theory A theory in biology stating that (1) all organisms are composed of one or more cells, (2) the cell is the smallest unit that retains a capacity for independent life, and (3) all cells arise from preexisting cells.

cell wall A semirigid, permeable structure that helps a cell hold its shape and resist rupturing if internal fluid pressure rises.

central nervous system The brain and spinal cord of vertebrates.

central vacuole A fluid-filled organelle in mature, living plant cells that stores amino acids, sugars, ions, and toxic wastes. As it enlarges, it forces increases in cell surface area that improve nutrient uptake.

centriole (SEN-tree-ohl) A cylinder of triplet microtubules that gives rise to microtubules of cilia and flagella.

centromere (SEN-troh-meer) [Gk. *kentron*, center, + *meros*, a part] A small, constricted region of a chromosome having attachment sites for the microtubules that move the chromosome during nuclear division.

cephalization (sef-ah-lah-ZAY-shun) [Gk. *kephalikos*, head] During the evolution of bilateral animals, the concentration of sensory structures and nerve cells in a head.

cerebellum (ser-ah-BELL-um) [L. diminutive of *cerebrum*, brain] Hindbrain region with reflex centers for maintaining posture and smoothing out limb movements.

cerebral cortex Thin surface layer of the cerebral hemispheres. Some parts receive sensory input, others integrate information and coordinate suitable responses.

cerebrospinal fluid Clear extracellular fluid surrounding and cushioning the brain and spinal cord.

cerebrum (suh-REE-bruhm) Forebrain region that first evolved to integrate olfactory input and select motor responses to it. In mammals, it evolved into the most complex integrating center.

channel protein A transport protein that acts as a channel through which specific ions and other water-soluble substances cross the plasma membrane. Some channels remain open; others are gated, and these open and close in controlled ways.

chemical bond A union between the electron structures of two or more atoms or ions.

chemical synapse (SIN-aps) [Gk. *synapsis*, union] A small cleft between a presynaptic neuron and a postsynaptic cell (another neuron, a muscle cell, or a gland cell) that is bridged by neurotransmitter molecules released from the presynaptic neuron.

chemiosmotic theory (kim-ee-OZ-MOT-ik) An electrochemical gradient across a cell membrane drives ATP formation. Hydrogen ions accumulate in a compartment formed by the membrane. The combined force of the H+ concentration and electric gradients propels ions through transport proteins (ATP synthases) spanning the membrane. By enzyme action at these proteins, ADP and inorganic phosphate combine to form ATP.

chemoreceptor (KEE-moe-ree-sep-tur) A sensory receptor that detects chemical energy (ions or molecules) dissolved in the fluid that bathes it.

chemosynthetic autotroph (KEE-moe-sin-THET-ik) One of a few kinds of bacteria able to synthesize its own organic compounds by using carbon dioxide as the carbon source and certain inorganic substances (such as sulfur) as the energy source.

chlorofluorocarbon (KLORE-oh-FLOOR-oh-car-bun), or **CFC** One of the odorless, invisible compounds of chlorine, fluorine, and carbon, widely used in commercial products, that are contributing to the thinning of the ozone layer above the Earth's surface.

chlorophyll (KLOR-uh-fills) [Gk. *chloros*, green, + *phyllon*, leaf] A light-sensitive pigment that absorbs violet-to-blue and red wavelengths but that transmits green. The main pigments in all but one small group of photoautotrophs. Certain chlorophylls donate electrons to the light-dependent reactions of photosynthesis.

chloroplast (KLOR-uh-plast) An organelle that specializes in photosynthesis in plants and photosynthetic protistans.

chordate An animal having a notochord, a dorsal hollow nerve cord, a pharynx, and gill slits in the pharynx wall for at least part of the life cycle.

chorion (CORE-ee-on) Of placental mammals, one of four extraembryonic membranes; it becomes a key component of the placenta. Absorptive structures (villi) develop at its surface and enhance the rapid exchange of substances between the embryo and mother.

chromatid (CROW-mah-tid) Of a duplicated eukaryotic chromosome, one of two DNA molecules (and associated proteins) that remain attached at their centromere region until separated by mitosis or meiosis; after this, each is a separate chromosome.

chromosome (CROW-moe-some) [Gk. *chroma*, color, + *soma*, body] Of eukaryotes, a DNA molecule with many associated proteins. Of prokaryotes, a DNA molecule without a comparable profusion of proteins.

chromosome number The sum total of chromosomes in cells of a given type. *See* haploidy; diploidy.

cilium (SILL-ee-um), plural **cilia** [L. *cilium*, eyelid] Of eukaryotic cells, a short, hairlike projection with an internal, regular array of microtubules. Cilia can serve as motile or sensory structures or help create currents of fluids. Typically more profuse than flagella.

circadian rhythm (ser-KAYD-ee-un) [L. *circa*, about, + *dies*, day] A cycle of physiological events that is completed every twenty-four hours or so independently of environmental change.

circulatory system An organ system having a muscular pump (heart, most often), blood vessels, and blood; the system transports materials to and from cells and often helps stabilize body temperature and pH.

cladogram [Gk. *clad-*, branch] Evolutionary tree diagram that arranges groups by branch points to show their relative relationships. Groups closer together share a more recent common ancestor than those farther apart.

classification system A way of organizing and retrieving information about species.

cleavage Third stage of animal embryonic development. Mitotic cell divisions divide the volume of egg cytoplasm into a number of smaller, nucleated cells (blastomeres). The number of cells increases, but the original volume of egg cytoplasm does not.

cleavage furrow A ringlike depression that forms during cytoplasmic division of an animal cell and that defines the cleavage plane for the cell. Microfilaments attached to the plasma membrane contract and draw it inward to cut the cell in two.

cleavage reaction A molecule splits into two smaller ones. Hydrolysis is an example.

climate Prevailing weather conditions for an ecosystem, such as temperature, humidity, wind speed, cloud cover, and rainfall.

climax community A self-perpetuating, stable array of species in equilibrium with one another and with their habitat.

climax pattern model Idea that one climax community may extend into another along gradients of environmental conditions, such as variations in climate, topography, and species interactions.

cloaca Of some vertebrates, the last part of a gut that receives feces, urine, and sperm or eggs; of some invertebrates, an excretory, respiratory, or reproductive duct.

cloned DNA Multiple, identical copies of restriction fragments that have been inserted into plasmids or some other cloning vector.

club fungus A fungus with reproductive structures having microscopic, club-shaped cells that produce and bear spores.

cnidarian A radial invertebrate at the tissue level of organization and the only organism to produce nematocysts. Two body forms (medusae and polyps) are common.

coal A nonrenewable source of energy that formed more than 280 million years ago from submerged, undecayed plant remains.

codominance A pair of nonidentical alleles that specify two phenotypes are expressed at the same time in heterozygotes.

codon One of the base triplets in an mRNA molecule, the linear sequence of which corresponds to a linear sequence of amino acids in a polypeptide chain. Of 64 codons, 61 specify different amino acids, and 3 of these also are start signals for translation; 1 serves as a stop signal for translation.

coelom (SEE-lum) [Gk. *koilos*, hollow] A cavity, lined with peritoneum, between the gut and body wall of most animals.

coenzyme A nucleotide; an enzyme helper that accepts electrons and hydrogen atoms stripped from substrates at a reaction site and transfers them elsewhere.

coevolution The joint evolution of two or more closely interacting species; when one species evolves, the change affects selection pressures operating between the two, so the other also evolves.

cofactor A metal ion or coenzyme; it helps an enzyme catalyze a reaction or transfers electrons, atoms, or functional groups from one substrate to another.

cohesion Capacity to resist rupturing when placed under tension (stretched).

cohesion theory of water transport Theory that water moves up through plants due to hydrogen bonding among water molecules confined as narrow columns in xylem. The collective cohesive strength of the bonds allows water to be pulled up in response to transpiration (evaporation from leaves).

collenchyma (coll-ENG-kih-mah) A simple plant tissue that offers flexible support for primary growth, as in lengthening stems.

colon (CO-lun) The large intestine.

commensalism [L. *com*, together, + *mensa*, table] An ecological interaction between species that directly benefits one but does not affect other much, if at all.

communication display A pattern of behavior, often ritualized with intended changes in the function of common behavior patterns, that serves as a social signal.

communication signal A social cue encoded in stimuli that holds unambiguous meaning for individuals of the same species. Specific odors, sounds, coloration and patterning, postures, and movements are examples.

community All populations living in the same habitat. Also, a group of organisms with similar life-styles in a habitat, such as a community of birds.

companion cell A specialized parenchyma cell that helps load organic compounds into conducting cells of phloem.

comparative morphology [Gk. *morph*, form] Study of comparable body parts of adults or embryonic stages of major lineages.

competitive exclusion Theory that species that require identical resources cannot coexist indefinitely.

complement system A set of about twenty proteins circulating in inactive form within vertebrate blood; different kinds induce lysis of pathogens, promote inflammation, and stimulate phagocytes to act during both nonspecific defenses and immune responses.

compound A substance consisting of two or more elements in unvarying proportions.

concentration gradient A difference in the number of molecules or ions of a substance between adjoining regions. Energy inherent in their constant molecular motion makes them collide and career outward from the region of higher to lower concentration. Barring other forces, all substances tend to diffuse down their concentration gradient.

condensation reaction Through covalent bonding, two molecules combine to form a larger molecule, often with the formation of water as a by-product.

cone cell In a vertebrate eye, a photoreceptor that responds to intense light and contributes to sharp daytime vision and color perception.

conifer A pollen- and seed-bearing plant of the dominant group of gymnosperms; mostly evergreen, woody trees and shrubs with needle-like or scale-like leaves.

conjugation, bacterial Of bacteria only, a mechanism by which a donor cell transfers plasmid DNA to a recipient cell.

connective tissue proper A category of animal tissues, all having mostly the same components but in different proportions. They incorporate fibroblasts and other cells, the secretions of which form fibers (mostly of collagen and elastin) and a ground substance of modified polysaccharides.

consumer [L. *consumere*, to take completely] A heterotroph that obtains energy and carbon by feeding on the tissues of other organisms. Herbivores, carnivores, and parasites are examples.

continuous variation Of a population, a more or less continuous range of small differences in a given trait among all of its individuals.

contractile vacuole (kun-TRAK-till VAK-you-ohl) [L. *contractus*, to draw together] Of some protistans, such as a paramecium, an organelle that takes up excess water in the cell body, then contracts; the contractile force is enough to expel the water outside the cell through a pore to its surface.

control group Of an experimental test, a group used to evaluate possible side effects of a test involving an experimental group. Ideally, the control group is identical to the experimental group in all respects except for the variable being studied.

cork cambium A lateral meristem that gives rise to a corky replacement for the epidermis of woody plant parts.

corpus callosum (CORE-pus ka-LOW-sum) A band of axons (200 million in humans) that functionally link two cerebral hemispheres.

corpus luteum (CORE-pus LOO-tee-um) A glandular structure that develops from cells of a ruptured ovarian follicle and secretes progesterone and estrogen.

cortex [L. *cortex*, bark] In general, a rindlike layer such as the kidney or adrenal cortex. In vascular plants, the ground tissue that makes up most of the primary plant body, supports plant parts, and stores food.

cotyledon A seed leaf, which develops as part of the embryo of monocots and dicots; cotyledons provide nourishment for the seedling at the time of germination and initial growth.

courtship display A pattern of ritualized social behavior between potential mates. It may include frozen postures as well as movements that are exaggerated and yet simplified. It may include visual signals, such as body parts that are conspicuously enlarged, distinctively colored or patterned, or some combination of these.

covalent bond (koe-VAY-lunt) [L. *con*, together, + *valere*, to be strong] A sharing of one or more electrons between atoms or groups of atoms. If electrons are shared equally, the bond is nonpolar. If shared unequally, it is polar (slightly positive at one end, slightly negative at the other).

continuous variation Of the individuals of a population, a range of small differences in one or more traits.

cross-bridge formation Of a muscle cell, a reversible interaction between its many actin and myosin filaments that is the basis of contraction.

crossing over During prophase I of meiosis, the breakage and exchange of corresponding segments between nonsister chromatids of a pair of homologous chromosomes; a form of genetic recombination that breaks up old combinations of alleles and puts new ones together in chromosomes.

culture The sum of behavior patterns of a social group, passed between generations by learning and by symbolic behavior, especially language.

cuticle (KEW-tih-kull) A body covering. Of land plants, a transparent cover of waxes and lipid-rich cutin deposited on the outer surface of epidermal cell walls. Of annelids, a thin, flexible surface coat. Of arthropods, a hardened, lightweight cover with protein and chitin components that functions as an exoskeleton.

cyclic AMP (SIK-lik) A nucleotide; cyclic adenosine monophosphate. It functions in intercellular communication, as when it is a second messenger (a cytoplasmic mediator of a cell's response to signaling molecules).

cyclic pathway of ATP formation Ancient photosynthetic pathway occurring at the plasma membrane of some bacteria and at the thylakoid membrane of chloroplasts. A photosystem embedded in the membrane gives up electrons to a transport system, which gives them back to the photosystem. The electron flow sets up concentration and electric gradients across the membrane that drive ATP formation at nearby membrane sites.

cyst Of many microorganisms, a resistant resting stage with thick, tough outer layers that typically forms in response to adverse conditions; of skin, any abnormal, fluid-filled sac without an external opening.

cytochrome (SIGH-toe-krome) [Gk. *kytos*, hollow vessel, + *chrōma*, color] Iron-containing protein molecule; a component of the electron transport systems used in photosynthesis and aerobic respiration.

cytokinesis (SIGH-toe-kih-NEE-sis) [Gk. *kinesis*, motion] Cytoplasmic division; the splitting of a parent cell into daughter cells.

cytokinin (SIGH-tow-KY-nin) Any of the class of plant hormones that stimulate cell division, promote leaf expansion, and retard leaf aging.

cytological marker One or more observable, unusual differences between chromosomes of the same type.

cytomembrane system [Gk. *kytos*, hollow vessel] Organelles functioning as a system to modify, package, and distribute newly formed proteins and lipids. Endoplasmic reticulum, Golgi bodies, lysosomes, and a variety of vesicles are its components.

cytoplasm (SIGH-toe-plaz-um) [Gk. *plassein*, to mold] All cellular parts, particles, and semifluid substances enclosed within the plasma membrane except for the nucleus (or nucleoid, in bacterial cells).

cytoplasmic localization When cleavage divides an animal zygote, each resulting blastomere receives a localized portion of maternal messages in the egg cytoplasm.

cytosine (SIGH-toe-seen) A pyrimidine; one of the nitrogen-containing bases in nucleotides.

cytoskeleton The internal "skeleton" of eukaryotic cells. Its microtubules and other components structurally support the cell and organize and move its internal components. The cytoskeleton also helps free-living cells move through their environment.

cytotoxic T cell A T lymphocyte that uses touch-killing to eliminate infected body cells or tumor cells. When it contacts targets, it delivers cell-killing chemicals into them.

decomposer [partly fr. L. *dis-*, to pieces, + *companere*, arrange] Of ecosystems, a heterotroph that gets energy and carbon by chemically breaking down the remains, products, or wastes of other organisms and helps cycle nutrients back to producers. Certain fungi and bacteria are examples.

deforestation The removal of all trees from a large tract of land, such as the Amazon Basin and the Pacific Northwest.

degradative pathway A metabolic pathway by which organic compounds are broken down in stepwise reactions that lead to products of lower energy.

deletion Loss of a chromosome segment.

denaturation (deh-NAY-chur-AY-shun) Of any molecule, the loss of three-dimensional shape following disruption of hydrogen bonds and other weak bonds.

dendrite (DEN-drite) [Gk. *dendron*, tree] A short, slender extension from the cell body of a neuron; commonly an input zone.

denitrification (DEE-nite-rih-fih-KAY-shun) Conversion of nitrate or nitrite by certain bacteria to gaseous nitrogen (N_2) and a small amount of nitrous oxide (N_2O).

density-dependent control A factor that limits population growth by reducing the birth rate, increasing the rates of death and dispersal, or all of these. Predation, parasitism, disease, and competition for resources are examples.

density-independent factor A factor that tends to cause a population's death rate to increase independently of its density. Storms and floods are examples.

dentition (den-TIH-shun) The type, size, and number of an animal's teeth.

derived trait A novel feature that evolved only once and is shared only by descendants of the ancestral species in which it evolved.

dermal tissue system All the tissues that cover and protect the surfaces of a plant.

dermis The layer of skin underlying the epidermis; consists primarily of dense connective tissue.

desert A biome that typically forms where the potential for evaporation greatly exceeds rainfall and vegetation cover is limited.

desertification (dez-urt-ih-fih-KAY-shun) Conversion of a grassland or an irrigated or rain-fed cropland to a desertlike condition, with a drop in agricultural productivity of 10 percent or more.

detrital food web A network of food chains in which energy flows mainly from plants through arrays of detritivores and decomposers.

detritivore (dih-TRY-tih-vorez) [L. *detritus*; after *deterere*, to wear down] A heterotroph that consumes decomposing particles of organic matter. Earthworms, crabs, and nematodes are examples.

deuterostome (DUE-ter-oh-stome) [Gk. *deuteros*, second, + *stoma*, mouth] A bilateral animal for which the first indentation that forms in the early embryo develops into an anus. An echinoderm or a chordate.

development Of multicelled organisms, the programmed emergence of specialized, morphologically different body parts.

diaphragm (DIE-uh-fram) [Gk. *diaphragma*, to partition] Muscular partition between the thoracic and abdominal cavities; its contraction and relaxation contribute to breathing. Also, a contraceptive device used temporarily to prevent sperm from entering the uterus during sexual intercourse.

dicot (DIE-kot) [Gk. *di*, two, + *kotylēdōn*, cup-shaped vessel] A dicotyledon. In general, a flowering plant characterized by seeds having embryos with two cotyledons (seed leaves); net-veined leaves; and floral parts arranged in fours, fives, or multiples of these.

diffusion Net movement of like molecules (or ions) down their concentration gradient. In the absence of other forces, the energy inherent in molecules makes them move constantly and collide at random. Their collisions are most frequent where they are most crowded together; thus they show a net outward movement from regions of higher to lower concentration.

digestive system An internal sac or tube from which ingested food is absorbed into the internal environment.

dihybrid cross An experimental cross in which true-breeding F_1 offspring inherit two gene pairs, each consisting of two nonidentical alleles.

diploidy (DIP-loyd-ee) The presence of two of each type of chromosome (that is, pairs of homologous chromosomes) in the interphase nucleus of somatic cells and germ cells. *Compare* haploidy.

directional selection A mode of natural selection by which the range of variation for some trait shifts in a consistent direction in response to directional change in the environment or to new environmental conditions.

disaccharide (die-SAK-uh-ride) [Gk. *di*, two, + *sakcharon*, sugar] A simple carbohydrate; one of the oligosaccharides consisting of two covalently bonded sugar monomers.

disease Outcome of an infection when the body's defenses cannot be mobilized fast enough; the pathogen's activities interfere with normal body functions.

disruptive selection A mode of natural selection by which forms of a trait at both ends of a range of variation are favored and intermediate forms are selected against.

distal tubule The tubular portion of a nephron farthest from the glomerulus; a region of water and sodium reabsorption.

distance effect Only species adapted for long-distance dispersal are potential colonists of islands far from their home range.

diversity, organismic Sum total of all the variations in form, function, and behavior that have accumulated in different lineages. Variations in traits generally are adaptive to prevailing conditions or were adaptive to conditions that existed in the past.

DNA Deoxyribonucleic acid (dee-OX-ee-RYE-bow-new-CLAY-ik). For all cells and many viruses, a nucleic acid that is the molecule of inheritance. It consists of two nucleotide strands twisted together helically and held together by numerous hydrogen bonds. The nucleotide sequence encodes instructions for synthesizing proteins and, ultimately, new individuals of a particular species.

DNA amplification Any of several methods by which a DNA library is copied again and again to yield multiple, identical copies of DNA fragments (cloned DNA).

DNA-DNA hybridization *See* nucleic acid hybridization.

DNA fingerprint A unique array of RFLPs, inherited in a Mendelian pattern from each parent, that gives each individual a unique identity.

DNA library A collection of DNA fragments produced by restriction enzymes and later incorporated into plasmids.

DNA ligase (LYE-gaze) An enzyme that seals together the new base-pairings during DNA replication; also used by technologists to seal base-pairings between DNA fragments and cut plasmid DNA.

DNA polymerase (poe-LIM-uh-raze) An enzyme that assembles a new strand on a parent DNA strand during replication; also takes part in DNA repair.

DNA probe A short DNA sequence that is synthesized from radioactively labeled nucleotides. Part of the probe is designed to base-pair with part of a gene under study.

DNA repair Following an alteration in the base sequence of a DNA strand, a process that may restore the original sequence, as carried out by DNA polymerases, DNA ligases, and other enzymes.

DNA replication Of cells, the process by which hereditary material is duplicated for distribution to daughter nuclei. Occurs prior to mitosis and meiosis in eukaryotic cells and during prokaryotic fission in bacterial cells.

dominance hierarchy A social organization in which some members of the group have adopted a subordinate status to others.

dominant allele In a diploid cell, an allele that masks the expression of its partner on the homologous chromosome.

dormancy [L. *dormire*, to sleep] A hormone-mediated time of inactivity during which metabolic activities idle. Perennials, seeds, many spores, cysts, and some animals go through dormancy.

double fertilization Of flowering plants only, the fusion of one sperm nucleus with the egg nucleus (to produce a zygote), *and* the fusion of a second sperm nucleus with nuclei of the endosperm mother cell, which gives rise to a nutritive tissue (endosperm).

doubling time The length of time it takes for a population to double in size.

drug addiction Chemical dependence on a drug following habituation and tolerance of it; in time the drug assumes an "essential" biochemical role in the body.

dry shrubland A biome that typically forms where annual rainfall is less than 25 to 60 centimeters; short, multibranched woody shrubs (e.g., chaparral) predominate.

dry woodland A biome that typically forms where annual rainfall is about 40 to 100 centimeters; there may be tall trees, but these do not form a dense canopy.

duplication A repeat of the same linear stretch of an individual's DNA in the same chromosome or in a different one.

ecdysone A hormone with major influence over the development of many insects.

echinoderm A type of invertebrate that has calcified spines, needles, or plates on the body wall. Although radially symmetrical, it has some bilateral features. Sea stars and sea urchins are examples.

ecology [Gk. *oikos*, home, + *logos*, reason] Study of the interactions of organisms with one another and with their physical and chemical environment.

ecosystem [Gk. *oikos*, home] An array of organisms and their physical environment, all of which are interacting through a flow of energy and a cycling of materials.

ecosystem modeling An analytical method, based on computer programs and models, of predicting unforeseen effects of specific disturbances to an ecosystem.

ectoderm [Gk. *ecto*, outside, + *derma*, skin] The first-formed, outermost primary tissue layer of animal embryos; forerunner of cell lineages that give rise to nervous system tissues and the integument's outer layer.

effector Of homeostatic systems, a muscle (or gland) that responds to signals from an integrator, such as the brain, by producing movement (or chemical change) that helps adjust the body to changing conditions.

effector cell A differentiated cell of one of the subpopulations of lymphocytes that form during an immune response; it acts at once to engage and destroy the antigen-bearing agent that triggered the response.

egg A type of mature female gamete; also called an ovum.

El Niño A recurring, massive eastward displacement of warm surface waters of the western equatorial Pacific, which displaces cooler waters off the South American coast. Causes global disruptions in climate.

electromagnetic spectrum The entire range of wavelengths, from the forms of radiant energy less than 10^{-5} nanometer long to radio waves more than 10 kilometers long.

electron A negatively charged unit of matter, with both particulate and wavelike properties, that occupies one of the orbitals around the atomic nucleus. Atoms can gain, lose, or share electrons with other atoms.

electron transfer The donation of one or more electrons stripped from one molecule to another molecule.

electron transport phosphorylation (FOSS-for-ih-LAY-shun) Final stage of aerobic respiration, when electrons from reaction intermediates flow through a membrane transport system that gives them up to oxygen. The flow sets up electrochemical gradients that drive ATP formation at other sites in the membrane.

electron transport system Organized array of enzymes and cofactors, bound in a cell membrane, that accept and donate electrons in series. When it operates, hydrogen ions flow across the membrane, and the flow drives ATP formation and other reactions.

element A substance that cannot be broken down to substances with different properties.

embryo (EM-bree-oh) [Gk. *en*, in, + probably *bryein*, to swell] Of animals generally, a stage formed by cleavage, gastrulation, and other early developmental events. Of seed plants, the young sporophyte, from the first cell divisions after fertilization until germination.

embryonic induction A change in the developmental fate of an embryonic tissue, as brought about by exposure to a gene product released from an adjacent tissue.

embryo sac Common name of the female gametophyte of flowering plants.

emerging pathogen A deadly pathogen, either a newly mutated strain of an existing species or one that evolved long ago and is only now taking great advantage of the increased presence of human hosts.

emulsification Of the chyme in the small intestine, a suspension of droplets of fat coated with bile salts.

end product A substance present at the end of a metabolic pathway.

endangered species A species at the brink of extinction owing to the extremely small size and severely limited genetic diversity of its remaining populations.

endergonic reaction (en-dur-GONE-ik) A chemical reaction having a net gain in energy.

endocrine gland A ductless gland that secretes hormones, which usually enter interstitial fluid and then the bloodstream.

endocrine system System of cells, tissues, and organs, functionally linked to the nervous system, that exerts control by way of its hormones and other chemical secretions.

endocytosis (EN-doe-sigh-TOE-sis) Transport of a substance into a cell by a vesicle, the membrane of which is a patch of plasma membrane that forms around the substance and sinks into the cytoplasm. Phagocytes also engulf prey or pathogens this way.

endoderm [Gk. *endon*, within, + *derma*, skin] The innermost primary tissue layer of animal embryos; gives rise to the inner lining of the gut and organs derived from it.

endodermis A sheetlike wrapping of single cells around the vascular cylinder of a root that functions in controlling the uptake of water and dissolved nutrients.

endometrium (EN-doh-MEET-ree-um) [Gk. *metrios*, of the womb] Innermost lining of the uterus, consisting of connective tissues, glands, and blood vessels.

endoplasmic reticulum or **ER** (EN-doe-PLAZ-mik reh-TIK-yoo-lum) An organelle that begins at the nucleus and curves through the cytoplasm. In rough ER (with many ribosomes on its cytoplasmic side), many new polypeptide chains acquire specialized side chains. Smooth ER (with no attached ribosomes) is a site of lipid synthesis.

endoskeleton [Gk. *endon*, within, + *skleros*, hard, stiff] An internal framework of bone, cartilage, or both in chordates. Together with skeletal muscle, supports and protects other body parts, helps maintain posture, and moves the body.

endosperm (EN-doe-sperm) Nutritive tissue that surrounds a flowering plant embryo and becomes food for the young seedling.

endospore A resting structure that forms around a copy of the chromosome and part of the cytoplasm of certain bacteria.

endosymbiosis In general, a mutually beneficial interdependence between two species, one of which resides permanently inside the other's body.

energy A capacity to do work.

energy carrier A molecule that delivers energy from one metabolic reaction site to another. ATP is the most common energy carrier in all cells.

energy flow pyramid A pyramid-shaped representation of an ecosystem's trophic structure, illustrating the energy losses at each transfer to a different trophic level.

enhancer A base sequence in DNA that is a binding site for an activator protein.

entropy (EN-trow-pee) A measure of the degree of disorder in a system (how much energy has become so disorganized and dispersed, usually as heat, that it is no longer readily available to do work). Any organized system tends toward entropy without energy inputs to make up for the flow of energy out of it.

enzyme (EN-zime) One of a class of proteins that enormously speed (catalyze) reactions between specific substances, usually at their functional groups.

eosinophil Fast-acting white blood cell; its enzyme secretions digest holes in parasitic worms during an inflammatory response.

epidermis The outermost tissue layer of a multicelled plant and of nearly all animals.

epinephrine (ep-ih-NEF-rin) Hormone of the adrenal medulla; raises blood levels of sugar and fatty acids; increases heart rate and the force of contraction.

epiglottis A flaplike structure at the start of the larynx, the position of which directs the movement of air into the trachea or of food into the esophagus.

epistasis (eh-PISS-tah-sis) An interaction between gene pairs. Two alleles of one gene mask expression of another gene's alleles, so expected phenotypes may not appear.

epithelium (EP-ih-THEE-lee-um) An animal tissue of one or more layers of adhering cells that covers the body's external surfaces and lines its internal cavities and tubes. It has one free surface; the opposite surface rests on a basement membrane between it and an underlying connective tissue. Epidermis is an example.

equilibrium, dynamic [Gk. *aequus*, equal, + *libra*, balance] The point at which a chemical reaction runs forward as fast as in reverse; the concentrations of reactant molecules and product molecules show no net change.

erosion The movement of land under the force of wind, running water, and ice.

erythrocyte (eh-RITH-row-site) [Gk. *erythros*, red, + *kytos*, vessel] Red blood cell.

esophagus (ee-SOF-uh-gus) Tubular portion of a digestive system that receives ingested food and leads to the stomach.

essential amino acid An amino acid that an organism cannot synthesize for itself and must obtain from a food source.

essential fatty acid A fatty acid that an organism cannot synthesize for itself and must obtain from food source.

estrogen (ESS-trow-jen) A sex hormone that helps oocytes mature, induces changes in the uterine lining during the menstrual cycle and pregnancy, and helps maintain secondary sexual traits; also influences bodily growth and development.

estrus (ESS-truss) [Gk. *oistrus*, frenzy] For mammals generally, the cyclic period of a female's sexual receptivity to the male.

estuary (EST-you-ehr-ee) A partly enclosed coastal region where seawater mixes with freshwater and runoff from the surrounding land, as by streams and rivers.

ethylene (ETH-il-een) Plant hormone that stimulates fruit ripening and abscission.

Eubacteria Kingdom of the most common species of bacterial cells.

eukaryotic cell (yoo-CARRY-oh-tic) [Gk. *eu*, good, + *karyon*, kernel] A cell having a "true nucleus" and other distinguishing membrane-bound organelles. *Compare* prokaryotic cell.

eutrophication Nutrient enrichment of a body of water, such as a lake, that typically results in reduced transparency and a phytoplankton-dominated community.

evaporation [L. *e-*, out, + *vapor*, steam] Heat energy converts a substance from the liquid to the gaseous state.

evolution, biological [L. *evolutio*, unrolling] Genetic change in a line of descent over time; brought about by microevolutionary processes (gene mutation, natural selection, genetic drift, and gene flow).

evolutionary systematics The branch of biology that applies evolutionary theory to the task of identifying patterns of diversity over time and in the environment.

evolutionary tree A treelike diagram in which the branches represent separate lines of descent from a common ancestor and branch points represent divergences.

excitatory postsynaptic potential (or EPSP) One of two competing signals at an input zone of a neuron; a graded potential that brings the neuron's plasma membrane closer to threshold.

excretion Any of several processes by which excess water, excess or harmful solutes, or waste materials leave the body by way of a urinary system or certain glands.

exergonic reaction (EX-ur-GONE-ik) A chemical reaction that shows a net loss in energy.

exocrine gland (EK-suh-krin) [Gk. *es*, out of, + *krinein*, to separate] Glandular structure that secretes products, usually through ducts or tubes, to a free epithelial surface.

exocytosis (EK-so-sigh-TOE-sis) Transport of a substance out of a cell by means of a vesicle, the membrane of which fuses with the plasma membrane, so that the vesicle's contents are released outside.

exodermis Layer of cells just inside the root epidermis of most flowering plants; helps control the uptake of water and solutes.

exon Any of the nucleotide sequences of a pre-mRNA molecule that become spliced together to form a mature mRNA transcript and ultimately get translated into protein.

exoskeleton [Gk. *exo*, out, + *skleros*, hard, stiff] An external skeleton, as in arthropods.

experiment A test of potentially falsifiable hypotheses about some aspect of nature. Its premise is that any aspect of the natural world has one or more underlying causes.

exponential growth (EX-po-NEN-shul) A pattern of population growth in which the population size expands by ever increasing increments during successive time intervals because the reproductive base becomes ever larger. The plot of population size against time has a characteristic J-shaped curve.

extinction, background A steady rate of species turnover that characterizes lineages through most of their histories.

extinction, mass An abrupt increase in the rate at which major taxa disappear, with several taxa being affected simultaneously.

extracellular fluid In animals generally, all the fluid not inside cells; includes plasma (the liquid portion of blood) and interstitial fluid (occupying the spaces between cells and tissues).

extracellular matrix A matrix that helps impart shape to many animal tissues; its ground substance contains fibrous proteins and other materials (mostly cell secretions).

FAD Flavin adenine dinucleotide, one of the nucleotide coenzymes that transfers electrons and unbound protons (H^+) from one reaction site to another. At such times it is abbreviated $FADH_2$.

fall overturn The vertical mixing of a body of water in autumn. Its upper layer cools, increases in density, and sinks; dissolved oxygen moves down and nutrients from bottom sediments move up.

family pedigree A chart of the genetic relationship of the individuals in a family through successive generations.

fat A lipid with a glycerol head and one, two, or three fatty acid tails. Tryglycerides (neutral fats) have three. Unsaturated tails have single covalent bonds in their carbon backbone; saturated tails also have one or more double bonds.

fate map A map of the surface of an animal embryo that shows the origin of each kind of differentiated cell in the adult.

fatty acid A molecule with a backbone of up to thirty-six carbon atoms, a carboxyl group (—COOH) at one end, and hydrogen atoms at most or all of the remaining bonding sites.

feedback inhibition Of cells or multicelled organisms, a control mechanism by which the output of a substance changes a specific condition or activity, which then triggers a decrease in further output of the substance or further activity.

fermentation [L. *fermentum*, yeast] A type of anaerobic pathway of ATP formation. It starts with glycolysis, ends with a transfer of electrons back to one of the breakdown products or intermediates, and regenerates NAD^+ required for the reaction. Its has a net yield of two ATP per glucose molecule.

fertilization [L. *fertilis*, to carry, to bear] The fusion of a sperm nucleus with the nucleus of an egg, which thus becomes a zygote.

fever A body temperature higher than a set point in the hypothalamic region that acts as the body's thermostat.

fibrous root system Of most monocots, all the lateral branchings of adventitious roots, which arose earlier from the young stem.

filter feeder An animal that filters food from a current of water that is directed through a body part, such as a sea squirt's pharynx.

filtration Of vertebrates, a process by which blood pressure forces water and solutes from capillaries into interstitial fluid. Filtration occurs in Bowman's capsule of a nephron.

fin Of fishes generally, an appendage that helps propel, stabilize, and guide the body through water.

first law of thermodynamics [Gk. *therme*, heat, + *dynamikos*, powerful] A law of nature stating the total amount of energy in the universe remains constant. Energy cannot be created from nothing and existing energy cannot be destroyed.

fish An aquatic animal of the most ancient, diverse vertebrate lineage, which includes jawless, cartilaginous, and bony fishes.

fitness An increase in adaptation to the environment, as brought about by genetic change.

fixation Of the individuals of a population, only one kind of allele remains at a specified locus; all individuals are homozygous for it.

fixed action pattern Of animals, a program of coordinated, stereotyped muscle activity that runs to completion independently of feedback from the environment.

flagellum (fluh-JELL-um), plural **flagella** [L. whip] Tail-like motile structure of many free-living eukaryotic cells; its core has a 9 + 2 array of microtubules.

flower A reproductive structure that distinguishes angiosperms from other seed plants and often attracts pollinators.

fluid mosaic model All cell membranes consist of a lipid bilayer and proteins. The lipids (phospholipids, mainly) impart basic structure, impermeability to water-soluble molecules, and (through packing variations and movements) fluidity. Diverse proteins spanning the bilayer or attached to one of its surfaces perform most of the membrane functions, including transport, enzyme activity, and reception of molecular signals or substances.

follicle (FOLL-ih-kul) A small sac, pit, or cavity, as around a hair; also a mammalian oocyte with its surrounding layer of cells.

food chain A straight-line sequence of who eats whom in an ecosystem.

food web A network of cross-connecting, interlinked food chains with some number of producers and consumers, as well as decomposers, detritivores, or both.

forebrain Most complex part of a vertebrate brain; it includes the cerebrum (and cerebral cortex), olfactory lobes, and hypothalamus.

forest A biome where tall trees grow together closely enough to form a fairly continuous canopy over a broad region.

fossil Recognizable, physical evidence of an organism that lived in the distant past.

fossil fuel Coal, petroleum, or natural gas; a nonrenewable source of energy formed in sediments by the compression of plant remains over hundreds of millions of years.

fossilization How fossils form. First an organism or traces of it becomes buried in sediments or volcanic ash. Water infiltrates the remains, which become infused with dissolved inorganic compounds. Sediments accumulate and exert pressure above the burial site. Over great spans of time, the pressure and chemical changes transform the remains to stony hardness.

founder effect A form of bottlenecking. By chance, allele frequencies of founders of a new population may not be the same as those of the original population. If there is no gene flow between the two, then natural selection will influence gene frequencies in drastically different ways through its interaction with genetic drift.

free radical A highly reactive, unbound molecular fragment that has the wrong number of electrons.

fruit [L. after *frui*, to enjoy] Of flowering plants, the expanded and ripened ovary of one or more carpels, some with accessory floral structures incorporated.

FSH Follicle-stimulating hormone; one of the hormones produced and secreted by the anterior lobe of the pituitary gland; serves reproductive roles in both sexes.

functional group An atom or group of atoms that is covalently bonded to the carbon backbone of an organic compound and that influences its chemical behavior.

functional-group transfer Donation of a functional group by one molecule to another.

Fungi The kingdom of fungi which, as a group, are major decomposers.

fungus A eukaryotic heterotroph that uses extracellular digestion and absorption; it secretes enzymes that break down an external food source into molecules small enough to be absorbed by its cells. Saprobes feed on nonliving organic matter, parasites feed on living organisms.

gall bladder Organ that stores bile secreted from the liver and that is connected by way of a duct to the small intestine.

gamete (GAM-eet) [Gk. *gametēs*, husband, and *gametē*, wife] A haploid cell, formed by meiosis and cytoplasmic division of a germ cell; required for sexual reproduction. Eggs and sperm are examples.

gamete formation Of animals, the first stage of development, in which sperm or eggs form and mature within reproductive tissues or organs of parents.

gametophyte (gam-EET-oh-fite) [Gk. *phyton*, plant] A haploid, multicelled, gamete-producing body that forms during the life cycle of most plants.

ganglion (GANG-lee-un), plural **ganglia** [Gk. *ganglion*, a swelling] A distinct clustering of cell bodies of neurons in regions other than the brain or spinal cord.

gastrulation (gas-tru-LAY-shun) The fourth stage of animal development; a time of major cellular reorganization when newly formed cells become arranged into two or three primary tissues, or germ layers.

gene [short for German *pangan*, after Gk. *pan*, all + *genes*, to be born] A unit of information about a heritable trait that is passed on from parents to offspring. Each gene has a specific location (locus) on a chromosome.

gene flow A microevolutionary process; the movement of alleles into and out of populations as a result of immigration and emigration.

gene frequency More precisely, allele frequency; the relative abundances of all the different alleles at a given gene locus that are carried by all individuals of a population.

gene locus The particular location of a gene along the length of a chromosome.

gene mutation [L. *mutatus*, a change] A heritable change in a DNA molecule by the deletion, addition, or substitution of one to several bases in its nucleotide sequence.

gene pair Of diploid cells, the two alleles at a particular gene locus (that is, on a pair of homologous chromosomes).

gene pool Sum total of all genotypes in a population.

gene therapy Generally, the transfer of one or more normal genes into an organism to correct or lessen the adverse effects of a genetic disorder.

genetic code [After L. *genesis*, to be born] The correspondence between the nucleotide triplets in DNA (then in RNA) and the specific sequences of amino acids in the resulting polypeptide chains; the basic language of protein synthesis in cells.

genetic disease An illness in which the expression of one or more genes increased the susceptibility of an individual to an infection or weakened its immune response.

genetic disorder An inherited condition that results in mild to severe medical problems.

genetic divergence Build-up of differences in gene pools of two or more populations of a species after a geographic barrier arises and separates them, because gene mutation, natural selection, and genetic drift are free to operate independently in each one.

genetic drift A random change in allele frequencies over the generations brought about by chance alone. The magnitude of its effect on genetic diversity and on the range of phenotypes relates to population size.

genetic engineering Altering the information content of DNA through use of recombinant DNA technology.

genetic equilibrium A state in which a population is not evolving. It occurs only if there is no mutation, if the population is large and isolated from other populations of the species, and if there is no natural selection (all members survive and reproduce equally by random mating).

genetic recombination A nonparental combination of some number of alleles in offspring; an outcome of gene mutation, crossing over, changes in chromosome structure or number, or recombinant DNA technology.

genome All the DNA in a haploid number of chromosomes of a given species.

genotype (JEEN-oh-type) Genetic constitution of an individual; a single gene pair or the sum total of an individual's genes. *Compare* phenotype.

genus, plural **genera** (JEEN-US, JEN-er-ah) [L. *genus*, race, origin] A taxon into which all species with phenotypic similarities and evolutionary relationship are grouped.

geographic dispersal Directional movement in which some residents of an established community leave their home range and take up residence elsewhere; they are considered to be exotic species in the new location.

geologic time scale A time scale for Earth history, the subdivisions of which are based on boundaries marked by episodes of mass extinction and which have been refined by radiometric dating.

germ cell Of animals, a cell of a lineage set aside for sexual reproduction; germ cells give rise to gametes. *Compare* somatic cell.

germ layer Of animal embryos, one of two or three primary tissue layers that form at gastrulation as forerunners of adult tissues. *Compare* ectoderm; endoderm; mesoderm.

germination (jur-min-AY-shun) Of resting spores and seeds, a resumption of activity following a period of arrested development.

gibberellin (JIB-er-ELL-un) A type of plant hormone that promotes stem elongation.

gill A respiratory organ, typically with a moist, thin vascularized layer of epidermis that functions in gas exchange.

gland A secretory cell or structure derived from epithelium and often connected to it.

glomerular capillaries Set of blood capillaries inside Bowman's capsule of the nephron.

glomerulus (glow-MARE-you-luss) [L. *glomus*, ball] First portion of the nephron, where water and solutes are filtered from blood.

glucagon (GLUE-kuh-gone) A hormone that stimulates cells to convert glycogen and amino acids to glucose; secreted by alpha cells of the pancreas when the blood level of glucose decreases.

glyceride (GLISS-er-eyed) One of the fats or oils; a molecule with one, two, or three fatty acid tails attached to a glycerol backbone.

glycerol (GLISS-er-oh) [Gk. *glykys*, sweet, + L. *oleum*, oil] A three-carbon molecule with three hydroxyl groups attached; one of the components of fats and oils.

glycogen (GLY-kuh-jen) A highly branched polysaccharide that is the main storage carbohydrate of animals; cleavage reactions at its many branchings yield an abundance of glucose monomers when required.

glycocalyx A sticky mesh of polysaccharides, polypeptides, or both around the cell wall of many bacteria.

glycolysis (gly-CALL-ih-sis) [Gk. *glykys*, sweet, + *lysis*, loosening or breaking apart] Initial energy-releasing reactions of aerobic and anaerobic pathways by which enzymes break down glucose (or some other organic compound) to pyruvate. It proceeds in the cytoplasm of all cells, it has a net yield of two ATP, and oxygen has no role in it.

glycoprotein A protein having linear or branched oligosaccharides covalently bonded to it. Nearly all surface proteins of animal cells and many proteins circulating in blood are glycoproteins.

gnetophyte Only gymnosperm known to have vessels in its xylem.

Golgi body (GOHL-gee) Organelle of lipid assembly, polypeptide chain modification, and packaging of both in vesicles for export or for transport to locations in the cytoplasm.

gonad (GO-nad) Primary reproductive organ in which animal gametes are produced.

graded potential Of neurons, a local signal that slightly alters the voltage difference across a patch of plasma membrane and that varies in magnitude according to the stimulus. With prolonged or intense stimulation, such signals may spread to a trigger zone of the membrane and initiate an action potential.

granum, plural **grana** In many chloroplasts, any of the stacks of flattened, membranous compartments with chlorophyll and other light-trapping pigments and reaction sites for ATP formation.

grassland A biome with flat or rolling land, 25–100 centimeters of annual rainfall, warm summers, grazing animals, and periodic fires that regenerate dominant species.

gravitropism (GRAV-ih-TROPE-izm) [L. *gravis*, heavy, + Gk. *trepein*, to turn] Tendency of a plant to grow directionally in response to the Earth's gravitational force.

gray matter Inside the brain and spinal cord, the unmyelinated axons, dendrites, and nerve cell bodies and neuroglial cells.

grazing food web A network of food chains in which energy flows from plants to an array of herbivores, then carnivores.

green alga An aquatic protistan with chlorophylls *a* and *b*; early members of its lineage may have given rise to plants.

green revolution In developing countries, the use of improved crop varieties, modern agricultural practices (including massive inputs of fertilizers and pesticides), and equipment to increase crop yields.

greenhouse effect Warming of the lower atmosphere as a result of the presence of increasing levels of greenhouse gases (such as carbon dioxide and methane).

ground meristem (MARE-ih-stem) [Gk. *meristos*, divisible] A primary meristem that produces the ground tissue system, hence the bulk of the plant body.

ground substance Of certain animal tissues, intercellular material made of cell secretions and other noncellular components.

ground tissue system Tissues making up the bulk of a plant body, the most common being parenchyma.

growth Of multicelled organisms, increases in the number, size, and volume of cells.

guanine A nitrogen-containing base; present in one of the four nucleotide building blocks of DNA and RNA.

guard cell Either of two adjacent cells with roles in the movement of carbon dioxide, oxygen, and water vapor across leaf or stem epidermis. An opening (stoma) forms when both swell with water and move apart; it closes when they lose water and collapse against one another.

gut A body region where food is digested and absorbed; of complete digestive systems, the gastrointestinal tract (the portions from the stomach onward).

gymnosperm (JIM-noe-sperm) [Gk. *gymnos*, naked, + *sperma*, seed] A vascular plant that bears seeds at exposed surfaces of reproductive structures, such as cone scales.

habitat [L. *habitare*, to live in] The type of place where an organism normally lives, as characterized by physical and chemical features and by its array of species.

hair cell A mechanoreceptor that may give rise to action potentials when bent or tilted.

half-life The time it takes for half of a given quantity of any radioisotope to decay into a different, less unstable daughter isotope.

halophile A type of archaebacterium that lives in extremely saline habitats.

haploidy (HAP-loyd) The presence of half the parental number of chromosomes in a gamete, as brought about by meiosis; the gamete has one of each pair of homologous chromosomes. *Compare* diploidy.

Hardy-Weinberg rule Allele frequencies will stay the same through the generations if there is no mutation, if the population is infinitely large and is isolated from other populations of the same species, if mating is random, and if all individuals survive and reproduce equally.

HCG Human chorionic gonadotropin. A hormone that helps maintain the lining of the uterus during the menstrual cycle and during the first trimester of pregnancy.

heart Muscular pump that keeps blood circulating through the animal body.

helper T cell A T lymphocyte which, when activated, produces and secretes chemicals that induce responsive T or B lymphocytes to divide and give rise to large populations of effector cells and memory cells.

hemoglobin (HEEM-oh-glow-bin) [Gk. *haima*, blood, + L. *globus*, ball] Iron-containing, oxygen-transporting protein that gives red blood cells their color.

hemostasis (HEE-mow-STAY-sis) [Gk. *haima*, blood, + *stasis*, standing] Stopping of blood loss from a damaged blood vessel through coagulation, blood vessel spasm, platelet plug formation, and other mechanisms.

herbivore [L. *herba*, grass, + *vovare*, to devour] Plant-eating animal.

hermaphrodite An individual with both male and female gonads; two individuals sexually reproduce by the mutual transfer of sperm.

heterocyst (HET-er-oh-sist) A self-modified cyanobacterial cell that makes a nitrogen-fixing enzyme when nitrogen is scarce.

heterotroph (HET-er-oh-trofe) [Gk. *heteros*, other, + *trophos*, feeder] Organism unable to synthesize its own organic compounds; it feeds on autotrophs, other heterotrophs, or organic wastes. *Compare* autotroph.

heterozygous condition (HET-er-oh-ZYE-guss) [Gk. *zygoun*, join together] Of a specified trait, having a pair of nonidentical alleles at a gene locus (on a pair of homologous chromosomes).

higher taxon (plural, **taxa**) One of the ever more inclusive groupings meant to reflect relationships among species. Family, order, class, phylum, and kingdom are examples.

hindbrain One of three divisions of the vertebrate brain; the medulla oblongata, cerebellum, and pons. Has reflex centers for respiration, blood circulation, and other basic functions; also coordinates motor responses and many complex reflexes.

histone Any of a class of proteins that are intimately associated with eukaryotic DNA and largely responsible for the organization of eukaryotic chromosomes.

homeostasis (HOE-me-oh-STAY-sis) [Gk. *homo*, same, + *stasis*, standing] Of animals, a physiological state in which physical and chemical aspects of the internal environment (blood and interstitial fluid) are maintained within ranges suitable for cell activities.

homeotic gene One of a class of master genes that specify the development of specific body part in animals.

hominid [L. *homo*, man] All species on or near the evolutionary road leading to modern humans.

hominoid Apes, humans, and their recent ancestors.

homologous chromosome (huh-MOLL-uh-gus) [Gk. *homologia*, correspondence] Of cells having a diploid chromosome number, one of a pair of chromosomes that are identical in size, shape, and gene sequence, and that interact during meiosis. Nonidentical sex chromosomes in a cell also interact during meiosis and are considered homologues also.

homology Similarity in one or more body parts in different species that is attributable to their descent from a common ancestor.

homologous structures The same body parts, modified in different ways, in different lines of descent from a common ancestor.

homozygous condition (HOE-moe-ZYE-guss) For a specified trait, having a pair of identical alleles at a gene locus (on a pair of homologous chromosomes).

homozygous dominant condition Having a pair of dominant alleles at a gene locus (on a pair of homologous chromosomes).

homozygous recessive condition Having a pair of recessive alleles at a gene locus (on a pair of homologous chromosomes).

hormone [Gk. *hormon*, to stir up, set in motion] Signaling molecule that stimulates or inhibits gene transcription in nonadjacent target cells (any cell having receptors for it).

horsetail A seedless vascular plant with photosynthetic stems that look like horsetails.

human genome project Worldwide basic research project to sequence the estimated 3 billion nucleotides present in the DNA of human chromosomes.

humus Decomposing organic matter in soil.

hydrogen bond A weak interaction between a small, highly electronegative atom of a molecule and a neighboring hydrogen atom that is taking part in a polar covalent bond.

hydrogen ion A free (unbound) proton; a hydrogen atom that has lost its electron and so bears a positive charge (H^+).

hydrologic cycle A biogeochemical cycle, driven by solar energy, in which water moves slowly through the atmosphere, on or through surface layers of land masses, to the ocean, and back again.

hydrolysis (high-DRAWL-ih-sis) [L. *hydro*, water, + Gk. *lysis*, loosening or breaking apart] Cleavage reaction in which covalent bonds break, splitting a molecule into two or more parts. Often H^+ and OH^- (derived from a water molecule) become attached to the exposed bonding sites.

hydrophilic substance [Gk. *philos*, loving] A polar substance that is attracted to the polar water molecule and dissolves easily in water. Sugars are examples.

hydrophobic substance [Gk. *phobos*, dreading] A nonpolar substance that is repelled by the polar water molecule and thus resists being dissolved in water. Oil is an example.

hydrosphere All liquid or frozen water on or near the Earth's surface.

hydrostatic pressure Any volume of fluid that exerts a force directed against a wall, membrane, or another structure enclosing that fluid. The greater its concentration of solutes, the greater will be the pressure that the fluid exerts.

hydrothermal vent ecosystem Ecosystem, near a fissure in the ocean floor, based on chemosynthetic bacteria that use dissolved minerals as their energy source.

hypha (HIGH-fuh), plural **hyphae** [Gk. *hyphe*, web] Of fungi, a filament with chitin-reinforced walls and, often, reinforcing cross-walls; component of the mycelium.

hypodermis A subcutaneous layer having stored fat that helps insulate the body; although not part of skin, it anchors skin and allows it some freedom of movement.

hypothalamus [Gk. *hypo*, under, + *thalamos*, inner chamber or possibly *tholos*, rotunda] Of the forebrain, a major center for homeostatic control of visceral activities (such as salt-water balance, temperature control, and reproduction), related forms of behavior (as in hunger, thirst, and sex), and emotional expression, such as sweating with fear.

hypothesis In science, a possible explanation of a specific phenomenon in nature, one that has the potential to be proved false by experimental tests.

immune response Events by which B and T lymphocytes recognize antigen, undergo cell divisions that form huge populations of lymphocytes, which differentiate into subpopulations of effector and memory cells. The effector cells destroy cells bearing antigen-MHC complexes. Memory cells are not activated until subsequent encounters with the same antigen.

immunization Various processes, including vaccination, that promote increased immunity against specific diseases.

immunoglobulin (Ig) One of five classes of antibodies, each with antigen-binding sites and other sites with specialized functions.

implantation Process by which a blastocyst adheres to the endometrium and establishes connections by which the mother and embryo will exchange substances during pregnancy.

imprinting A time-dependent form of learning that is triggered by exposure to sign stimuli and that usually occurs during a sensitive period when the animal is young.

inbreeding Nonrandom mating among close relatives, which have many identical alleles in common. Inbreeding is a form of genetic drift in a small group of relatives that are preferentially interbreeding.

incomplete dominance One allele of a pair is not fully dominant over its partner, so a heterozygous phenotype in between the two homozygous phenotypes emerges.

independent assortment, theory of By the end of meiosis in a germ cell, each pair of homologous chromosomes—hence the genes that they carry—have been sorted for shipment into gametes independently of how the other pairs were sorted out.

indirect selection A theory in evolutionary biology that self-sacrificing individuals can pass on their genes indirectly by helping relatives survive and reproduce.

induced-fit model A substrate induces change in the shape of an enzyme's active site when bound to it, the result being a more precise molecular fit between the two that promotes reactivity.

infection Invasion and multiplication of a pathogen in host cells or tissues. Disease follows if defenses cannot be mobilized fast enough to prevent the pathogen's activities from interfering with normal functions.

inflammation, acute Important aspect of nonspecific defenses and immune responses when cells of a local tissue are damaged or killed, as by infection; requires action of fast-acting phagocytes and plasma proteins, including complement proteins.

inheritance The transmission, from parents to offspring, of structural and functional patterns that have a genetic basis and are characteristic of their species.

inhibiting hormone A signaling molecule produced and secreted by the hypothalamus that suppresses a particular secretion by the anterior lobe of the pituitary gland.

inhibitor A substance that can bind with an enzyme and interfere with its functioning.

inhibitory postsynaptic potential (IPSP) Of neurons, one of two competing types of graded potentials at an input zone; it tends to drive the resting membrane potential away from threshold.

instinctive behavior A behavior performed without having been learned by experience in the environment. The nervous system of a newly born or hatched animal is prewired to recognize one or two sign stimuli (simple, well-defined cues in the environment) that can trigger a suitable response.

insulin Hormone secreted by beta cells of the pancreas that lowers the glucose level in blood by stimulating cells to take up glucose; also promotes protein and fat synthesis and inhibits protein conversion to glucose.

integration, neural [L. *integrare*, coordinate] Moment-by-moment summation of all the excitatory and inhibitory synapses acting on the neuron; it takes place at each level of synapsing in a nervous system.

integrator Of homeostatic systems, a control point such as a brain where information is pulled together in the selection of responses to stimuli.

integument Of animals, a protective body cover such as skin. Of seed-bearing plants, one or more layers around an ovule that harden, thicken, and form a seed coat.

integumentary exchange (in-teg-you-MEN-tuh-ree) A mode of respiration in which oxygen and carbon dioxide diffuse across a thin, moist, vascularized layer at the body surface of certain animals.

interleukin One of several chemical mediator molecules secreted by helper T cells that fan mitotic cell divisions and differentiation of responsive T and B cells.

intermediate A compound formed between the start and end of a metabolic pathway.

intermediate filament In different types of animal cells, a cytoskeletal element made of particular proteins.

interneuron Any neuron of the vertebrate brain and spinal cord.

internode In vascular plants, the stem region between two successive nodes.

interphase Of the cell cycle, an interval in between nuclear divisions when a cell increases in mass, roughly doubles the number of its cytoplasmic components, and then duplicates its chromosomes (replicates its DNA). The interval differs among species.

interspecific competition Individuals of different species compete with one another for a share of resources in their habitat.

interstitial fluid (IN-ter-STISH-ul) [L. *interstitus*, to stand in the middle of something] That portion of extracellular fluid occupying the spaces between cells and tissues of complex animals.

intertidal zone Generally, part of a rocky or sandy shoreline above the low water mark and below the high water mark; organisms that inhabit it are alternately submerged, then exposed, by tides.

intervertebral disk One of a number of disk-shaped structures containing cartilage that act as shock absorbers and flex points between bony segments of the vertebral column.

intraspecific competition All individuals of a population compete with one another for a share of resources in their habitat.

intron A noncoding portion of a newly formed mRNA molecule.

inversion A linear stretch of DNA within a chromosome that has become oriented in the reverse direction, with no molecular loss.

invertebrate Animal without a backbone.

in vitro fertilization Conception outside the body (literally, "in glass" petri dishes or test tubes).

ion, negatively charged (EYE-on) An atom or a compound that acquired an overall negative charge by gaining one or more electrons.

ion, positively charged An atom or a compound that acquired an overall positive charge by losing one or more electrons.

ionic bond Ions of opposite charge have attracted each other and are staying together.

isotonic condition Equality in the relative concentrations of solutes in two fluids; for two fluids separated by a cell membrane, there is no net osmotic (water) movement across the membrane.

isotope (EYE-so-tope) Of an element, an atom with more or fewer neutrons than the atoms having the most common number.

J-shaped curve The type of curve that emerges when population size is plotted against time; it represents unrestricted, exponential growth.

joint An area of contact or near-contact between bones.

juvenile Of many animals, a miniaturized form between the embryo and adult; it simply changes in size and proportion until it reaches sexual maturity.

karyotype (CARRY-oh-type) For an individual (or a species), a preparation of metaphase chromosomes sorted by length, centromere location, and other defining features.

keratin A tough, water-insoluble protein made by most epidermal cells that becomes concentrated in skin's outermost layers.

keratinization (care-AT-in-iz-AY-shun) Process by which keratin-producing epidermal cells die and accumulate as keratinized bags at the skin surface to form a barrier against dehydration, bacteria, many toxins, and ultraviolet radiation.

keystone species A species that dominates a community and dictates its structure.

key innovation A modified structure or function that allows a lineage to exploit the environment in more efficient or novel ways.

kidney One of a pair of vertebrate organs that filter mineral ions, organic wastes, and other substances from the blood; it controls the amounts returned to blood and thereby helps maintain the volume and solute levels of extracellular fluid.

kilocalorie 1,000 calories of heat energy; the amount of energy required to raise the temperature of 1 kilogram of water by 1°C; unit of measure for the caloric value of foods.

kinase One of a class of enzymes that catalyze phosphate-group transfers.

kinetochore Group of proteins and DNA at a chromosome's centromere where spindle microtubules become attached at mitosis or meiosis. Each chromatid of a duplicated chromosome has its own kinetochore.

Krebs cycle A cyclic pathway that occurs in mitochondria; together with a few preparatory steps, the stage of aerobic respiration in which pyruvate is completely broken down to carbon dioxide and water. Coenzymes accept unbound protons (H^+) and electrons stripped from intermediates and deliver them to the next stage.

lactate fermentation Anaerobic pathway of ATP formation; pyruvate from glycolysis is converted to the three-carbon compound lactate, and NAD^+ is regenerated. The net energy yield is two ATP.

lactation Milk production by hormone-primed mammary glands.

lake A body of standing freshwater with littoral, limnetic, and profundal zones.

large intestine Colon; a gut region that receives unabsorbed food residues from the small intestine and concentrates and stores feces until they are expelled from the body.

larva, plural larvae Of many animals, an immature developmental stage between the embryo and adult.

larynx (LARE-inks) A tubular airway to and from the lungs. In humans, it contains vocal cords.

lateral meristem A type of meristem in plants that show secondary growth; either vascular cambium or cork cambium.

lateral root Of taproot systems, a lateral branching from the first, primary root.

leaching The removal of some nutrients in soil as water percolates through it.

leaf For most vascular plants, a structure having chlorophyll-containing tissue that is the major region of photosynthesis.

learned behavior Variation or change in responses to stimuli after an animal has processed and integrated information gained from specific experiences.

lek Of some birds and other animals, a type of communal display ground occupied during times of courtship.

lethal mutation A mutation with drastic effects on phenotype that usually cause the individual's death.

LH Luteinizing hormone. Hormone secreted by the anterior lobe of the pituitary gland, with roles in male and female reproduction.

lichen (LY-kun) A symbiotic interaction between a fungus and a photoautotroph, such as a green alga.

life cycle A recurring pattern of genetically programmed events by which individuals grow, develop, maintain themselves, and reproduce.

life table Tabulation of age-specific patterns of birth and death for a population.

ligament A strap of dense connective tissue that bridges a joint.

light-dependent reactions The first stage of photosynthesis, in which sunlight energy is trapped and converted to the chemical energy of ATP alone (by a cyclic pathway) or ATP and NADPH (by a noncyclic pathway).

light-independent reactions The second stage of photosynthesis, in which ATP makes phosphate-group transfers required to build sugar phosphates. Often NADPH delivers electrons and hydrogen atoms for the synthesis reactions, which also require carbon from carbon dioxide. The sugar phosphates enter other reactions by which starch, cellulose, and other end products of photosynthesis are assembled.

lignification Of land plants, a process by which lignin is deposited in secondary cell walls. Lignin imparts strength and rigidity by anchoring cellulose strands in the walls, stabilizes and protects other components of the wall, and forms a waterproof barrier around the cellulose. A key factor in the evolution of vascular plants.

lignin A complex organic compound that strengthens and waterproofs cell walls in certain tissues of vascular plants.

limbic system Brain centers, located in the middle of the cerebral hemispheres, that interact to govern emotions and influence memory. Distantly related to the olfactory lobes; it still deals with the sense of smell.

limiting factor Any essential resource that, in short supply, limits population growth.

lineage (LIN-ee-age) A line of descent.

linkage group A quantitative measure of the relative positions of the genes along the length of a chromosome.

linkage mapping An investigative approach by which the relative linear positions of a chromosome's genes are deduced by tracking the percentage of recombinants in gametes.

lipid Mainly a hydrocarbon, greasy or oily, that strongly resists dissolving in water but quickly dissolves in nonpolar substances. In nearly all cells, certain lipids are the main reservoirs of stored energy. Other lipids are structural materials (as in membranes) and cell products (such as surface coatings).

lipid bilayer Structural basis of all cell membranes, with two layers of mostly phospholipid molecules. The hydrophobic tails of the molecules are sandwiched in between the hydrophilic heads, which are dissolved in the fluid that bathes them.

lipoprotein A molecule that forms when proteins in blood combine with cholesterol, triglycerides, and phospholipids that were absorbed from the gut, after a meal.

liver A large gland in vertebrates and many invertebrates that stores and interconverts organic compounds and helps maintain their blood concentrations. It additionally inactivates most hormone molecules that have served their functions as well as ammonia and other compounds that can be toxic at high concentrations.

local signaling molecules Secretions from cells of many local tissues that quickly alter chemical conditions in the vicinity, then are degraded before they can travel elsewhere.

logistic population growth (low-JISS-tik) A pattern of population growth in which a low-density population slowly increases in size, then enters a rapid growth phase that levels off once the carrying capacity has been reached.

loop of Henle The hairpin-shaped, tubular region of a nephron that functions in the reabsorption of water and solutes.

lung An internal respiratory surface in the shape of a cavity or sac that can be filled with air; a pair occur in a few fishes and in amphibians, birds, reptiles, and mammals.

lycophyte A seedless vascular plant of mostly wet or shaded habitats; requires free water to complete its life cycle.

lymph (LIMF) [L. lympha, water] Tissue fluid that has drained into the vessels of the lymphatic system.

lymph capillary A small-diameter vessel of the lymph vascular system that has no pronounced entrance; tissue fluid moves inward by passing between overlapping endothelial cells at the vessel's tip.

lymph node A lymphoid organ that is a battleground for immune responses; packed, organized arrays of lymphocytes inside cleanse lymph before it reaches the blood.

lymph vascular system [L. *lympha*, water, + *vasculum*, a small vessel] Of the lymphatic system, vessels that take up and transport excess tissue fluid and reclaimable solutes as well as fats absorbed from the digestive tract.

lymphatic system An organ system that supplements the circulatory system. Its vessels deliver fluid and solutes from interstitial fluid to the bloodstream; its lymphoid organs have roles in immunity.

lymphocyte Any of various T and B cells with roles in immune responses that follow recognition of antigen.

lysis [Gk. *lysis*, a loosening] Gross damage to a plasma membrane, cell wall, or both that allows the cytoplasm to leak out and thereby leads to cell death.

lysogenic pathway A common pathway in which a viral replication cycle enters a latent period and viral genes are integrated with a host chromosome. The recombinant molecule is a miniature time bomb that is passed on to all of the host cell's progeny.

lysosome (LYE-so-sohm) The main organelle of intracellular digestion, with enzymes that can break down nearly all polysaccharides, proteins, and nucleic acids, and some lipids.

lysozyme An infection-fighting enzyme in mucous membranes.

lytic pathway A common pathway in which a viral replication cycle proceeds rapidly and ends with lysis of the host cell.

macroevolution The large-scale patterns, trends, and rates of change among families and other more inclusive groups of species.

macrophage A leukocyte (white blood cell) that phagocytizes worn-out cells and tissue debris and anything bearing antigen. It may become an antigen-presenting cell and thus trigger immune responses.

mammal The only vertebrate whose females nourish offspring by milk, produced by and secreted from mammary glands.

mass extinction A sudden rise in rates of extinction above the background level; a catastrophic, global event during which a number of major taxa are wiped out simultaneously.

mass number The total number of protons and neutrons in an atom's nucleus.

mast cell In connective tissues, a basophil-like cell that releases histamine during an inflammatory response.

mechanoreceptor Sensory cell or nearby cell that detects mechanical energy (changes in pressure, position, or acceleration).

medulla oblongata Part of the hindbrain; it has reflex centers for vital tasks, such as respiration and circulation. It coordinates motor responses with complex reflexes, such as coughing. It influences brain regions concerned with sleep and arousal.

medusa (meh-DOO-sah) [Gk. *Medousa*, one of three sisters in Greek mythology having snake-entwined hair] Free-swimming, bell-shaped stage in cnidarian life cycles; oral arms and tentacles extend from the bell.

megaspore A haploid spore that forms in the ovary of seed-bearing plants; one of its cellular descendants develops into an egg.

meiosis (my-OH-sis) [Gk. *meioun*, to diminish] Two-stage nuclear division process that reduces the chromosome number of a germ cell by half, to the haploid number; each daughter nucleus receives one of each type of chromosome. It is the basis of gamete formation and (in plants) spore formation. *Compare* mitosis.

membrane excitability A special membrane property of any cell that can generate action potentials in response to stimulation.

memory The capacity of the brain to store and retrieve information about past sensory experience.

memory cell One of the subpopulations of B or T cells that form during an immune response to antigen but that do not act at once; it enters a resting phase, from which it is released during a secondary immune response.

menopause (MEN-uh-pozz) [L. *mensis*, month, + *pausa*, stop] End of the period of a human female's reproductive potential.

menstrual cycle Cyclic release of a secondary oocyte and priming of the endometrium to receive it, should it become fertilized; the complete cycle averages about twenty-eight days in female humans.

menstruation Cyclic sloughing of the blood-enriched endometrium (lining of the uterus) when pregnancy does not occur.

mesoderm (MEH-zoe-derm) [Gk. *mesos*, middle, + *derm*, skin] Of most animals, intermediate primary tissue layer that forms in the embryo and gives rise to muscle, most of the skeleton; circulatory, reproductive, and excretory organs; and connective tissue layers of the gut and integument.

mesophyll (MEH-zoe-fill) Of vascular plants, a tissue of photosynthetic parenchyma cells with an abundance of air spaces.

messenger RNA (mRNA) A single strand of ribonucleotides transcribed from DNA and translated into a polypeptide chain; the only RNA with protein-building instructions.

metabolic pathway (MEH-tuh-BALL-ik) One of many orderly sequences of enzyme-mediated reactions by which cells maintain, increase, or decrease the concentrations of substances. Different pathways are linear or circular, and one typically interconnects with others.

metabolism (meh-TAB-oh-lizm) [Gk. *meta*, change] All of the controlled, enzyme-mediated chemical reactions by which cells acquire and use energy to synthesize, store, degrade, and eliminate substances in ways that contribute to growth, survival, and reproduction.

metamorphosis (me-tuh-MOR-foe-sis) [Gk. *meta*, change, + *morphe*, form] Major change in body form during the transition from an embryo to the adult owing to hormonally controlled increases in size, reorganization of tissues, and remodeling of body parts.

metaphase Of meiosis I, the stage when all pairs of homologous chromosomes have become positioned midway between the spindle poles. Of mitosis or meiosis II, a stage when each duplicated chromosome has become positioned midway between the spindle poles.

metazoan A multicelled animal.

methanogen A type of archaebacterium that lives in oxygen-free habitats and produces methane gas as a metabolic by-product.

MHC marker Any of a variety of proteins that are self-markers. Some occur on all body cells of an individual; others are unique to the macrophages and lymphocytes.

micelle (my-SELL) Of fat digestion, a small droplet that is assembled from bile salts, fatty acids, and monoglycerides and that assists in fat absorption from the small intestine.

microevolution Change in allele frequencies resulting from mutation, genetic drift, gene flow, and natural selection.

microfilament [Gk. *mikros*, small, + L. *filum*, thread] Thin cytoskeletal element of two twisted polypeptide chains; it has roles in cell movement, especially at the cell surface, and in producing and maintaining cell shapes.

microorganism An organism, usually single-celled, that is far too small to be observed without the aid of a microscope.

microspore Of gymnosperms and flowering plants, a walled haploid spore that develops into a pollen grain.

microtubular spindle Of eukaryotic cells, a temporary bipolar structure composed of organized arrays of microtubules that forms during nuclear division and that moves the chromosomes to prescribed destinations.

microtubule (my-crow-TUBE-yool) A hollow, cylindrical cytoskeletal element, consisting mainly of tubulin, with roles in cell shape, motion, and growth and in the structure of cilia and flagella.

microtubule organizing center (MTOC) In the cytoplasm of eukaryotic cells, a small mass of proteins and other substances. The number, type, and location of MTOCs dictate the particular organization and orientation of microtubules inside the cell.

microvillus (MY-crow-VILL-us) [L. *villus*, shaggy hair] Of many cells, one of a number of very slender cylindrical extensions of the plasma membrane that serve in absorption or secretion.

midbrain Part of the vertebrate brain with coordination centers for reflex responses to visual and auditory input; also swiftly relays sensory signals to forebrain.

migration A recurring pattern of movement between two or more locations in response to environmental rhythms, such as circadian rhythms and seasonal changes in daylength. It requires activation or suppression of internal timing mechanisms that govern physiological and behavioral functions.

mimicry (MIM-ik-ree) In form, behavior, or both, a prey organism (the mimic) closely resembles a dangerous, unpalatable, or hard-to-catch species (its model).

mineral An element or inorganic compound formed by natural geologic processes and required for normal cell functioning.

mitochondrion (MY-toe-KON-dree-on), plural **mitochondria** Of eukaryotic cells, the organelle that specializes in formation of ATP; the second and third stages of aerobic respiration, an oxygen-requiring pathway, occur only in mitochondria.

mitosis (my-TOE-sis) [Gk. *mitos*, thread] Type of nuclear division that maintains the parental chromosome number for daughter cells; for eukaryotes, the basis of growth in size (and often of asexual reproduction).

mixture Two or more elements intermingled in proportions that can and usually do vary.

molar One of the cheek teeth; a tooth with a platform having cusps (surface bumps) that help crush, grind, and shear food.

molecular clock An accumulation of neutral mutations in a lineage that can be measured as a series of predictable "ticks" back through time; a way of calculating the time of origin of one lineage or species relative to others.

molecule A unit of matter in which chemical bonding holds together two or more atoms of the same or different elements.

mollusk An invertebrate with a tissue fold (mantle) draped around a soft, fleshy body; snails, clams, and squids are examples.

molting Shedding of hair, feathers, horns, epidermis, a shell, or cuticle in a process of increases in size or periodic renewal.

Monera In some traditional classification schemes, a kingdom that encompasses both archaebacteria and eubacteria.

monocot (MON-oh-kot) A monocotyledon; a flowering plant with seeds bearing one cotyledon, floral parts generally in threes (or multiples of three), and leaves typically parallel-veined. *Compare* dicot.

monohybrid cross [Gk. *monos*, alone] An experimental cross in which offspring inherit a pair of nonidentical alleles for a single trait being studied, so that they are heterozygous.

monomer Small molecule that is commonly a subunit of polymers, such as the sugar monomers of starch.

monosaccharide (MON-oh-SAK-ah-ride) [Gk. *monos*, alone, single, + *sakharon*, sugar] A simple carbohydrate, with only one sugar monomer. Glucose is an example.

monosomy A chromosome abnormality; the presence of a chromosome that has no homologue in a diploid cell.

morphogenesis (MORE-foe-JEN-ih-sis) [Gk. *morphe*, form, + *genesis*, origin] A program of orderly changes in an animal embryo's size, shape, and proportions, the outcome being specialized tissues and early organs. As part of the program, cells divide, grow, migrate, and change in size. Tissues expand and fold, and cells in some die in controlled ways at prescribed locations.

morphological convergence Lineages only remotely related evolved in response to similar environmental pressures, and they became similar in appearance, functions, or both. Analogous structures are evidence of this macroevolutionary pattern.

morphological divergence Of two or more lineages, change in appearance, functions, or both as they evolve from a shared ancestor. Homologous structures are evidence of this macroevolutionary pattern.

motor neuron A type of neuron that swiftly delivers signals from the brain or spinal cord to muscle cells, gland cells, or both.

multicelled organism An organism with differentiated cells arranged into tissues, organs, and often organ systems.

multiple allele system Three or more different molecular forms of the same gene (alleles) among individuals of a population.

muscle fatigue A decline in tension of a muscle kept in a state of tetanic contraction as a result of continuous, high-frequency stimulation.

muscle tension A mechanical force exerted by a contracting muscle; it resists opposing forces such as gravity and the weight of an object being lifted.

muscle tissue A tissue having cells able to contract under stimulation, then passively lengthen and return to the resting position.

mutagen (MEW-tuh-jen) Any environmental agent that can alter the molecular structure of DNA. Ultraviolet radiation and certain viruses are examples.

mutation [L. *mutatus*, a change, + *-ion*, result or a process or an act] A heritable change in the molecular structure of DNA. Mutations are the source of all alleles and, ultimately, of life's diversity. *See also* lethal mutation; neutral mutation.

mutation frequency The number of times a gene mutation has occurred in a population, as in 1 million gametes from which 500,000 individuals were produced.

mutation rate Of a gene, the probability of it undergoing spontaneous mutation during some specified interval, such as each DNA replication cycle.

mutualism [L. *mutuus*, reciprocal] A two-species, symbiotic ecological interaction that directly benefits both participants.

mycelium (my-SEE-lee-um), plural **mycelia** [Gk. *mykes*, fungus, mushroom, + *helos*, callus] A mesh of tiny, branching filaments (hyphae) that is the food-absorbing part of most fungi.

mycorrhiza (MY-coe-RIZE-uh) "Fungus-root"; a form of mutualism between the hyphae of a fungus and young roots of many plants. The fungus obtains carbohydrates from the plant and in turn releases dissolved mineral ions that plant roots can take up.

myelin sheath Of many sensory and motor neurons, an axonal sheath that enhances the propagation of action potentials; the plasma membranes of Schwann cells are wrapped repeatedly around the axon, and only a small node separates one from the other.

myofibril (MY-oh-FY-brill) One of the many threadlike structures inside a muscle cell; it is divided into many sarcomeres, the basic units of contraction.

myosin (MY-uh-sin) A motor protein with roles in contraction. In muscles, it interacts with actin to bring about contraction.

NAD$^+$ Nicotinamide adenine dinucleotide; a nucleotide coenzyme. When carrying electrons and unbound protons (H$^+$) between reaction sites, it is abbreviated NADH.

NADP$^+$ Nicotinamide adenine dinucleotide phosphate; a phosphorylated nucleotide coenzyme. When carrying electrons and unbound protons (H$^+$) between reaction sites, it is abbreviated NADPH$_2$.

natural killer cell A cytotoxic lymphocyte that reconnoiters for tumor cells and virus-infected cells, then touch-kills them.

natural selection A microevolutionary process; outcome of differences in survival and reproduction among individuals that show variation in heritable traits. Over the generations, it can lead to increased fitness (increased adaptation to the environment).

necrosis (neh-CROW-sis) Of multicelled organisms, the passive death of many cells that results from severe tissue damage.

negative feedback mechanism Homeostatic feedback mechanism in which an activity changes some condition in the internal environment; the altered condition triggers a response that reverses the change.

nematocyst (NEM-ad-uh-sist) [Gk. *nema*, thread, + *kystis*, pouch] A capsule formed only by cnidarians; it houses dischargeable, tubular threads that have prey-piercing barbs and that dispense toxins or sticky substances.

nephridium (neh-FRID-ee-um), plural **nephridia** Of earthworms and some other invertebrates, a system of regulating water and solute levels.

nephron (NEFF-ron) [Gk. *nephros*, kidney] Of the vertebrate kidney, a slender tubule into which water and solutes are filtered from blood, then selectively reabsorbed, and in which urine forms.

nerve Cordlike communication line of the peripheral nervous system, composed of axons of sensory neurons, motor neurons, or both bundled in connective tissue. In the brain and spinal cord, similar cord-like bundles are called nerve tracts.

nerve cord Of many animals, a cordlike communication line of axons of neurons.

nerve impulse *See* action potential.

nerve net A simple nervous system of a diffuse mesh of nerve cells that interact with contractile and often sensory cells in an epithelium.

nervous system An organ system in which nerve cells, such as neurons, are oriented relative to one another in signal-conducting and information-processing pathways. It detects and processes information about changes outside and inside the body and elicits responses from muscle and gland cells.

nervous tissue A type of connective tissue composed of neurons.

net energy Of energy resources available to the human population, the amount of energy that is left over after subtracting the energy used to locate, extract, transport, store, and deliver energy to consumers.

net population growth rate per individual (*r*) Of population growth equations, a single variable combining birth and death rates, which are assumed to remain constant.

neural tube Embryonic forerunner of the brain and spinal cord.

neuroglial cell (NUR-oh-GLEE-uhl) One of the cells that structurally and metabolically support neurons. Neuroglia represent about half the volume of the vertebrate nervous system.

neuromodulator A signaling molecule that can magnify or reduce the effects of a given neurotransmitter on neighboring or distant neurons.

neuromuscular junction A site of chemical synapsing between the axonal endings of a motor neuron and a muscle cell.

neuron (NUR-on) A nerve cell; the basic unit of communication in nervous systems. *See* motor neuron; interneuron; sensory neuron.

neurotransmitter Any of a class of signaling molecules that are secreted from neurons and act on immediately adjacent cells, then are rapidly degraded or recycled.

neutral mutation A mutation with little or no effect on phenotype; natural selection cannot increase or decrease its frequency in a population, for it cannot influence an individual's survival and reproduction. *Compare* molecular clock.

neutron A unit of matter, one or more of which occupies the atomic nucleus and has mass but no electric charge.

neutrophil The most abundant, fast-acting white blood cell; it phagocytizes bacteria during an inflammatory response.

niche (NITCH) [L. *nidas*, nest] The sum total of all activities and relationships in which individuals of a species engage as they secure and use the resources necessary to survive and reproduce.

nitrification (nye-trih-fih-KAY-shun) Of the nitrogen cycle, a process by which certain bacteria strip electrons from ammonia or ammonium in soil. The end product, nitrite (NO_2^-), is broken down to nitrate (NO_3^-) by other bacteria.

nitrogen cycle An atmospheric cycle; a biogeochemical cycle in which the largest nitrogen reservoir is the atmosphere.

nitrogen fixation Process by which a few kinds of bacteria convert gaseous nitrogen (N_2) to ammonia. This dissolves rapidly in their cytoplasm to form ammonium, which can be used in biosynthetic pathways.

nociceptor (NO-SEE-sep-tur) A pain receptor, such as a free nerve ending that detects tissue damage as by burns or distortions.

node A site where one or more leaves are attached to a plant stem.

noncyclic pathway of ATP formation (non-SIK-lik) [L. *non*, not, + Gk. *kylos*, circle] Photosynthetic pathway of ATP formation in which electrons from water molecules flow through two photosystems and two electron transport chains, and end up in NADPH.

nondisjunction The failure of two sister chromatids or of a pair of homologous chromosomes to separate at meiosis or mitosis, so that daughter cells end up with too many or too few chromosomes.

notochord (KNOW-toe-kord) Of chordates, a rod of stiffened tissue (not cartilage or bone) that serves as a supporting structure for the body.

nuclear envelope A double membrane (two lipid bilayers and associated proteins) that is the outermost portion of a cell nucleus.

nucleic acid (new-CLAY-ik) A single-stranded or double-stranded chain of four different kinds of nucleotides joined one after the other at their phosphate groups. They differ in which nucleotide base follows the next in sequence. DNA and RNA are examples.

nucleic acid hybridization Single strands of DNA or RNA from different species are recombined into double-stranded, hybrid molecules, which are then heated to break hydrogen bonds between them. The amount of heat required to separate the strands is a measure of biochemical similarity, which is greatest among closely related species and weakest among distantly related ones.

nucleoid (NEW-KLEE-oid) Of bacteria, the region of the cell in which the bacterial chromosome is physically organized (but not bounded by membrane, as with the nuclear envelope of eukaryotic cells).

nucleolus (new-KLEE-oh-lus) [L. *nucleolus*, a little kernel] In the nucleus of a nondividing cell, a site where the protein and RNA subunits of ribosomes are assembled.

nucleosome (NEW-KLEE-oh-sohm) One of the organizational units of the eukaryotic chromosome; a stretch of DNA looped twice around a "spool" of histone molecules.

nucleotide (NEW-KLEE-oh-tide) A small organic compound with a five-carbon sugar (deoxyribose), nitrogen-containing base, and phosphate group. Nucleotides are the structural units of adenosine phosphates, nucleotide coenzymes, and nucleic acids.

nucleotide coenzyme A protein that assists an enzyme by transporting electrons and hydrogen atoms released at one reaction site to another site.

nucleus (NEW-klee-us) [L. *nucleus*, a kernel] Of atoms, the central core of one or more positively charged protons and (in all but hydrogen) electrically neutral neutrons. In eukaryotic cells, a membranous organelle that physically isolates and organizes DNA out of the way of cytoplasmic machinery.

numerical taxonomy Practice by which traits of an unidentified organism are compared with those of a known group. The greater the total number of traits that it has in common with the known group, the closer is their inferred relatedness.

nutrient An element essential for growth and survival of a given organism; directly or indirectly, it has one or more roles in metabolism that no other element fulfills.

nutrition All of those processes by which food is selectively taken in, digested, absorbed, and later converted to the body's own organic compounds.

obesity An excess of fat in the body's adipose tissues, caused by imbalances between caloric intake and energy output.

oligosaccharide A carbohydrate consisting of a short chain of two or more covalently bonded sugar monomers. The type called a disaccharide has two sugar monomers. *Compare* monosaccharide; polysaccharide.

omnivore [L. *omnis*, all, + *vovare*, to devour] An animal able to obtain energy and carbon from more than one food source rather than being limited to one trophic level.

oncogene (ON-koe-jeen) Any gene having the potential to induce cancerous transformation.

oocyte An immature egg.

oogenesis (oo-oh-JEN-uh-sis) Formation of a female gamete, from a germ cell to a haploid ovum (mature egg).

operator A short base sequence intervening between a promoter and bacterial genes.

operon A promoter-operator sequence that services more than one bacterial gene.

orbital One of a number of volumes of space around the atomic nucleus in which electrons are likely to be at any instant.

organ A structure having definite form and function that consists of more than one tissue.

organ formation A stage of embryonic development in animals in which primary tissue layers split into subpopulations of cells, and different lines of cells become unique in structure and function; basis of growth and tissue specialization.

organ system Two or more organs that interact chemically, physically, or both in performing a common task.

organelle Of cells, an internal, membrane-bounded sac or compartment having one or more specific, specialized metabolic functions.

organic compound A molecule consisting of one or more elements covalently bonded to some number of carbon atoms.

osmoreceptor A sensory receptor that detects changes in water volume (hence solute concentration) in fluid that bathes it.

osmosis (oss-MOE-sis) [Gk. *osmos*, act of pushing] The diffusion of water in response to a water concentration gradient between two regions that are separated from each other by a selectively permeable membrane. The greater the number of molecules and ions dissolved in a solution, the lower its water concentration will be.

osmotic pressure At some point following the development of hydrostatic pressure in a cell, an amount of force that counters the inward diffusion of water and prevents any further increase in fluid volume.

ovary (oh-vuh-ree) A type of primary reproductive organ in which eggs form. In seed-bearing plants, the part of the carpel where eggs develop, fertilization takes place, and seeds mature; at maturity, it and sometimes other plant parts is a fruit.

oviduct (OH-vih-dukt) One of a pair of ducts through which eggs travel from an ovary to the uterus. Formerly called Fallopian tube.

ovulation (AHV-you-LAY-shun) The release of a secondary oocyte from an ovary during a menstrual cycle.

ovule (OHV-youl) [L. *ovum*, egg] Of seed-bearing plants, a female gametophyte with egg cell, surrounding tissue, and one or two protective layers (integuments). After fertilization, it matures into a seed.

ovum (OH-vum) Of vertebrates, the mature female gamete (egg).

oxidation-reduction reaction An electron transfer from one atom or molecule to another. Often hydrogen is also transferred along with an electron or electrons.

ozone thinning A pronounced seasonal thinning of the ozone layer in the lower stratosphere, especially above polar regions.

pancreas (PAN-cree-us) Gland that secretes enzymes and bicarbonate into the small intestine during digestion, and that secretes the hormones insulin and glucagon.

pancreatic islet Any of the 2 million clusters of endocrine cells in the pancreas, including alpha cells, beta cells, and delta cells.

parapatric speciation Neighbor populations become distinct species while maintaining contact along their common border.

parasite [Gk. *para*, alongside, + *sitos*, food] An organism that lives in or on other living organisms (hosts), feeds on specific tissues during part of its life cycle, and usually does not kill its host outright.

parasitism A two-species interaction in which one species lives in or on a host species and uses its tissues for nutrients. The parasite benefits; the host does not.

parasitoid An insect larva that grows and develops inside a host organism (usually another insect), eventually consuming the soft tissues and killing it.

parasympathetic nerve Of the autonomic nervous system, any of the nerves carrying signals that tend to slow down the body's activities and divert energy to basic tasks; such nerves work continually in opposition with sympathetic nerves to make small adjustments in internal organ activity.

parenchyma (par-ENG-kih-mah) A simple tissue that makes up the bulk of the plant body. Its cells take part in photosynthesis, storage, secretion, and other tasks.

parthenogenesis Development of an embryo from an unfertilized egg.

passive transport A transport protein that spans the lipid bilayer of a cell membrane passively allows a solute to diffuse through its interior, down its concentration gradient.

pathogen (PATH-oh-jen) [Gk. *pathos*, suffering, + *genēs*, origin] A virus, bacterium, fungus, protozoan, parasitic worm, or some other infectious agent that can invade organisms, multiply in them, and cause disease.

pattern formation Of animals, a sculpting of nondescript clumps of embryonic cells into specialized tissues and organs according to an ordered, spatial pattern.

PCR Polymerase chain reaction; a method of enormously amplifying the quantity of DNA fragments cut by restriction enzymes.

peat bog An accumulation of the remains of peat mosses in an compressed, excessively moist, and highly acidic mat.

pedigree A chart of genetic connections among related individuals, constructed according to standardized methods.

pelagic province The full volume of ocean water; it is subdivided into a neritic zone (relatively shallow water over continental shelves) and an oceanic zone (water over ocean basins).

penis A male copulatory organ by which sperm can be deposited inside a female reproductive tract.

peptide hormone A protein hormone or some other water-soluble hormone unable to cross the lipid bilayer of a target cell. It enters by receptor-mediated endocytosis or activates cell surface receptors.

perennial [L. *per-*, throughout, + *annus*, year] A flowering plant that lives for more than two growing seasons.

pericycle (PARE-ih-sigh-kul) [Gk. *peri-*, around, + *kyklos*, circle] Of a root vascular cylinder, one or more layers just inside the endodermis that gives rise to lateral roots and contributes to secondary growth.

periderm Of plants that show extensive secondary growth, a protective covering that replaces epidermis.

peripheral nervous system (per-IF-ur-uhl) [Gk. *peripherein*, to carry around] The nerves leading into and out from a spinal cord and brain and the ganglia along those communication lines.

peristalsis (pare-ih-STAL-sis) Waves of contraction and relaxation of muscles in the wall of a tubular or saclike organ.

peritoneum (pare-ih-tah-NEE-um) A lining of the coelom that also covers and helps maintain the position of internal organs.

peritubular capillary One of the set of blood capillaries around the tubular parts of a nephron; it functions in reabsorption of water and solutes back into the body and in secretion of hydrogen ions and some other substances in the forming urine.

permafrost A permanently frozen, water-impenetrable layer beneath the soil surface in arctic tundra.

peroxisome Enzyme-filled vesicle in which fatty acids and amino acids are digested into hydrogen peroxide and converted to harmless products.

pest resurgence Insecticide resistance arises among a population of pests; the insecticide acts as an agent of directional selection.

PGA Phosphoglycerate (FOSS-foe-GLISS-er-ate) A key intermediate of glycolysis and of the Calvin-Benson cycle.

PGAL Phosphoglyceraldehyde. A key intermediate of glycolysis and of the Calvin-Benson cycle.

pH scale A measure of the concentration of free hydrogen ions in blood, water, and other solutions; pH 0 is the most acidic, 14 the most basic, and 7, neutral.

phagocyte (FAG-uh-sight) [Gk. *phagein*, to eat, + *kytos*, hollow vessel] A cell that obtains nutrients or destroys foreign cells and cellular debris by phagocytosis. The macrophages and amoeboid protozoans are examples.

phagocytosis (FAG-uh-sigh-TOE-sis) [Gk. *phagein*, to eat, + *kytos*, hollow vessel] The engulfment of foreign cells or substances by means of pseudopod formation and then endocytosis.

pharynx (FARE-inks) A muscularized tube by which food enters the gut. Also an organ of gas exchange in some chordates; in land vertebrates, a dual entrance for the tubular part of the digestive tract and windpipe.

phenotype (FEE-no-type) [Gk. *phainein*, to show, + *typos*, image] Observable trait or traits of an individual that arise from the interactions between genes, and between genes and the environment.

pheromone (FARE-oh-moan) [Gk. *phero*, to carry, + *-mone*, as in hormone] A nearly odorless, hormone-like, exocrine gland secretion that is a communication signal between individuals of the same species and that helps integrate social behavior.

phloem (FLOW-um) Of vascular plants, a tissue with living cells that interconnect and form the tubes through which sugars and other dissolved organic compounds are conducted.

phospholipid An organic compound with a glycerol backbone, two fatty acid tails, and a hydrophilic head (a phosphate group and another polar group). Phospholipids are the main structural materials of cell membrane.

phosphorus cycle Movement of phosphorus from rock or soil through organisms, then back to soil.

phosphorylation (FOSS-for-ih-LAY-shun) The attachment of unbound (inorganic) phosphate to a molecule; also the transfer of a phosphate group from one molecule to another, as when ATP phosphorylates glucose.

photoautotroph An organism able to synthesize all organic molecules it requires using carbon dioxide as the carbon source and sunlight as the energy source. All plants, some protistans, and a few bacteria are photoautotrophs.

photolysis (foe-TALL-ih-sis) [Gk. *photos*, light, + *-lysis*, breaking apart] A reaction sequence of the noncyclic pathway of photosynthesis, triggered by photon energy, in which water is split into oxygen, hydrogen, and electrons.

photoperiodism A biological response to a change in the relative length of daylight and darkness.

photoreceptor Light-sensitive sensory cell.

photosynthesis The trapping of energy from the sun and its conversion to chemical energy (ATP, NADPH, or both), followed by synthesis of sugar phosphates that become converted to sucrose, cellulose, starch, and other products. The main pathway by which energy and carbon enter the web of life.

photosystem One of the clusters of light-trapping pigments that are embedded in photosynthetic membranes and that donate electrons to transport systems that are involved in the formation of ATP, NADPH, or both. Examples are photosystems II and I of the thylakoid membrane of chloroplasts.

phototropism [Gk. *photos*, light, + *trope*, turning, direction] An adjustment in the direction and rate of plant growth in response to a source of light.

photovoltaic cell A device that can convert sunlight energy into electricity.

phycobilin (FIE-koe-BY-lins) One of the light-sensitive, accessory pigments that transfer energy they absorb to chlorophylls. The phycobilins are especially abundant in red algae and cyanobacteria.

phylogeny Evolutionary relationships among species, starting with most ancestral forms and including the branches leading to their descendants.

phytochrome Light-sensitive pigment, the activation and inactivation of which triggers plant hormone activities governing leaf expansion, stem branching, stem length and often seed germination and flowering.

phytoplankton (FIE-toe-PLANK-tun) [Gk. *phyton*, plant, + *planktos*, wandering] A community of floating or weakly swimming photoautotrophs in saltwater and freshwater habitats.

pigment A light-absorbing molecule.

pioneer species An opportunistic colonizer of barren or disturbed habitats, notable for its high dispersal rate and rapid growth.

pituitary gland An endocrine gland that interacts with the hypothalamus to control diverse physiological functions, including activity of many other endocrine glands. The posterior lobe stores and secretes hypothalamic hormones; the anterior lobe produces and secretes its own hormones.

placenta (play-SEN-tuh) A blood-engorged organ, consisting of endometrial tissue and extraembryonic membranes, that develops during pregnancy in the female placental mammal. It permits exchanges between the mother and her fetus without intermingling their bloodstreams, thus sustaining the new individual and allowing its blood vessels to develop apart from the mother's.

plankton [Gk. *planktos*, wandering] Any community of floating or weakly swimming organisms, mostly microscopic, living in freshwater and saltwater habitats. *See also* phytoplankton; zooplankton.

plant One of the eukaryotic organisms, nearly all of which are photoautrophic, that have chlorophylls *a* and *b* and that generally have well-developed root and shoot systems.

Plantae The kingdom of plants.

plasma (PLAZ-muh) Liquid component of blood; water and dissolved ions, diverse proteins, sugars, gases, and other substances.

plasma membrane Of cells, the outermost membranous boundary between cytoplasm and external fluid bathing the cell. Its lipid bilayer is the basic structural part of the membrane; diverse proteins embedded in the bilayer or attached to its surfaces carry out most functions, such as transport and reception of extracellular signals.

plasmid Of many bacteria, a small, circular molecule of extra DNA that carries only a few genes and that replicates independently of the bacterial chromosome.

plasmodesma, plural **plasmodesmata** (PLAZ-moe-DEZ-muh) A plant cell junction; a membrane-lined channel across walls of adjacent cells that connects their cytoplasm.

plasmolysis An osmotically induced shrinkage of a cell's cytoplasm.

plate tectonics Theory that slabs, or plates, of the Earth's outer layer (lithosphere) float on a hot, plastic layer of the underlying mantle. All plates are in motion and they raft the continents to new positions over great spans of geologic time.

platelet (PLAYT-let) Any of the fragments of megakaryocytes that release substances necessary for clot formation.

pleiotropy (PLEE-oh-troe-pee) [Gk. *pleon*, more, + *trope*, direction] As a result of expression of alleles at a single gene locus, positive or negative effects on two or more traits. The effects may or may not be manifest at the same time in the individual.

polar body One of four cells that form during the meiotic cell division of an oocyte but that does not become the ovum.

pollen grain [L. *pollen*, fine dust] Depending on the species, the immature or mature, sperm-bearing male gametophyte of gymnosperms and angiosperms.

pollen sac In anthers of flowers, any of the chambers in which pollen grains develop.

pollen tube A tube formed after a pollen grain germinates; grows through carpel tissues, and carries sperm to the ovule.

pollination Of flowering plants, the arrival of a pollen grain on the landing platform (stigma) of a carpel.

pollutant Any substance with which an ecosystem has had no prior evolutionary experience, in terms of kinds or amounts, and that can accumulate to disruptive or harmful levels. Can be naturally occurring or synthetic.

polymer (POH-lih-mur) [Gk. *polus*, many, + *meris*, part] A large molecule with three to millions of subunits.

polymerase (puh-LIM-ur-aze) An enzyme that catalyzes a polymerization reaction. Examples are DNA polymerase (of DNA replication and repair), RNA polymerase (of gene transcription), and the enzyme that catalyzes cellulose synthesis.

polypeptide chain Organic compound with nitrogen atoms arrayed in a regular pattern in its backbone (-N-C-C-N-C-C-). A protein consists of one or more of these chains.

polymorphism (poly-MORE-fizz-um) [Gk. *polus*, many, + *morphe*, form] Of populations, the persistence of two or more distinctive forms of a trait (morphs).

polyp (POH-lip) Vase-shaped, sedentary stage of cnidarian life cycles.

polypeptide chain Three or more amino acids joined by peptide bonds.

polyploidy (POL-ee-PLOYD-ee) Having three or more of each type of chromosome in the interphase nucleus, compared to a diploid chromosome number.

polysaccharide [Gk. *polus*, many, + *sakcharon*, sugar] A straight or branched chain of many covalently linked sugar units, of the same or different kinds. The most common are cellulose, starch, and glycogen.

polysome A number of ribosomes translating the same mRNA molecule one after the other, at the same time, during protein synthesis.

pons Part of the hindbrain; a traffic center for signals passing between the cerebellum and integrating centers of the forebrain.

population Individuals of the same species occupying the same area.

population density A count of individuals of a population occupying a specified area or volume of a habitat.

population distribution The general pattern of dispersion of individuals of a population throughout their habitat.

population size The number of individuals that make up the gene pool of a population.

positive feedback mechanism Homeostatic mechanism by which a chain of events is set in motion that intensifies a change from an original condition; after a limited time, the intensification reverses the change.

predation An ecological interaction in which a predatory species eats (and so directly harms) a prey species.

predator [L. *prehendere*, to grasp, seize] An organism that feeds on other living organisms (its prey), that does not live in or on them, and that may or may not kill them.

prediction A statement about what you can expect to observe in nature if a theory or hypothesis is not false.

pressure flow theory Of vascular plants, a theory that organic compounds move through phloem because of gradients in solute concentrations and pressure between sources (such as photosynthetically active leaves) and sinks (such as growing parts).

primary growth Plant growth originating at root tips and shoot tips and resulting in increases in length.

primary immune response Defensive acts by white blood cells and their secretions, as elicited by a first-time encounter with an antigen; includes both antibody-mediated and cell-mediated activities.

primary productivity, gross Of ecosystems, the rate at which producers capture and store a given amount of energy in their cells and tissues during a specified interval.

primary productivity, net Of ecosystems, the rate of energy storage in the cells and tissues of producers in excess of their rate of aerobic respiration.

primate A mammalian lineage that includes the prosimians, tarsioids, and anthropoids (monkeys, apes, and humans).

primer A short nucleotide sequence that will base-pair with any complementary DNA sequence; later, DNA polymerases recognize them as start tags for replication.

prion A small protein linked with eight rare, fatal degenerative diseases of the nervous system. Prions are altered products of a gene that is present in normal as well as infected individuals.

probability The chance that each outcome of a given event will occur is proportional to the number of ways it can be reached.

procambium (pro-KAM-bee-um) Of vascular plants, a primary meristem that gives rise to primary vascular tissues.

producer Of ecosystems, an autotrophic (self-feeding) organism; it nourishes itself using sources of energy and carbon from its physical environment. Photoautorophs and chemoautotrophs are examples.

progesterone (pro-JESS-tuh-rown) A type of sex hormone secreted by the pair of ovaries in female mammals.

prokaryotic cell (pro-CARRY-oh-tic) [L. *pro*, before, + Gk. *karyon*, kernel] A bacterium; a single-celled organism, usually walled, that does not have the profusion of membrane-bound organelles seen in eukaryotic cells.

prokaryotic fission Division mechanism by which a bacterial cell reproduces.

promoter A short base sequence in DNA that signals the start of a gene; the site where RNA polymerase binds to start transcription.

prophase Of mitosis, the stage when each duplicated chromosome starts to condense, microtubules form a spindle apparatus, and the nuclear envelope starts to break up.

prophase I Of meiosis, the stage when the microtubular spindle starts to form, the nuclear envelope starts to break up, and each duplicated chromosome condenses and pairs with its homologous partner. Nonsister chromatids typically undergo crossing over and genetic recombination.

prophase II Brief first stage of meiosis II when chromosomes start moving toward spindle equator; each has already been separated from its homologue but still consists of two chromatids.

protein Organic compound of one or more chains of amino acids, which peptide bonds join in a linear sequence that is the basis of the protein's three-dimensional structure and chemical behavior.

Protista The kingdom of protistans.

protistan (pro-TISS-tun) [Gk. *prōtistos*, primal, very first] One of the staggeringly diverse eukaryotes, single-celled species of which may resemble the first eukaryotic cells. At present categorized in part by what they are *not* (not bacteria, fungi, plants, or animals).

proto-oncogene A gene sequence similar to an oncogene but that codes for a protein required for normal cell function. When a mutation alters its structure, it may trigger cancerous transformation.

proton Positively charged particle, one or more of which resides in the nucleus of each type of atom; when unbound, a proton is the same as a hydrogen ion (H^+).

protostome (PRO-toe-stome) [Gk. *proto*, first, + *stoma*, mouth] One of the lineage of coelomate, bilateral animals in which the first indentation to form in the embryo develops into a mouth. Includes mollusks, annelids, and anthropods.

protozoan A group of protistans, some predatory, others parasitic, that in certain respects resemble single-celled heterotrophs that presumably gave rise to animals. The amoeboid and animal-like protozoans, as well as sporozoans, are examples.

proximal tubule Of a nephron, the tubular portion into which water and solutes enter after being filtered from the blood at the preceding portion (Bowman's capsule).

pulmonary circuit Blood circulation route leading to and from the lungs.

Punnett-square method A way to predict the probable outcome of a genetic cross in simple diagrammatic form.

purine A type of nucleotide base with a double ring structure. Examples are adenine and guanine.

pyrimidine (phi-RIM-ih-deen) A type of nucleotide bases with a single ring structure. Examples are cytosine and thymine.

pyruvate (PIE-roo-vate) A small organic compound with a backbone of three carbon atoms. Glycolysis produces two molecules of pyruvate as end products.

r A variable used in population growth equations to signify net population growth rate; the assumption is that birth and death rates remain constant and therefore may be combined into this one variable.

radial symmetry Of animals, a body plan having four or more roughly equivalent parts arranged around a central axis.

radiometric dating A method of measuring the proportions of (1) a radioisotope in a mineral trapped long ago in a newly formed rock and (2) a daughter isotope that formed from it by radioactive decay in the same rock sample. Used to assign absolute dates to fossil-containing rocks and to a geologic time scale.

radioisotope An unstable atom, with an uneven number of protons and neutrons. It spontaneously emits particles and energy, and over a predictable time span becomes transformed (decays) into a different atom.

rain shadow A reduction in rainfall on the leeward side of high mountains, resulting in arid or semiarid conditions.

reabsorption In kidneys, diffusion or active transport of water and reclaimable solutes from a nephron into peritubular capillaries under the control of ADH and aldosterone. At a capillary bed, osmotic movement of some interstitial fluid into a capillary when there is a difference in the concentration of water between plasma and interstitial fluid.

rearrangement, molecular A conversion of one type of organic compound to another through a juggling of its internal bonds.

receptor, molecular Of cells, a membrane protein that binds a hormone or some other extracellular substance that triggers change in cell activities.

receptor, sensory Of nervous systems, a sensory cell or specialized cell adjacent to it that can detect a particular stimulus.

recessive allele [L. *recedere*, to recede] In heterozygotes, an allele whose expression is fully or partially masked by expression of its partner; only in the homozygous recessive condition is it fully expressed.

recognition protein Of mammalian cell membranes, a glycolipid or glycoprotein that functions as a molecular fingerprint; it identifies a cell as being of a specific type. Self-proteins of the body's own cells are an example.

recombinant technology Procedures by which DNA (genes) from different species may be isolated, cut, spliced together, and the new recombinant molecules multiplied in quantity, as by PCR or by a population of rapidly dividing bacterial cells.

red alga A protistan, most of which are multicelled, aquatic, and photosynthetic, with an abundance of phycobilins.

blood cell An erythrocyte; an oxygen-transporting cell in blood.

red marrow A site of blood cell formation in the spongy tissue of many bones.

reflex [L. *reflectere*, to bend back] A simple, stereotyped movement elicited directly by sensory stimulation.

reflex pathway [L. *reflectere*, to bend back] The simplest route by which stimulation of nerve cells leads to contractile or glandular action. In complex animals, sensory neurons directly stimulate or inhibit motor neurons, without intervention by interneurons.

refractory period Of neurons, the period following an action potential at a given patch of membrane when sodium gates are shut and potassium gates are open, so that the patch is insensitive to stimulation.

regulatory protein A protein involved in a positive or negative control system that operates during transcription or translation or after translation. The protein interacts with DNA, RNA, or a gene product.

releasing hormone A signaling molecule produced by the hypothalamus that enhances or slows down secretions from target cells in the anterior lobe of the pituitary gland.

repressor Regulatory protein that can block gene transcription.

reproduction Of nature, any of a number of processes by which a new generation of cells or multicelled individuals is produced. For eukaryotes, it may be sexual or it may be asexual (as by binary fission, budding, or vegetative propagation). For bacteria only, it occurs by prokaryotic fission.

reproductive isolating mechanism Any heritable feature of body form, functioning, or behavior that prevents interbreeding between one or more genetically divergent populations.

reproductive success The survival and production of offspring by an individual.

reptile A type of carnivorous vertebrate; its ancestors were the first vertebrates to escape dependency on standing water, largely by means of internal fertilization and amniote eggs. Examples are dinosaurs (extinct), crocodiles, lizards, and snakes.

resource partitioning Coexistence among competing species; they share the same resource differently or at different times.

respiration [L. *respirare*, to breathe] Of animals, the exchange of oxygen from the environment for carbon dioxide wastes from cells by way of integumentary exchange, a respiratory system, or both. *Compare* aerobic respiration.

respiratory surface An epithelium or some other surface tissue that functions in gas exchange in animals; it is thin enough to allow oxygen to diffuse into the body and carbon dioxide to diffuse out of it.

resting membrane potential Of a neurons and other excitable cells, a steady voltage difference across the plasma membrane in the absence of outside stimulation.

restoration ecology Attempts to reestablish biodiversity in abandoned farmlands and other large areas altered by agriculture, mining, and other severe disturbances.

restriction enzyme Of bacteria, one of a class of enzymes that can cut apart foreign DNA injected into the cell body, as by viruses; also used in recombinant DNA technology.

reticular formation A mesh of interneurons extending from the upper spinal cord, through the brain stem, and into the cerebral cortex; a low-level pathway through the vertebrate nervous system.

reverse transcriptase A viral enzyme that uses viral RNA as a template to synthesize either DNA or mRNA in a host cell.

reverse transcription A process by which reverse transcriptase assembles DNA on an RNA template. Used by RNA viruses as well as by technologists who use mature mRNA transcripts as the templates for synthesizing cDNA.

RFLP Short for restriction fragment length polymorphism. Such fragments are formed by using restriction enzymes to cut DNA; for each individual, they show slight but unique differences in their banding pattern.

Rh blood typing A laboratory method of characterizing red blood cells on the basis of a protein that serves as a self-marker at their surface; Rh^+ signifies its presence, and Rh^- its absence.

rhizoid A rootlike absorptive structure of some fungi and nonvascular plants.

ribosomal RNA (rRNA) A type of RNA molecule that combines with proteins to form ribosomes, on which the polypeptide chains of proteins are assembled.

ribosome In all cells, the structure at which amino acids are strung together in specified sequence to form the polypeptide chains of proteins. An intact ribosome consists of two subunits, each composed of ribosomal RNA and protein molecules.

RNA Ribonucleic acid. A general category of single-stranded nucleic acids that function in processes by which genetic instructions encoded in DNA are used to build proteins.

rod cell A vertebrate photoreceptor that is sensitive to very dim light and contributes to coarse perception of movement.

root hair Of vascular plants, an extension of a specialized root epidermal cell; root hairs collectively enhance the surface area available for absorbing water and solutes.

root nodule A localized swelling on roots of certain legumes and other plants that contain symbiotic, nitrogen-fixing bacteria.

root A part of a plant that typically grows belowground, absorbs water and dissolved nutrients, helps anchor aboveground parts, and often stores food.

RuBP Ribulose bisphosphate. A compound with a backbone of five carbon atoms; used for carbon fixation by C3 plants and also regenerated in the Calvin-Benson cycle of photosynthesis.

S-shaped curve A curve, obtained when population size is plotted against time, that is characteristic of logistic growth.

salination A salt buildup in soil as a result of evaporation, poor drainage, and heavy irrigation.

salivary gland A gland that secretes saliva, a fluid that initially mixes with food in the mouth and starts the breakdown of starch.

salt A compound that releases ions other than H^+ and OH^- in solution.

saltatory conduction Of myelinated neurons, a rapid form of action potential propagation by a hopping of excitation from one node to another between the jellyrolled membranes of cells making up the myelin sheath.

sampling error A rule of probability: The fewer times that a chance event occurs, the greater will be the variance from the expected outcome of that occurrence. Of experimental groups, large sampling tends to lessen the chance that differences among individuals will distort the results.

saprobe A heterotroph that obtains energy and carbon from nonliving organic matter and so causes its decay. Saprobic fungi are examples.

sarcomere (SAR-koe-meer) Of muscles, the basic unit of contraction; a region of myosin and actin filaments organized in parallel arrays between two Z lines of one of the many myofibrils inside a muscle cell.

sarcoplasmic reticulum (sar-koe-PLAZ-mik reh-TIK-you-lum) A membrane system around a muscle cell's myofibrils that takes up, stores, and releases the calcium ions required for cross-bridge formation in sarcomeres, hence for contraction.

Schwann cell One of a series of neuroglial cells wrapped like jellyrolls around an axon in a nerve and forming a myelin sheath.

sclerenchyma (skler-ENG-kih-mah) A simple plant tissue in which most cells have thick, lignin-impregnated walls. Sclerenchyma supports mature plant parts and commonly protects seeds.

sea-floor spreading Molten rock erupts from great, continuous ridges on the ocean floor, flows laterally in both directions, and hardens to form new crust. As the seafloor spreads out, it forces older crust down into great trenches elsewhere in the seafloor.

second law of thermodynamics A law of nature stating that the spontaneous direction of energy flow is from organized to less organized forms. Overall, the total amount of energy in the universe is spontaneously flowing from forms of higher to lower quality; with each conversion, some energy gets randomly dispersed in a form (usually heat) not as readily available to do work.

second messenger A molecule within a cell that mediates a hormonal signal by triggering a response to it.

secondary immune response An immune response to previously encountered antigen that is more rapid and prolonged than the first owing to the swift participation of memory cells.

secondary sexual trait A trait associated with maleness or femaleness but one with no direct role in reproduction.

secretion A product released across the plasma membrane of a cell that may act singly or as part of glandular tissue.

sedimentary cycle A biogeochemical cycle with no gaseous phase; the element moves from land to the seafloor, then returns only through long-term geological uplifting.

seed Of gymnosperms and angiosperms, a fully mature ovule (contains the plant embryo) with integuments that form the seed coat.

segmentation Of animals, a body plan with a repeating series of units that may or may not be similar to one another in appearance. Also, an oscillating movement created by rings of circular muscle in a tubular wall. In a small intestine, such movement constantly mixes and forces the contents of the lumen against the absorptive wall surface.

segregation, theory of [L. *se-*, apart, + *grex*, herd] A Mendelian theory that diploid organisms inherit pairs of genes for traits (on pairs of homologous chromosomes) and that the two genes segregate during meiosis and end up in separate gametes.

selective gene expression Of a multicelled organism, the activation and suppression of some fraction of the same inherited genes in different populations of cells; it leads to cell differentiation.

selective permeability Of a cell membrane, a capacity to allow some substances but not others to cross it at certain sites, at certain times, owing to its molecular structure.

selfish behavior A behavior by which an individual protects or increases its own chance of producing offspring, regardless of the consequences to the group to which it belongs.

selfish herd A society held together simply by reproductive self-interest.

semen (SEE-mun) [L. *serere*, to sow] Sperm-bearing fluid expelled from a penis during male orgasm.

semiconservative replication [Gk. *hēmi*, half, + L. *conservare*, to keep] How a DNA molecule duplicates itself. A DNA double helix unzips and a complementary strand is assembled on the exposed bases of each strand. Each conserved strand and its new partner are wound together into a double helix. Thus the outcome is two "half-old, half-new" molecules.

senescence (sen-ESS-cents) [L. *senescere*, to grow old] Sum of processes leading to the natural death of an organism or some of its parts.

sensation A conscious awareness of a stimulus. Not the same thing as perception, which is an understanding of the meaning of a sensation.

sensory neuron Any of the nerve cells or cells adjacent to them that detect specific stimuli (such as light energy) and relay signals to the brain and spinal cord.

sensory system Of animals, the "front door" of the nervous system; the components that detect external and internal stimuli and relay information about them into the central nervous system.

sessile animal (SESS-ihl) An animal that remains attached to a substrate during some stage (often the adult) of its life cycle.

sex chromosome Of animals and some plants, one of two types of chromosomes, the combinations of which govern gender. *Compare* autosome.

sexual dimorphism Of a species, having individuals with distinctively male or female phenotypes.

sexual reproduction Production of offspring by way of meiosis, gamete formation, and fertilization.

sexual selection A microevolutionary process; natural selection favors a trait that gives the individual a competitive edge in attracting and holding onto mates, hence in reproductive success.

shifting cultivation The cutting and burning of trees, followed by tilling of ashes into the soil; once called slash-and-burn agriculture.

shell model Model of electron distribution in atoms, in which all orbitals available to electrons occupy a nested series of shells.

shoot system The aboveground parts of a plant, such as stems, leaves, and flowers.

sieve-tube member Of flowering plants, a cellular component of the interconnecting sugar-conducting tubes in phloem.

sign stimulus A simple but important cue in the environment that triggers a suitable response to a stimulus that the nervous system is prewired to recognize.

sink region Of plants, any region using or stockpiling organic compounds for growth and development.

sister chromatid One of two DNA molecules (and associated proteins) of a duplicated chromosome that remain attached at their centromere region until separated at mitosis or meiosis; each ends up as a chromosome in a different daughter nucleus.

skeletal muscle An organ with hundreds to many thousands of muscle cells arrayed in bundles. Connective tissue surrounds the bundles and extends beyond the muscle, in the form of tendons that attach it to bone.

sliding filament model Model of muscle contraction in which each myosin filament in a sarcomere physically slides along and pulls actin filaments toward the center of the sarcomere, which shortens. The sliding requires ATP energy and formation of cross-bridges between the actin and myosin.

small intestine Of vertebrates, the portion of the digestive system where digestion is completed and most nutrients absorbed.

smog, industrial Polluted air, gray colored, that predominates in industrialized cities with cold, wet winters.

smog, photochemical Polluted air, brown and smelly, that occurs in large cities with many gas-burning vehicles and a warm climate.

social behavior Intraspecific interactions that require mixes of the instinctive and learned behaviors by which individuals send and respond to communication signals.

social parasite An animal that exploits the social behavior of another species to gain food, care for young, or some other factor necessary to survive, reproduce, or both.

sodium-potassium pump A membrane transport protein; when activated by ATP, it selectively transports potassium ions across the membrane, against its concentration gradient, and allows the crossing of sodium ions in the opposite direction.

soil A variable mixture of mineral particles and decomposing organic material; air and water occupy spaces between the particles.

solute (SOL-yoot) [L. *solvere*, to loosen] Any substance dissolved in a solution. In water, this means that hydration spheres surround the charged parts of individual ions or molecules and keep them dispersed.

solvent A fluid, such as water, in which one or more substances is dissolved.

somatic cell (so-MAT-ik) [Gk. *somā*, body] Of animals, any body cell that is not a germ cell (which gives rise to gametes).

somatic nervous system Nerves leading from the central nervous system to skeletal muscles.

somite One of many paired segments in a vertebrate embryo that will give rise to most bones, skeletal muscles of the head and trunk, and the overlying dermis.

source region Of plants, a site where organic compounds form by photosynthesis.

speciation (spee-cee-AY-shun) Evolutionary process by which a daughter species forms from a population or subpopulation of the parent species. The process can vary in its details and in the length of time it takes before the required reproductive isolation from each other is complete.

species (SPEE-sheez) [L. *species*, a kind] In general, one kind of organism. Of sexually reproducing organisms only, one or more populations of individuals that are interbreeding under natural conditions and producing fertile offspring, and that are reproductively isolated from other such groups.

sperm [Gk. *sperma*, seed] A type of mature male gamete.

spermatogenesis (sperm-AT-oh-JEN-ih-sis) Process by which mature sperm form from a germ cell in males.

sphere of hydration Through positive or negative interactions, a clustering of water molecules around individual molecules of a substance placed in water. *Compare* solute.

spinal cord The portion of the central nervous system that threads through a canal inside the vertebral column and that affords direct reflex connections between sensory and motor neurons, and that has tracts to and from the brain.

spleen One of the lymphoid organs; it is a filtering station for blood, a reservoir of red blood cells, and a reservoir of macrophages.

spore A single-celled reproductive structure or resistant, resting body; often walled, and often adapted to survive adverse conditions. An asexual spore may directly give rise to a new stage of the life cycle; a sexual spore must unite with another sexual spore to produce a new stage.

sporophyte [Gk. *phyton*, plant] Of plant life cycles, a vegetative body that grows by way of mitosis from a zygote and that produces spore-bearing structures.

spring overturn Of some bodies of water, movement during spring of dissolved oxygen from the surface layer to the depths and movement of nutrients from bottom sediments to the surface.

stabilizing selection A mode of natural selection by which intermediate phenotypes are favored, and extremes at both ends of the range of variation are eliminated.

stamen (STAY-mun) A male reproductive structure in a flower, commonly consisting of a pollen-bearing structure (anther) on a single stalk (filament).

start codon A base triplet in a strand of mRNA that serves as the start signal for the translation stage of protein synthesis.

stem cell A type of self-perpetuating cell that remains unspecialized. Some of its daughter cells also are self-perpetuating; others differentiate into cells of specialized structure and function. Stem cells in bone marrow that give rise to daughter stem cells and to blood cells are an example.

sterol (STAIR-all) A type of lipid with a rigid backbone of four fused carbon rings. Sterols differ in the number, position, and type of functional groups. They occur in eukaryotic cell membranes; cholesterol is the main type in animal tissues.

steroid hormone A lipid-soluble hormone, synthesized from cholesterol, that diffuses directly across the lipid bilayer of a target cell's plasma membrane and binds with a receptor inside the cell.

stigma Of many flowers, a sticky or hairy surface tissue on upper portion of ovary; it captures pollen grains and favor their germination.

stimulus [L. *stimulus*, goad] A specific form of energy, such as mechanical energy, light, and heat, that activates a sensory receptor having the capacity to detect it.

stoma (STOW-muh), plural **stomata** [Gk. *stoma*, mouth] A controllable gap between two guard cells in stems and leaves; any of the tiny passageways across the epidermis by which carbon dioxide moves into a plant and water vapor and oxygen move out.

stomach A muscular, stretchable sac that receives ingested food; of vertebrates, an organ between the esophagus and intestine in which much protein digestion occurs.

stop codon A base triplet in a strand of mRNA that serves as a stop signal during the translation stage of protein synthesis; it blocks further additions of amino acids to a new polypeptide chain.

strain One of two or more organisms with differences that are too minor to classify it as a separate species. An example is a strain of *Escherichia coli* or some other bacterium.

stratification Ancient layers of sedimentary rock resulting from the slow deposition of volanic ash, silt, and other materials, one above the other.

stream A flowing-water ecosystem that starts out as a freshwater spring or seep.

stroma [Gk. *strōma*, bed] The semifluid interior between a thylakoid membrane system and the two outer membranes of a chloroplast; the chloroplast zone where sucrose, starch, cellulose, and other end products of photosynthesis are assembled.

stromatolite A formation of sediments and matted remains of photosynthetic species that lived shallow seas and gradually accumulated in thin, cakelike layers.

substrate A reactant or precursor for a metabolic reaction; a specific molecule or molecules that an enzyme can chemically recognize, bind reversibly to itself, and modify rapidly in a predictable way.

substrate-level phosphorylation A direct, enzyme-mediated transfer of a phosphate group from a substrate of a reaction to a different molecule, as when an intermediate of glycolysis gives up a phosphate group to ADP, thus forming ATP.

succession, primary (suk-SESH-un) [L. *succedere*, to follow after] An ecological pattern by which a community develops in sequence, from the time pioneer species colonize a new, barren habitat to the climax community (an end array of species that remain in equilibrium over the region).

succession, secondary A pattern by which some disturbed area within a community recovers and moves back toward the climax state; typical of abandoned fields, burned forests, and storm-battered intertidal zones.

surface-to-volume ratio A mathematical relationship in which volume increases with the cube of the diameter, but surface area increases only with the square. Of growing cells, the volume of cytoplasm increases faster than the surface area of the plasma membrane that must service the cytoplasm. This constraint generally keeps cells small, elongated, or with membrane foldings.

survivorship curve A plot of age-specific survival of a group of individuals in a given environment, from the time of their birth until the last one dies.

swim bladder An adjustable flotation device that changes in volume as it exchanges gases with blood and that allows many fishes to maintain neutral buoyancy in water.

symbiosis (sim-by-OH-sis) [Gk. *sym*, together, + *bios*, life, mode of life] Literally, living together. For at least part of the life cycle, individuals of one species live near, in, or on individuals of another species. Commensalism, mutualism, and parasitism are examples of symbiotic interactions.

sympathetic nerve Of autonomic nervous systems, one of the nerves dealing mainly with increasing overall body activities at times of heightened awareness, excitement, or danger; also work continually in opposition with parasympathetic nerves to make minor adjustments in internal organ activity.

sympatric speciation [Gk. *sym*, together, + *patria*, native land] Species form within the home range of an existing species, in the absence of a physical barrier. It may happen instantaneously, as by polyploidy.

synaptic integration (sin-AP-tik) Of an individual neuron, the moment-by-moment combining of all excitatory and inhibitory signals arriving at its trigger zone.

syndrome A set of symptoms that may not individually be a telling clue but that collectively characterize a particular genetic disorder or disease.

systemic circuit (sis-TEM-ik) Of a closed circulatory system, a route by which oxygen-enriched blood flows from the lungs to the left half of the heart, through the rest of the body (where it gives up oxygen and takes up carbon dioxide), then back to the right half of the heart.

T lymphocyte One of a class of white blood cells that carry out immune responses. The helper T and cytotoxic T cells are examples.

taproot system A primary root together with its lateral branchings.

target cell Of a signaling molecule such as a hormone, any cell having the type of receptor that can bind with that molecule.

tectum The midbrain's roof. In fishes and amphibians, a center for coordinating most sensory input and initiating motor responses. In most vertebrates (not mammals), a reflex center that swiftly relays sensory input to higher integrative centers in the forebrain.

telophase (TEE-low-faze) Of meiosis I, the stage when one of each pair of homologous chromosomes has arrived at a spindle pole. Of mitosis and of meiosis II, the final stage when chromosomes decondense into threadlike structures and two daughter nuclei form.

temperature A measure of how rapidly the molecules or ions of a substance are moving.

tendon A cord or strap of dense connective tissue that attaches muscle to bones.

territory An area that one or more animals defend against competitors for food, water, mates, or suitable living space.

test A way to determine the accuracy of a prediction, as by conducting experimental or observational tests and by developing models. In science, such predictions must be based on potentially falsifiable hypotheses; that is, they must be tested in the natural world in ways that might disprove them.

testcross For an individual of unknown genotype that shows dominance for a trait, a type of experimental cross that may reveal whether the individual is homozygous dominant or heterozygous.

testis, plural **testes** Male gonad; a primary reproductive organ in which male gametes and sex hormones are produced.

testosterone (tess-TOSS-tuh-rown) Of male vertebrates, a sex hormone with key roles in the development and functioning of the male reproductive system.

tetanus Of muscles, a large contraction in which repeated stimulation of a motor unit causes muscle twitches to mechanically run together. In a disease by the same name, toxins block muscles from relaxing.

thalamus Of the forebrain, a coordinating center for sensory input and a relay station for signals to the cerebrum.

theory, scientific A testable explanation of a broad range of related phenomena, one that has been extensively tested and can be used with a high degree of confidence. A scientific theory remains open to tests, revision, and tentative acceptance or rejection.

thermal inversion The trapping of a layer of dense, cool air beneath a layer of warm air; can cause air pollutants to accumulate to high levels close to the ground.

thermophile A type of archaebacterium of unusually hot aquatic habitats, as in hot springs or near hydrothermal vents.

thermoreceptor A sensory cell or specialized cell by it that detects radiant energy (heat).

thigmotropism (thig-MOE-truh-pizm) [Gk. *thigm*, touch] An orientation of the direction of growth in response to physical contact with a solid object, as when a vine curls around a fencepost.

threshold Of a neuron and other excitable cells, the minimum amount by which the resting membrane potential must change to trigger an action potential.

thylakoid membrane system Of chloroplasts, an internal membrane system commonly folded into flattened channels and disks (grana) that have light-absorbing pigments and enzymes used in the formation of ATP, NADPH, or both during photosynthesis.

thymine A nitrogen-containing base of one of the nucleotides in DNA.

thymus gland A lymphoid organ that has endocrine functions; lymphocytes of the immune system multiply, differentiate, and mature in its tissues, and its hormone secretions affect their functioning.

thyroid gland An endocrine gland, the hormones of which affect overall metabolic rates, growth, and development.

tissue Of multicelled organisms, a group of cells and intercellular substances that function together in the performance of one or more specialized tasks.

tonicity The relative concentrations of solutes in two fluids, such as inside and outside a cell. When solute concentrations are isotonic (equal in both fluids), water shows no net osmotic movement in either direction. When one fluid is hypotonic (has less solutes), the other is hypertonic (has more solutes) and water tends to move into it.

toxin A normal metabolic product of one species with chemical effects that can harm or kill individuals of a different species that encounter it.

trace element Any element that represents less than 0.01 percent of body weight.

tracer A substance that has a radioisotope attached to it so that its pathway or destination in a cell, organism, ecosystem, or some other system can be tracked, as by scintillation counters that detect its emissions.

trachea (TRAY-kee-uh), plural **tracheae** An air-conducting tube of respiratory systems; of land vertebrates, the windpipe, which carries air between the larynx and bronchi.

tracheal respiration Of insects, spiders, and some other animals, a respiratory system consisting of finely branching tracheae that extend from openings in the integument and that dead-end in body tissues.

tracheid (TRAY-kid) Of flowering plants, one of two types of cells in xylem that conduct water and dissolved minerals.

tract A communication line inside the brain and spinal cord; comparable to a nerve of the peripheral nervous system.

transcription [L. *trans*, across, + *scribere*, to write] The first stage of protein synthesis, when an RNA strand is assembled on one of the two strands of a DNA double helix; the base sequence of the resulting transcript is complementary to the DNA template.

transfer RNA (tRNA) An RNA molecule that binds and delivers an amino acid to a ribosome and that pairs with an mRNA codon during the translation stage of protein synthesis.

translation The stage of protein synthesis when the encoded sequence of information in mRNA becomes converted to a sequence of particular amino acids, the outcome being a polypeptide chain; rRNA, tRNA, and mRNA interact to bring this about.

translocation Of cells, a stretch of DNA that physically moved to a different location in the same chromosome or in a different one, with no molecular loss. Of vascular plants, the process by which organic compounds are distributed by way of phloem.

transpiration Evaporative water loss from aboveground plant parts, leaves especially.

transport protein A membrane protein that allows water-soluble substances to move through their interior, which spans the lipid bilayer of a cell membrane.

transposable element A DNA region that moves spontaneously from one location to another in the same DNA molecule or a different one. Often such regions inactivate genes into which they become inserted and cause changes in phenotype.

triglyceride (neutral fat) A lipid having three fatty acid tails attached to a glycerol backbone. Triglycerides are the body's most abundant lipids and richest energy source.

trisomy (TRY-so-mee) Of cells, the presence of three chromosomes of a given type rather than the two characteristic of the parental diploid chromosome number.

trophic level (TROE-fik) [Gk. *trophos*, feeder] All of the organisms in an ecosystem that are the same number of transfer steps away from the energy input into the system.

tropical rain forest A biome where rainfall is regular and heavy, the annual mean temperature is 25°C, humidity is 80 percent or more, and biodiversity is spectacular.

tropism (TROE-pizm) Of plants, a directional growth response to an environmental factor, such as growth toward light.

true breeding Of a sexually reproducing species, a lineage in which the offspring of successive generations are just like the parents in one or more traits being studied.

tumor A tissue mass composed of cells that are dividing at an abnormally high rate.

turgor pressure (TUR-gore) [L. *turgere*, to swell] Internal fluid pressure applied to a cell wall when water moves into the cell by way of osmosis.

ultrafiltration Bulk flow of a small amount of protein-free plasma from a blood capillary when the outward-directed force of blood pressure is greater than the inward-directed osmotic force of interstitial fluid.

uniformity Various theories that mountain building, erosion, and other forces of nature have worked over the Earth's surface in the same repetitive ways through time.

upwelling An upward movement of deep, nutrient-rich water along coasts; it replaces surface waters that move away from shore when the direction of prevailing wind shifts.

uracil (YUR-uh-sill) Nitrogen-containing base of a nucleotide found in RNA molecules; like thymine, it can base-pair with adenine.

ureter A tubular channel for urine flow between the kidney and urinary bladder.

urethra A tubular channel for urine flow between the urinary bladder and an opening at the body surface.

urinary bladder A distensible sac in which urine is stored before being excreted.

urinary excretion A mechanism by which excess water and solutes are removed by way of a urinary system.

urinary system An organ system that adjusts the volume and composition of blood, and so helps maintain extracellular fluid.

urine A fluid formed in kidneys by filtration, reabsorption, and secretion; urine consists of wastes, excess water, and solutes.

uterus (YOU-tur-us) [L. *uterus*, womb] A chamber in which the developing embryo is contained and nurtured during pregnancy.

vaccination An immunization procedure against a specific pathogen.

vaccine An antigen-containing preparation, swallowed or injected, that increases immunity to certain diseases by inducing formation of armies of effector and memory B and T cells.

vagina Part of a reproductive system of mammalian females that receives sperm, forms part of the birth canal, and channels menstrual flow to the exterior.

variable Of an experimental test, a specific aspects of an object or event that may differ over time and among individuals.

vascular bundle Of vascular plants, the arrangement of primary xylem and phloem into multistranded, sheathed cords that extend lengthwise through the ground tissue system.

vascular cambium A lateral meristem that increases stem or root diameter (girth).

vascular cylinder Arrangement of vascular tissues as a central cylinder in roots.

vascular plant A plant with xylem and phloem, and well-developed roots, stems, and leaves.

vascular tissue system Xylem and phloem; conducting tissues that distribute water and solutes through the body of vascular plants.

vein Of the circulatory system, any of the large-diameter vessels that lead back to the heart; of leaves, one of the vascular bundles that thread through photosynthetic tissues.

ventricle (VEN-tri-kuhl) Of the vertebrate heart, one of two chambers from which blood is pumped out. Blood circulation is driven by ventricular contraction.

venule A small blood vessel that accepts blood from capillaries and delivers it to a vein.

vernalization Of flowering plants, stimulation of flowering by exposure to low temperatures.

vertebra, plural **vertebrae** One of a series of hard bones organized, with intervertebral disks, into a backbone.

vertebrate Animal with a backbone.

vesicle (VESS-ih-kul) [L. *vesicula*, little bladder] In the cytoplasm of cells, one of a variety of small membrane-bound sacs that function in the transport, storage, or digestion of substances or in some other activity.

vessel member A cell in xylem; although dead at maturity, its wall becomes part of a water-conducting pipeline.

villus (VIL-us), plural **villi** Any of several fingerlike absorptive structures projecting from the free surface of an epithelium.

viroid An infectious particle consisting only of very short, tightly folded strands or circles of RNA. Viroids may have evolved from introns, which they resemble.

virus A noncellular infectious agent that consists of DNA or RNA and a protein coat; it can replicate only after its genetic material enters a host cell and subverts the host's metabolic machinery.

vision Perception of visual stimuli; requires focusing of light precisely onto a layer of photoreceptive cells that is dense enough to sample details of a light stimulus, followed by image formation in a brain.

visual signal An observable action or cue that functions as a communication signal.

vitamin Any of more than a dozen organic substances that animals require in small amounts for metabolism but that they generally cannot synthesize for themselves.

vocal cord One of the thickened, muscular folds of the larynx that help produce sound waves for speech.

water potential Sum of two opposing forces (osmosis and turgor pressure) that can cause a directional movement of water into or out of a walled cell.

water table The upper limit at which the ground in a specified area has become fully saturated with water.

watershed Any specified region in which all precipitation drains into a single stream or river.

wavelength A wavelike form of energy in motion. The horizontal distance between two crests of every two successive waves.

wax A molecule with long-chain fatty acids packed together and linked to long-chain alcohols or to carbon rings. Waxes have a firm consistency and repel water.

white blood cell One of the eosinophils, neutrophils, macrophages, T and B cells, and other leukocytes which, together with their chemical weapons and mediators, defend the vertebrate body against attacks by pathogens and against tissue damage.

white matter Inside the brain and spinal cord, axons that have glistening white sheaths and that specialize in rapid signal transmission.

wild-type allele For a given gene locus, the allele that occurs normally or with the greatest frequency among individuals of a population.

wing A body part that functions in flight, as among birds, bats, and many insects. A bird wing is a forelimb with feathers, strong muscles, and extremely lightweight bones. An insect wing develops as a lateral fold of the exoskeleton.

X chromosome A sex chromosome with genes that, in humans, causes an embryo to develop into a female, provided that it inherits a pair of these.

X-linked gene Any gene located on an X chromosome.

X-linked recessive inheritance Recessive condition in which the responsible, mutated gene occurs on the X chromosome.

xylem (ZYE-lum) [Gk. *xylon*, wood] Of vascular plants, a tissue that transports water and solutes through the plant body.

Y chromosome A distinctive chromosome present in males or females of many species, but not both. In human males, one is paired with an X chromosome (XY); and in human females, the pairing is XX.

Y-linked gene One of the genes located on a Y chromosome.

yellow marrow A fatty tissue in the cavities of most mature bones that produces red blood cells when blood loss from the body is severe.

yolk sac Of land vertebrates, one of four extraembryonic membranes. In most shelled eggs, it holds nutritive yolk; in humans, part becomes a site of blood cell formation and some of its cells give rise to the forerunners of gametes.

zero population growth A population for which the number of births is balanced by the number of deaths over a specified period, assuming that immigration and emigration also are balanced.

zooplankton A freshwater or marine community of floating or weakly swimming heterotrophs, mostly microscopic, such as rotifers and copepods.

zygote (ZYE-goat) The first cell of a new individual, formed by the fusion of a sperm nucleus with the nucleus of an egg at fertilization; also called a fertilized egg.

CREDITS AND ACKNOWLEDGMENTS

CHAPTER 33 33.1 David Macdonald. 33.2 (a) Focus on Sports;inset, Manfred Kage/Bruce Coleman Ltd.; art, Palay/Beaubois; (b)left, Lennart Nilsson from Behold Man, © 1974 by Albert Bonniers Forlag and Little, Brown and Company, Boston; (center) Manfred Kage/ Bruce Coleman Ltd.; right, Ed Reschke/Peter Arnold, Inc. 33.3 Art, R. Ciemma. 33.4 Gregory Dimijian/Photo Researchers; art, R. Ciemma adapted from C. P. Hickman, Jr., L. S. Roberts, and A. Larson, Integrated Principles of Zoology, Ninth edition, Wm. C. Brown, 1995. 33.5 (a-c, e, f) Ed Reschke; (d) Fred Hossler/Visuals Unlimited. 33.6 Roger K. Burnard;left art, Joel Ito; right art, L. Calver. 33.7, 33.8 Ed Reschke. 33.9 Robert Demarest. 33.10 (a) Lennart Nilsson from Behold Man, © 1974 Albert Bonniers Forlag and Little, Brown and Company, Boston; (b) Kim Taylor/Bruce Coleman Ltd. 33.11 Art, L. Calver. 33.14 Fred Bruemmer

CHAPTER 34 34.1 Adrian Warren/Ardea London. Page 559 Art, L. Calver. 34.2 (d) Manfred Kage/ Peter Arnold, Inc. / 34.3, 34.5 R. Ciemma. 34.7 Ed Reschke; art, Kevin Somerville. 34.8 (b) Dr. Constantino Sotelo from International Cell Biology, page 83, 1977. Used by copyright permission of the Rockefeller University Press. 34.10 R. G. Kessel and R. H. Kardon from Tissues and Organs: A Text-Atlas of Scanning Electron Microscopy, W. H. Freeman and Company, 1979. Used by permission of R. G. Kessel; art, Robert Demarest. 34.12 (b) Robert Demarest. 34.13 Painting by Sir Charles Bell, 1809, courtesy of Royal College of Surgeons, Edinburgh.

CHAPTER 35 35.1 Comstock, Inc. 35.2 Kjell B. Sandved. 35.3, 35.5 (b) R. Ciemma. 35.6 Kevin Somerville. 35.9 (a) Robert Demarest; (b) Manfred Kage/Peter Arnold, Inc. 35.10-35.12 Kevin Somerville. 35.13 (a) Colin Chumbley/ Science Source/Photo Researchers. (b) C. Yokochi, J. Rohen, Photographic Anatomy of the Human Body, Second edition, Igaku-Shoin Ltd., 1979. 35.14 Left, Marcus Raichle, Washington University School of Medicine. 35.15 Left, Palay/Beaubois after Penfield and Rasmussen, The Cerebral Cortex of Man, copyright © 1950 Macmillan Publishing Company, Inc. Renewed 1978 by Theodore Rasmussen; right, Colin Chumbley/Science Source/Photo Researchers. 35.22 (a,b) From Edythe D. London et al., Archives of General Psychiatry, 47:567-574 (1990); right, Ogden Gigli/Photo Researchers.

CHAPTER 36 36.1 (a) Eric A. Newman; (b) Merlin D. Tuttle, Bat Conservation International. 36.2 Art, Kevin Somerville. 36.3 From Hensel and Bowman, Journal of Physiology, 23:564-568, 1960. 36.4 Left, Palay/Beaubois after Penfield and Rasmussen, The Cerebral Cortex of Man, copyright © 1950 Macmillan Publishing Company, Inc. Renewed 1978 by Theodore Rasmussen; right, Colin Chumbley/Science Source/Photo Researchers. 36.5, 36.6 R. Ciemma. 36.7 Omikron/SPL/Photo Researchers; art, Robert Demarest. 36.9 Edward W. Bower/© 1991 TIB/West; art, Kevin Somerville. 36.10 (b,c) Kevin Somerville . 36.12 Robert Demarest. 36.13 (a,b) Robert E. Preston, courtesy Joseph E. Hawkins, Kresge Hearing Research Institute,

University of Michigan Medical School. 36.14 Art, R. Ciemma; (d) Keith Gillett/Tom Stack & Associates. 36.15 E. R. Degginger; sketch after M. Gardiner, The Biology of Vertebrates, McGraw-Hill, 1972 . 36.16 G. A. Mazohkin-Porshnykov (1958). Reprinted with permission from Insect Vision, © 1969 Plenum Press. 36.17 Chris Newbert. 36.18 Robert Demarest. 36.19 Chase Swift. 36.20, 36.21 Kevin Somerville. 36.22 Lennart Nilsson © Boehringer Ingelheim International GmbH. 36.24 Palay/Beaubois after S. Kuffler and J. Nicholls, From Neuron to Brain, Sinauer, 1977. 36.25, 36.26 Gerry Ellis/ The Wildlife Collection. 36.27 Douglas Faulkner/ Sally Faulkner Collection.

CHAPTER 37 37.1 Hugo van Lawick. 37.2 Kevin Somerville. 37.5, 37.6 Robert Demarest. 37.7 (a) Mitchell Layton; (b) Syndication International (1986) Ltd.; (c) courtesy of Dr. William H. Daughaday, Washington University School of Medicine, from A. I. Mendelhoff and D. E. Smith, eds., American Journal of Medicine, 20:133 (1956). 37.8 Leonard Morgan. 37.9 (a, b) Art, R. Ciemma; (c) Corbis-Bettmann. 37.11 Biophoto Associates/SPL/Photo Researchers. 37.12 Leonard Morgan. 37.13 (a) John S. Dunning/Ardea London; (b) Evan Cerasoli. 37.14 (a, b) From R. C. Brusca and G. J. Brusca, Invertebrates, © 1990 Sinauer Associates. Used by permission; (c) Frans Lanting/Bruce Coleman Ltd.; (d) Robert & Linda Mitchell. 37.15 Roger K. Burnard.

CHAPTER 38 38.1 John Brandenberg/Minden Pictures. 38.2 L. Calver. 38.3 (a) Robert Demarest; (b) CNRI/SPL/Photo Researchers; (c) Gregory Dimijian/Photo Researchers. 38.4 (a) Tom McHugh; (b) M.P.L. Fogden/Bruce Coleman Ltd.; (c) Chaumeton-Lanceau/Agence Nature; sketches from The Life of Birds, Fourth edition, by Luis Baptista and Joel Carl Welty, © 1988 by Saunders College Publishing. Reproduced by permission of the publisher. 38.5 Ed Reschke. 38.6 Michael Keller/FPG. 38.7 Linda Pitkin/Planet Earth Pictures. 38.8 D. A. Parry, Journal of Experimental Biology, 36:654, 1959. 38.9 (a) Stephen Dalton/Photo Researchers; (b) R. Ciemma. 38.10 D. & V. Hennings. 38.11 (a) R. Ciemma; (b) Ed Reschke, art, Joel Ito. 38.12 K. Kasnot. 38.13 National Osteoporosis Foundation. 38.14 C. Yokochi and J. Rohen, Photographic Anatomy of the Human Body, Second edition, Igaku-Shoin Ltd., 1979. 38.15, 38.16 R. Ciemma. 38.17 (a) N.H.P.A./A.N.T. Photo Library; (b) Robert Demarest. 38.18 Kevin Somerville. 38.19 (b) John D. Cunningham; (c) D. W. Fawcett, The Cell, Philadelphia: W. B. Saunders Co., 1966; art, Robert Demarest. 38.20 Nadine Sokol. 38.21 Robert Demarest. 38.22 After Stephen L. Wolfe, Molecular and Cellular Biology, Wadsworth, 1993. 38.25 Michael Neveux.

CHAPTER 39 39.1 (a) A. D. Waller, Physiology, The Servant of Medicine, Hitchcock Lectures, University of London Press, 1910; (b) courtesy The New York Academy of Medicine Library. 39.3 (c) After M. Labarbera and S. Vogel, American Scientist, 70:54-60, 1982. 39.4 Precision Graphics. 39.5 Palay/Beaubois. 39.6 CNRI/

SPL/Photo Researchers. 39.7 R. Ciemma. 39.8 (a,b) Lester V. Bergman & Associates, Inc.; (c) after F. Ayala and J. Kiger, Modern Genetics, © 1980 Benjamin-Cummings. 39.9 Nadine Sokol after Gerard J. Tortora and Nicholas P. Anagnostakos, Principles of Anatomy and Physiology, Sixth edition, Copyright © 1990 by Biological Sciences Textbooks, Inc., A & P Textbooks, Inc. and Elia-Sparta, Inc. Reprinted by permission of HarperCollins Publishers. 39.11 Kevin Somerville. 39.12 (a) C. Yokochi and J. Rohen, Photographic Anatomy of the Human Body, Second edition, Igaku-Shoin Ltd, 1979; (b, c) R. Ciemma. 39.14 Michael Abbey/ Photo Researchers; art, R. Ciemma. 39.15 R. Ciemma. 39.16 Robert Demarest based on A. Spence, Basic Human Anatomy, Benjamin-Cummings, 1982 . 39.18 Sheila Terry/SPL/ Photo Researchers. 39.19 R. Ciemma. 39.20 Kevin Somerville. 39.21 (a) Ed Reschke; (b) F. Sloop and W. Ober/Visuals Unlimited. 39.24 Lennart Nilsson © Boehringer Ingelheim International GmbH. 39.25, 39.26 R. Ciemma. 39.27 Lennart Nilsson from Behold Man, © 1974 by Albert Bonniers Forlag and Little, Brown and Company, Boston

CHAPTER 40 40.1 (a) The Granger Collection, New York; (b) Lennart Nilsson © Boehringer Ingelheim International GmbH. 40.2 Nadine Sokol. 40.3 Robert R. Dourmashkin, courtesy of Clinical Research Centre, Harrow, England. 40.4 Lennart Nilsson © Boehringer Ingelheim International GmbH. 40.5, 40.8 R. Ciemma. 40.9 Morton H. Nielsen and Ole Werdlin, University of Copenhagen; art, R. Ciemma. 40.10 courtesy Don C. Wiley, Harvard University; art, R. Ciemma. 40.11 Hans & Cassady, Inc. 40.13, 40.14 Palay/Beaubois after B. Alberts et al., Molecular Biology of the Cell, Garland Publishing Company, 1983. 40.16 Above, Lowell Georgia/Science Source/ Photo Researchers; (below) Matt Meadows/ Peter Arnold, Inc. 40.17 (left) David Scharf/ Peter Arnold, Inc.; right, Kent Wood/Photo Researchers. 40.18 Ted Thai/Time Magazine. 40.19 After Stephen L. Wolfe, Molecular B iology of the Cell, Wadsworth, 1993. 40.20 Z. Salahuddin, National Institutes of Health.

CHAPTER 41 41.1 Galen Rowell/Peter Arnold, Inc. 41.4 (a) Peter Parks/Oxford Scientific Films; (b) Hervé Chaumeton/Agence Nature. 41.5 Ed Reschke . 41.6 R. Ciemma. 41.7 After C. P. Hickman et al., Integrated Principles of Zoology, Sixth edition, St. Louis: C. V. Mosby Co., 1979. 41.9 H. R. Duncker, Justus-Liebig University, Giessen, Germany. 41.10 Kevin Somerville . 41.11 CNRI/SPL/Photo Researchers. 41.12 SIU/Visuals Unlimited; art, K. Kosnot. 41.13 From L. G. Mitchell, J. A. Mutchmor, and W. D. Dolphin, Zoology, copyright © 1988 by The Benjamin-Cummings Publishing Company; reprinted by permission. 41.14 Modified from A. Spence and E. Mason, Human Anatomy and Physiology, Fourth edition, 1992, West Publishing. 41.16 Leonard Morgan. 41.17 O. Auerbach/Visuals Unlimited. 41.18 Lennart Nilsson from Behold Man, © 1974 by Albert Bonniers Forlag and Little, Brown and

Company, Boston. **41.19** Giorgio Gualco/ Bruce Coleman Ltd. **41.20** From Tom Garrison, *Oceanography: An Invitation to Marine Science*, Wadsworth, 1993. **41.21** Christian Zuber/ Bruce Coleman Ltd. **41.22** Steve Lissau/Rainbow.

CHAPTER 42 **42.1** Gary Head. **42.3** R. Ciemma. **42.4** (a) D. Robert Franz/Planet Earth Pictures; (c) adapted from A. Romer and T. Parsons, *The Verterbrate Body*. Sixth edition, Saunders Publishing Company, 1986. **42.5** Art, Kevin Somerville. **42.7** Omikron/SPL/Photo Researchers; art, Robert Demarest. **42.8** After A. Vander et al., *Human Physiology: Mechanisms of Body Function*, Fifth edition, McGraw-Hill, 1990, used by permission; (c) Redrawn from *Human Anatomy and Physiology*, Fourth edition, by A. Spence and E. Mason. Copyright © 1992 by West Publishing Company. All rights reserved. **42.9** (a) R. Ciemma; (b,c)*right*, Lennart Nilsson © Boehringer Ingelheim International GmbH; (c) *left*, Biophoto Associates/SPL/Photo Researchers; art, Robert Demarest. **42.10** R. Ciemma. **Page 719** (in-text art) After A. Vander et al., *Human Physiology: Mechanisms of Body Function*, Fifth edition, McGraw-Hill, 1990. Used by permission. **42.11** Ralph Pleasant/ FPG. **42.15** Gary Head. **42.16** Dr. Douglas Coleman, The Jackson Laboratory.

CHAPTER 43 **43.1** *Above*, David Noble/ FPG; *below*, Claude Steelman/Tom Stack & Associates. **43.3, 43.4** Robert Demarest. **43.5** *Above*, Robert Demarest;*below*, Precision

Graphics. **43.7** (a, b) From Tom Garrison, *Oceanography: An Invitation to Marine Science*, Wadsworth, 1993; (c) Thomas D. Mangelsen/ Images of Nature. **43.8** (a) Colin Monteath, Hedgehog House, New Zealand; (b) Bob McKeever/Tom Stack & Associates. **43.9** Everett C. Johnson. **43.10** *Left*, Robert Demarest;*right*, Kevin Somerville. **43.11** David Jennings/Image Works / **43.12** Corbis-Bettmann

CHAPTER 44 **44.1** (a) Hans Pfletschinger; (b) R. Ciemma; (c-f) John H. Gerard. **44.2** (a) Frieder Sauer/Bruce Coleman Ltd.; (b) Wisniewski/ZEFA; (c) Zig Leszczynski/ Animals Animals; (d) Carolina Biological Supply Company; (e) FredMcKinney/FPG; (f) Gary Head. **44.4** Carolina Biological Supply Company. **44.7** (a) R. Ciemma after V. E. Foe and B. M. Alberts, *Journal of Cell Science*, 61:32, © The Company of Biologists 1983; (b) J. B. Morrill. **44.8** *Left*, Carolina Biological Supply Company;*far right*, Peter Parks/Oxford Scientific Films/Animals Animals. **44.9** R. Ciemma;*right*, after B. Burnside, *Developmental Biology*, 26:416-441, 1971. Used by permission of Academic Press. **44.10** F. R. Turner; art, R. Ciemma. **44.11** Carolina Biological Supply Company. **44.12** (a) Palay/Beaubois after Robert F. Weaver and Philip W. Hedrick, *Genetics*. Copyright © 1989 Wm. C. Brown Publishers; (b) R. Ciemma after Scott Gilbert, *Developmental Biology*, 4/E. Sinauer. **44.13** R. Ciemma after Scott Gilbert, *Developmental Biology*, 4/E. Sinauer. **44.14** (a) Peter Parks/

Oxford Scientific Films/Animal (b-d) Hans & Cassady, Inc. adapt W. Freeman and Brian Bracegirdl *of Embryology*, Third edition, Heine Educational Books, 1978; (e) S. R. H J. W. Yang, *The Anatomical Record*, 197:﹏‑4᠎ 1980. **Page 759** Dennis Green/Bruce Coleman Ltd. **44.15** Palay/Beaubois adapted from R. G. Ham and M. J. Veomett, *Mechanisms of Development*, St. Louis, C. V. Mosby Co., 1980.

CHAPTER 45 **45.1** Lennart Nilsson from *A Child Is Born*, © 1966, 1977 Dell Publishing Company, Inc. **45.2** R. Ciemma. **45.3** (a) R. Ciemma; (b) Ed Reschke. **45.4-45.6** R. Ciemma. **45.7***Above*, Robert Demarest;*below*, R. Ciemma. **45.8** Lennart Nilsson from *A Child Is Born*, © 1966, 1977 Dell Publishing Company, Inc. **45.9** Robert Demarest. **45.10-45.13** R. Ciemma. **45.14** Lennart Nilsson, *A Child Is Born*, © 1966, 1977 Dell Publishing Company, Inc.; art, R. Ciemma. **45.15** (a) R. Ciemma modified from Keith L. Moore, *The Developing Human: Clinically Oriented Embryology*, Fourth edition, Philadelphia: W. B. Saunders Co., 1988. **45.16** (a) From Lennart Nilsson, *A Child Is Born*, © 1966, 1977 Dell Publishing Company, Inc.; (b) James W. Hanson, M.D. **45.17** Robert Demarest. **45.18** R. Ciemma. **45.19** R. Ciemma adapted from L. B. Arey, *Developmental Anatomy*, Philadelphia, W. B. Saunders Co., 1965. **45.21** CNRI/SPL/Photo Researchers. **45.22** (a) John D. Cunningham/Visuals Unlimited; (b) David M. Phillips/Visuals Unlimited. **Page 791** Alan and Sandy Carey.

Mole rat, 628i
Mollusk, 650
Molt-inhibiting hormone (MIH), in arthropods, 623i
Monoclonal antibody, 681
Monocyte, 652i, 653, 653i, 675
Mononucleosis, infectious, 654
Montagu, M., 670
Morphogen, 757
Morphogenesis, 745, 753, 753i, 760
Morula, 751, 774i, 783t
Motion sickness, 595
Motor cortex, 581i
Motor neuron, 567i, 568
Motor unit, of nerve and muscle, 644, 646
Mouth, 714i, 715, 715i, 728, 728t
Mucous membrane, as barrier against infection, 672t
Mucus, secretion by stomach, 716
Muirhead, P., 758
Multiple sclerosis, 566
Mumps, 684i
Muscle
  cardiac, 627, 638, 658, 668
  contraction, 642–642, 642i, 643, 643i
  as effector, 555i
  effects of aging, 644
  effects of exercise, 644
  fatigue, 644
  intercostal, 696i
  sense, 592
  skeletal, 640–641, 640i, 641i, 642i
  smooth, 627, 638
  structure and function, 640–644
  system, 552i, 627, 627i
  tension, 644, 644i, 646
  tissue, 550, 550i, 551i, 556
  twitch, 644, 644i, 646
Mutation resulting in aging, 758
Myasthenia gravis, 685
Myelin sheath, 566, 566i
Myocardium, 658
Myofibril, 627, 640, 640i, 642i, 646
Myoglobin, 705
Myometrium, 769i, 782
Myopia, 604, 604i
Myosin, 640–641, 640i, 641i

Nervous system, human
  autonomic, 571, 576–577, 576i, 586
  bilateral and cephalized, 586
  central, 571
  divisions of, 574
  general arrangement, 552i
  gray matter, 575
  motor, 575i
  parasympathetic, 575i, 576, 576i, 586
  peripheral, 571, 576–577, 576i, 577i
  sensory, 575i
  somatic, 571, 576, 586
  sympathetic, 575i, 576i, 577, 586
  white matter, 575
Nervous system, of invertebrates snd lower vertebrates
  invertebrate, 572–573, 573i
  vertebrate, 574–575, 574i, 575i
Net protein utilization (NPU), 723
Neural fold, 776i
Neural groove, 752i, 776i
Neural plate, of frog embryo, 749i
Neural tube, 574, 574i, 578, 578i, 579, 586, 749i, 752, 752i, 762, 776i, 778i
Neuroglia, 551, 571
Neuromodulator, 564, 565i
Neuromuscular junction, 564i
Neuron, 551, 551i, 559–568, 559i, 560i, 567i, 571
Neurotransmitter, 564, 564i, 565, 565i, 567i, 568, 610
Neutrophil, 652i, 653, 653i, 674, 675, 675i, 688t
Newborn, 782–783, 783t
Niacin, 724t
Nicotine, 584, 664
Nitrogen and decompression sickness, 705
Nocireceptor (pain receptor), 674t
Nonshivering heat production, 740
Norepinephrine, 564, 585, 617t, 618
Notochord, 574, 749i, 776
Nuclease, 717t
Nutrition, 710–711, 728
Nutritional requirements, 722–723, 722i, 723i

Ommatidium, 599, 599i
Oocyte, 750, 768, 769t, 770, 770i, 771i, 773, 773i
Optic lobe, 579
Optic nerve, 576i, 580i, 600i, 600t, 602i, 603
Oral cavity, 696i, 715i
Oral contraception, 785, 785i
Organ, formation of, 748, 748i, 752, 752i, 760
Organ system
  animal, 552–553, 552i, 553i
  defined, 545
  human, 552i, 553i
  integumentary, 552i
  muscular, 552i
  skeletal, 552i
Orgasm, 773
Osmoreceptor, 590, 590t, 606, 735
Osteoarthritis, 636
Osteoblast, 635
Osteocyte, 635
Osteoporosis, 635, 635i
Otolith, of inner ear, 595, 595i
Oval window, 597i
Ovary, 611i, 617t, 763, 768, 768t, 769i, 770
Oviduct, 762, 768, 768t, 769i, 770i, 774i
Oviparous, 746, 747i
Ovoviviparous, 746, 747i
Ovulation, 768, 769i, 769t, 770, 770i, 771i, 772i, 773i, 788
Ovum, 773
Oxygen
  debt, 643
  diffusion into animal body, 691
Oxyhemoglobin, 700
Oxytocin, 555, 614, 614t, 624, 782, 783

Parturition (human birth), 782
Pasteur, L., 671
Pasteurization, 671
Patella, 637i
Pathogen, 672, 672t
Pattern formation, 756–757, 756i, 757i, 760
Peacock, 629i, 645
Pectoral girdle, 637i
Pectoralis major muscle, 639i
Pelvic
  cavity, 552i
  girdle, 637i
  inflammatory disease (PID), 786, 787
  nerve, 576i
Pelvis, renal, 733i
Penguin, 645
Penis, 764, 764t, 765i, 766i
Pepsin, 716, 717t
Pepsinogen, 716
Peptic ulcer, 716
Peptide hormone, 612t, 613, 613i
Peptidoglycan, 673
Perception, 590, 606
Perforin, 678, 679i
Pericardium, 658i
Periodontal disease, 715
Periodontal membrane, of tooth, 715i
Peripheral nervous system, 575, 575i
Peristalsis, stomach, 716
Peritoneum, 732i
Peritubular capillaries, 733i
PET scan, 580, 581i, 585i
pH, 736
Phagocyte, 675i
Phagocytosis, 671, 674, 674i
Phalanges, 637i
Pharyngeal arches, 776i, 778i
Pharynx, 696i, 697, 714i, 715, 728, 728t
Pheromone, 594, 610, 624
Phosphorus, 725t
Photoreceptor, 590, 590t, 598, 598i, 599, 599i, 602–603, 602i, 606
Pilomotor response, 740
Pineal gland, 578i, 580i, 611i, 617t, 622, 622i
Pituitary dwarfism, 616, 616i
Pituitary gigantism, 616, 616i
Pituitary gland, 578i, 610, 610i, 611i, 614–615, 614t, 615i, 616, 616i, 618i, 619i, 624, 771i, 772i
Placenta, 777, 777i, 778, 779i, 782i, 788–789
Plane of symmetry, 552i, 553i
Planula, 573, 573i
Plasma, of blood, 652, 652i, 668
Plasma protein, 675
Platelet, of blood, 652, 652i, 653, 653i, 666, 666i
Pleural membrane, of lung, 696i, 697
Pneumocystis carinii, cause of one type of pneumonia, 686
Polar body, 770, 770i